# Measurement and Instrumentation

*Theory and Application*

Alan S. Morris
Reza Langari

AMSTERDAM • BOSTON • HEIDELBERG • LONDON
NEW YORK • OXFORD • PARIS • SAN DIEGO
SAN FRANCISCO • SINGAPORE • SYDNEY • TOKYO
Academic Press is an imprint of Elsevier

Academic Press is an imprint of Elsevier
225 Wyman Street, Waltham, MA 02451, USA
525 B Street, Suite 1800, San Diego, California 92101-4495, USA
84 Theobald's Road, London WC1X 8RR, UK



**Library of Congress Cataloging-in-Publication Data**
Morris, Alan S., 1948-
Measurement and instrumentation : theory and application / Alan S. Morris ; Reza Langari, contributor.
p. cm.
ISBN 978-0-12-381960-4 (pbk.)
1. Measurement. 2. Automatic control. 3. Engineering instrument. I. Langari, Reza. II. Title.
T50.M644 2011
681–dc23
2011018587

**British Library Cataloguing-in-Publication Data**
A catalogue record for this book is available from the British Library.

ISBN: 978-0-12-381960-4

For information on all Academic Press publications
visit our Web site at www.elsevierdirect.com

Printed in the United States of America
11 12 13 14 9 8 7 6 5 4 3 2 1

# Measurement and Instrumentation

# Contents

# Acknowledgement

The authors gratefully acknowledge permission by John Wiley and Sons, Ltd, to reproduce some material that was previously published in *Measurement and Calibration Requirements for Quality Assurance to ISO9000* by A.S. Morris (published 1997). The second author would like to acknowledge the help of many graduate teaching assistants who helped revise the laboratory material that served as the basis for Chapters 5 and 6 of the present text.

# Preface

The foundations of this book lie in several earlier texts by Alan Morris: *Principles of Measurement and Instrumentation* (Prentice-Hall, 1988, 1993), *Essence of Measurement* (Prentice Hall, 1996) and, *Measurement and Instrumentation Principles* (Butterworth-Heinemann, 2001). Since the most recent of these was published in 2001, there have been significant developments in the field of measurement, particularly in smart sensors, intelligent instruments, microsensors, digital signal processing, digital recorders, digital fieldbuses, and new methods of signal transmission. It is therefore timely to create an up-to-date book that incorporates all of the latest developments in measurement. Whilst doing this, the opportunity has been taken to strengthen the book by bringing in a second author, Professor Reza Langari of Texas A & M University, who has made significant contributions especially in the areas of data acquisition and signal processing and the implementation of these using industry standard LabView software.

The overall aim of the book is to present the topics of sensors and instrumentation, and their use within measurement systems, as an integrated and coherent subject. Measurement systems, and the instruments and sensors used within them, are of immense importance in a wide variety of domestic and industrial activities. The growth in the sophistication of instruments used in industry has been particularly significant as advanced automation schemes have been developed. Similar developments have also been evident in military and medical applications.

Unfortunately, the crucial part that measurement plays in all of these systems tends to get overlooked, and measurement is therefore rarely given the importance that it deserves. For example, much effort goes into designing sophisticated automatic control systems, but little regard is given to the accuracy and quality of the raw measurement data that such systems use as their inputs. This disregard of measurement system quality and performance means that such control systems will never achieve their full potential, as it is very difficult to increase their performance beyond the quality of the raw measurement data on which they depend.

Ideally, the principles of good measurement and instrumentation practice should be taught throughout the duration of engineering courses, starting at an elementary level and moving on to more advanced topics as the course progresses. With this in mind, the material contained in this book is designed both to support introductory courses in measurement and instrumentation, and also to provide in-depth coverage of advanced topics for higher-level courses. In addition, besides its role as a student course text, it is also anticipated that the book will be useful to practicing engineers, both to update their knowledge of the latest developments in measurement theory and practice, and also to serve as a guide to the typical characteristics and capabilities of the range of sensors and instruments that are currently in use.

As is normal with measurement texts, the principles and theory of measurement are covered in the first set of chapters and then subsequent chapters cover the ranges of instruments and sensors that are available for measuring various physical quantities. This order of coverage means that the general characteristics of measuring instruments, and their behaviour in different operating environments, are well established before the reader is introduced to the procedures involved in choosing a measurement device for a particular application. This ensures that the reader will be properly equipped to appreciate and critically appraise the various merits and characteristics of different instruments when faced with the task of choosing a suitable instrument.

It should be noted that, while measurement theory inevitably involves some mathematics, the mathematical content of the book has deliberately been kept to the minimum necessary for the reader to be able to design and build measurement systems that perform to a level commensurate with the needs of the automatic control scheme or other system that they support. Where mathematical procedures are necessary, worked examples are provided as necessary throughout the book to illustrate the principles involved. Self-assessment questions are also provided at the end of each chapter to enable readers to test their level of understanding.

The earlier chapters are organized such that all of the elements in a typical measurement system are presented in a logical order, starting with data acquisition from sensors and then proceeding through the stages of signal processing, sensor output transducing, signal transmission, and signal display or recording. Ancillary issues, such as calibration and measurement system reliability, are also covered. Discussion starts with a review of the different classes of instruments and sensors available, and the sorts of applications in which these different types are typically used. This opening discussion includes analysis of the static and dynamic characteristics of instruments and exploration of how these affect instrument usage.
A comprehensive discussion of measurement uncertainty then follows, with appropriate procedures for quantifying and reducing measurement errors being presented. The importance of calibration procedures in all aspects of measurement systems, and particularly to satisfy the requirements of standards such as ISO9000 and ISO14000, is recognized by devoting a full chapter to the issues involved. The principles of computer data acquisition are covered next, and a comprehensive explanation is given about how to implement this using industry-standard LabVIEW software. After this, signal processing using both analogue and digital filters is explained, with implementation using LabVIEW software in the digital case. Then, after a

chapter covering the range of electrical indicating and test instruments that are used to monitor electrical measurement signals, the following chapter discusses the display, recording, and presentation of measurement data. A chapter is then devoted to presenting the range of variable conversion elements (transducers) and techniques that are used to convert non-electrical sensor outputs into electrical signals, with particular emphasis on electrical bridge circuits. The problems of measurement signal transmission are considered next, and various means of improving the quality of transmitted signals are presented. This is followed by a discussion of intelligent devices, including the digital computation principles used within them, and the methods used to create communication mechanisms between such devices and the other components within a measurement system. To conclude the set of chapters covering measurement theory, a chapter is provided that discusses the issues of measurement system reliability, and the effect of unreliability on plant safety systems. This discussion also includes the important subject of software reliability, since computational elements are now embedded in most measurement systems.

The subsequent chapters covering the measurement of particular physical variables are introduced by a chapter that reviews the various different technologies that are used in measurement sensors. The chapters that follow provide comprehensive coverage of the main types of sensors and instruments that exist for measuring all the physical quantities that a practicing engineer is likely to meet in normal situations. However, while the coverage is as comprehensive as possible, the distinction is emphasized between (a) instruments that are current and in common use, (b) instruments that are current but not widely used except in special applications, for reasons of cost or limited capabilities, and (c) instruments that are largely obsolete as regards new industrial implementations, but are still encountered on older plant that was installed some years ago. As well as emphasizing this distinction, some guidance is given about how to go about choosing an instrument for a particular measurement application.

Resources for Instructors: A solutions manual and image bank containing electronic versions of figures from the book are available by registering at www.textbooks.elsevier.com

CHAPTER 1

# *Fundamentals of Measurement Systems*



## 1.1 Introduction

Measurement techniques have been of immense importance ever since the start of human civilization, when measurements were first needed to regulate the transfer of goods in barter trade in order to ensure that exchanges were fair. The industrial revolution during the 19th century brought about a rapid development of new instruments and measurement techniques to satisfy the needs of industrialized production techniques. Since that time, there has been a large and rapid growth in new industrial technology. This has been particularly evident during the last part of the 20th century because of the many developments in electronics in general and computers in particular. In turn, this has required a parallel growth in new instruments and measurement techniques.

The massive growth in the application of computers to industrial process control and monitoring tasks has greatly expanded the requirement for instruments to measure, record, and control process variables. As modern production techniques dictate working to ever tighter accuracy limits, and as economic forces to reduce production costs become more severe, so the requirement for instruments to be both accurate and inexpensive becomes ever harder to satisfy. This latter problem is at the focal point of the research and development efforts of all instrument manufacturers. In the past few years, the most cost-effective means of improving instrument accuracy has been found in many cases to be the inclusion of digital computing power within instruments themselves. These intelligent instruments therefore feature prominently in current instrument manufacturers' catalogues.

This opening chapter covers some fundamental aspects of measurement. First, we look at how standard measurement units have evolved from the early units used in barter trade to the

**Measurement and Instrumentation: Theory and Application**

more exact units belonging to the Imperial and metric measurement systems. We then go on to study the major considerations in designing a measurement system. Finally, we look at some of the main applications of measurement systems.

## 1.2 Measurement Units

The very first measurement units were those used in barter trade to quantify the amounts being exchanged and to establish clear rules about the relative values of different commodities. Such early systems of measurement were based on whatever was available as a measuring unit. For purposes of measuring length, the human torso was a convenient tool and gave us units of the hand, the foot, and the cubit. Although generally adequate for barter trade systems, such measurement units are, of course, imprecise, varying as they do from one person to the next. Therefore, there has been a progressive movement toward measurement units that are defined much more accurately.

The first improved measurement unit was a unit of length (the meter) defined as $10^{-7}$ times the polar quadrant of the earth. A platinum bar made to this length was established as a standard of length in the early part of the 19th century. This was superseded by a superior quality standard bar in 1889, manufactured from a platinum–iridium alloy. Since that time, technological research has enabled further improvements to be made in the standard used for defining length. First, in 1960, a standard meter was redefined in terms of $1.65076373 \times 10^6$ wavelengths of the radiation from krypton-86 in vacuum. More recently, in 1983, the meter was redefined yet again as the length of path traveled by light in an interval of 1/299,792,458 seconds. In a similar fashion, standard units for the measurement of other physical quantities have been defined and progressively improved over the years. The latest standards for defining the units used for measuring a range of physical variables are given in Table 1.1.

The early establishment of standards for the measurement of physical quantities proceeded in several countries at broadly parallel times; in consequence, several sets of units emerged for measuring the same physical variable. For instance, length can be measured in yards, meters, or several other units. Apart from the major units of length, subdivisions of standard units exist such as feet, inches, centimeters, and millimeters, with a fixed relationship between each fundamental unit and its subdivisions.

Yards, feet, and inches belong to the Imperial system of units, which is characterized by having varying and cumbersome multiplication factors relating fundamental units to subdivisions such as 1760 (miles to yards), 3 (yards to feet), and 12 (feet to inches). The metric system is an alternative set of units, which includes, for instance, the unit of the meter and its centimeter and millimeter subdivisions for measuring length. All multiples and subdivisions of basic metric units are related to the base by factors of 10 and such units are therefore much easier to

**Table 1.1 Definitions of Standard Units**

| Physical Quantity | Standard Unit | Definition |
|---|---|---|
| Length | Meter | Length of path traveled by light in an interval of 1/299,792,458 seconds |
| Mass | Kilogram | Mass of a platinum–iridium cylinder kept in the International Bureau of Weights and Measures, Sevres, Paris |
| Time | Second | $9.192631770 \times 10^9$ cycles of radiation from vaporized cesium 133 (an accuracy of 1 in $10^{12}$ or one second in 36,000 years) |
| Temperature | Degrees | Temperature difference between absolute zero Kelvin and the triple point of water is defined as 273.16 K |
| Current | Amphere | One ampere is the current flowing through two infinitely long parallel conductors of negligible cross section placed 1 meter apart in vacuum and producing a force of $2 \times 10^{-7}$ newtons per meter length of conductor |
| Luminous intensity | Candela | One candela is the luminous intensity in a given direction from a source emitting monochromatic radiation at a frequency of 540 terahertz (Hz $\times$ $10^{12}$) and with a radiant density in that direction of 1.4641 mW/steradian (1 steradian is the solid angle, which, having its vertex at the centre of a sphere, cuts off an area of the sphere surface equal to that of a square with sides of length equal to the sphere radius) |
| Matter | Mole | Number of atoms in a 0.012-kg mass of carbon 12 |

use than Imperial units. However, in the case of derived units such as velocity, the number of alternative ways in which these can be expressed in the metric system can lead to confusion.

As a result of this, an internationally agreed set of standard units (SI units or Systèmes internationales d'unités) has been defined, and strong efforts are being made to encourage the adoption of this system throughout the world. In support of this effort, the SI system of units is used exclusively in this book. However, it should be noted that the Imperial system is still widely used in the engineering industry, particularly in the United States.

The full range of fundamental SI measuring units and the further set of units derived from them are given in Tables 1.2 and 1.3. Conversion tables relating common Imperial and metric units to their equivalent SI units can also be found in Appendix 1.

## 1.3 Measurement System Design

This section looks at the main considerations in designing a measurement system. First, we learn that a measurement system usually consists of several separate components, although only one component might be involved for some very simple measurement tasks. We then go on to look at how measuring instruments and systems are chosen to satisfy the requirements of particular measurement situations.

**Table 1.2 Fundamental SI Units**

| (a) Fundamental Units | | |
|---|---|---|
| **Quantity** | **Standard Unit** | **Symbol** |
| Length | meter | m |
| Mass | kilogram | kg |
| Time | second | s |
| Electric current | ampere | A |
| Temperature | kelvin | K |
| Luminous | candela | cd |
| Matter | mole | mol |
| **(b) Supplementary Fundamental Units** | | |
| **Quantity** | **Standard Unit** | **Symbol** |
| Plane angle | radian | rad |
| Solid angle | steradian | sr |

### *1.3.1 Elements of a Measurement System*

A *measuring system* exists to provide information about the physical value of some variable being measured. In simple cases, the system can consist of only a single unit that gives an output reading or signal according to the magnitude of the unknown variable applied to it. However, in more complex measurement situations, a measuring system consists of several separate elements as shown in Figure 1.1. These components might be contained within one or more boxes, and the boxes holding individual measurement elements might be either close together or physically separate. The term *measuring instrument* is used commonly to describe a measurement system, whether it contains only one or many elements, and this term is widely used throughout this text.

The first element in any measuring system is the primary *sensor*: this gives an output that is a function of the measurand (the input applied to it). For most but not all sensors, this function is at least approximately linear. Some examples of primary sensors are a liquid-in-glass thermometer, a thermocouple, and a strain gauge. In the case of a mercury-in-glass thermometer, because the output reading is given in terms of the level of the mercury, this particular primary sensor is also a complete measurement system in itself. However, in general, the primary sensor is only part of a measurement system. The types of primary sensors available for measuring a wide range of physical quantities are presented in the later chapters of this book.

Variable conversion elements are needed where the output variable of a primary transducer is in an inconvenient form and has to be converted to a more convenient form. For instance, the displacement-measuring strain gauge has an output in the form of a varying resistance. Because the resistance change cannot be measured easily, it is converted to a change in voltage by a *bridge circuit*, which is a typical example of a variable conversion element. In some cases, the

**Table 1.3 Derived SI Units**

| Quantity | Standard Unit | Symbol | Derivation Formula |
|---|---|---|---|
| Area | square meter | $m^2$ | |
| Volume | cubic meter | $m^3$ | |
| Velocity | metre per second | m/s | |
| Acceleration | metre per second squared | $m/s^2$ | |
| Angular velocity | radian per second | rad/s | |
| Angular acceleration | radian per second squared | $rad/s^2$ | |
| Density | kilogram per cubic meter | $kg/m^3$ | |
| Specific volume | cubic meter per kilogram | $m^3/kg$ | |
| Mass flow rate | kilogram per second | kg/s | |
| Volume flow rate | cubic meter per second | $m^3/s$ | |
| Force | newton | N | $kg\text{-}m/s^2$ |
| Pressure | pascal | Pa | $N/m^2$ |
| Torque | newton meter | N-m | |
| Momentum | kilogram meter per second | kg-m/s | |
| Moment of inertia | kilogram meter squared | $kg\text{-}m^2$ | |
| Kinematic viscosity | square meter per second | $m^2/s$ | |
| Dynamic viscosity | newton second per sq metre | $N\text{-}s/m^2$ | |
| Work, energy, heat | joule | J | N-m |
| Specific energy | joule per cubic meter | $J/m^3$ | |
| Power | watt | W | J/s |
| Thermal conductivity | watt per meter Kelvin | W/m-K | |
| Electric charge | coulomb | C | A-s |
| Voltage, e.m.f., pot diff | volt | V | W/A |
| Electric field strength | volt per meter | V/m | |
| Electric resistance | ohm | Ω | V/A |
| Electric capacitance | farad | F | A-s/V |
| Electric inductance | henry | H | V-s/A |
| Electric conductance | siemen | S | A/V |
| Resistivity | ohm meter | Ω-m | |
| Permittivity | farad per meter | F/m | |
| Permeability | henry per meter | H/m | |
| Current density | ampere per square meter | $A/m^2$ | |
| Magnetic flux | weber | Wb | V-s |
| Magnetic flux density | tesla | T | $Wb/m^2$ |
| Magnetic field strength | ampere per meter | A/m | |
| Frequency | hertz | Hz | $s^{-1}$ |
| Luminous flux | lumen | lm | cd-sr |
| Luminance | candela per square meter | $cd/m^2$ | |
| Illumination | lux | lx | $lm/m^2$ |
| Molar volume | cubic meter per mole | $m^3/mol$ | |
| Molarity | mole per kilogram | mol/kg | |
| Molar energy | joule per mole | J/mol | |

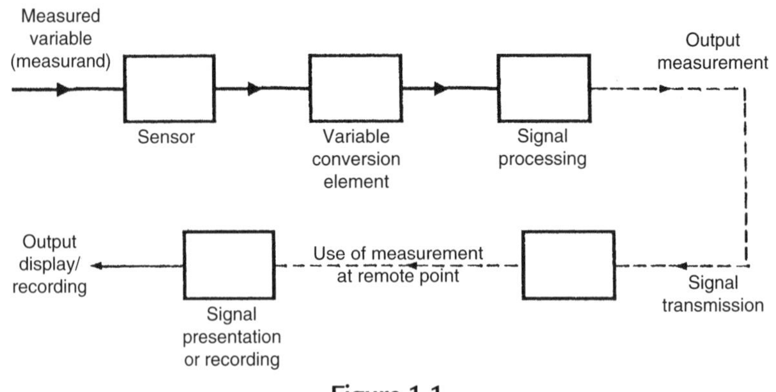


**Figure 1.1**
Elements of a measuring system.

primary sensor and variable conversion element are combined; this combination is known as a *transducer*.*

Signal processing elements exist to improve the quality of the output of a measurement system in some way. A very common type of signal processing element is the electronic amplifier, which amplifies the output of the primary transducer or variable conversion element, thus improving the sensitivity and resolution of measurement. This element of a measuring system is particularly important where the primary transducer has a low output. For example, thermocouples have a typical output of only a few millivolts. Other types of signal processing elements are those that filter out induced noise and remove mean levels, etc. In some devices, signal processing is incorporated into a transducer, which is then known as a *transmitter*.*

In addition to these three components just mentioned, some measurement systems have one or two other components, first to transmit the signal to some remote point and second to display or record the signal if it is not fed automatically into a feedback control system. Signal transmission is needed when the observation or application point of the output of a measurement system is some distance away from the site of the primary transducer. Sometimes, this separation is made solely for purposes of convenience, but more often, it follows from the physical inaccessibility or environmental unsuitability of the site of the primary transducer for mounting the signal presentation/recording unit. The signal transmission element has traditionally consisted of single or multicored cable, which is often screened to minimize signal corruption by induced electrical noise. However, fiber-optic cables are being used in ever-increasing numbers in modern installations, in part because of their low transmission loss and imperviousness to the effects of electrical and magnetic fields.

* In some cases, the word "sensor" is used generically to refer to both transducers and transmitters.

The final optional element in a measurement system is the point where the measured signal is utilized. In some cases, this element is omitted altogether because the measurement is used as part of an automatic control scheme, and the transmitted signal is fed directly into the control system. In other cases, this element in the measurement system takes the form either of a signal presentation unit or of a signal-recording unit. These take many forms according to the requirements of the particular measurement application, and the range of possible units is discussed more fully in Chapter 8.

### 1.3.2 Choosing Appropriate Measuring Instruments

The starting point in choosing the most suitable instrument to use for measurement of a particular quantity in a manufacturing plant or other system is specification of the instrument characteristics required, especially parameters such as desired measurement accuracy, resolution, sensitivity, and dynamic performance (see the next chapter for definitions of these). It is also essential to know the environmental conditions that the instrument will be subjected to, as some conditions will immediately either eliminate the possibility of using certain types of instruments or else will create a requirement for expensive protection of the instrument. It should also be noted that protection reduces the performance of some instruments, especially in terms of their dynamic characteristics (e.g., sheaths protecting thermocouples and resistance thermometers reduce their speed of response). Provision of this type of information usually requires the expert knowledge of personnel who are intimately acquainted with the operation of the manufacturing plant or system in question. Then, a skilled instrument engineer, having knowledge of all instruments available for measuring the quantity in question, will be able to evaluate the possible list of instruments in terms of their accuracy, cost, and suitability for the environmental conditions and thus choose the most appropriate instrument. As far as possible, measurement systems and instruments should be chosen that are as insensitive as possible to the operating environment, although this requirement is often difficult to meet because of cost and other performance considerations. The extent to which the measured system will be disturbed during the measuring process is another important factor in instrument choice. For example, significant pressure loss can be caused to the measured system in some techniques of flow measurement.

Published literature is of considerable help in the choice of a suitable instrument for a particular measurement situation. Many books are available that give valuable assistance in the necessary evaluation by providing lists and data about all instruments available for measuring a range of physical quantities (e.g., the later chapters of this text). However, new techniques and instruments are being developed all the time, and therefore a good instrumentation engineer must keep abreast of the latest developments by reading the appropriate technical journals regularly.

The instrument characteristics discussed in the next chapter are the features that form the technical basis for a comparison between the relative merits of different instruments. Generally, the better the characteristics, the higher the cost. However, in comparing the cost and relative suitability of different instruments for a particular measurement situation, considerations of durability, maintainability, and constancy of performance are also very important because the instrument chosen will often have to be capable of operating for long periods without performance degradation and a requirement for costly maintenance. In consequence of this, the initial cost of an instrument often has a low weighting in the evaluation exercise.

Cost is correlated very strongly with the performance of an instrument, as measured by its static characteristics. Increasing the accuracy or resolution of an instrument, for example, can only be done at a penalty of increasing its manufacturing cost. Instrument choice therefore proceeds by specifying the minimum characteristics required by a measurement situation and then searching manufacturers' catalogues to find an instrument whose characteristics match those required. To select an instrument with characteristics superior to those required would only mean paying more than necessary for a level of performance greater than that needed.

As well as purchase cost, other important factors in the assessment exercise are instrument durability and maintenance requirements. Assuming that one had $20,000 to spend, one would not spend $15,000 on a new motor car whose projected life was 5 years if a car of equivalent specification with a projected life of 10 years was available for $20,000. Likewise, durability is an important consideration in the choice of instruments. The projected life of instruments often depends on the conditions in that the instrument will have to operate. Maintenance requirements must also be taken into account, as they also have cost implications.

As a general rule, a good assessment criterion is obtained if the total purchase cost and estimated maintenance costs of an instrument over its life are divided by the period of its expected life. The figure obtained is thus a cost per year. However, this rule becomes modified where instruments are being installed on a process whose life is expected to be limited, perhaps in the manufacture of a particular model of car. Then, the total costs can only be divided by the period of time that an instrument is expected to be used for, unless an alternative use for the instrument is envisaged at the end of this period.

To summarize therefore, instrument choice is a compromise among performance characteristics, ruggedness and durability, maintenance requirements, and purchase cost. To carry out such an evaluation properly, the instrument engineer must have a wide knowledge of the range of instruments available for measuring particular physical quantities, and he/she must also have a deep understanding of how instrument characteristics are affected by particular measurement situations and operating conditions.

## 1.4 Measurement System Applications

Today, the techniques of measurement are of immense importance in most facets of human civilization. Present-day applications of measuring instruments can be classified into three major areas. The first of these is their use in regulating trade, applying instruments that measure physical quantities such as length, volume, and mass in terms of standard units. The particular instruments and transducers employed in such applications are included in the general description of instruments presented in the later chapters of this book.

The second application area of measuring instruments is in monitoring functions. These provide information that enables human beings to take some prescribed action accordingly. The gardener uses a thermometer to determine whether he should turn the heat on in his greenhouse or open the windows if it is too hot. Regular study of a barometer allows us to decide whether we should take our umbrellas if we are planning to go out for a few hours. While there are thus many uses of instrumentation in our normal domestic lives, the majority of monitoring functions exist to provide the information necessary to allow a human being to control some industrial operation or process. In a chemical process, for instance, the progress of chemical reactions is indicated by the measurement of temperatures and pressures at various points, and such measurements allow the operator to make correct decisions regarding the electrical supply to heaters, cooling water flows, valve positions, and so on. One other important use of monitoring instruments is in calibrating the instruments used in the automatic process control systems described here.

Use as part of automatic feedback control systems forms the third application area of measurement systems. Figure 1.2 shows a functional block diagram of a simple temperature control system in which temperature $T_a$ of a room is maintained at reference value $T_d$. The value

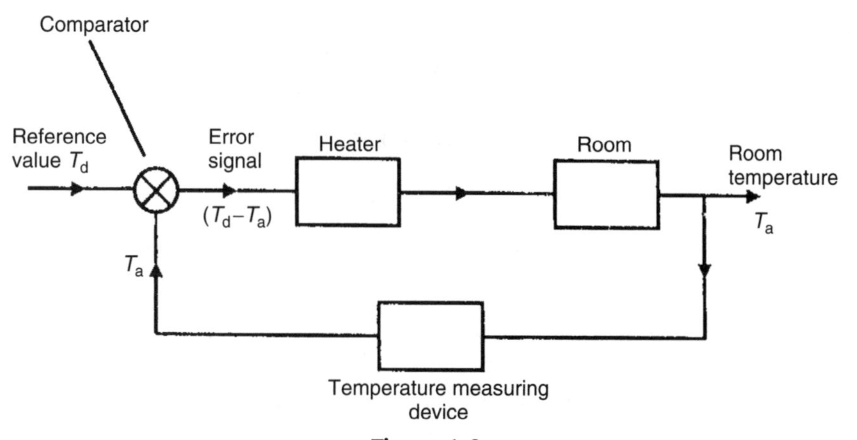


**Figure 1.2**
Elements of a simple closed-loop control system.

of the controlled variable, $T_a$, as determined by a temperature-measuring device, is compared with the reference value, $T_d$, and the difference, $e$, is applied as an error signal to the heater. The heater then modifies the room temperature until $T_a = T_d$. The characteristics of the measuring instruments used in any feedback control system are of fundamental importance to the quality of control achieved. The accuracy and resolution with which an output variable of a process is controlled can never be better than the accuracy and resolution of the measuring instruments used. This is a very important principle, but one that is often discussed inadequately in many texts on automatic control systems. Such texts explore the theoretical aspects of control system design in considerable depth, but fail to give sufficient emphasis to the fact that all gain and phase margin performance calculations are entirely dependent on the quality of the process measurements obtained.

## 1.5 Summary

This opening chapter covered some fundamental aspects of measurement systems. First, we looked at the importance of having standard measurement units and how these have evolved into the Imperial and metric systems of units. We then went on to look at the main aspects of measurement system design and, in particular, what the main components in a measurement system are and how these are chosen for particular measurement requirements. Finally, we had a brief look at the range of applications of measurement systems.

## 1.6 Problems

1.1. How have systems of measurement units evolved over the years?

1.2. What are the main elements in a measurement system and what are their functions? Which elements are not needed in some measurement systems and why are they not needed?

1.3. What are the main factors governing the choice of a measuring instrument for a given application?

1.4. Name and discuss three application areas for measurement systems.

CHAPTER 2

# Instrument Types and Performance Characteristics



## 2.1 Introduction

Two of the important aspects of measurement covered in the opening chapter concerned how to choose appropriate instruments for a particular application and a review of the main applications of measurement. Both of these activities require knowledge of the characteristics of different classes of instruments and, in particular, how these different classes of instrument perform in different applications and operating environments.

**Measurement and Instrumentation: Theory and Application**

We therefore start this chapter by reviewing the various classes of instruments that exist. We see first of all that instruments can be divided between active and passive ones according to whether they have an energy source contained within them. The next distinction is between null-type instruments that require adjustment until a datum level is reached and deflection-type instruments that give an output measurement in the form of either a deflection of a pointer against a scale or a numerical display. The third distinction covered is between analogue and digital instruments, which differ according to whether the output varies continuously (analogue instrument) or in discrete steps (digital instrument). Fourth, we look at the distinction between instruments that are merely indicators and those that have a signal output. Indicators give some visual or audio indication of the magnitude of the measured quantity and are commonly found in the process industries. Instruments with a signal output are commonly found as part of automatic control systems. The final distinction we consider is between smart and nonsmart instruments. Smart, often known as intelligent, instruments are very important today and predominate in most measurement applications. Because of their importance, they are given more detailed consideration later in Chapter 11.

The second part of this chapter looks at the various attributes of instruments that determine their performance and suitability for different measurement requirements and applications. We look first of all at the static characteristics of instruments. These are their steady-state attributes (when the output measurement value has settled to a constant reading after any initial varying output) such as accuracy, measurement sensitivity, and resistance to errors caused by variations in their operating environment. We then go on to look at the dynamic characteristics of instruments. This describes their behavior following the time that the measured quantity changes value up until the time when the output reading attains a steady value. Various kinds of dynamic behavior can be observed in different instruments ranging from an output that varies slowly until it reaches a final constant value to an output that oscillates about the final value until a steady reading is obtained. The dynamic characteristics are a very important factor in deciding on the suitability of an instrument for a particular measurement application. Finally, at the end of the chapter, we also briefly consider the issue of instrument calibration, although this is considered in much greater detail later in Chapter 4.

## 2.2 Review of Instrument Types

Instruments can be subdivided into separate classes according to several criteria. These subclassifications are useful in broadly establishing several attributes of particular instruments such as accuracy, cost, and general applicability to different applications.

### 2.2.1 Active and Passive Instruments

Instruments are divided into active or passive ones according to whether instrument output is produced entirely by the quantity being measured or whether the quantity being measured simply modulates the magnitude of some external power source. This is illustrated by examples.

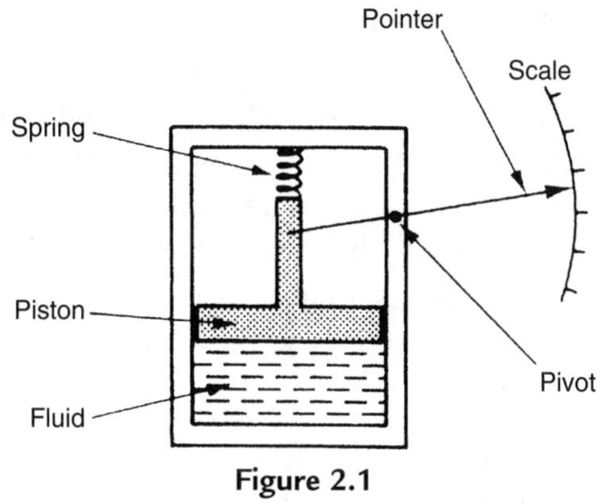


**Figure 2.1**
Passive pressure gauge.

An example of a passive instrument is the pressure-measuring device shown in Figure 2.1. The pressure of the fluid is translated into movement of a pointer against a scale. The energy expended in moving the pointer is derived entirely from the change in pressure measured: there are no other energy inputs to the system.

An example of an active instrument is a float-type petrol tank level indicator as sketched in Figure 2.2. Here, the change in petrol level moves a potentiometer arm, and the output signal consists of a proportion of the external voltage source applied across the two ends of the potentiometer. The energy in the output signal comes from the external power source: the primary transducer float system is merely modulating the value of the voltage from this external power source.

In active instruments, the external power source is usually in electrical form, but in some cases, it can be other forms of energy, such as a pneumatic or hydraulic one.

One very important difference between active and passive instruments is the level of measurement resolution that can be obtained. With the simple pressure gauge shown, the

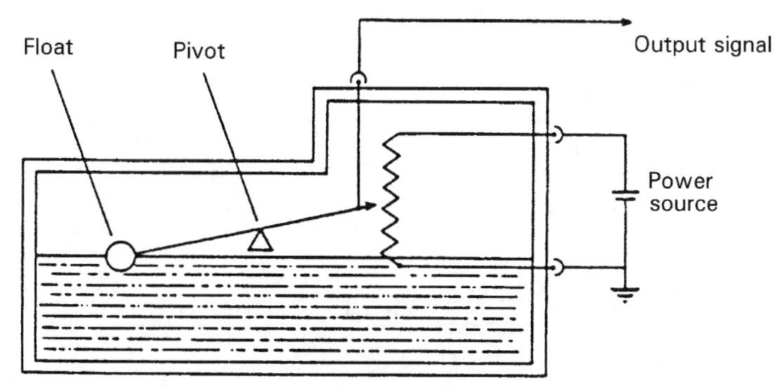


**Figure 2.2**
Petrol-tank level indicator.

amount of movement made by the pointer for a particular pressure change is closely defined by the nature of the instrument. While it is possible to increase measurement resolution by making the pointer longer, such that the pointer tip moves through a longer arc, the scope for such improvement is clearly restricted by the practical limit of how long the pointer can conveniently be. In an active instrument, however, adjustment of the magnitude of the external energy input allows much greater control over measurement resolution. While the scope for improving measurement resolution is much greater incidentally, it is not infinite because of limitations placed on the magnitude of the external energy input, in consideration of heating effects and for safety reasons.

In terms of cost, passive instruments are normally of a more simple construction than active ones and are therefore less expensive to manufacture. Therefore, a choice between active and passive instruments for a particular application involves carefully balancing the measurement resolution requirements against cost.

### 2.2.2 Null-Type and Deflection-Type Instruments

The pressure gauge just mentioned is a good example of a deflection type of instrument, where the value of the quantity being measured is displayed in terms of the amount of movement of a pointer. An alternative type of pressure gauge is the dead-weight gauge shown in Figure 2.3, which is a null-type instrument. Here, weights are put on top of the piston until the downward force balances the fluid pressure. Weights are added until the piston reaches a datum level, known as the null point. Pressure measurement is made in terms of the value of the weights needed to reach this null position.

The accuracy of these two instruments depends on different things. For the first one it depends on the linearity and calibration of the spring, whereas for the second it relies on calibration of the weights. As calibration of weights is much easier than careful choice and calibration of a linear-characteristic spring, this means that the second type of instrument will normally be the more accurate. This is in accordance with the general rule that null-type instruments are more accurate than deflection types.

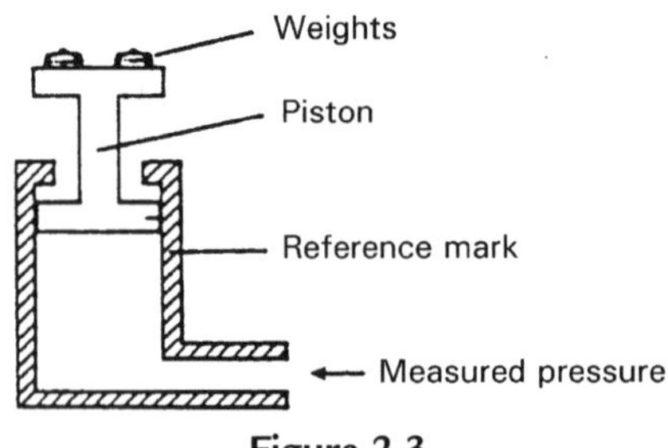


**Figure 2.3**
Dead-weight pressure gauge.

In terms of usage, a deflection-type instrument is clearly more convenient. It is far simpler to read the position of a pointer against a scale than to add and subtract weights until a null point is reached. A deflection-type instrument is therefore the one that would normally be used in the workplace. However, for calibration duties, a null-type instrument is preferable because of its superior accuracy. The extra effort required to use such an instrument is perfectly acceptable in this case because of the infrequent nature of calibration operations.

### 2.2.3 Analogue and Digital Instruments

An analogue instrument gives an output that varies continuously as the quantity being measured changes. The output can have an infinite number of values within the range that the instrument is designed to measure. The deflection-type of pressure gauge described earlier in this chapter (Figure 2.1) is a good example of an analogue instrument. As the input value changes, the pointer moves with a smooth continuous motion. While the pointer can therefore be in an infinite number of positions within its range of movement, the number of different positions that the eye can discriminate between is strictly limited; this discrimination is dependent on how large the scale is and how finely it is divided.

A digital instrument has an output that varies in discrete steps and so can only have a finite number of values. The rev counter sketched in Figure 2.4 is an example of a digital instrument. A cam is attached to the revolving body whose motion is being measured, and on each revolution the cam opens and closes a switch. The switching operations are counted by an electronic counter. This system can only count whole revolutions and cannot discriminate any motion that is less than a full revolution.

The distinction between analogue and digital instruments has become particularly important with rapid growth in the application of microcomputers to automatic control systems. Any digital computer system, of which the microcomputer is but one example, performs its computations in digital form. An instrument whose output is in digital form is therefore particularly advantageous in such applications, as it can be interfaced directly to the control

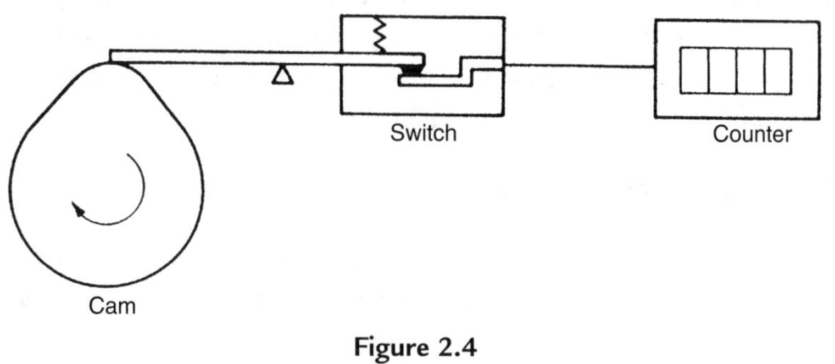


**Figure 2.4**
Rev counter.

computer. Analogue instruments must be interfaced to the microcomputer by an analogue-to-digital (A/D) converter, which converts the analogue output signal from the instrument into an equivalent digital quantity that can be read into the computer. This conversion has several disadvantages. First, the A/D converter adds a significant cost to the system. Second, a finite time is involved in the process of converting an analogue signal to a digital quantity, and this time can be critical in the control of fast processes where the accuracy of control depends on the speed of the controlling computer. Degrading the speed of operation of the control computer by imposing a requirement for A/D conversion thus impairs the accuracy by which the process is controlled.

### 2.2.4 Indicating Instruments and Instruments with a Signal Output

The final way in which instruments can be divided is between those that merely give an audio or visual indication of the magnitude of the physical quantity measured and those that give an output in the form of a measurement signal whose magnitude is proportional to the measured quantity.

The class of indicating instruments normally includes all null-type instruments and most passive ones. Indicators can also be further divided into those that have an analogue output and those that have a digital display. A common analogue indicator is the liquid-in-glass thermometer. Another common indicating device, which exists in both analogue and digital forms, is the bathroom scale. The older mechanical form of this is an analogue type of instrument that gives an output consisting of a rotating pointer moving against a scale (or sometimes a rotating scale moving against a pointer). More recent electronic forms of bathroom scales have a digital output consisting of numbers presented on an electronic display. One major drawback with indicating devices is that human intervention is required to read and record a measurement. This process is particularly prone to error in the case of analogue output displays, although digital displays are not very prone to error unless the human reader is careless.

Instruments that have a signal-type output are used commonly as part of automatic control systems. In other circumstances, they can also be found in measurement systems where the output measurement signal is recorded in some way for later use. This subject is covered in later chapters. Usually, the measurement signal involved is an electrical voltage, but it can take other forms in some systems, such as an electrical current, an optical signal, or a pneumatic signal.

### 2.2.5 Smart and Nonsmart Instruments

The advent of the microprocessor has created a new division in instruments between those that do incorporate a microprocessor (smart) and those that don't. Smart devices are considered in detail in Chapter 11.

## 2.3 Static Characteristics of Instruments

If we have a thermometer in a room and its reading shows a temperature of 20°C, then it does not really matter whether the true temperature of the room is 19.5 or 20.5°C. Such small variations around 20°C are too small to affect whether we feel warm enough or not. Our bodies cannot discriminate between such close levels of temperature and therefore a thermometer with an inaccuracy of ±0.5°C is perfectly adequate. If we had to measure the temperature of certain chemical processes, however, a variation of 0.5°C might have a significant effect on the rate of reaction or even the products of a process. A measurement inaccuracy much less than ±0.5°C is therefore clearly required.

Accuracy of measurement is thus one consideration in the choice of instrument for a particular application. Other parameters, such as sensitivity, linearity, and the reaction to ambient temperature changes, are further considerations. These attributes are collectively known as the static characteristics of instruments and are given in the data sheet for a particular instrument. It is important to note that values quoted for instrument characteristics in such a data sheet only apply when the instrument is used under specified standard calibration conditions. Due allowance must be made for variations in the characteristics when the instrument is used in other conditions.

The various static characteristics are defined in the following paragraphs.

### *2.3.1 Accuracy and Inaccuracy (Measurement Uncertainty)*

The *accuracy* of an instrument is a measure of how close the output reading of the instrument is to the correct value. In practice, it is more usual to quote the *inaccuracy* or *measurement uncertainty* value rather than the accuracy value for an instrument. Inaccuracy or measurement uncertainty is the extent to which a reading might be wrong and is often quoted as a percentage of the full-scale (f.s.) reading of an instrument.

The aforementioned example carries a very important message. Because the maximum measurement error in an instrument is usually related to the full-scale reading of the instrument, measuring quantities that are substantially less than the full-scale reading means that the possible measurement error is amplified. For this reason, it is an important system design rule that instruments are chosen such that their range is appropriate to the spread of values being measured in order that the best possible accuracy is maintained in instrument readings. Clearly, if we are measuring pressures with expected values between 0 and 1 bar, we would not use an instrument with a measurement range of 0–10 bar.

## Example 2.1

A pressure gauge with a measurement range of 0–10 bar has a quoted inaccuracy of ±1.0% f.s. (±1% of full-scale reading).

(a) What is the maximum measurement error expected for this instrument?
(b) What is the likely measurement error expressed as a percentage of the output reading if this pressure gauge is measuring a pressure of 1 bar?

■

## Solution

(a) The maximum error expected in any measurement reading is 1.0% of the full-scale reading, which is 10 bar for this particular instrument. Hence, the maximum likely error is 1.0% × 10 bar = 0.1 bar.
(b) The maximum measurement error is a constant value related to the full-scale reading of the instrument, irrespective of the magnitude of the quantity that the instrument is actually measuring. In this case, as worked out earlier, the magnitude of the error is 0.1 bar. Thus, when measuring a pressure of 1 bar, the maximum possible error of 0.1 bar is 10% of the measurement value.

■

### *2.3.2 Precision/Repeatability/Reproducibility*

*Precision* is a term that describes an instrument's degree of freedom from random errors. If a large number of readings are taken of the same quantity by a high-precision instrument, then the spread of readings will be very small. Precision is often, although incorrectly, confused with accuracy. High precision does not imply anything about measurement accuracy. A high-precision instrument may have a low accuracy. Low accuracy measurements from a high-precision instrument are normally caused by a bias in the measurements, which is removable by recalibration.

The terms repeatability and reproducibility mean approximately the same but are applied in different contexts, as given later. *Repeatability* describes the closeness of output readings when the same input is applied repetitively over a short period of time, with the same measurement conditions, same instrument and observer, same location, and same conditions of use maintained throughout. *Reproducibility* describes the closeness of output readings for the same input when there are changes in the method of measurement, observer, measuring instrument, location, conditions of use, and time of measurement. Both terms thus describe the spread of output readings for the same input. This spread is referred to as repeatability if the measurement conditions are constant and as reproducibility if the measurement conditions vary.

The degree of repeatability or reproducibility in measurements from an instrument is an alternative way of expressing its precision. Figure 2.5 illustrates this more clearly by showing results of tests on three industrial robots programmed to place components at a particular point on a table. The target point was at the center of the concentric circles shown, and black dots represent points where each robot actually deposited components at each attempt. Both the accuracy and the precision of Robot 1 are shown to be low in this trial. Robot 2 consistently puts the component down at approximately the same place but this is the wrong point. Therefore, it has high precision but low accuracy. Finally, Robot 3 has both high precision and high accuracy because it consistently places the component at the correct target position.

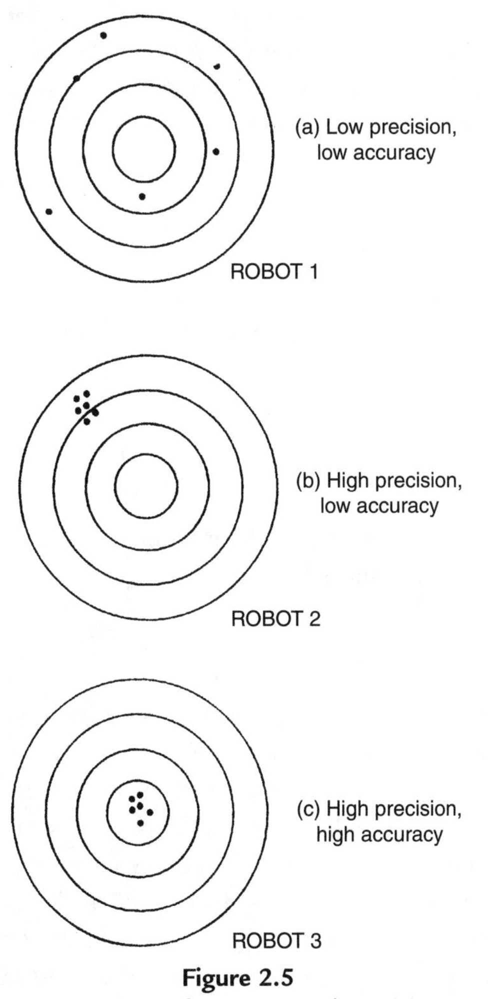


**Figure 2.5**
Comparison of accuracy and precision.

### 2.3.3 Tolerance

Tolerance is a term that is closely related to accuracy and defines the maximum error that is to be expected in some value. While it is not, strictly speaking, a static characteristic of measuring instruments, it is mentioned here because the accuracy of some instruments is sometimes quoted as a tolerance value. When used correctly, tolerance describes the maximum deviation of a manufactured component from some specified value. For instance, crankshafts are machined with a diameter tolerance quoted as so many micrometers ($10^{-6}$ m), and electric circuit components such as resistors have tolerances of perhaps 5%.

## Example 2.2

A packet of resistors bought in an electronics component shop gives the nominal resistance value as 1000 Ω and the manufacturing tolerance as ±5%. If one resistor is chosen at random from the packet, what is the minimum and maximum resistance value that this particular resistor is likely to have?

## Solution

The minimum likely value is 1000 Ω – 5% = 950 Ω.
The maximum likely value is 1000 Ω + 5% = 1050 Ω.

### 2.3.4 Range or Span

The range or span of an instrument defines the minimum and maximum values of a quantity that the instrument is designed to measure.

### 2.3.5 Linearity

It is normally desirable that the output reading of an instrument is linearly proportional to the quantity being measured. The Xs marked on Figure 2.6 show a plot of typical output readings of an instrument when a sequence of input quantities are applied to it. Normal procedure is to draw a good fit straight line through the Xs, as shown in Figure 2.6. (While this can often be done with reasonable accuracy by eye, it is always preferable to apply a mathematical least-squares line-fitting technique, as described in Chapter 8.) Nonlinearity is then defined as the maximum deviation of any of the output readings marked X from this straight line. Nonlinearity is usually expressed as a percentage of full-scale reading.

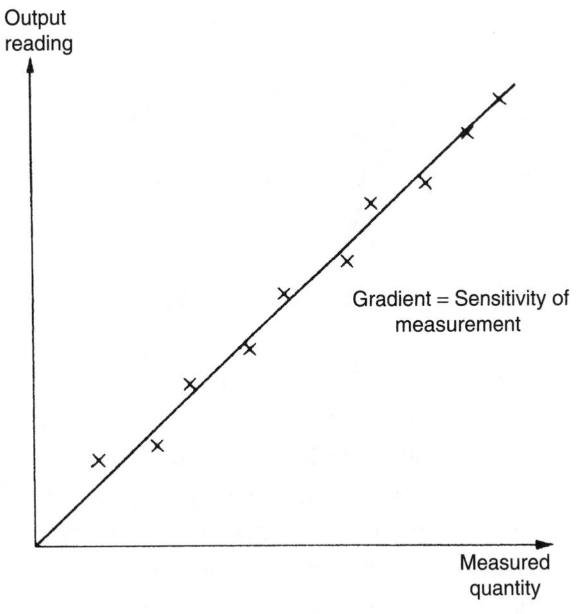


**Figure 2.6**
Instrument output characteristic.

### *2.3.6 Sensitivity of Measurement*

The sensitivity of measurement is a measure of the change in instrument output that occurs when the quantity being measured changes by a given amount. Thus, sensitivity is the ratio:

$$\frac{\text{scale deflection}}{\text{value of measurand producing deflection}}$$

The sensitivity of measurement is therefore the slope of the straight line drawn on Figure 2.6. If, for example, a pressure of 2 bar produces a deflection of 10 degrees in a pressure transducer, the sensitivity of the instrument is 5 degrees/bar (assuming that the deflection is zero with zero pressure applied).

## Example 2.3

The following resistance values of a platinum resistance thermometer were measured at a range of temperatures. Determine the measurement sensitivity of the instrument in ohms/°C.

| Resistance (Ω) | Temperature (°C) |
|---|---|
| 307 | 200 |
| 314 | 230 |
| 321 | 260 |
| 328 | 290 |

## Solution

If these values are plotted on a graph, the straight-line relationship between resistance change and temperature change is obvious.

For a change in temperature of 30°C, the change in resistance is 7 Ω. Hence the measurement sensitivity = 7/30 = 0.233 Ω/°C.

■

### 2.3.7 Threshold

If the input to an instrument is increased gradually from zero, the input will have to reach a certain minimum level before the change in the instrument output reading is of a large enough magnitude to be detectable. This minimum level of input is known as the threshold of the instrument. Manufacturers vary in the way that they specify threshold for instruments. Some quote absolute values, whereas others quote threshold as a percentage of full-scale readings. As an illustration, a car speedometer typically has a threshold of about 15 km/h. This means that, if the vehicle starts from rest and accelerates, no output reading is observed on the speedometer until the speed reaches 15 km/h.

### 2.3.8 Resolution

When an instrument is showing a particular output reading, there is a lower limit on the magnitude of the change in the input measured quantity that produces an observable change in the instrument output. Like threshold, resolution is sometimes specified as an absolute value and sometimes as a percentage of f.s. deflection. One of the major factors influencing the resolution of an instrument is how finely its output scale is divided into subdivisions. Using a car speedometer as an example again, this has subdivisions of typically 20 km/h. This means that when the needle is between the scale markings, we cannot estimate speed more accurately than to the nearest 5 km/h. This value of 5 km/h thus represents the resolution of the instrument.

### 2.3.9 Sensitivity to Disturbance

All calibrations and specifications of an instrument are only valid under controlled conditions of temperature, pressure, and so on. These standard ambient conditions are usually defined in the instrument specification. As variations occur in the ambient temperature, certain static instrument characteristics change, and the sensitivity to disturbance is a measure of the magnitude of this change. Such environmental changes affect instruments in two main ways, known as *zero drift* and *sensitivity drift*. Zero drift is sometimes known by the alternative term, *bias*.

Zero drift or bias describes the effect where the zero reading of an instrument is modified by a change in ambient conditions. This causes a constant error that exists over the full range of measurement of the instrument. The mechanical form of a bathroom scale is a common example of an instrument prone to zero drift. It is quite usual to find that there is a reading of perhaps 1 kg with no one on the scale. If someone of known weight 70 kg were to get on the scale, the reading would be 71 kg, and if someone of known weight 100 kg were to get on the scale, the reading would be 101 kg. Zero drift is normally removable by calibration. In the case of the bathroom scale just described, a thumbwheel is usually provided that can be turned until the reading is zero with the scales unloaded, thus removing zero drift.

The typical unit by which such zero drift is measured is volts/°C. This is often called the *zero drift coefficient* related to temperature changes. If the characteristic of an instrument is sensitive to several environmental parameters, then it will have several zero drift coefficients, one for each environmental parameter. A typical change in the output characteristic of a pressure gauge subject to zero drift is shown in Figure 2.7a.

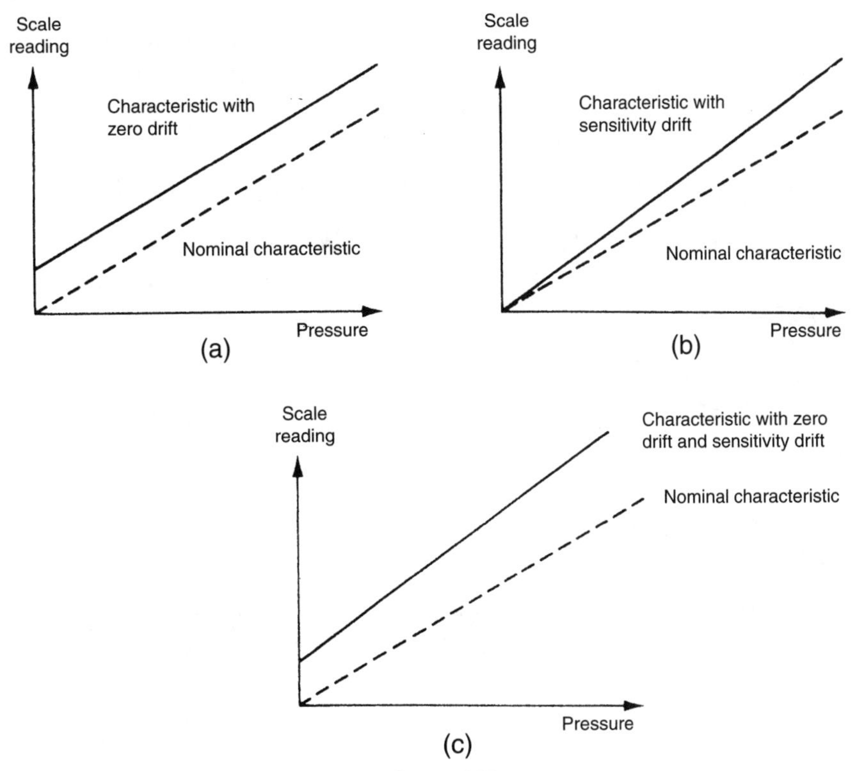


**Figure 2.7**
Effects of disturbance: (a) zero drift, (b) sensitivity drift, and (c) zero drift plus sensitivity drift.

*Sensitivity drift* (also known as *scale factor drift*) defines the amount by which an instrument's sensitivity of measurement varies as ambient conditions change. It is quantified by sensitivity drift coefficients that define how much drift there is for a unit change in each environmental parameter that the instrument characteristics are sensitive to. Many components within an instrument are affected by environmental fluctuations, such as temperature changes: for instance, the modulus of elasticity of a spring is temperature dependent. Figure 2.7b shows what effect sensitivity drift can have on the output characteristic of an instrument. Sensitivity drift is measured in units of the form (angular degree/bar)/°C. If an instrument suffers both zero drift and sensitivity drift at the same time, then the typical modification of the output characteristic is shown in Figure 2.7c.

## Example 2.4

The following table shows output measurements of a voltmeter under two sets of conditions:

(a) Use in an environment kept at 20°C which is the temperature that it was calibrated at.
(b) Use in an environment at a temperature of 50°C.

| Voltage readings at calibration temperature of 20°C (assumed correct) | Voltage readings at temperature of 50°C |
|---|---|
| 10.2 | 10.5 |
| 20.3 | 20.6 |
| 30.7 | 40.0 |
| 40.8 | 50.1 |

Determine the zero drift when it is used in the 50°C environment, assuming that the measurement values when it was used in the 20°C environment are correct. Also calculate the zero drift coefficient.

■

## Solution

Zero drift at the temperature of 50°C is the constant difference between the pairs of output readings, that is, 0.3 volts.

The zero drift coefficient is the magnitude of drift (0.3 volts) divided by the magnitude of the temperature change causing the drift (30°C). Thus the zero drift coefficient is $0.3/30 = 0.01$ volts/°C.

■

## Example 2.5

A spring balance is calibrated in an environment at a temperature of 20°C and has the following deflection/load characteristic:

| Load (kg) | 0 | 1 | 2 | 3 |
|---|---|---|---|---|
| Deflection (mm) | 0 | 20 | 40 | 60 |

It is then used in an environment at a temperature of 30°C, and the following deflection/load characteristic is measured:

| Load (kg) | 0 | 1 | 2 | 3 |
|---|---|---|---|---|
| Deflection (mm) | 5 | 27 | 49 | 71 |

Determine the zero drift and sensitivity drift per °C change in ambient temperature. ■

## Solution

At 20°C, deflection/load characteristic is a straight line. Sensitivity = 20 mm/kg.
At 30°C, deflection/load characteristic is still a straight line. Sensitivity = 22 mm/kg.
Zero drift (bias) = 5 mm (the no-load deflection)
Sensitivity drift = 2 mm/kg
Zero drift/°C = 5/10 = 0.5 mm/°C
Sensitivity drift/°C = 2/10 = 0.2 (mm/kg)/°C

■

### *2.3.10 Hysteresis Effects*

Figure 2.8 illustrates the output characteristic of an instrument that exhibits *hysteresis*. If the input measured quantity to the instrument is increased steadily from a negative value, the output reading varies in the manner shown in curve A. If the input variable is then decreased steadily, the output varies in the manner shown in curve B. The noncoincidence between these loading and unloading curves is known as *hysteresis*. Two quantities are defined, *maximum input hysteresis* and *maximum output hysteresis*, as shown in Figure 2.8. These are normally expressed as a percentage of the full-scale input or output reading, respectively.

Hysteresis is found most commonly in instruments that contain springs, such as a passive pressure gauge (Figure 2.1) and a Prony brake (used for measuring torque). It is also evident when friction forces in a system have different magnitudes depending on the direction of movement, such as in the pendulum-scale mass-measuring device. Devices such as the mechanical flyball (a device for measuring rotational velocity) suffer hysteresis from both of the aforementioned sources because they have friction in moving parts and also contain a spring. Hysteresis can also

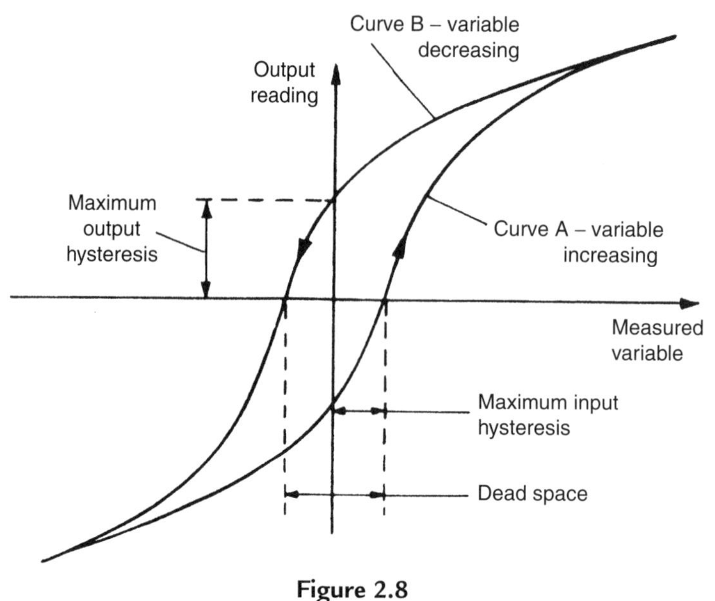


**Figure 2.8**
Instrument characteristic with hysteresis.

occur in instruments that contain electrical windings formed round an iron core, due to magnetic hysteresis in the iron. This occurs in devices such as the variable inductance displacement transducer, the linear variable differential transformer, and the rotary differential transformer.

### *2.3.11 Dead Space*

Dead space is defined as the range of different input values over which there is no change in output value. Any instrument that exhibits hysteresis also displays dead space, as marked on Figure 2.8. Some instruments that do not suffer from any significant hysteresis can still exhibit a dead space in their output characteristics, however. Backlash in gears is a typical cause of dead space and results in the sort of instrument output characteristic shown in Figure 2.9. Backlash is commonly experienced in gear sets used to convert between translational and rotational motion (which is a common technique used to measure translational velocity).

## 2.4 Dynamic Characteristics of Instruments

The static characteristics of measuring instruments are concerned only with the steady-state reading that the instrument settles down to, such as accuracy of the reading.

The dynamic characteristics of a measuring instrument describe its behavior between the time a measured quantity changes value and the time when the instrument output attains a steady value in response. As with static characteristics, any values for dynamic characteristics quoted in instrument

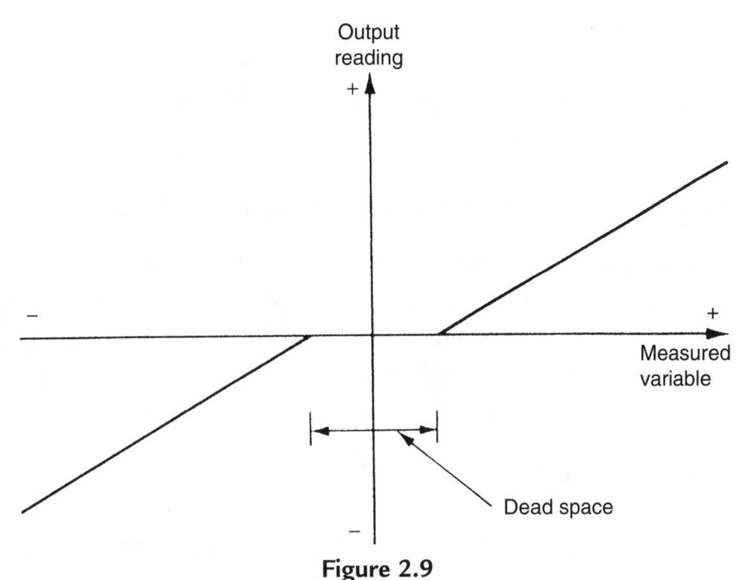


**Figure 2.9**
Instrument characteristic with dead space.

data sheets only apply when the instrument is used under specified environmental conditions. Outside these calibration conditions, some variation in the dynamic parameters can be expected.

In any linear, time-invariant measuring system, the following general relation can be written between input and output for time $(t) > 0$:

$$a_n \frac{d^n q_o}{dt^n} + a_{n-1} \frac{d^{n-1} q_o}{dt^{n-1}} + \cdots + a_1 \frac{dq_o}{dt} + a_0 q_o = b_m \frac{d^m q_i}{dt^m} + b_{m-1} \frac{d^{m-1} q_i}{dt^{m-1}} + \cdots + b_1 \frac{dq_i}{dt} + b_0 q_i, \tag{2.1}$$

where $q_i$ is the measured quantity, $q_o$ is the output reading, and $a_o \ldots a_n$, $b_o \ldots b_m$ are constants.

The reader whose mathematical background is such that Equation (2.1) appears daunting should not worry unduly, as only certain special, simplified cases of it are applicable in normal measurement situations. The major point of importance is to have a practical appreciation of the manner in which various different types of instruments respond when the measurand applied to them varies.

If we limit consideration to that of step changes in the measured quantity only, then Equation (2.1) reduces to

$$a_n \frac{d^n q_o}{dt^n} + a_{n-1} \frac{d^{n-1} q_o}{dt^{n-1}} + \cdots + a_1 \frac{dq_o}{dt} + a_0 q_o = b_0 q_i. \tag{2.2}$$

Further simplification can be made by taking certain special cases of Equation (2.2), which collectively apply to nearly all measurement systems.

### 2.4.1 Zero-Order Instrument

If all the coefficients $a_1 \ldots a_n$ other than $a_0$ in Equation (2.2) are assumed zero, then

$$a_0 q_o = b_0 q_i \quad or \quad q_o = b_0 q_i / a_0 = K q_i, \tag{2.3}$$

where $K$ is a constant known as the instrument sensitivity as defined earlier.

Any instrument that behaves according to Equation (2.3) is said to be of a zero-order type. Following a step change in the measured quantity at time $t$, the instrument output moves immediately to a new value at the same time instant $t$, as shown in Figure 2.10. A potentiometer, which measures motion, is a good example of such an instrument, where the output voltage changes instantaneously as the slider is displaced along the potentiometer track.

### 2.4.2 First-Order Instrument

If all the coefficients $a_2 \ldots a_n$ except for $a_o$ and $a_1$ are assumed zero in Equation (2.2) then

$$a_1 \frac{dq_o}{dt} + a_0 q_o = b_0 q_i. \tag{2.4}$$

Any instrument that behaves according to Equation (2.4) is known as a first-order instrument. If $d/dt$ is replaced by the $D$ operator in Equation (2.4), we get

$$a_1 D q_o + a_0 q_o = b_0 q_i$$

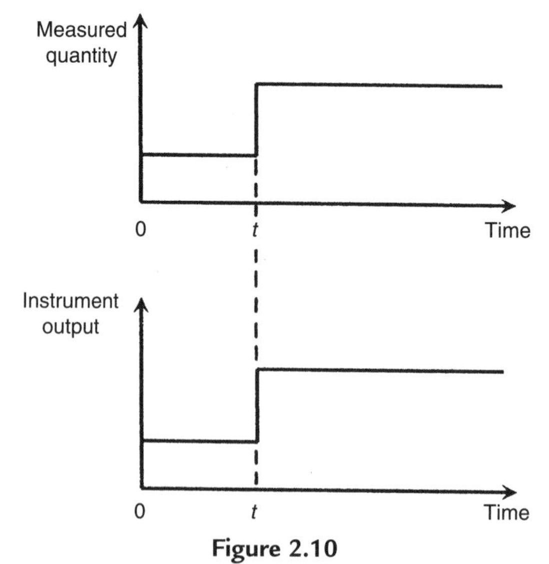


**Figure 2.10**
Zero-order instrument characteristic.

and rearranging this then gives

$$q_o = \frac{(b_0/a_0)q_i}{[1 + (a_1/a_0)D]}. \tag{2.5}$$

Defining $K = b_0/a_0$ as the static sensitivity and $\tau = a_1/a_0$ as the time constant of the system, Equation (2.5) becomes

$$q_o = \frac{Kq_i}{1 + \tau D}. \tag{2.6}$$

If Equation (2.6) is solved analytically, the output quantity $q_0$ in response to a step change in $q_i$ at time $t$ varies with time in the manner shown in Figure 2.11. The time constant $\tau$ of the step response is time taken for the output quantity $q_0$ to reach 63% of its final value.

The thermocouple (see Chapter 14) is a good example of a first-order instrument. It is well known that if a thermocouple at room temperature is plunged into boiling water, the output e.m.f. does not rise instantaneously to a level indicating 100°C, but instead approaches a reading indicating 100°C in a manner similar to that shown in Figure 2.11.

A large number of other instruments also belong to this first-order class: this is of particular importance in control systems where it is necessary to take account of the time lag that occurs between a measured quantity changing in value and the measuring instrument indicating

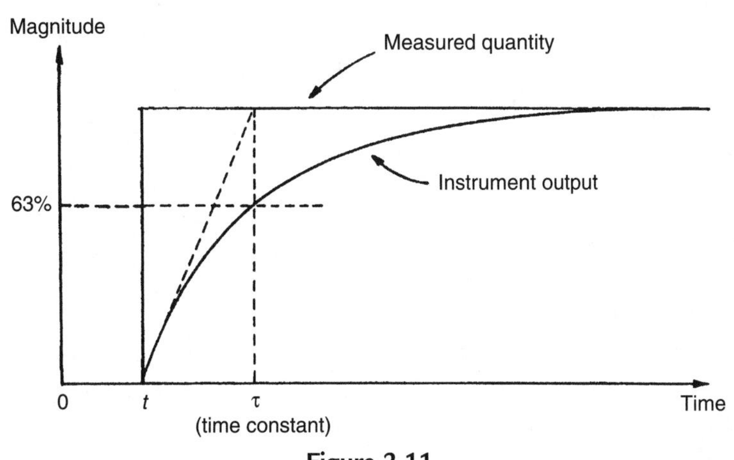


**Figure 2.11**
First-order instrument characteristic.

the change. Fortunately, because the time constant of many first-order instruments is small relative to the dynamics of the process being measured, no serious problems are created.

## Example 2.6

A balloon is equipped with temperature- and altitude-measuring instruments and has radio equipment that can transmit the output readings of these instruments back to the ground. The balloon is initially anchored to the ground with the instrument output readings in steady state. The altitude-measuring instrument is approximately zero order, and the temperature transducer is first order with a time constant of 15 seconds. The temperature on the ground, $T_0$, is 10°C and the temperature $T_x$ at an altitude of $x$ meters is given by the relation: $T_x = T_0 - 0.01x$.

(a) If the balloon is released at time zero, and thereafter rises upward at a velocity of 5 meters/second, draw a table showing the temperature and altitude measurements reported at intervals of 10 seconds over the first 50 seconds of travel. Show also in the table the error in each temperature reading.

(b) What temperature does the balloon report at an altitude of 5000 meters?

■

## Solution

In order to answer this question, it is assumed that the solution of a first-order differential equation has been presented to the reader in a mathematics course. If the reader is not so equipped, the following solution will be difficult to follow.

Let the temperature reported by the balloon at some general time $t$ be $T_r$. Then $T_x$ is related to $T_r$ by the relation:

$$T_r = \frac{T_x}{1+\tau D} = \frac{T_o - 0.01x}{1+\tau D} = \frac{10 - 0.01x}{1+15D}.$$

It is given that $x = 5t$, thus

$$T_r = \frac{10 - 0.05t}{1+15D}.$$

The transient or complementary function part of the solution ($T_x = 0$) is given by $T_{r_{cf}} = Ce^{-t/15}$

The particular integral part of the solution is given by $T_{r_{pi}} = 10 - 0.05(t-15)$

Thus, the whole solution is given by $T_r = T_{r_{cf}} + T_{r_{pi}} = Ce^{-t/15} + 10 - 0.05(t-15)$

Applying initial conditions: At $t = 0$, $T_r = 10$, that is, $10 = Ce^{-0} + 10 - 0.05(-15)$
Thus $C = -0.75$ and the solution can be written as $T_r = 10 - 0.75e^{-t/15} - 0.05(t - 15)$

Using the aforementioned expression to calculate $T_r$ for various values of $t$, the following table can be constructed:

| Time | Altitude | Temperature reading | Temperature error |
|---|---|---|---|
| 0 | 0 | 10 | 0 |
| 10 | 50 | 9.86 | 0.36 |
| 20 | 100 | 9.55 | 0.55 |
| 30 | 150 | 9.15 | 0.65 |
| 40 | 200 | 8.70 | 0.70 |
| 50 | 250 | 8.22 | 0.72 |

(c) At 5000 m, $t = 1000$ seconds. Calculating $T_r$ from the aforementioned expression:

$$T_r = 10 - 0.75e^{-1000/15} - 0.05(1000 - 15).$$

The exponential term approximates to zero and so $T_r$ can be written as

$$T_r \approx 10 - 0.05(985) = -39.25°\text{C}.$$

This result might have been inferred from the table given earlier where it can be seen that the error is converging toward a value of 0.75. For large values of $t$, the transducer reading lags the true temperature value by a period of time equal to the time constant of 15 seconds. In this time, the balloon travels a distance of 75 meters and the temperature falls by 0.75°. Thus for large values of $t$, the output reading is always 0.75° less than it should be.

■

### 2.4.3 Second-Order Instrument

If all coefficients $a_3 \ldots a_n$ other than $a_0$, $a_1$, and $a_2$ in Equation (2.2) are assumed zero, then we get

$$a_2 \frac{d^2 q_o}{dt^2} + a_1 \frac{dq_o}{dt} + a_0 q_o = b_0 q_i. \tag{2.7}$$

Applying the $D$ operator again:

$$a_2 D^2 q_o + a_1 D q_o + a_0 q_o = b_0 q_i,$$

and rearranging:

$$q_o = \frac{b_0 q_i}{a_0 + a_1 D + a_2 D^2}. \tag{2.8}$$

It is convenient to reexpress the variables $a_0$, $a_1$, $a_2$, and $b_0$ in Equation (2.8) in terms of three parameters: $K$ (static sensitivity), $\omega$ (undamped natural frequency), and $\xi$ (damping ratio), where

$$K = b_0/a_0 \quad ; \quad \omega = \sqrt{a_0/a_2} \quad ; \quad \xi = a_1/2\sqrt{a_0 a_2}$$

$\xi$ can be written as

$$\xi = \frac{a_1}{2a_0\sqrt{a_2/a_0}} = \frac{a_1\omega}{2a_0}.$$

If Equation (2.8) is now divided through by $a_0$, we get

$$q_o = \frac{(b_0/a_0)q_i}{1 + (a_1/a_0)D + (a_2/a_0)D^2}. \tag{2.9}$$

The terms in Equation (2.9) can be written in terms of $\omega$ and $\xi$ as follows:

$$\frac{b_o}{a_0} = K \quad ; \quad \left(\frac{a_1}{a_0}\right)D = \frac{2\xi D}{\omega} \quad ; \quad \left(\frac{a_2}{a_0}\right)D^2 = \frac{D^2}{\omega^2}.$$

Hence, dividing Equation (2.9) through by $q_i$ and substituting for $a_0$, $a_1$, and $a_2$ gives

$$\frac{q_o}{q_i} = \frac{K}{D^2/\omega^2 + 2\xi D/\omega + 1}. \tag{2.10}$$

This is the standard equation for a second-order system, and any instrument whose response can be described by it is known as a second-order instrument. If Equation (2.9) is solved analytically, the shape of the step response obtained depends on the value of the damping ratio parameter $\xi$. The output responses of a second-order instrument for various values of $\xi$ following a step change in the value of the measured quantity at time $t$ are shown in Figure 2.12. For case A, where $\xi = 0$, there is no damping and the instrument output exhibits constant amplitude oscillations when disturbed by any change in the physical quantity measured. For light damping of $\xi = 0.2$, represented by case B, the response to a step change in input is still oscillatory but the oscillations die down gradually. A further increase in the value of $\xi$ reduces oscillations and overshoots still more, as shown by curves C and D, and finally the response becomes very overdamped, as shown by curve E, where the output reading creeps up slowly toward the correct reading. Clearly, the extreme response curves A and E are grossly unsuitable for any measuring instrument. If an instrument were to be only ever subjected to step inputs, then the design strategy would be to aim toward a damping ratio of 0.707, which gives the critically damped response (C). Unfortunately, most of the physical quantities that instruments are required to measure do not change in the mathematically convenient form of steps, but rather in the form of ramps of varying slopes. As the

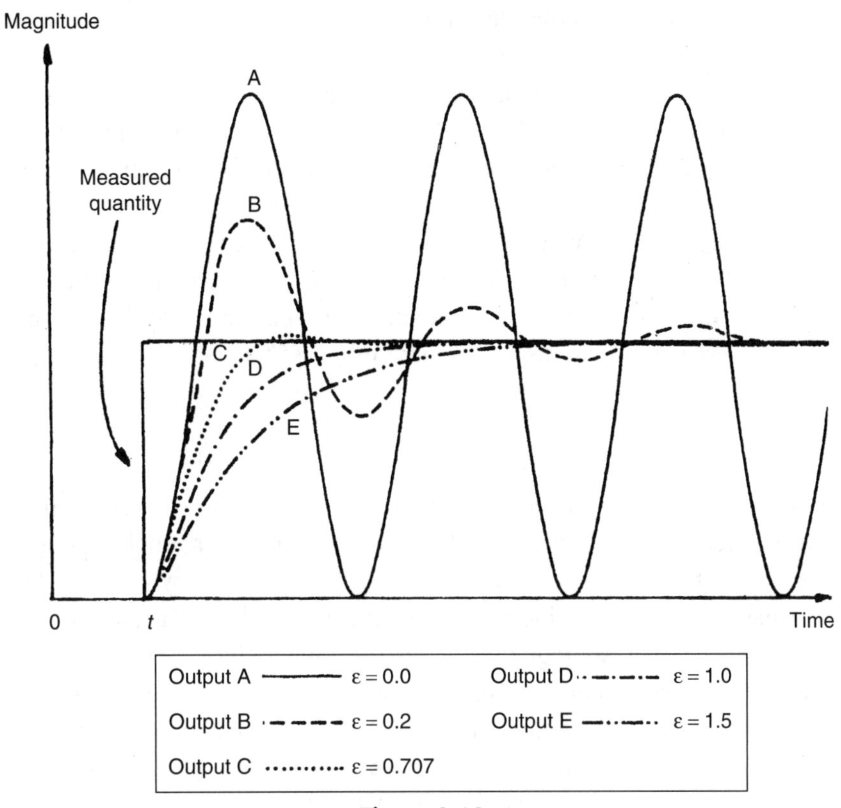


**Figure 2.12**
Response characteristics of second-order instruments.

form of the input variable changes, so the best value for ξ varies, and choice of ξ becomes one of compromise between those values that are best for each type of input variable behavior anticipated. Commercial second-order instruments, of which the accelerometer is a common example, are generally designed to have a damping ratio (ξ) somewhere in the range of 0.6–0.8.

## 2.5 Necessity for Calibration

The foregoing discussion has described the static and dynamic characteristics of measuring instruments in some detail. However, an important qualification that has been omitted from this discussion is that an instrument only conforms to stated static and dynamic patterns of behavior after it has been calibrated. It can normally be assumed that a new instrument will have been calibrated when it is obtained from an instrument manufacturer and will therefore initially behave according to the characteristics stated in the specifications. During use, however, its behavior will gradually diverge from the stated specification for a variety of reasons. Such reasons include mechanical wear and the effects of dirt, dust, fumes, and chemicals in the operating environment. The rate of divergence from standard specifications varies according to the type of instrument, the frequency of usage, and the severity of the operating conditions. However, there will come a time, determined by practical knowledge, when the characteristics of the instrument will have drifted from the standard specification by an unacceptable amount. When this situation is reached, it is necessary to recalibrate the instrument back to standard specifications. Such recalibration is performed by adjusting the instrument at each point in its output range until its output readings are the same as those of a second standard instrument to which the same inputs are applied. This second instrument is one kept solely for calibration purposes whose specifications are accurately known. Calibration procedures are discussed more fully in Chapter 4.

## 2.6 Summary

This chapter began by reviewing various different classes of instruments and considering how these differences affect their typical usage. We saw, for example, that null-type instruments are favored for calibration duties because of their superior accuracy, whereas deflection-type instruments are easier to use for routine measurements. We also looked at the distinction between active and passive instruments, analogue and digital instruments, indicators and signal output-type instruments, and, finally, smart and nonsmart instruments. Following this, we went on to look at the various static characteristics of instruments. These define the quality of measurements when an instrument output has settled to a steady reading. Several important lessons arose out of this coverage. In particular, we saw the important distinction between accuracy and precision, which are often equated incorrectly as meaning the same thing. We saw that high precision does not promise anything at all about measurement accuracy; in fact, a high-precision instrument can sometimes give very poor measurement accuracy. The final topic covered in this chapter was the dynamic characteristics of instruments. We saw that there are three kinds of dynamic characteristics: zero order, first order, and second order. Analysis of these showed that both first- and second-order instruments take time to settle to a steady-state reading when the measured quantity changes. It is therefore necessary to wait until the dynamic motion has ended before a reading is recorded. This places a serious limitation on the use of first- and second-order instruments to make repeated measurements. Clearly, the frequency of repeated measurements is limited by the time taken by the instrument to settle to a steady-state reading.

## 2.7 Problems

2.1. Briefly explain four ways in which measuring instruments can be subdivided into different classes according to their mode of operation, giving examples of instruments that fall into each class.

2.2. Explain what is meant by
(a) active instruments
(b) passive instruments
Give examples of each and discuss the relative merits of these two classes of instruments.

2.3. Discuss the advantages and disadvantages of null and deflection types of measuring instruments. What are null types of instruments mainly used for and why?

2.4. What are the differences between analogue and digital instruments? What advantages do digital instruments have over analogue ones?

2.5. Explain the difference between static and dynamic characteristics of measuring instruments.

2.6. Briefly define and explain all the static characteristics of measuring instruments.

2.7. How is the accuracy of an instrument usually defined? What is the difference between accuracy and precision?

2.8. Draw sketches to illustrate the dynamic characteristics of the following:
(a) zero-order instrument
(b) first-order instrument
(c) second-order instrument
In the case of a second-order instrument, indicate the effect of different degrees of damping on the time response.

2.9. State briefly how the dynamic characteristics of an instrument affect its usage.

2.10. A tungsten resistance thermometer with a range of –270 to +1100°C has a quoted inaccuracy of ±1.5% of full-scale reading. What is the likely measurement error when it is reading a temperature of 950°C?

2.11. A batch of steel rods is manufactured to a nominal length of 5 meters with a quoted tolerance of ±2%. What is the longest and shortest length of rod to be expected in the batch?

2.12. What is the measurement range for a micrometer designed to measure diameters between 5.0 and 7.5 cm?

2.13. A tungsten/5% rhenium–tungsten/26% rhenium thermocouple has an output e.m.f. as shown in the following table when its hot (measuring) junction is at the temperatures shown. Determine the sensitivity of measurement for the thermocouple in mV/°C.

| mV | 4.37 | 8.74 | 13.11 | 17.48 |
|---|---|---|---|---|
| °C | 250 | 500 | 750 | 1000 |

2.14. Define sensitivity drift and zero drift. What factors can cause sensitivity drift and zero drift in instrument characteristics?

2.15. (a) An instrument is calibrated in an environment at a temperature of 20°C and the following output readings $y$ are obtained for various input values $x$:

| $y$ | 13.1 | 26.2 | 39.3 | 52.4 | 65.5 | 78.6 |
|---|---|---|---|---|---|---|
| $x$ | 5 | 10 | 15 | 20 | 25 | 30 |

Determine the measurement sensitivity, expressed as the ratio $y/x$.

(b) When the instrument is subsequently used in an environment at a temperature of 50°C, the input/output characteristic changes to the following:

| $y$ | 14.7 | 29.4 | 44.1 | 58.8 | 73.5 | 88.2 |
|---|---|---|---|---|---|---|
| $x$ | 5 | 10 | 15 | 20 | 25 | 30 |

Determine the new measurement sensitivity. Hence determine the sensitivity drift due to the change in ambient temperature of 30°C.

2.16. The following temperature measurements were taken with an infrared thermometer that produced biased measurements due to the instrument being out of calibration. Calculate the bias in the measurements.

| Values measured by uncalibrated instrument (°C) | Correct value of temperature (°C) |
|---|---|
| 20 | 21.5 |
| 35 | 36.5 |
| 50 | 51.5 |
| 65 | 66.5 |

2.17. A load cell is calibrated in an environment at a temperature of 21°C and has the following deflection/load characteristic:

| Load (kg) | 0 | 50 | 100 | 150 | 200 |
|---|---|---|---|---|---|
| Deflection (mm) | 0.0 | 1.0 | 2.0 | 3.0 | 4.0 |

When used in an environment at 35°C, its characteristic changes to the following:

| Load (kg) | 0 | 50 | 100 | 150 | 200 |
|---|---|---|---|---|---|
| Deflection (mm) | 0.2 | 1.3 | 2.4 | 3.5 | 4.6 |

(a) Determine the sensitivity at 21 and 35°C.
(b) Calculate the total zero drift and sensitivity drift at 35°C.
(c) Hence determine the zero drift and sensitivity drift coefficients (in units of μm/°C and (μm per kg)/(°C).

2.18. An unmanned submarine is equipped with temperature- and depth-measuring instruments and has radio equipment that can transmit the output readings of these instruments back to the surface. The submarine is initially floating on the surface of the sea with the instrument output readings in steady state. The depth-measuring instrument is approximately zero order and the temperature transducer first order with a time constant of 50 seconds. The water temperature on the sea surface, $T_0$, is 20°C and the temperature $T_x$ at a depth of $x$ meters is given by the relation:

$$T_x = T_0 - 0.01x$$

(a) If the submarine starts diving at time zero, and thereafter goes down at a velocity of 0.5 meters/second, draw a table showing the temperature and depth measurements reported at intervals of 100 seconds over the first 500 seconds of travel. Show also in the table the error in each temperature reading.

(b) What temperature does the submarine report at a depth of 1000 meters?

2.19. Write down the general differential equation describing the dynamic response of a second-order measuring instrument and state the expressions relating the static sensitivity, undamped natural frequency, and damping ratio to the parameters in this differential equation. Sketch the instrument response for cases of heavy damping, critical damping, and light damping and state which of these is the usual target when a second-order instrument is being designed.

CHAPTER 3

# Measurement Uncertainty




**Measurement and Instrumentation: Theory and Application**

## 3.1 Introduction

We have already been introduced to the subject of measurement uncertainty in the last chapter, in the context of defining the accuracy characteristic of a measuring instrument. The existence of measurement uncertainty means that we would be entirely wrong to assume (although the uninitiated might assume this) that the output of a measuring instrument or larger measurement system gives the exact value of the measured quantity. Measurement errors are impossible to avoid, although we can minimize their magnitude by good measurement system design accompanied by appropriate analysis and processing of measurement data.

We can divide errors in measurement systems into those that arise during the measurement process and those that arise due to later corruption of the measurement signal by induced noise during transfer of the signal from the point of measurement to some other point. This chapter considers only the first of these, with discussion on induced noise being deferred to Chapter 6.

It is extremely important in any measurement system to reduce errors to the minimum possible level and then to quantify the maximum remaining error that may exist in any instrument output reading. However, in many cases, there is a further complication that the final output from a measurement system is calculated by combining together two or more measurements of separate physical variables. In this case, special consideration must also be given to determining how the calculated error levels in each separate measurement should be combined to give the best estimate of the most likely error magnitude in the calculated output quantity. This subject is considered in Section 3.7.

The starting point in the quest to reduce the incidence of errors arising during the measurement process is to carry out a detailed analysis of all error sources in the system. Each of these error sources can then be considered in turn, looking for ways of eliminating or at least reducing the magnitude of errors. Errors arising during the measurement process can be divided into two groups, known as systematic errors and random errors.

*Systematic errors* describe errors in the output readings of a measurement system that are consistently on one side of the correct reading, that is, either all errors are positive or are all negative. (Some books use alternative name bias errors for systematic errors, although this is not entirely incorrect, as systematic errors include errors such as sensitivity drift that are not biases.) Two major sources of systematic errors are system disturbance during measurement and the effect of environmental changes (sometimes known as *modifying inputs*), as discussed in Sections 3.4.1 and 3.4.2. Other sources of systematic error include bent meter needles, use of uncalibrated instruments, drift in instrument characteristics, and poor cabling practices. Even when systematic errors due to these factors have been reduced or eliminated, some errors remain that are inherent in the manufacture of an instrument. These are quantified by the accuracy value quoted in the published specifications contained in the instrument data sheet.

*Random errors*, which are also called *precision errors* in some books, are perturbations of the measurement either side of the true value caused by random and unpredictable effects, such that positive errors and negative errors occur in approximately equal numbers for a series of measurements made of the same quantity. Such perturbations are mainly small, but large perturbations occur from time to time, again unpredictably. Random errors often arise when measurements are taken by human observation of an analogue meter, especially where this involves interpolation between scale points. Electrical noise can also be a source of random errors. To a large extent, random errors can be overcome by taking the same measurement a number of times and extracting a value by averaging or other statistical techniques, as discussed in Section 3.5. However, any quantification of the measurement value and statement of error bounds remains a statistical quantity. Because of the nature of random errors and the fact that large perturbations in the measured quantity occur from time to time, the best that we can do is to express measurements in probabilistic terms: we may be able to assign a 95 or even 99% confidence level that the measurement is a certain value within error bounds of, say, $\pm 1\%$, but we can never attach a 100% probability to measurement values that are subject to random errors. In other words, even if we say that the maximum error is $\leq \pm 0.5\%$ of the measurement reading, there is still a 1% chance that the error is greater than $\pm 0.5\%$.

Finally, a word must be said about the distinction between systematic and random errors. Error sources in the measurement system must be examined carefully to determine what type of error is present, systematic or random, and to apply the appropriate treatment. In the case of manual data measurements, a human observer may make a different observation at each attempt, but it is often reasonable to assume that the errors are random and that the mean of these readings is likely to be close to the correct value. However, this is only true as long as the human observer is not introducing a parallax-induced systematic error as well by persistently reading the position of a needle against the scale of an analogue meter from one side rather than from directly above. A human-induced systematic error is also introduced if an instrument with a first-order characteristic is read before it has settled to its final reading. Wherever a systematic error exists alongside random errors, correction has to be made for the systematic error in the measurements before statistical techniques are applied to reduce the effect of random errors.

## 3.2 Sources of Systematic Error

The main sources of systematic error in the output of measuring instruments can be summarized as

- effect of environmental disturbances, often called modifying inputs
- disturbance of the measured system by the act of measurement
- changes in characteristics due to wear in instrument components over a period of time
- resistance of connecting leads

These various sources of systematic error, and ways in which the magnitude of the errors can be reduced, are discussed here.

### *3.2.1 System Disturbance due to Measurement*

Disturbance of the measured system by the act of measurement is a common source of systematic error. If we were to start with a beaker of hot water and wished to measure its temperature with a mercury-in-glass thermometer, then we would take the thermometer, which would initially be at room temperature, and plunge it into the water. In so doing, we would be introducing a relatively cold mass (the thermometer) into the hot water and a heat transfer would take place between the water and the thermometer. This heat transfer would lower the temperature of the water. While the reduction in temperature in this case would be so small as to be undetectable by the limited measurement resolution of such a thermometer, the effect is finite and clearly establishes the principle that, in nearly all measurement situations, the process of measurement disturbs the system and alters the values of the physical quantities being measured.

A particularly important example of this occurs with the orifice plate. This is placed into a fluid-carrying pipe to measure the flow rate, which is a function of the pressure that is measured either side of the orifice plate. This measurement procedure causes a permanent pressure loss in the flowing fluid. The disturbance of the measured system can often be very significant.

Thus, as a general rule, the process of measurement always disturbs the system being measured. The magnitude of the disturbance varies from one measurement system to the next and is affected particularly by the type of instrument used for measurement. Ways of minimizing disturbance of measured systems are important considerations in instrument design. However, an accurate understanding of the mechanisms of system disturbance is a prerequisite for this.

*Measurements in electric circuits*

In analyzing system disturbance during measurements in electric circuits, Thévenin's theorem (see Appendix 2) is often of great assistance. For instance, consider the circuit shown in Figure 3.1a in which the voltage across resistor $R_5$ is to be measured by a voltmeter with resistance $R_m$. Here, $R_m$ acts as a shunt resistance across $R_5$, decreasing the resistance between points $AB$ and so disturbing the circuit. Therefore, the voltage $E_m$ measured by the meter is not the value of the voltage $E_o$ that existed prior to measurement. The extent of the disturbance can be assessed by calculating the open-circuit voltage $E_o$ and comparing it with $E_m$.

Thévenin's theorem allows the circuit of Figure 3.1a comprising two voltage sources and five resistors to be replaced by an equivalent circuit containing a single resistance and one

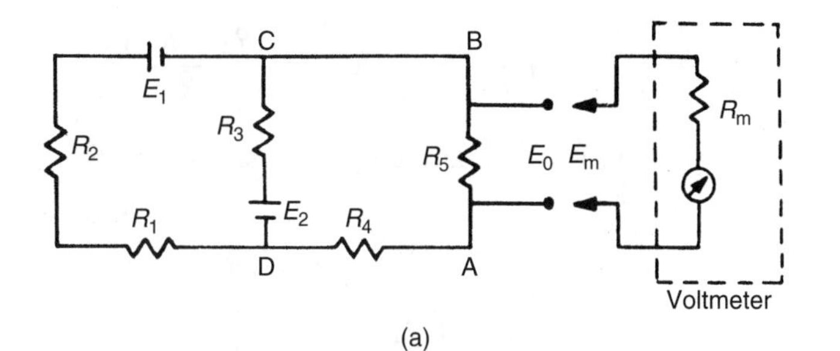


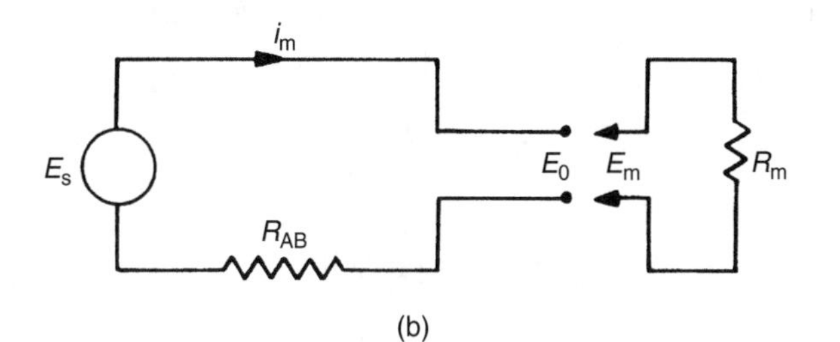


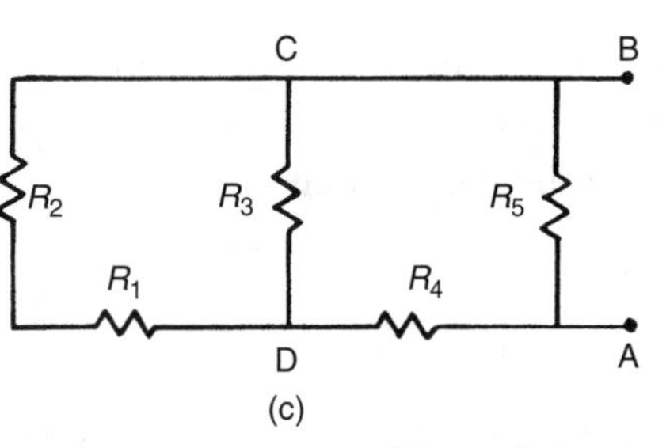


**Figure 3.1**
Analysis of circuit loading: (a) circuit in which the voltage across $R_5$ is to be measured, (b) equivalent circuit by Thévenin's theorem, and (c) circuit used to find the equivalent single.

voltage source, as shown in Figure 3.1b. For the purpose of defining the equivalent single resistance of a circuit by Thévenin's theorem, all voltage sources are represented just by their internal resistance, which can be approximated to zero, as shown in Figure 3.1c. Analysis proceeds by calculating the equivalent resistances of sections of the circuit and building these up until the required equivalent resistance of the whole of the circuit is obtained. Starting at $C$ and $D$, the circuit to the left of $C$ and $D$ consists of a series pair of resistances ($R_1$ and $R_2$) in parallel with $R_3$, and the equivalent resistance can be written as

$$\frac{1}{R_{CD}} = \frac{1}{R_1 + R_2} + \frac{1}{R_3} \quad or \quad R_{CD} = \frac{(R_1 + R_2)R_3}{R_1 + R_2 + R_3}.$$

Moving now to $A$ and $B$, the circuit to the left consists of a pair of series resistances ($R_{CD}$ and $R_4$) in parallel with $R_5$. The equivalent circuit resistance $R_{AB}$ can thus be written as

$$\frac{1}{R_{AB}} = \frac{1}{R_{CD} + R_4} + \frac{1}{R_5} \quad or \quad R_{AB} = \frac{(R_4 + R_{CD})R_5}{R_4 + R_{CD} + R_5}.$$

Substituting for $R_{CD}$ using the expression derived previously, we obtain

$$R_{AB} = \frac{\left[\frac{(R_1 + R_2)R_3}{R_1 + R_2 + R_3} + R_4\right]R_5}{\frac{(R_1 + R_2)R_3}{R_1 + R_2 + R_3} + R_4 + R_5}. \tag{3.1}$$

Defining $I$ as the current flowing in the circuit when the measuring instrument is connected to it, we can write

$$I = \frac{E_o}{R_{AB} + R_m},$$

and the voltage measured by the meter is then given by

$$E_m = \frac{R_m E_o}{R_{AB} + R_m}.$$

In the absence of the measuring instrument and its resistance $R_m$, the voltage across $AB$ would be the equivalent circuit voltage source whose value is $E_o$. The effect of measurement is therefore to reduce the voltage across $AB$ by the ratio given by

$$\frac{E_m}{E_o} = \frac{R_m}{R_{AB} + R_m}. \tag{3.2}$$

It is thus obvious that as $R_m$ gets larger, the ratio $E_m/E_o$ gets closer to unity, showing that the design strategy should be to make $R_m$ as high as possible to minimize disturbance of the measured system. (Note that we did not calculate the value of $E_o$, as this is not required in quantifying the effect of $R_m$.)

At this point, it is interesting to note the constraints that exist when practical attempts are made to achieve a high internal resistance in the design of a moving-coil voltmeter. Such an instrument consists of a coil carrying a pointer mounted in a fixed magnetic field. As current flows through the coil, the interaction between the field generated and the fixed field causes the pointer it carries to turn in proportion to the applied current (for further details, see Chapter 7). The simplest way of increasing the input impedance (the resistance) of the meter is either to increase the number of turns in the coil or to construct the same number of coil turns with a higher resistance material. However, either of these solutions decreases the current flowing in the coil, giving less magnetic torque and thus decreasing the measurement sensitivity of the instrument (i.e., for a given applied voltage, we get less deflection of the pointer). This problem can be overcome by changing the spring constant of the restraining springs of the instrument, such that less torque is required to turn the pointer by a given amount. However, this reduces the ruggedness of the instrument and also demands better pivot design to reduce friction. This highlights a very important but tiresome principle in instrument design: any attempt to improve the performance of an instrument in one respect generally decreases the performance in some other aspect. This is an inescapable fact of life with passive instruments such as the type of voltmeter mentioned and is often the reason for the use of alternative active instruments such as digital voltmeters, where the inclusion of auxiliary power improves performance greatly.

Bridge circuits for measuring resistance values are a further example of the need for careful design of the measurement system. The impedance of the instrument measuring the bridge output voltage must be very large in comparison with the component resistances in the bridge circuit. Otherwise, the measuring instrument will load the circuit and draw current from it. This is discussed more fully in Chapter 9.

## Example 3.1

Suppose that the components of the circuit shown in Figure 3.1a have the following values:

$$R_1 = 400\ \Omega;\ R_2 = 600\ \Omega;\ R_3 = 1000\ \Omega;\ R_4 = 500\ \Omega;\ R_5 = 1000\ \Omega.$$

The voltage across $AB$ is measured by a voltmeter whose internal resistance is 9500 Ω. What is the measurement error caused by the resistance of the measuring instrument?

■

## Solution

Proceeding by applying Thévenin's theorem to find an equivalent circuit to that of Figure 3.1a of the form shown in Figure 3.1b, and substituting the given component values into the equation for $R_{AB}$ (3.1), we obtain

$$R_{AB} = \frac{[(1000^2/2000) + 500]1000}{(1000^2/2000) + 500 + 1000} = \frac{1000^2}{2000} = 500\ \Omega.$$

From Equation (3.2), we have

$$\frac{E_m}{E_o} = \frac{R_m}{R_{AB} + R_m}.$$

The measurement error is given by $(E_o - E_m)$:

$$E_o - E_m = E_o\left(1 - \frac{R_m}{R_{AB} + R_m}\right).$$

Substituting in values:

$$E_o - E_m = E_o\left(1 - \frac{9500}{10{,}000}\right) = 0.95E_o.$$

Thus, the error in the measured value is 5%.

■

### 3.2.2 Errors due to Environmental Inputs

An environmental input is defined as an apparently real input to a measurement system that is actually caused by a change in the environmental conditions surrounding the measurement system. The fact that the static and dynamic characteristics specified for measuring instruments are only valid for particular environmental conditions (e.g., of temperature and pressure) has already been discussed at considerable length in Chapter 2. These specified conditions must be reproduced as closely as possible during calibration exercises because, away from the specified calibration conditions, the characteristics of measuring instruments vary to some extent and cause measurement errors. The magnitude of this environment-induced variation is quantified by the two constants known as sensitivity drift and zero drift, both of which are generally included in the published specifications for an instrument. Such variations of environmental conditions away from the calibration conditions are sometimes described as *modifying inputs* to the measurement system because they modify the output of the system. When such modifying inputs are present, it is often difficult to determine how much of the output change in a measurement system is due to a change in the measured variable and how much is due to a change in environmental conditions. This is illustrated by the following example. Suppose we are given a small closed box and told that it may contain either a mouse or a rat. We are also told that the box weighs 0.1 kg when empty. If we put the box onto a bathroom scale and observe a reading of 1.0 kg, this does not immediately tell us what is in the box because the reading may be due to one of three things:

(a) a 0.9 kg rat in the box (real input)
(b) an empty box with a 0.9 kg bias on the scale due to a temperature change (environmental input)
(c) a 0.4 kg mouse in the box together with a 0.5 kg bias (real + environmental inputs)

Thus, the magnitude of any environmental input must be measured before the value of the measured quantity (the real input) can be determined from the output reading of an instrument.

In any general measurement situation, it is very difficult to avoid environmental inputs, as it is either impractical or impossible to control the environmental conditions surrounding the measurement system. System designers are therefore charged with the task of either reducing the susceptibility of measuring instruments to environmental inputs or, alternatively, quantifying the effects of environmental inputs and correcting for them in the instrument output reading. The techniques used to deal with environmental inputs and minimize their effects on the final output measurement follow a number of routes as discussed later.

### 3.2.3 Wear in Instrument Components

Systematic errors can frequently develop over a period of time because of wear in instrument components. Recalibration often provides a full solution to this problem.

### 3.2.4 Connecting Leads

In connecting together the components of a measurement system, a common source of error is the failure to take proper account of the resistance of connecting leads (or pipes in the case of pneumatically or hydraulically actuated measurement systems). For instance, in typical applications of a resistance thermometer, it is common to find that the thermometer is separated from other parts of the measurement system by perhaps 100 meters. The resistance of such a length of 20-gauge copper wire is 7 Ω, and there is a further complication that such wire has a temperature coefficient of 1 mΩ/°C.

Therefore, careful consideration needs to be given to the choice of connecting leads. Not only should they be of adequate cross section so that their resistance is minimized, but they should be screened adequately if they are thought likely to be subject to electrical or magnetic fields that could otherwise cause induced noise. Where screening is thought essential, then the routing of cables also needs careful planning. In one application in the author's personal experience involving instrumentation of an electric-arc steelmaking furnace, screened signal-carrying cables between transducers on the arc furnace and a control room at the side of the furnace were initially corrupted by high-amplitude 50-Hz noise. However, by changing the route of the cables between the transducers and the control room, the magnitude of this induced noise was reduced by a factor of about ten.

## 3.3 Reduction of Systematic Errors

The prerequisite for the reduction of systematic errors is a complete analysis of the measurement system that identifies all sources of error. Simple faults within a system, such as bent meter needles and poor cabling practices, can usually be rectified readily and inexpensively once they have been identified. However, other error sources require more detailed analysis and treatment. Various approaches to error reduction are considered next.

### *3.3.1 Careful Instrument Design*

Careful instrument design is the most useful weapon in the battle against environmental inputs by reducing the sensitivity of an instrument to environmental inputs to as low a level as possible. For instance, in the design of strain gauges, the element should be constructed from a material whose resistance has a very low temperature coefficient (i.e., the variation of the resistance with temperature is very small). However, errors due to the way in which an instrument is designed are not always easy to correct, and a choice often has to be made between the high cost of redesign and the alternative of accepting the reduced measurement accuracy if redesign is not undertaken.

### *3.3.2 Calibration*

Instrument calibration is a very important consideration in measurement systems and therefore calibration procedures are considered in detail in Chapter 4. All instruments suffer drift in their characteristics, and the rate at which this happens depends on many factors, such as the environmental conditions in which instruments are used and the frequency of their use. Error due to an instrument being out of calibration is never zero, even immediately after the instrument has been calibrated, because there is always some inherent error in the reference instrument that a working instrument is calibrated against during the calibration exercise. Nevertheless, the error immediately after calibration is of low magnitude. The calibration error then grows steadily with the drift in instrument characteristics until the time of the next calibration. The maximum error that exists just before an instrument is recalibrated can therefore be made smaller by increasing the frequency of recalibration so that the amount of drift between calibrations is reduced.

### *3.3.3 Method of Opposing Inputs*

The method of opposing inputs compensates for the effect of an environmental input in a measurement system by introducing an equal and opposite environmental input that cancels it out. One example of how this technique is applied is in the type of millivoltmeter shown in Figure 3.2. This consists of a coil suspended in a fixed magnetic field produced by a permanent magnet. When an unknown voltage is applied to the coil, the magnetic field due to the current interacts with the fixed field and causes the coil (and a pointer attached to the coil) to turn. If the coil resistance $R_{coil}$ is sensitive to temperature, then any environmental input to the system in the form of a temperature change will alter the value of the coil current for a given applied voltage and so alter the pointer output reading. Compensation for this is made by introducing a compensating resistance $R_{comp}$ into the circuit, where $R_{comp}$ has a temperature coefficient equal in magnitude but opposite in sign to that of the coil. Thus, in response to an increase in temperature, $R_{coil}$ increases but $R_{comp}$ decreases, and so the total resistance remains approximately the same.

### *3.3.4 High-Gain Feedback*

The benefit of adding high-gain feedback to many measurement systems is illustrated by considering the case of the voltage-measuring instrument whose block diagram is shown in Figure 3.3. In this system, unknown voltage $E_i$ is applied to a motor of torque constant $K_m$, and the induced torque turns a pointer against the restraining action of a spring with spring

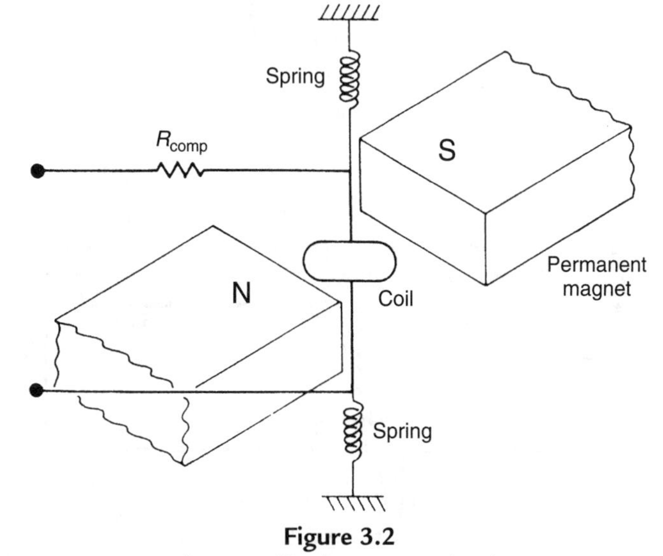


**Figure 3.2**
Analogue millivoltmeter mechanism.

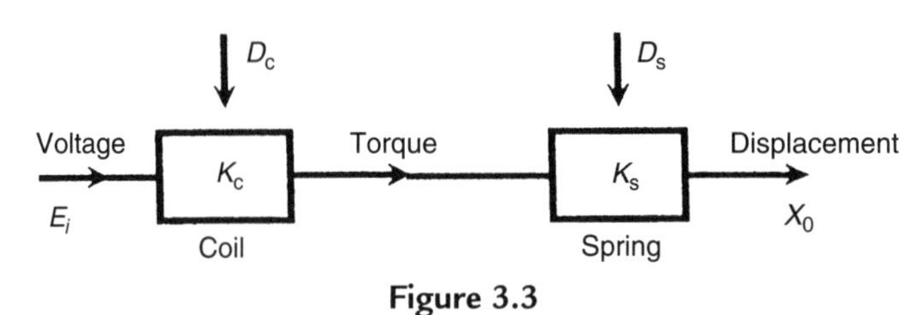


**Figure 3.3**
Block diagram for voltage-measuring instrument.

constant $K_s$. The effect of environmental inputs on the motor and spring constants is represented by variables $D_m$ and $D_s$. In the absence of environmental inputs, the displacement of the pointer $X_o$ is given by $Xo = K_m K_s E_i$. However, in the presence of environmental inputs, both $K_m$ and $K_s$ change, and the relationship between $X_o$ and $E_i$ can be affected greatly. Therefore, it becomes difficult or impossible to calculate $E_i$ from the measured value of $X_o$. Consider now what happens if the system is converted into a high-gain, closed-loop one, as shown in Figure 3.4, by adding an amplifier of gain constant $K_a$ and a feedback device with gain constant $K_f$. Assume also that the effect of environmental inputs on the values of $K_a$ and $K_f$ are represented by $D_a$ and $D_f$. The feedback device feeds back a voltage $E_o$ proportional to the pointer displacement $X_o$. This is compared with the unknown voltage $E_i$ by a comparator and the error is amplified. Writing down the equations of the system, we have

$$E_o = K_f X_o \, ; X_o = (E_i - E_o) K_a K_m K_s = \left(E_i - K_f X_o\right) K_a K_m K_s .$$

Thus

$$E_i K_a K_m K_s = \left(1 + K_f K_a K_m K_s\right) X_o,$$

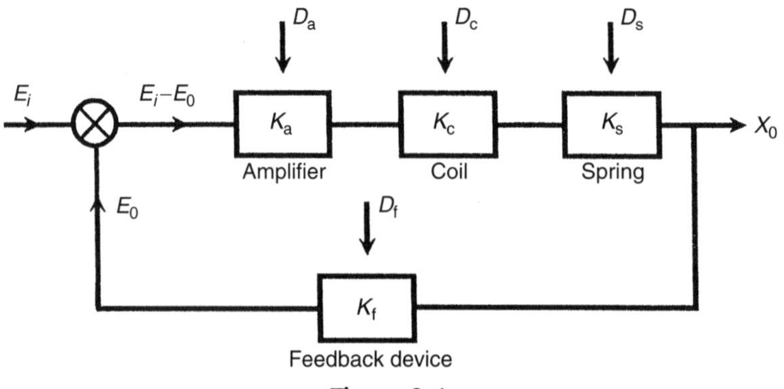


**Figure 3.4**
Block diagram of voltage-measuring instrument with high-gain feedback.

that is,

$$X_o = \frac{K_a K_m K_s}{1 + K_f K_a K_m K_s} E_i. \tag{3.3}$$

Because $K_a$ is very large (it is a high-gain amplifier), $K_f \cdot K_a \cdot K_m \cdot K_s >> 1$, and Equation (3.3) reduces to

$$X_o = E_i / K_f.$$

This is a highly important result because we have reduced the relationship between $X_o$ and $E_i$ to one that involves only $K_f$. The sensitivity of the gain constants $K_a$, $K_m$, and $K_s$ to the environmental inputs $D_a$, $D_m$, and $D_s$ has thereby been rendered irrelevant, and we only have to be concerned with one environmental input, $D_f$. Conveniently, it is usually easy to design a feedback device that is insensitive to environmental inputs: this is much easier than trying to make a motor or spring insensitive. Thus, high-gain feedback techniques are often a very effective way of reducing a measurement system's sensitivity to environmental inputs. However, one potential problem that must be mentioned is that there is a possibility that high-gain feedback will cause instability in the system. Therefore, any application of this method must include careful stability analysis of the system.

### 3.3.5 Signal Filtering

One frequent problem in measurement systems is corruption of the output reading by periodic noise, often at a frequency of 50 Hz caused by pickup through the close proximity of the measurement system to apparatus or current-carrying cables operating on a mains supply. Periodic noise corruption at higher frequencies is also often introduced by mechanical oscillation or vibration within some component of a measurement system. The amplitude of all such noise components can be substantially attenuated by the inclusion of filtering of an appropriate form in the system, as discussed at greater length in Chapter 6. Band-stop filters can be especially useful where corruption is of one particular known frequency, or, more generally, low-pass filters are employed to attenuate all noise in the frequency range of 50 Hz and above. Measurement systems with a low-level output, such as a bridge circuit measuring a strain-gauge resistance, are particularly prone to noise, and Figure 3.5 shows typical corruption of a bridge output by 50-Hz pickup. The beneficial effect of putting a simple passive RC low-pass filter across the output is shown in Figure 3.5.

### 3.3.6 Manual Correction of Output Reading

In the case of errors that are due either to system disturbance during the act of measurement or to environmental changes, a good measurement technician can substantially reduce errors at the output of a measurement system by calculating the effect of such systematic errors and

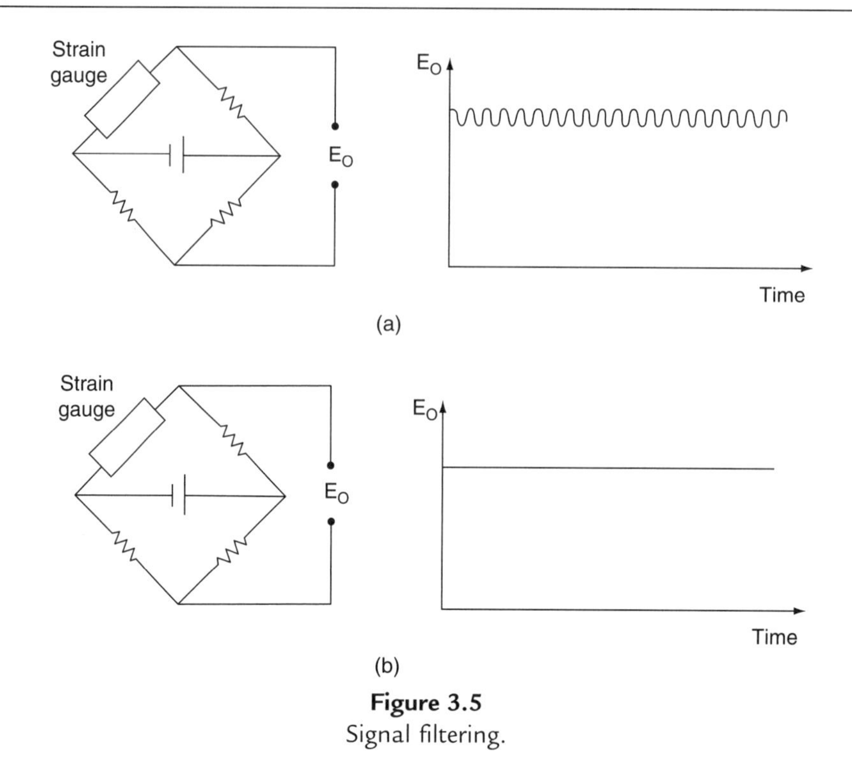


**Figure 3.5**
Signal filtering.

making appropriate correction to the instrument readings. This is not necessarily an easy task and requires all disturbances in the measurement system to be quantified. This procedure is carried out automatically by intelligent instruments.

### *3.3.7 Intelligent Instruments*

Intelligent instruments contain extra sensors that measure the value of environmental inputs and automatically compensate the value of the output reading. They have the ability to deal very effectively with systematic errors in measurement systems, and errors can be attenuated to very low levels in many cases. A more detailed coverage of intelligent instruments can be found in Chapter 11.

## 3.4 Quantification of Systematic Errors

Once all practical steps have been taken to eliminate or reduce the magnitude of systematic errors, the final action required is to estimate the maximum remaining error that may exist in a measurement due to systematic errors. This quantification of the maximum likely systematic error in a measurement requires careful analysis.

### 3.4.1 Quantification of Individual Systematic Error Components

The first complication in the quantification of systematic errors is that it is not usually possiblė to specify an exact value for a component of systematic error, and the quantification has to be in terms of a "best estimate." Once systematic errors have been reduced as far as reasonably possible using the techniques explained in Section 3.3, a sensible approach to estimate the various kinds of remaining systematic error would be as follows.

*Environmental condition errors*

If a measurement is subject to unpredictable environmental conditions, the usual course of action is to assume midpoint environmental conditions and specify the maximum measurement error as $\pm x\%$ of the output reading to allow for the maximum expected deviation in environmental conditions away from this midpoint. Of course, this only refers to the case where the environmental conditions remain essentially constant during a period of measurement but vary unpredictably on perhaps a daily basis. If random fluctuations occur over a short period of time from causes such as random draughts of hot or cold air, this is a random error rather than a systematic error that has to be quantified according to the techniques explained in Section 3.5.

*Calibration errors*

All measuring instruments suffer from drift in their characteristics over a period of time. The schedule for recalibration is set so that the frequency at which an instrument is calibrated means that the drift in characteristics by the time just before the instrument is due for recalibration is kept within an acceptable limit. The maximum error just before the instrument is due for recalibration becomes the basis for estimating the maximum likely error. This error due to the instrument being out of calibration is usually in the form of a bias. The best way to express this is to assume some midpoint value of calibration error and compensate all measurements by this midpoint error. The maximum measurement error over the full period of time between when the instrument has just been calibrated and time just before the next calibration is due can then be expressed as $\pm x\%$ of the output reading.

## Example 3.2

The recalibration frequency of a pressure transducer with a range of 0 to 10 bar is set so that it is recalibrated once the measurement error has grown to $+1\%$ of the full-scale reading. How can its inaccuracy be expressed in the form of a $\pm x\%$ error in the output reading?

■

## Solution

Just before the instrument is due for recalibration, the measurement error will have grown to +0.1 bar (1% of 10 bar). An amount of half this maximum error, that is, 0.05 bar, should be subtracted from all measurements. Having done this, the error just after the instrument has been calibrated will be −0.05 bar (−0.5% of full-scale reading) and the error just before the next recalibration will be +0.05 bar (+0.5% of full-scale reading). Inaccuracy due to calibration error can then be expressed as ±0.05% of full-scale reading.

■

*System disturbance errors*

Disturbance of the measured system by the act of measurement itself introduces a systematic error that can be quantified for any given set of measurement conditions. However, if the quantity being measured and/or the conditions of measurement can vary, the best approach is to calculate the maximum likely error under worst-case system loading and then to express the likely error as a plus or minus value of half this calculated maximum error, as suggested for calibration errors.

*Measurement system loading errors*

These have a similar effect to system disturbance errors and are expressed in the form of $\pm x\%$ of the output reading, where $x$ is half the magnitude of the maximum predicted error under the most adverse loading conditions expected.

### 3.4.2 Calculation of Overall Systematic Error

The second complication in the analysis to quantify systematic errors in a measurement system is the fact that the total systemic error in a measurement is often composed of several separate components, for example, measurement system loading, environmental factors, and calibration errors. A worst-case prediction of maximum error would be to simply add up each separate systematic error. For example, if there are three components of systematic error with a magnitude of ±1% each, a worst-case prediction error would be the sum of the separate errors, that is, ±3%. However, it is very unlikely that all components of error would be at their maximum or minimum values simultaneously. The usual course of action is therefore to combine separate sources of systematic error using a *root-sum-squares method*. Applying this method for $n$ systematic component errors of magnitude $\pm x_1\%$, $\pm x_2\%$, $\pm x_3\%$, ….. $\pm x_n\%$, the best prediction of likely maximum systematic error by the root-sum-squares method is

$$\text{error} = \pm\sqrt{x_1^2 + x_2^2 + x_3^2 + \cdots\cdots + x_n^2}.$$

Before closing this discussion on quantifying systematic errors, a word of warning must be given about the use of manufacturers' data sheets. When instrument manufacturers provide data sheets with an instrument that they have made, the measurement uncertainty or inaccuracy value quoted in the data sheets is the best estimate that the manufacturer can give about the way that the instrument will perform when it is new, used under specified conditions, and recalibrated at the recommended frequency. Therefore, this can only be a starting point in estimating the measurement accuracy that will be achieved when the instrument is actually used. Many sources of systematic error may apply in a particular measurement situation that are not included in the accuracy calculation in the manufacturer's data sheet, and careful quantification and analysis of all systematic errors are necessary, as described earlier.

### Example 3.3

Three separate sources of systematic error are identified in a measurement system and, after reducing the magnitude of these errors as much as possible, the magnitudes of the three errors are estimated to be

System loading: $\pm 1.2\%$
Environmental changes: 0.8%
Calibration error: 0.5%

Calculate the maximum possible total systematic error and the likely system error by the root-mean-square method.

### Solution

The maximum possible system error is $\pm(1.2 + 0.8 + 0.5)\% = \pm 2.5\%$

Applying the root-mean-square method,

$$\text{likely error} = \pm\sqrt{1.2^2 + 0.8^2 + 0.5^2} = \pm 1.53\%.$$

## 3.5 Sources and Treatment of Random Errors

Random errors in measurements are caused by unpredictable variations in the measurement system. In some books, they are known by the alternative name *precision errors*. Typical sources of random error are

- measurements taken by human observation of an analogue meter, especially where this involves interpolation between scale points.
- electrical noise.
- random environmental changes, for example, sudden draught of air.

Random errors are usually observed as small perturbations of the measurement either side of the correct value, that is, positive errors and negative errors occur in approximately equal numbers for a series of measurements made of the same constant quantity. Therefore, random errors can largely be eliminated by calculating the average of a number of repeated measurements. Of course, this is only possible if the quantity being measured remains at a constant value during the repeated measurements. This averaging process of repeated measurements can be done automatically by intelligent instruments, as discussed in Chapter 11.

While the process of averaging over a large number of measurements reduces the magnitude of random errors substantially, it would be entirely incorrect to assume that this totally eliminates random errors. This is because the mean of a number of measurements would only be equal to the correct value of the measured quantity if the measurement set contained an infinite number of values. In practice, it is impossible to take an infinite number of measurements. Therefore, in any practical situation, the process of averaging over a finite number of measurements only reduces the magnitude of random error to a small (but nonzero) value. The degree of confidence that the calculated mean value is close to the correct value of the measured quantity can be indicated by calculating the standard deviation or variance of data, these being parameters that describe how the measurements are distributed about the mean value (see Sections 3.6.1 and 3.6.2). This leads to a more formal quantification of this degree of confidence in terms of the standard error of the mean in Section 3.6.6.

## 3.6 Statistical Analysis of Measurements Subject to Random Errors

### 3.6.1 Mean and Median Values

The average value of a set of measurements of a constant quantity can be expressed as either the mean value or the median value. Historically, the median value was easier for a computer to compute than the mean value because the median computation involves a series of logic operations, whereas the mean computation requires addition and division. Many years ago, a computer performed logic operations much faster than arithmetic operations, and there were computational speed advantages in calculating average values by computing the median rather than the mean. However, computer power increased rapidly to a point where this advantage disappeared many years ago.

As the number of measurements increases, the difference between mean and median values becomes very small. However, the average calculated in terms of the mean value is always slightly closer to the correct value of the measured quantity than the average calculated as the median value for any finite set of measurements. Given the loss of any computational speed advantage because of the massive power of modern-day computers, this means that there is now little argument for calculating average values in terms of the median.

For any set of $n$ measurements $x_1, x_2 \cdots x_n$ of a constant quantity, the most likely true value is the *mean* given by

$$x_{mean} = \frac{x_1 + x_2 + \cdots x_n}{n}. \tag{3.4}$$

This is valid for all data sets where the measurement errors are distributed equally about the zero error value, that is, where positive errors are balanced in quantity and magnitude by negative errors.

The *median* is an approximation to the mean that can be written down without having to sum the measurements. The median is the middle value when measurements in the data set are written down in ascending order of magnitude. For a set of $n$ measurements $x_1, x_2 \cdots x_n$ of a constant quantity, written down in ascending order of magnitude, the median value is given by

$$x_{median} = x_{n+1/2}. \tag{3.5}$$

Thus, for a set of nine measurements $x_1, x_2 \cdots x_9$ arranged in order of magnitude, the median value is $x_5$. For an even number of measurements, the median value is midway between the two center values, that is, for 10 measurements $x_1 \cdots x_{10}$, the median value is given by $(x_5 + x_6)/2$.

Suppose that the length of a steel bar is measured by a number of different observers and the following set of 11 measurements are recorded (units millimeter). We will call this measurement set A.

398 420 394 416 404 408 400 420 396 413 430 (Measurement set A)

Using Equations (3.4) and (3.5), mean = 409.0 and median = 408. Suppose now that the measurements are taken again using a better measuring rule and with the observers taking more care to produce the following measurement set B:

409 406 402 407 405 404 407 404 407 407 408 (Measurement set B)

For these measurements, mean = 406.0 and median = 407. Which of the two measurement sets, A and B, and the corresponding mean and median values should we have the most confidence in? Intuitively, we can regard measurement set B as being more reliable because the measurements are much closer together. In set A, the spread between the smallest (396) and largest (430) value is 34, while in set B, the spread is only 6.

- *Thus, the smaller the spread of the measurements, the more confidence we have in the mean or median value calculated.*

Let us now see what happens if we increase the number of measurements by extending measurement set B to 23 measurements. We will call this measurement set C.

409 406 402 407 405 404 407 404 407 407 408 406 410
406 405 408 406 409 406 405 409 406 407 (Measurement set C)

Now, mean = 406.5 and median = 406

- *This confirms our earlier statement that the median value tends toward the mean value as the number of measurements increases.*

### 3.6.2 Standard Deviation and Variance

Expressing the spread of measurements simply as a range between the largest and the smallest value is not, in fact, a very good way of examining how measurement values are distributed about the mean value. A much better way of expressing the distribution is to calculate the variance or standard deviation of the measurements. The starting point for calculating these parameters is to calculate the deviation (error) $d_i$ of each measurement $x_i$ from the mean value $x_{mean}$ in a set of measurements $x_1, x_2, \cdots\cdots x_n$:

$$d_i = x_i - x_{mean}. \tag{3.6}$$

The *variance* ($V_s$) of the set of measurements is defined formally as the mean of the squares of deviations:

$$V_s = \frac{d_1^2 + d_2^2 \cdots d_n^2}{n}. \tag{3.7}$$

The *standard deviation* ($\sigma_s$) of the set of measurements is defined as the square root of the variance:

$$\sigma = \sqrt{V_s} = \sqrt{\frac{d_1^2 + d_2^2 \cdots d_n^2}{n}}. \tag{3.8}$$

Unfortunately, these formal definitions for the variance and standard deviation of data are made with respect to an infinite population of data values whereas, in all practical situations, we can only have a finite set of measurements. We have made the observation previously that the mean value $x_m$ of a finite set of measurements will differ from the true mean $\mu$ of the theoretical infinite population of measurements that the finite set is part of. This means that there is an error in the mean value $x_{mean}$ used in the calculation of $d_i$ in Equation (3.6). Because of this, Equations (3.7) and (3.8) give a biased estimate that tends to underestimate the variance and standard deviation of the infinite set of measurements. A better prediction of the variance of the infinite population can be obtained by applying the Bessel correction factor $(n/n-1)$ to the formula for $V_s$ in Equation (3.7):

$$V = \left(\frac{n}{n-1}\right)V_s = \frac{d_1^2 + d_2^2 \cdots d_n^2}{n-1}, \tag{3.9}$$

where $V_s$ is the variance of the finite set of measurements and V is the variance of the infinite population of measurements.

This leads to a similar better prediction of the standard deviation by taking the square root of the variance in Equation (3.9):

$$\sigma = \sqrt{V} = \sqrt{\frac{d_1^2 + d_2^2 \cdots d_n^2}{n-1}}. \tag{3.10}$$

## Example 3.4

Calculate σ and *V* for measurement sets A, B, and C given earlier.

## Solution

First, draw a table of measurements and deviations for set A (mean = 409 as calculated earlier):

| Measurement | 398 | 420 | 394 | 416 | 404 | 408 | 400 | 420 | 396 | 413 | 430 |
|---|---|---|---|---|---|---|---|---|---|---|---|
| Deviation from mean | −11 | +11 | −15 | +7 | −5 | −1 | −9 | +11 | −13 | +4 | +21 |
| (deviations)$^2$ | 121 | 121 | 225 | 49 | 25 | 1 | 81 | 121 | 169 | 16 | 441 |

$\sum(deviations^2) = 1370$; $n$ = number of measurements = 11.

Then, from Equations (3.9) and (3.10), $V = \sum(deviations^2)/n - 1 = 1370/10 = 137$; $\sigma = \sqrt{V} = 11.7$.

Measurements and deviations for set B are (mean = 406 as calculated earlier):

| Measurement | 409 | 406 | 402 | 407 | 405 | 404 | 407 | 404 | 407 | 407 | 408 |
|---|---|---|---|---|---|---|---|---|---|---|---|
| Deviation from mean | +3 | 0 | −4 | +1 | −1 | −2 | +1 | −2 | +1 | +1 | +2 |
| (deviations)$^2$ | 9 | 0 | 16 | 1 | 1 | 4 | 1 | 4 | 1 | 1 | 4 |

From these data, using Equations (3.9) and (3.10), $V = 4.2$ and $\sigma = 2.05$.

Measurements and deviations for set C are (mean = 406.5 as calculated earlier):

| Measurement | 409 | 406 | 402 | 407 | 405 | 404 | 407 | 404 | 407 | 407 | 408 |
|---|---|---|---|---|---|---|---|---|---|---|---|
| Deviation from mean | +2.5 | −0.5 | −4.5 | +0.5 | −1.5 | −2.5 | +0.5 | −2.5 | +0.5 | +0.5 | +1.5 |
| (deviations)$^2$ | 6.25 | 0.25 | 20.25 | 0.25 | 2.25 | 6.25 | 0.25 | 6.25 | 0.25 | 0.25 | 2.25 |

| Measurement | 406 | 410 | 406 | 405 | 408 | 406 | 409 | 406 | 405 | 409 | 406 | 407 |
|---|---|---|---|---|---|---|---|---|---|---|---|---|
| Deviation from mean | −0.5 | +3.5 | −0.5 | −1.5 | +1.5 | −0.5 | +2.5 | −0.5 | −1.5 | +2.5 | −0.5 | +0.5 |
| (deviations)$^2$ | 0.25 | 12.25 | 0.25 | 2.25 | 2.25 | 0.25 | 6.25 | 0.25 | 2.25 | 6.25 | 0.25 | 0.25 |

From this data, using Equations (3.9) and (3.10), $V = 3.53$ and $\sigma = 1.88$.

Note that the smaller values of $V$ and $\sigma$ for measurement set B compared with A correspond with the respective size of the spread in the range between maximum and minimum values for the two sets.

- *Thus, as $V$ and $\sigma$ decrease for a measurement set, we are able to express greater confidence that the calculated mean or median value is close to the true value, that is, that the averaging process has reduced the random error value close to zero.*
- *Comparing $V$ and $\sigma$ for measurement sets B and C, $V$ and $\sigma$ get smaller as the number of measurements increases, confirming that confidence in the mean value increases as the number of measurements increases.*

We have observed so far that random errors can be reduced by taking the average (mean or median) of a number of measurements. However, although the mean or median value is close to the true value, it would only become exactly equal to the true value if we could average an infinite number of measurements. As we can only make a finite number of measurements in a practical situation, the average value will still have some error. This error can be quantified as the *standard error of the mean*, which is discussed in detail a little later. However, before that, the subject of graphical analysis of random measurement errors needs to be covered.

■

### 3.6.3 Graphical Data Analysis Techniques—Frequency Distributions

Graphical techniques are a very useful way of analyzing the way in which random measurement errors are distributed. The simplest way of doing this is to draw a *histogram*, in which bands of equal width across the range of measurement values are defined and the number of measurements within each band is counted. The bands are often given the name *data bins*. A useful rule for defining the number of bands (bins) is known as the Sturgis rule, which calculates the number of bands as

$$\text{Number of bands} = 1 + 3.3\log_{10}(n),$$

where $n$ is the number of measurement values.

## ■ Example 3.5

Draw a histogram for the 23 measurements in set C of length measurement data given in Section 3.5.1.

■

## Solution

For 23 measurements, the recommended number of bands calculated according to the Sturgis rule is $1 + 3.3 \log10(23) = 5.49$. This rounds to five, as the number of bands must be an integer number.

To cover the span of measurements in data set C with five bands, data bands need to be 2 mm wide. The boundaries of these bands must be chosen carefully so that no measurements fall on the boundary between different bands and cause ambiguity about which band to put them in. Because the measurements are integer numbers, this can be accomplished easily by defining the range of the first band as 401.5 to 403.5 and so on. A histogram can now be drawn as in Figure 3.6 by counting the number of measurements in each band.

In the first band from 401.5 to 403.5, there is just one measurement, so the height of the histogram in this band is 1 unit.

In the next band from 403.5 to 405.5, there are five measurements, so the height of the histogram in this band is $1 = 5$ units.

The rest of the histogram is completed in a similar fashion.

When a histogram is drawn using a sufficiently large number of measurements, it will have the characteristic shape shown by truly random data, with symmetry about the mean value of the measurements. However, for a relatively small number of measurements, only approximate symmetry in the histogram can be expected about the mean value. It is a matter of judgment as to whether the shape of a histogram is close enough to symmetry to justify a conclusion that data on which it is based are truly

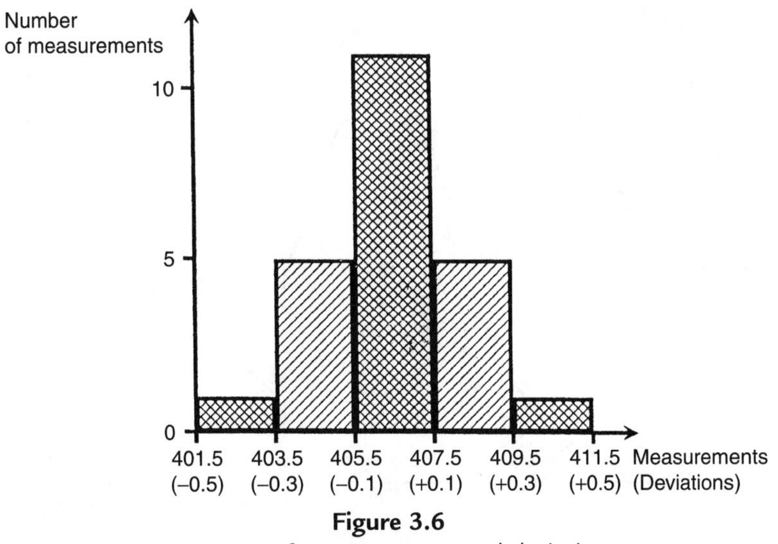

**Figure 3.6**
Histogram of measurements and deviations.

random. It should be noted that the 23 measurements used to draw the histogram in Figure 3.6 were chosen carefully to produce a symmetrical histogram but exact symmetry would not normally be expected for a measurement data set as small as 23.

As it is the actual value of measurement error that is usually of most concern, it is often more useful to draw a histogram of deviations of measurements from the mean value rather than to draw a histogram of the measurements themselves. The starting point for this is to calculate the deviation of each measurement away from the calculated mean value. Then a *histogram of deviations* can be drawn by defining deviation bands of equal width and counting the number of deviation values in each band. This histogram has exactly the same shape as the histogram of raw measurements except that scaling of the horizontal axis has to be redefined in terms of the deviation values (these units are shown in parentheses in Figure 3.6).

Let us now explore what happens to the histogram of deviations as the number of measurements increases. As the number of measurements increases, smaller bands can be defined for the histogram, which retains its basic shape but then consists of a larger number of smaller steps on each side of the peak. In the limit, as the number of measurements approaches infinity, the histogram becomes a smooth curve known as a *frequency distribution curve*, as shown in Figure 3.7. The ordinate of this curve is the frequency of occurrence of each deviation value, $F(D)$, and the abscissa is the magnitude of deviation, $D$.

The symmetry of Figures 3.6 and 3.7 about the zero deviation value is very useful for showing graphically that measurement data only have random errors. Although these figures cannot be used to quantify the magnitude and distribution of the errors

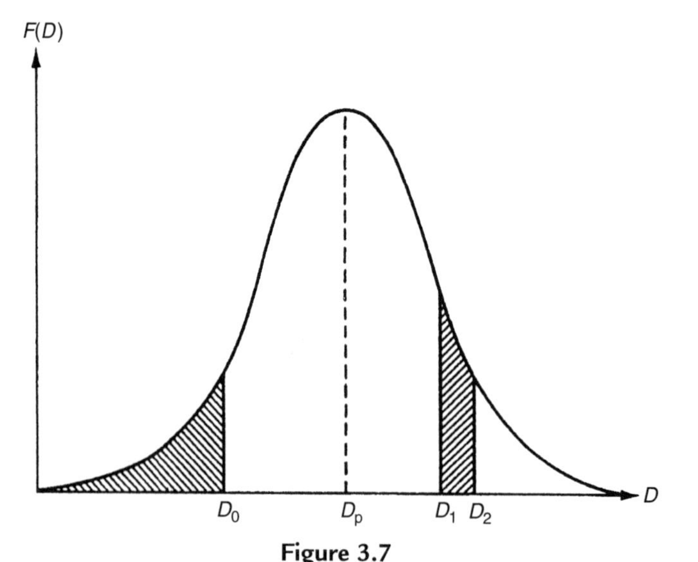


**Figure 3.7**
Frequency distribution curve of deviations.

easily, very similar graphical techniques do achieve this. If the height of the frequency distribution curve is normalized such that the area under it is unity, then the curve in this form is known as a *probability curve*, and the height $F(D)$ at any particular deviation magnitude $D$ is known as the *probability density function* (p.d.f.). The condition that the area under the curve is unity can be expressed mathematically as

$$\int_{-\infty}^{\infty} F(D)dD = 1.$$

The probability that the error in any one particular measurement lies between two levels $D_1$ and $D_2$ can be calculated by measuring the area under the curve contained between two vertical lines drawn through $D_1$ and $D_2$, as shown by the right-hand hatched area in Figure 3.7. This can be expressed mathematically as

$$P(D_1 \leq D \leq D_2) = \int_{D_1}^{D_2} F(D)dD. \qquad (3.11)$$

Of particular importance for assessing the maximum error likely in any one measurement is the *cumulative distribution function* (c.d.f.). This is defined as the probability of observing a value less than or equal to $D_o$ and is expressed mathematically as

$$P(D \leq D_0) = \int_{-\infty}^{D_0} F(D)dD. \qquad (3.12)$$

Thus, the c.d.f. is the area under the curve to the left of a vertical line drawn through $D_o$, as shown by the left-hand hatched area in Figure 3.7.

The deviation magnitude $D_p$ corresponding with the peak of the frequency distribution curve (Figure 3.7) is the value of deviation that has the greatest probability. If the errors are entirely random in nature, then the value of $D_p$ will equal zero. Any nonzero value of $D_p$ indicates systematic errors in data in the form of a bias that is often removable by recalibration.

■

### 3.6.4 Gaussian (Normal) Distribution

Measurement sets that only contain random errors usually conform to a distribution with a particular shape that is called *Gaussian*, although this conformance must always be tested (see the later section headed "Goodness of fit"). The shape of a Gaussian curve is such that the frequency of small deviations from the mean value is much greater than the frequency of large deviations. This coincides with the usual expectation in measurements subject to random errors that the number of measurements with a small error is much larger than the number of measurements with a large error. Alternative names for the Gaussian distribution are *normal distribution* or *bell-shaped distribution.* A Gaussian curve is defined formally as a normalized frequency distribution that is symmetrical about the line of zero error and in which the frequency and magnitude of quantities are related by the expression:

$$F(x) = \frac{1}{\sigma\sqrt{2\pi}} e^{\left[-(x-m)^2/2\sigma^2\right]}, \tag{3.13}$$

where $m$ is the mean value of data set $x$ and the other quantities are as defined before. Equation (3.13) is particularly useful for analyzing a Gaussian set of measurements and predicting how many measurements lie within some particular defined range. If measurement deviations $D$ are calculated for all measurements such that $D = x - m$, then the curve of deviation frequency $F(D)$ plotted against deviation magnitude $D$ is a Gaussian curve known as the *error frequency distribution curve*. The mathematical relationship between $F(D)$ and $D$ can then be derived by modifying Equation (3.13) to give

$$F(D) = \frac{1}{\sigma\sqrt{2\pi}} e^{\left[-D^2/2\sigma^2\right]}. \tag{3.14}$$

The shape of a Gaussian curve is influenced strongly by the value of σ, with the width of the curve decreasing as σ becomes smaller. As a smaller σ corresponds with typical deviations of measurements from the mean value becoming smaller, this confirms the earlier observation that the mean value of a set of measurements gets closer to the true value as σ decreases.

If the standard deviation is used as a unit of error, the Gaussian curve can be used to determine the probability that the deviation in any particular measurement in a Gaussian data set is greater than a certain value. By substituting the expression for $F(D)$ in Equation (3.14) into probability Equation (3.11), the probability that the error lies in a band between error levels $D_1$ and $D_2$ can be expressed as

$$P(D_1 \leq D \leq D_2) = \int_{D_1}^{D_2} \frac{1}{\sigma\sqrt{2\pi}} e^{\left(-D^2/2\sigma^2\right)} dD. \tag{3.15}$$

Solution of this expression is simplified by the substitution

$$z = D/\sigma. \tag{3.16}$$

The effect of this is to change the error distribution curve into a new Gaussian distribution that has a standard deviation of one ($\sigma = 1$) and a mean of zero. This new form, shown in Figure 3.8, is known as a *standard Gaussian curve* (or sometimes as a *z distribution*), and the dependent variable is now $z$ instead of $D$. Equation (3.15) can now be re-expressed as

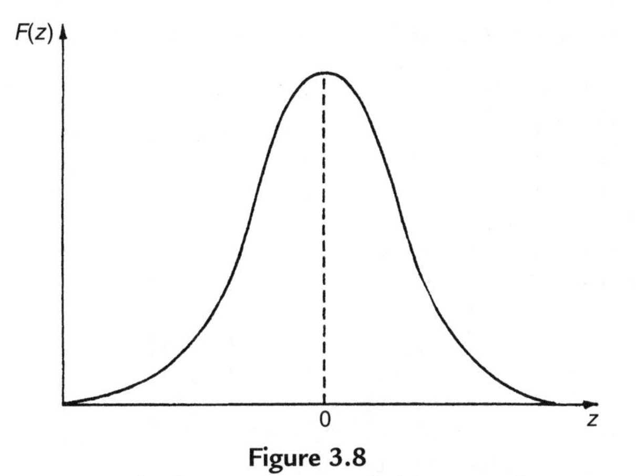


**Figure 3.8**
Standard Gaussian curve [$F(z)$ versus $z$].

$$P(D_1 \leq D \leq D_2) = P(z_1 \leq z \leq z_2) = \int_{z_1}^{z_2} \frac{1}{\sigma\sqrt{2\pi}} e^{\left(-z^2/2\right)} dz. \quad (3.17)$$

Unfortunately, neither Equation (3.15) nor Equation (3.17) can be solved analytically using tables of standard integrals, and numerical integration provides the only method of solution. However, in practice, the tedium of numerical integration can be avoided when analyzing data because the standard form of Equation (3.17), and its independence from the particular values of the mean and standard deviation of data, means that standard Gaussian tables that tabulate $F(z)$ for various values of $z$ can be used.

### *3.6.5 Standard Gaussian Tables (z Distribution)*

A standard Gaussian table (sometimes called *z distribution*), such as that shown in Table 3.1, tabulates the area under the Gaussian curve $F(z)$ for various values of $z$, where $F(z)$ is given by

$$F(z) = \int_{-\infty}^{z} \frac{1}{\sigma\sqrt{2\pi}} e^{\left(-z^2/2\right)} dz. \quad (3.18)$$

Thus, $F(z)$ gives the proportion of data values that are less than or equal to $z$. This proportion is the area under the curve of $F(z)$ against $z$ that is to the left of $z$. Therefore, the expression given in Equation (3.17) has to be evaluated as $[F(z_2) - F(z_1)]$. Study of Table 3.1 shows that $F(z) = 0.5$ for $z = 0$. This confirms that, as expected, the number of data values $\leq 0$ is 50% of the total. This must be so if data only have random errors. It will also be observed that Table 3.1, in common with most published standard Gaussian tables, only gives $F(z)$ for

**Table 3.1 Error Function Table (Area under a Gaussian Curve or $z$ Distribution)**

| z | 0.00 | 0.01 | 0.02 | 0.03 | 0.04 | 0.05 | 0.06 | 0.07 | 0.08 | 0.09 |
|---|---|---|---|---|---|---|---|---|---|---|
| 0.0 | 0.5000 | 0.5000 | 0.5080 | 0.5120 | 0.5160 | 0.5199 | 0.5239 | 0.5279 | 0.5319 | 0.5359 |
| 0.1 | 0.5398 | 0.5438 | 0.5478 | 0.5517 | 0.5557 | 0.5596 | 0.5636 | 0.5675 | 0.5714 | 0.5753 |
| 0.2 | 0.5793 | 0.5832 | 0.5871 | 0.5910 | 0.5948 | 0.5987 | 0.6026 | 0.6064 | 0.6103 | 0.6141 |
| 0.3 | 0.6179 | 0.6217 | 0.6255 | 0.6293 | 0.6331 | 0.6368 | 0.6406 | 0.6443 | 0.6480 | 0.6517 |
| 0.4 | 0.6554 | 0.6591 | 0.6628 | 0.6664 | 0.6700 | 0.6736 | 0.6772 | 0.6808 | 0.6844 | 0.6879 |
| 0.5 | 0.6915 | 0.6950 | 0.6985 | 0.7019 | 0.7054 | 0.7088 | 0.7123 | 0.7157 | 0.7190 | 0.7224 |
| 0.6 | 0.7257 | 0.7291 | 0.7324 | 0.7357 | 0.7389 | 0.7422 | 0.7454 | 0.7486 | 0.7517 | 0.7549 |
| 0.7 | 0.7580 | 0.7611 | 0.7642 | 0.7673 | 0.7703 | 0.7734 | 0.7764 | 0.7793 | 0.7823 | 0.7852 |
| 0.8 | 0.7881 | 0.7910 | 0.7939 | 0.7967 | 0.7995 | 0.8023 | 0.8051 | 0.8078 | 0.8106 | 0.8133 |
| 0.9 | 0.8159 | 0.8186 | 0.8212 | 0.8238 | 0.8264 | 0.8289 | 0.8315 | 0.8340 | 0.8365 | 0.8389 |
| 1.0 | 0.8413 | 0.8438 | 0.8461 | 0.8485 | 0.8508 | 0.8531 | 0.8554 | 0.8577 | 0.8599 | 0.8621 |
| 1.1 | 0.8643 | 0.8665 | 0.8686 | 0.8708 | 0.8729 | 0.8749 | 0.8770 | 0.8790 | 0.8810 | 0.8830 |
| 1.2 | 0.8849 | 0.8869 | 0.8888 | 0.8906 | 0.8925 | 0.8943 | 0.8962 | 0.8980 | 0.8997 | 0.9015 |
| 1.3 | 0.9032 | 0.9049 | 0.9066 | 0.9082 | 0.9099 | 0.9115 | 0.9131 | 0.9147 | 0.9162 | 0.9177 |
| 1.4 | 0.9192 | 0.9207 | 0.9222 | 0.9236 | 0.9251 | 0.9265 | 0.9279 | 0.9292 | 0.9306 | 0.9319 |
| 1.5 | 0.9332 | 0.9345 | 0.9357 | 0.9370 | 0.9382 | 0.9394 | 0.9406 | 0.9418 | 0.9429 | 0.9441 |
| 1.6 | 0.9452 | 0.9463 | 0.9474 | 0.9484 | 0.9495 | 0.9505 | 0.9515 | 0.9525 | 0.9535 | 0.9545 |
| 1.7 | 0.9554 | 0.9564 | 0.9573 | 0.9582 | 0.9591 | 0.9599 | 0.9608 | 0.9616 | 0.9625 | 0.9633 |
| 1.8 | 0.9641 | 0.9648 | 0.9656 | 0.9664 | 0.9671 | 0.9678 | 0.9686 | 0.9693 | 0.9699 | 0.9706 |
| 1.9 | 0.9713 | 0.9719 | 0.9726 | 0.9732 | 0.9738 | 0.9744 | 0.9750 | 0.9756 | 0.9761 | 0.9767 |
| 2.0 | 0.9772 | 0.9778 | 0.9783 | 0.9788 | 0.9793 | 0.9798 | 0.9803 | 0.9808 | 0.9812 | 0.9817 |
| 2.1 | 0.9821 | 0.9826 | 0.9830 | 0.9834 | 0.9838 | 0.9842 | 0.9846 | 0.9850 | 0.9854 | 0.9857 |
| 2.2 | 0.9861 | 0.9864 | 0.9868 | 0.9871 | 0.9875 | 0.9878 | 0.9881 | 0.9884 | 0.9887 | 0.9890 |
| 2.3 | 0.9893 | 0.9896 | 0.9898 | 0.9901 | 0.9904 | 0.9906 | 0.9909 | 0.9911 | 0.9913 | 0.9916 |
| 2.4 | 0.9918 | 0.9920 | 0.9922 | 0.9924 | 0.9926 | 0.9928 | 0.9930 | 0.9932 | 0.9934 | 0.9936 |
| 2.5 | 0.9938 | 0.9940 | 0.9941 | 0.9943 | 0.9945 | 0.9946 | 0.9948 | 0.9949 | 0.9951 | 0.9952 |
| 2.6 | 0.9953 | 0.9955 | 0.9956 | 0.9957 | 0.9959 | 0.9960 | 0.9961 | 0.9962 | 0.9963 | 0.9964 |
| 2.7 | 0.9965 | 0.9966 | 0.9967 | 0.9968 | 0.9969 | 0.9970 | 0.9971 | 0.9972 | 0.9973 | 0.9974 |
| 2.8 | 0.9974 | 0.9975 | 0.9976 | 0.9977 | 0.9977 | 0.9978 | 0.9979 | 0.9979 | 0.9980 | 0.9981 |
| 2.9 | 0.9981 | 0.9982 | 0.9982 | 0.9983 | 0.9984 | 0.9984 | 0.9985 | 0.9985 | 0.9986 | 0.9986 |
| 3.0 | 0.9986 | 0.9987 | 0.9987 | 0.9988 | 0.9988 | 0.9989 | 0.9989 | 0.9989 | 0.9990 | 0.9990 |
| 3.1 | 0.9990 | 0.9991 | 0.9991 | 0.9991 | 0.9992 | 0.9992 | 0.9992 | 0.9992 | 0.9993 | 0.9993 |
| 3.2 | 0.9993 | 0.9993 | 0.9994 | 0.9994 | 0.9994 | 0.9994 | 0.9994 | 0.9995 | 0.9995 | 0.9995 |
| 3.3 | 0.9995 | 0.9995 | 0.9995 | 0.9996 | 0.9996 | 0.9996 | 0.9996 | 0.9996 | 0.9996 | 0.9996 |
| 3.4 | 0.9997 | 0.9997 | 0.9997 | 0.9997 | 0.9997 | 0.9997 | 0.9997 | 0.9997 | 0.9997 | 0.9998 |
| 3.5 | 0.9998 | 0.9998 | 0.9998 | 0.9998 | 0.9998 | 0.9998 | 0.9998 | 0.9998 | 0.9998 | 0.9998 |
| 3.6 | 0.9998 | 0.9998 | 0.9998 | 0.9999 | 0.9999 | 0.9999 | 0.9999 | 0.9999 | 0.9999 | 0.9999 |

positive values of $z$. For negative values of $z$, we can make use of the following relationship because the frequency distribution curve is normalized:

$$F(-z) = 1 - F(z). \tag{3.19}$$

[$F(-z)$ is the area under the curve to the left of $(-z)$, i.e., it represents the proportion of data values $\leq -z$.]

## Example 3.6

How many measurements in a data set subject to random errors lie outside deviation boundaries of $+\sigma$ and $-\sigma$, that is, how many measurements have a deviation greater than $|\sigma|$?

■

## Solution

The required number is represented by the sum of the two shaded areas in Figure 3.9. This can be expressed mathematically as $P(E<-\sigma \text{ or } E>+\sigma) = P(E<-\sigma)+P(E>+\sigma)$. For $E = -\sigma$, $z = -1.0$ [from Equation (3.14)].

Using Table 3.1: $P(E<-\sigma)=F(-1)=1-F(1)=1-0.8413=0.1587$.

Similarly, for $E = +\sigma$, $z = +1.0$. Table 3.1 gives $P(E>+\sigma)=1-P(E<+\sigma)=1-F(1)= 1-0.8413=0.1587$. (This last step is valid because the frequency distribution curve is normalized such that the total area under it is unity.) Thus, $P[E<-\sigma] + P[E>+\sigma] = 0.1587 + 0.1587 = 0.3174 \sim 32\%$, that is, 32% of the measurements lie outside the $\pm\sigma$ boundaries, then 68% of the measurements lie inside.

The analysis just given shows that, for Gaussian-distributed data values, 68% of the measurements have deviations that lie within the bounds of $\pm\sigma$. Similar analysis shows that boundaries of $\pm 2\sigma$ contain 95.4% of data points, and extending the boundaries to $\pm 3\sigma$ encompasses 99.7% of data points. The probability of any data point lying outside particular deviation boundaries can therefore be expressed by the following table.

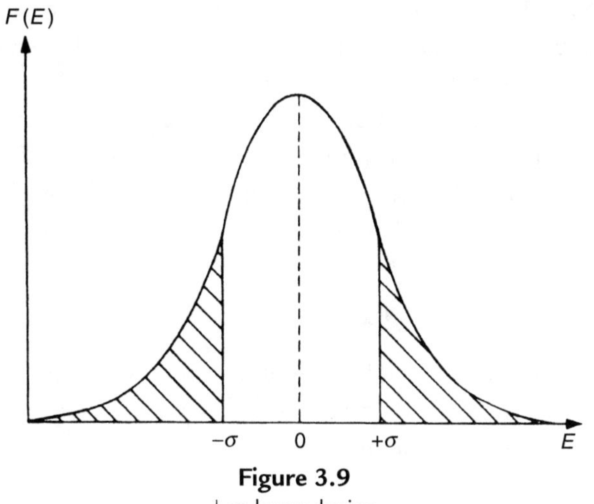


**Figure 3.9**
$\pm\sigma$ boundaries.

| Deviation boundaries | % of data points within boundary | Probability of any particular data point being outside boundary |
|---|---|---|
| $\pm\sigma$ | 68.0 | 32.0% |
| $\pm 2\sigma$ | 95.4 | 4.6% |
| $\pm 3\sigma$ | 99.7 | 0.3% |

■

### 3.6.6 Standard Error of the Mean

The foregoing analysis has examined the way in which measurements with random errors are distributed about the mean value. However, we have already observed that some error exists between the mean value of a finite set of measurements and the true value, that is, averaging a number of measurements will only yield the true value if the number of measurements is infinite. If several subsets are taken from an infinite data population with a Gaussian distribution, then, by the central limit theorem, the means of the subsets will form a Gaussian distribution about the mean of the infinite data set. The standard deviation of mean values of a series of finite sets of measurements relative to the true mean (the mean of the infinite population that the finite set of measurements is drawn from) is defined as the *standard error of the mean*, $\alpha$. This is calculated as

$$\alpha = \sigma/\sqrt{n}. \tag{3.20}$$

Clearly, $\alpha$ tends toward zero as the number of measurements ($n$) in the data set expands toward infinity.

The next question is how do we use the standard error of the mean to predict the error between the calculated mean of a finite set of measurements and the mean of the infinite population? In other words, if we use the mean value of a finite set of measurements to predict the true value of the measured quantity, what is the likely error in this prediction? This likely error can only be expressed in probabilistic terms. All we know for certain is the standard deviation of the error, which is expressed as $\alpha$ in Equation (3.20). We also know that a range of $\pm$ one standard deviation (i.e., $\pm\alpha$) encompasses 68% of the deviations of sample means either side of the true value. Thus we can say that the measurement value obtained by calculating the mean of a set of $n$ measurements, $x_1, x_2, \cdots x_n$, can be expressed as

$$x = x_{mean} \pm \alpha$$

with 68% certainty that the magnitude of the error does not exceed $|\alpha|$. For data set C of length measurements used earlier, $n = 23$, $\sigma = 1.88$, and $\alpha = 0.39$. The length can therefore be expressed as $406.5 \pm 0.4$ (68% confidence limit).

The problem of expressing the error with 68% certainty is that there is a 32% chance that the error is greater than $\alpha$. Such a high probability of the error being greater than $\alpha$ may not be acceptable in many situations. If this is the case, we can use the fact that a range of $\pm$ two standard deviations, that is, $\pm 2\alpha$, encompasses 95.4% of the deviations of sample means either side of the true value. Thus, we can express the measurement value as

$$x = x_{mean} \pm 2\alpha$$

with 95.4% certainty that the magnitude of the error does not exceed $|2\alpha|$. This means that there is only a 4.6% chance that the error exceeds $2\alpha$. Referring again to set C of length measurements, $2\sigma = 3.76$, $2\alpha = 0.78$, and the length can be expressed as $406.5 \pm 0.8$ (95.4% confidence limits).

If we wish to express the maximum error with even greater probability that the value is correct, we could use $\pm 3\alpha$ limits (99.7% confidence). In this case, for length measurements again, $3\sigma = 5.64$, $3\alpha = 1.17$, and the length should be expressed as $406.5 \pm 1.2$ (99.7% confidence limits). There is now only a 0.3% chance (3 in 1000) that the error exceeds this value of 1.2.

### *3.6.7 Estimation of Random Error in a Single Measurement*

In many situations where measurements are subject to random errors, it is not practical to take repeated measurements and find the average value. Also, the averaging process becomes invalid if the measured quantity does not remain at a constant value, as is usually the case when process variables are being measured. Thus, if only one measurement can be made, some means of estimating the likely magnitude of error in it is required. The normal approach to this is to calculate the error within 95% confidence limits, that is, to calculate the value of deviation $D$ such that 95% of the area under the probability curve lies within limits of $\pm D$. These limits correspond to a deviation of $\pm 1.96\sigma$. Thus, it is necessary to maintain the measured quantity at a constant value while a number of measurements are taken in order to create a reference measurement set from which $\sigma$ can be calculated. Subsequently, the maximum likely deviation in a single measurement can be expressed as Deviation $= \pm 1.96\sigma$. However, this only expresses the maximum likely deviation of the measurement from the calculated mean of the reference measurement set, which is not the true value as observed earlier. Thus the calculated value for the standard error of the mean has to be added to the likely maximum deviation value. To be consistent, this should be expressed to the same 95% confidence limits. Thus, the maximum likely error in a single measurement can be expressed as

$$\text{Error} = \pm 1.96(\sigma + \alpha). \tag{3.21}$$

Before leaving this matter, it must be emphasized that the maximum error specified for a measurement is only specified for the confidence limits defined. Thus, if the maximum error is specified as $\pm 1\%$ with 95% confidence limits, this means that there is still 1 chance in 20 that the error will exceed $\pm 1\%$.

## Example 3.7

Suppose that a standard mass is measured 30 times with the same instrument to create a reference data set, and the calculated values of $\sigma$ and $\alpha$ are $\sigma = 0.46$ and $\alpha = 0.08$. If the instrument is then used to measure an unknown mass and the reading is 105.6 kg, how should the mass value be expressed?

## Solution

Using Equation (3.21), $1.96(\sigma + \alpha) = 1.06$. The mass value should therefore be expressed as $105.6 \pm 1.1$ kg.

### *3.6.8 Distribution of Manufacturing Tolerances*

Many aspects of manufacturing processes are subject to random variations caused by factors similar to those that cause random errors in measurements. In most cases, these random variations in manufacturing, which are known as *tolerances*, fit a Gaussian distribution, and the previous analysis of random measurement errors can be applied to analyze the distribution of these variations in manufacturing parameters.

## Example 3.8

An integrated circuit chip contains $10^5$ transistors. The transistors have a mean current gain of 20 and a standard deviation of 2. Calculate the following:

(a) number of transistors with a current gain between 19.8 and 20.2
(b) number of transistors with a current gain greater than 17

## Solution

(a) The proportion of transistors where $19.8 < \text{gain} < 20.2$ is

$$P[X < 20] - P[X < 19.8] = P[z < 0.2] - P[z < -0.2] \quad (\text{for } z = (X - \mu)/\sigma)$$

For $X = 20.2$, $z = 0.1$, and for $X = 19.8$, $z = -0.1$

From tables, $P[z < 0.1] = 0.5398$ and thus $P[z < -0.1] = 1 - P[z < 0.1] = 1 - 0.5398 = 0.4602$

Hence, $P[z < 0.1] - P[z < -0.1] = 0.5398 - 0.4602 = 0.0796$

Thus, $0.0796 \times 10^5 = 7960$ transistors have a current gain in the range from 19.8 to 20.2.

(b) The number of transistors with gain $>17$ is given by

$$P[x > 17] = 1 - P[x < 17] = 1 - P[z < -1.5] = P[z < +1.5] = 0.9332$$

Thus, 93.32%, that is, 93,320 transistors have a gain $>17$.

■

### 3.6.9 Chi-Squared ($\chi^2$) Distribution

We have already observed the fact that, if we calculate the mean value of successive sets of samples of *N* measurements, the means of those samples form a Gaussian distribution about the true value of the measured quantity (the true value being the mean of the infinite data set that the set of samples are part of). The standard deviation of the distribution of the mean values was quantified as the standard error of the mean.

It is also useful for many purposes to look at distribution of the variance of successive sets of samples of *N* measurements that form part of a Gaussian distribution. This is expressed as the chi-squared distribution $F(\chi^2)$, where $\chi^2$ is given by

$$\chi^2 = k\sigma_x{}^2/\sigma^2, \tag{3.22}$$

where $\sigma_x{}^2$ is the variance of a sample of *N* measurements and $\sigma^2$ is the variance of the infinite data set that sets of *N* samples are part of. k is a constant known as the number of degrees of freedom and is equal to $(N-1)$.

The shape of the $\chi^2$ distribution depends on the value of k, with typical shapes being shown in Figure 3.10. The area under the $\chi^2$ distribution curve is unity but, unlike the Gaussian distribution, the $\chi^2$ distribution is not symmetrical. However, it tends toward the symmetrical shape of a Gaussian distribution as k becomes very large.

The $\chi^2$ distribution expresses the expected variation due to random chance of the variance of a sample away from the variance of the infinite population that the sample is part of. The magnitude of this expected variation depends on what level of "random chance" we set. The level of random chance is normally expressed as a *level of significance*, which is usually

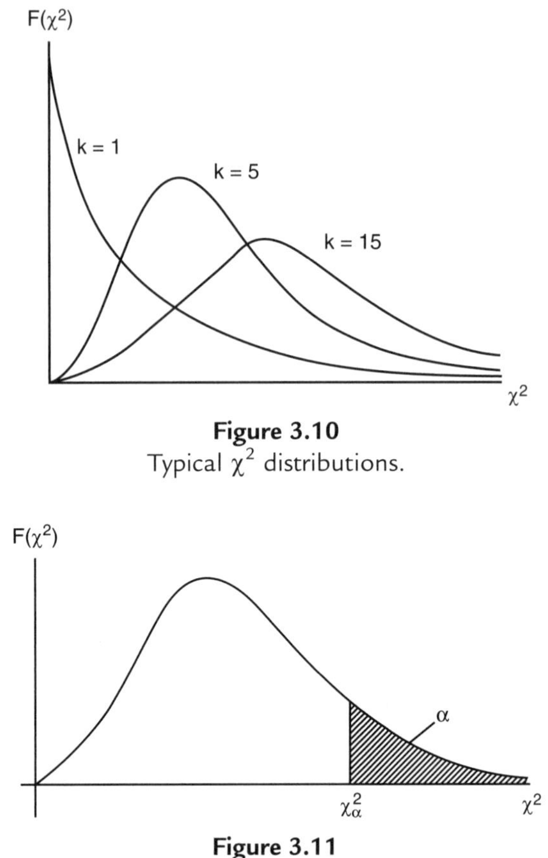


**Figure 3.10**
Typical $\chi^2$ distributions.

**Figure 3.11**
Meaning of symbol $\alpha$ for $\chi^2$ distribution.

denoted by the symbol $\alpha$. Referring to the $\chi^2$ distribution shown in Figure 3.11, value $\chi^2_\alpha$ denotes the $\chi^2$ value to the left of which lies $100(1-\alpha)\%$ of the area under the $\chi^2$ distribution curve. Thus, the area of the curve to the right of $\chi^2_\alpha$ is $\alpha$ and that to the left is $(1-\alpha)$.

Numerical values for $\chi^2$ are obtained from tables that express the value of $\chi^2$ for various degrees of freedom k and for various levels of significance $\alpha$. Published tables differ in the number of degrees of freedom and the number of levels of significance covered. A typical table is shown as Table 3.2.

One major use of the $\chi^2$ distribution is to predict the variance $\sigma^2$ of an infinite data set, given the measured variance $\sigma_x{}^2$ of a sample of $N$ measurements drawn from the infinite population. The boundaries of the range of $\chi^2$ values expected for a particular level of significance $\alpha$ can be expressed by the probability expression:

$$P\left[\chi^2_{1-\alpha/2} \le \chi^2 \le \chi^2_{\alpha/2}\right] = 1 - \alpha. \tag{3.23}$$

**Table 3.2 Chi-Squared ($\chi^2$) Distribution**

| | 0.005 | 0.025 | 0.050 | 0.900 | 0.950 | 0.975 | 0.990 | 0.995 | 0.999 |
|---|---|---|---|---|---|---|---|---|---|
| 1 | 3.93x 10–5 | $.0^3982$ | $.0^2393$ | 2.71 | 3.84 | 5.02 | 6.63 | 7.88 | 10.8 |
| 2 | .0100 | .0506 | .103 | 4.61 | 5.99 | 7.38 | 9.21 | 10.6 | 13.8 |
| 3 | .0717 | .216 | .352 | 6.25 | 7.81 | 9.35 | 11.3 | 12.8 | 16.3 |
| 4 | 0.207 | .484 | .711 | 7.78 | 9.49 | 11.1 | 13.3 | 14.9 | 18.5 |
| 5 | .412 | .831 | 1.15 | 9.24 | 11.1 | 12.8 | 15.1 | 16.7 | 20.5 |
| 6 | .676 | 1.24 | 1.64 | 10.6 | 12.6 | 14.4 | 16.8 | 18.5 | 22.5 |
| 7 | .989 | 1.69 | 2.17 | 12.0 | 14.1 | 16.0 | 18.5 | 20.3 | 24.3 |
| 8 | 1.34 | 2.18 | 2.73 | 13.4 | 15.5 | 17.5 | 20.1 | 22.0 | 26.1 |
| 9 | 1.73 | 2.70 | 3.33 | 14.7 | 16.9 | 19.0 | 21.7 | 23.6 | 27.9 |
| 10 | 2.16 | 3.25 | 3.94 | 16.0 | 18.3 | 20.5 | 23.2 | 25.2 | 29.6 |
| 11 | 2.60 | 3.82 | 4.57 | 17.3 | 19.7 | 21.9 | 24.7 | 26.8 | 31.3 |
| 12 | 3.07 | 4.40 | 5.23 | 18.5 | 21.0 | 23.3 | 26.2 | 28.3 | 32.9 |
| 13 | 3.57 | 5.01 | 5.89 | 19.8 | 22.4 | 24.7 | 27.7 | 29.8 | 34.5 |
| 14 | 4.07 | 5.63 | 6.57 | 21.1 | 23.7 | 26.1 | 29.1 | 31.3 | 36.1 |
| 15 | 4.60 | 6.26 | 7.26 | 22.3 | 25.0 | 27.5 | 30.6 | 32.8 | 37.7 |
| 16 | 5.14 | 6.91 | 7.96 | 23.5 | 26.3 | 28.8 | 32.0 | 34.3 | 39.3 |
| 17 | 5.70 | 7.56 | 8.67 | 24.8 | 27.6 | 30.2 | 33.4 | 35.7 | 40.8 |
| 18 | 6.26 | 8.23 | 9.39 | 26.0 | 28.9 | 31.5 | 34.8 | 37.2 | 42.3 |
| 19 | 6.84 | 8.91 | 10.1 | 27.2 | 30.1 | 32.9 | 36.2 | 38.6 | 43.8 |
| 20 | 7.43 | 9.59 | 10.9 | 28.4 | 31.4 | 34.2 | 37.6 | 40.0 | 45.3 |
| 21 | 8.03 | 10.3 | 11.6 | 29.6 | 32.7 | 35.5 | 38.9 | 41.4 | 46.8 |
| 22 | 8.64 | 11.0 | 12.3 | 30.8 | 33.9 | 36.8 | 40.3 | 42.8 | 48.3 |
| 23 | 9.26 | 11.7 | 13.1 | 32.0 | 35.2 | 38.1 | 41.6 | 44.2 | 49.7 |
| 24 | 9.89 | 12.4 | 13.8 | 33.2 | 36.4 | 39.4 | 43.0 | 45.6 | 51.2 |
| 25 | 10.5 | 13.1 | 14.6 | 34.4 | 37.7 | 40.6 | 44.3 | 46.9 | 52.6 |
| 26 | 11.2 | 13.8 | 15.4 | 35.6 | 38.9 | 41.9 | 45.6 | 48.3 | 54 J |
| 27 | 11.8 | 14.6 | 16.2 | 36.7 | 40.1 | 43.2 | 47.0 | 49.6 | 55.5 |
| 28 | 12.5 | 15.3 | 16.9 | 37.9 | 41.3 | 44.5 | 48.3 | 51.0 | 56.9 |
| 29 | 13.1 | 16.0 | 17.7 | 39.1 | 42.6 | 45.7 | 49.6 | 52.3 | 58.3 |
| 30 | 13.8 | 16.8 | 18.5 | 40.3 | 43.8 | 47.0 | 50.9 | 53.7 | 59.7 |
| 35 | 17.2 | 20.6 | 12.5 | 46.1 | 49.8 | 53.2 | 57.3 | 60.3 | 66.6 |
| 40 | 20.7 | 24.4 | 26.5 | 51.8 | 55.8 | 59.3 | 63.7 | 66.8 | 73.4 |
| 45 | 24.3 | 28.4 | 30.6 | 57.5 | 61.7 | 65.4 | 70.0 | 73.2 | 80.1 |
| 50 | 28.0 | 32.4 | 34.8 | 63.2 | 67.5 | 71.4 | 76.2 | 79.5 | 86.7 |
| 75 | 47.2 | 52.9 | 56.1 | 91.1 | 96.2 | 100.8 | 106.4 | 110.3 | 118.6 |
| 100 | 67.3 | 74.2 | 77.9 | 118.5 | 124.3 | 129.6 | 135.8 | 140.2 | 149.4 |

To put this in simpler terms, we are saying that there is a probability of $(1-\alpha)\%$ that $\chi^2$ lies within the range bounded by $\chi^2_{1-\alpha/2}$ and $\chi^2_{\alpha/2}$ for a level of significance of $\alpha$. For example, for a level of significance $\alpha = 0.5$, there is a 95% probability (95% confidence level) that $\chi^2$ lies between $\chi^2_{0.975}$ and $\chi^2_{0.025}$.

Substituting into Equation (3.23) using the expression for $\chi^2$ given in Equation (3.22):

$$P\left[\chi^2_{1-\alpha/2} \le \frac{k{\sigma_x}^2}{\sigma^2} \le \chi^2_{\alpha/2}\right] = 1 - \alpha.$$

This can be expressed in an alternative but equivalent form by inverting the terms and changing the "≤" relationships to "≥" ones:

$$P\left[\frac{1}{\chi^2_{1-\alpha/2}} \ge \frac{\sigma^2}{k{\sigma_x}^2} \ge \frac{1}{\chi^2_{\alpha/2}}\right] = 1 - \alpha.$$

Now multiplying the expression through by $k{\sigma_x}^2$ gives the following expression for the boundaries of the variance $\sigma^2$:

$$P\left[\frac{k{\sigma_x}^2}{\chi^2_{1-\alpha/2}} \ge \sigma^2 \ge \frac{k\sigma^2}{\chi^2_{\alpha/2}}\right] = 1 - \alpha. \tag{3.24}$$

The thing that is immediately evident in this solution is that the range within which the true variance and standard deviation lies is very wide. This is a consequence of the relatively small number of measurements (10) in the sample. It is therefore highly desirable wherever possible to use a considerably larger sample when making predictions of the true variance and standard deviation of some measured quantity.

The solution to Example 3.10 shows that, as expected, the width of the estimated range in which the true value of the standard deviation lies gets wider as we increase the confidence level from 90 to 99%. It is also interesting to compare the results in Examples 3.9 and 3.10 for the same confidence level of 95%. The ratio between maximum and minimum values of estimated variance is much greater for the 10 samples in Example 3.9 compared with the 25 samples in Example 3.10. This shows the benefit of having a larger sample size when predicting the variance of the whole population that the sample is drawn from.

## Example 3.9

The length of each rod in a sample of 10 brass rods is measured, and the variance of the length measurement in the sample is found to be 16.3 mm. Estimate the true variance and standard deviation for the whole batch of rods from which the sample of 10 was drawn, expressed to a confidence level of 95%.

■

## Solution

Degrees of freedom (k) $= N - 1 = 9$

For ${\sigma_x}^2 = 16.3$, $k{\sigma_x}^2 = 146.7$

For confidence level of 95%, level of significance, $\alpha$, $= 0.05$

Applying Equation (3.24), the true variance is bounded by the values of $146.7/\chi^2_{0.975}$ and $146.7/\chi^2_{0.025}$

Looking up the appropriate values in the $\chi^2$ distribution table for k = 9 gives

$\chi^2_{0.975} = 2.70$ ; $\chi^2_{0.025} = 19.02$ ; $146.7/\chi^2_{0.975} = 54.3$ ; $146.7/\chi^2_{0.025} = 7.7$

The true variance can therefore be expressed as $7.7 \leq \sigma^2 \leq 54.3$

The true standard deviation can be expressed as $\sqrt{7.7} \leq \sigma \leq \sqrt{54.3}$, that is, $2.8 \leq \sigma \leq 7.4$

■

## Example 3.10

The length of a sample of 25 bricks is measured and the variance of the sample is calculated as 6.8 mm. Estimate the true variance for the whole batch of bricks from which the sample of 25 was drawn, expressed to confidence levels of (a) 90%, (b) 95%, and (c) 99%.

■

## Solution

Degrees of freedom (k) = $N - 1 = 24$

For $\sigma_x{}^2 = 6.8$, $k\sigma_x{}^2 = 163.2$

(a) For confidence level of 90%, level of significance, $\alpha$, = 0.10 and $\alpha/2 = 0.05$

Applying Equation (3.24), the true variance is bounded by the values of $163.2/\chi^2_{0.95}$ and $163.2/\chi^2_{0.05}$

Looking up the appropriate values in the $\chi^2$ distribution table for k = 24 gives

$\chi^2_{0.95} = 13.85$ ; $\chi^2_{0.05} = 36.42$ ; $163.2/\chi^2_{0.95} = 11.8$ ; $163.2/\chi^2_{0.05} = 4.5$

The true variance can therefore be expressed as $4.5 \leq \sigma^2 \leq 11.8$

(b) For confidence level of 95%, level of significance, $\alpha$, = 0.05 and $\alpha/2 = 0.025$

Applying Equation (3.24), the true variance is bounded by the values of $163.2/\chi^2_{0.975}$ and $163.2/\chi^2_{0.025}$

Looking up the appropriate values in the $\chi^2$ distribution table for k = 24 gives

$\chi^2_{0.975} = 12.40$ ; $\chi^2_{0.025} = 39.36$ ; $163.2/\chi^2_{0.975} = 13.2$ ; $163.2/\chi^2_{0.025} = 4.1$

The true variance can therefore be expressed as $4.1 \leq \sigma^2 \leq 13.2$

(c) For confidence level of 99%, level of significance, $\alpha$, = 0.01 and $\alpha/2 = 0.005$

Applying Equation (3.24), the true variance is bounded by the values of $146.7/\chi^2_{0.995}$ and $146.7/\chi^2_{0.005}$

Looking up the appropriate values in the $\chi^2$ distribution table for k = 24 gives

$\chi^2_{0.995} = 9.89$ ; $\chi^2_{0.005} = 45.56$ ; $163.2/\chi^2_{0.995} = 16.5$ ; $163.2/\chi^2_{0.005} = 3.6$

The true variance can therefore be expressed as $3.6 \leq \sigma^2 \leq 16.5$

■

### 3.6.10 Goodness of Fit to a Gaussian Distribution

All of the analysis of random deviations presented so far only applies when data being analyzed belong to a Gaussian distribution. Hence, the degree to which a set of data fits a Gaussian distribution should always be tested before any analysis is carried out. This test can be carried out in one of three ways:

(a) *Inspecting the shape of the histogram*: The simplest way to test for Gaussian distribution of data is to plot a histogram and look for a "bell shape" of the form shown earlier in Figure 3.6. Deciding whether the histogram confirms a Gaussian distribution is a matter of judgment. For a Gaussian distribution, there must always be approximate symmetry about the line through the center of the histogram, the highest point of the histogram must always coincide with this line of symmetry, and the histogram must get progressively smaller either side of this point. However, because the histogram can only be drawn with a finite set of measurements, some deviation from the perfect shape of histogram as described previously is to be expected even if data really are Gaussian.

(b) *Using a normal probability plot*: A normal probability plot involves dividing data values into a number of ranges and plotting the cumulative probability of summed data frequencies against data values on special graph paper.* This line should be a straight line if the data distribution is Gaussian. However, careful judgment is required, as only a finite number of data values can be used and therefore the line drawn will not be entirely straight even if the distribution is Gaussian. Considerable experience is needed to judge whether the line is straight enough to indicate a Gaussian distribution. This will be easier to understand if data in measurement set C are used as an example. Using the same five ranges as used to draw the histogram, the following table is first drawn:

| **Range** | **401.5 to 403.5** | **403.5 to 405.5** | **405.5 to 407.5** | **407.5 to 409.5** | **409.5 to 411.5** |
|---|---|---|---|---|---|
| Number of data items in range | 1 | 5 | 11 | 5 | 1 |
| Cumulative number of data items | 1 | 6 | 17 | 22 | 23 |
| Cumulative number of data items as % | 4.3 | 26.1 | 73.9 | 95.7 | 100.0 |

The normal probability plot drawn from this table is shown in Figure 3.12. This is sufficiently straight to indicate that data in measurement set C are Gaussian.

---

* This is available from specialist stationery suppliers.

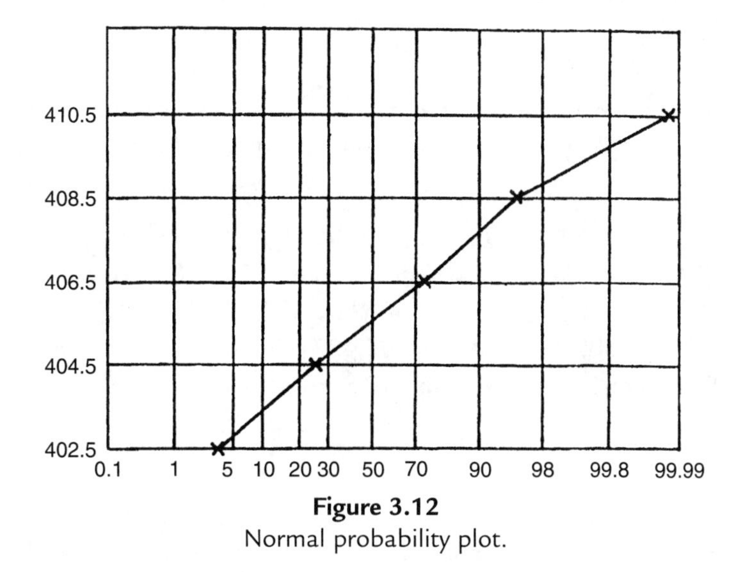


**Figure 3.12**
Normal probability plot.

(c) *The $\chi^2$ test*: The $\chi^2$ distribution provides a more formal method for testing whether data follow a Gaussian distribution. The principle of the $\chi^2$ test is to divide data into p equal width bins and to count the number of measurements $n_i$ in each bin, using exactly the same procedure as done to draw a histogram. The expected number of measurements $n_i'$ in each bin for a Gaussian distribution is also calculated. Before proceeding any further, a check must be made at this stage to confirm that at least 80% of the bins have a data count greater than a minimum number for both $n_i$ and $n_i'$. We will apply a minimum number of four, although some statisticians use the smaller minimum of three and some use a larger minimum of five. If this check reveals that too many bins have data counts less than the minimum number, it is necessary to reduce the number of bins by redefining their widths. The test for at least 80% of the bins exceeding the minimum number then has to be reapplied. Once the data count in the bins is satisfactory, a $\chi^2$ value is calculated for data according to the following formula:

$$\chi^2 = \sum_{i=1}^{p} \frac{\left(n_i - n_i'\right)^2}{n_i'} \tag{3.25}$$

The $\chi^2$ test then examines whether the calculated value of $\chi^2$ is greater than would be expected for a Gaussian distribution according to some specified level of chance. This involves reading off the expected value from the $\chi^2$ distribution table (Table 3.2) for the specified confidence level and comparing this expected value with that calculated in Equation (3.25). This procedure will become clearer if we work through an example.

## Example 3.11

A sample of 100 pork pies produced in a bakery is taken, and the mass of each pie (grams) is measured. Apply the $\chi^2$ test to examine whether the data set formed by the set of 100 mass measurements shown here conforms to a Gaussian distribution.

487 504 501 515 491 496 482 502 508 494 505 501 485 503 507 494 489 501 510 491
503 492 483 501 500 493 505 501 517 500 494 503 500 488 496 500 519 499 495 490
503 500 497 492 510 506 497 499 489 506 502 484 495 498 502 496 512 504 490 497
488 503 512 497 480 509 496 513 499 502 487 499 505 493 498 508 492 498 486 511
499 504 495 500 484 513 509 497 505 510 516 499 495 507 498 514 506 500 508 494

■

## Solution

Applying the Sturgis rule, the recommended number of data bins p for N data points is given by

$$p = 1 + 3.3\log_{10}N = 1 + (3.3)(2.0000) = 7.6.$$

This rounds to 8.

Mass measurements span the range from 480 to 519. Hence we will choose data bin widths of 5 grams, with bin boundaries set at 479.5, 484.5, 489.5, 494.5, 499.5, 504.5, 509.5, 514.5 and 519.5 (boundaries set so that there is no ambiguity about which bin any particular data value fits in). The next step involves counting the number of measurements in each bin. These are the $n_i$ values, $i = 1, \cdots 8$, for Equation (3.25). Results of this counting are set out in the following table.

| Bin number (i) | 1 | 2 | 3 | 4 | 5 | 6 | 7 | 8 |
|---|---|---|---|---|---|---|---|---|
| Data range | 479.5 to 484.5 | 484.5 to 489.5 | 489.5 to 494.5 | 494.5 to 499.5 | 499.5 to 504.5 | 504.5 to 509.5 | 509.5 to 514.5 | 514.5 to 519.5 |
| Measurements in range ($n_i$) | 5 | 8 | 13 | 23 | 24 | 14 | 9 | 4 |

Because none of the bins have a count less than our stated minimum threshold of four, we can now proceed to calculate $n_i'$ values. These are the expected numbers of measurements in each data bin for a Gaussian distribution. The starting point for this calculation is knowing the mean value ($\mu$) and standard deviation of the 100 mass measurements. These are calculated using Equations (3.4) and (3.10) as $\mu = 499.53$ and $\sigma = 8.389$. We now calculate the *z* values corresponding to the measurement values (*x*) at the upper end of each data bin using Equation (3.16) and then use the error function table (Table 3.1) to calculate *F*(*z*). *F*(*z*) gives the proportion of *z* values that are $\leq z$, which gives the proportion of measurements less than the corresponding *x* values. This then allows calculation of the expected number of measurements $(n_i')$ in each data bin. These calculations are shown in the following table.

| x | $z\left(\frac{x-\mu}{\sigma}\right)$ | *F*(*z*) | **Expected number of data in bin** $(n_i')$ |
|---|---|---|---|
| 484.5 | -1.792 | 0.037 | 3.7 |
| 489.5 | -1.195 | 0.116 | 7.9 |
| 494.5 | -0.600 | 0.274 | 15.8 |
| 499.5 | -0.004 | 0.498 | 22.4 |
| 504.5 | 0.592 | 0.723 | 22.5 |
| 509.5 | 1.188 | 0.883 | 16.0 |
| 514.5 | 1.784 | 0.963 | 8.0 |
| 519.5 | 2.381 | 0.991 | 2.8 |

In case there is any confusion about the calculation of numbers in the final column, let us consider rows 1 and 2. Row 1 shows that the proportion of data points less than 484.5 is 0.037. Because there are 100 data points in total, the actual estimated number of data points less than 484.5 is 3.7. Row 2 shows that the proportion of data points less than 489.5 is 0.116, and hence the total estimated number of data points less than 489.5 is 11.6. This total includes the 3.7 data points less than 484.5 calculated in the previous row. Hence, the number of data points in this bin between 484.5 and 489.5 is 11.6 minus 3.7, that is, 7.9.

We can now calculate the $\chi^2$ value for data using Equation (3.25). The steps of the calculation are shown in the following table.

| **Bin number (p)** | $n_i$ | $n_i'$ | $(n_i - n_i')$ | $(n_i - n_i')^2$ | $\frac{(n_i - n_i')^2}{n_i'}$ |
|---|---|---|---|---|---|
| 1 | 5 | 3.7 | 1.3 | 1.69 | 0.46 |
| 2 | 8 | 7.9 | 0.1 | 0.01 | 0.00 |
| 3 | 13 | 15.8 | -2.8 | 7.84 | 0.50 |
| 4 | 23 | 22.4 | 0.6 | 0.36 | 0.02 |
| 5 | 24 | 22.5 | 1.5 | 2.25 | 0.10 |
| 6 | 14 | 16.0 | -2.0 | 4.00 | 0.25 |
| 7 | 9 | 8.0 | 1.0 | 1.00 | 0.12 |
| 8 | 4 | 2.8 | 1.2 | 1.44 | 0.51 |

The value of $\chi^2$ is now found by summing the values in the final column to give $\chi^2 = 1.96$. The final step is to check whether this value of $\chi^2$ is greater than would be expected for a Gaussian distribution. This involves looking up $\chi^2$ in Table 3.2. Before doing this, we have to specify the number of degrees of freedom, k. In this case, k is the number of bins minus 2, because data are manipulated twice to obtain the $\mu$ and $\sigma$ statistical values used in the calculation of $n_i'$. Hence, $k = 8 - 2 = 6$.

Table 3.2 shows that, for $k = 6$, $\chi^2 = 1.64$ for a 95% confidence level and $\chi^2 = 2.20$ for a 90% confidence level. Hence, our calculated value for $\chi^2$ of 1.96 shows that the confidence level that data follow a Gaussian distribution is between 90% and 95%.

■

We will now look at a slightly different example where we meet the problem that our initial division of data into bins produces too many bins that do not contain the minimum number of data points necessary for the $\chi^2$ test to work reliably.

## Example 3.12

Suppose that the production machinery used to produce the pork pies featured in Example 3.11 is modified to try and reduce the amount of variation in mass. The mass of a new sample of 100 pork pies is then measured. Apply the $\chi^2$ test to examine whether the data set formed by the set of 100 new mass measurements shown here conforms to a Gaussian distribution.

503 509 495 500 504 491 496 499 501 489 507 501 486 497 500 493 499 505 501 495
499 515 505 492 499 502 507 500 498 507 494 499 506 501 493 498 505 499 496 512
498 502 508 500 497 485 504 499 502 496 483 501 510 494 498 505 491 499 503 495
502 481 498 503 508 497 511 490 506 500 508 504 517 494 487 505 499 509 492 484
500 507 501 496 510 503 498 490 501 492 497 489 502 495 491 500 513 499 494 498

■

## Solution

The recommended number of data bins for 100 measurements according to the Sturgis rule is eight, as calculated in Example 3.11. Mass measurements in this new data set span the range from 481 to 517. Hence, data bin widths of 5 grams are still suggested, with bin boundaries set at 479.5, 484.5, 489.5, 494.5, 499.5, 504.5, 509.5, 514.5, and 519.5. The number of measurements in each bin is then counted, with the counts given in the following table:

| Bin number (i) | 1 | 2 | 3 | 4 | 5 | 6 | 7 | 8 |
|---|---|---|---|---|---|---|---|---|
| Data range | 479.5 to 484.5 | 484.5 to 489.5 | 489.5 to 494.5 | 494.5 to 499.5 | 499.5 to 504.5 | 504.5 to 509.5 | 509.5 to 514.5 | 514.5 to 519.5 |
| Measurements in range ($n_i$) | 3 | 5 | 14 | 29 | 26 | 16 | 5 | 2 |

Looking at these counts, we see that there are two bins with a count less than four. This amounts to 25% of the data bins. We have said previously that not more than 20% of data bins can have a data count less than the threshold of four if the $\chi^2$ test is to operate reliably. Hence, we must combine the bins and count the measurements again. The usual approach is to combine pairs of bins, which in this case reduces the number of bins from eight to four. The boundaries of the new set of four bins are now 479.5, 489.5, 499.5, 509.5, and 519.5. New data ranges and counts are shown in the following table.

| Bin number (i) | 1 | 2 | 3 | 4 |
|---|---|---|---|---|
| Data range | 479.5 to 489.5 | 489.5 to 499.5 | 499.5 to 509.5 | 509.5 to 519.5 |
| Measurements in range ($n_i$) | 8 | 43 | 42 | 7 |

Now, none of the bins have a count less than our stated minimum threshold of four and so we can proceed to calculate $n_i'$ values as before. The mean value ($\mu$) and standard deviation of the new mass measurements are $\mu = 499.39$ and $\sigma = 6.979$. We now calculate the *z* values corresponding to the measurement values (*x*) at the upper end of each data bin, read off the corresponding *F*(*z*) values from Table 3.1, and so calculate the expected number of measurements $(n_i')$ in each data bin:

| x | z $\left(\frac{x-\mu}{\sigma}\right)$ | *F*(*z*) | **Expected number of data in bin $(n_i')$** |
|---|---|---|---|
| 489.5 | -1.417 | 0.078 | 7.8 |
| 499.5 | -0.016 | 0.494 | 41.6 |
| 509.5 | 1.449 | 0.926 | 43.2 |
| 519.5 | 2.882 | 0.998 | 7.2 |

We now calculate the $\chi^2$ value for data using Equation (3.25). The steps of the calculation are shown in the following table.

| **Bin number (p)** | $n_i$ | $n_i'$ | $(n_i - n_i')$ | $(n_i - n_i')^2$ | $\frac{(n_i - n_i')^2}{n_i'}$ |
|---|---|---|---|---|---|
| 1 | 8 | 7.8 | 0.2 | 0.04 | 0.005 |
| 2 | 43 | 41.6 | 1.4 | 1.96 | 0.047 |
| 3 | 42 | 43.2 | -1.2 | 1.44 | 0.033 |
| 4 | 7 | 7.2 | -0.2 | 0.04 | 0.006 |

The value of $\chi^2$ is now found by summing the values in the final column to give $\chi^2 = 0.091$. The final step is to check whether this value of $\chi^2$ is greater than would be expected for a Gaussian distribution. This involves looking up $\chi^2$ in Table 3.2. This time, k = 2, as there are four bins and k is the number of bins minus 2 (as explained in Example 3.11, data were manipulated twice to obtain the $\mu$ and $\sigma$ statistical values used in the calculation of $n_i'$).

Table 3.2 shows that, for k = 2, $\chi^2 = 0.10$ for a 95% confidence level. Hence, our calculated value for $\chi^2$ of 0.91 shows that the confidence level that data follow a Gaussian distribution is slightly better than 95%.

Out of interest, if the two bin counts less than four had been ignored and $\chi^2$ had been calculated for the eight original data bins, a value of $\chi^2 = 2.97$ would have been obtained. (It would be a useful exercise for the reader to check this for himself/herself.) For six degrees of freedom (k = 8 − 2), the predicted value of $\chi^2$ for a Gaussian population from Table 3.2 is 2.20 at a 90% confidence level. Thus the confidence that data fit a Gaussian distribution is substantially less than 90% given the $\chi^2$ value of 2.97 calculated for data. This result arises because of the unreliability associated with calculating $\chi^2$ from data bin counts of less than four.

■

### 3.6.11 Rogue Data Points (Data Outliers)

In a set of measurements subject to random error, measurements with a very large error sometimes occur at random and unpredictable times, where the magnitude of the error is much larger than could reasonably be attributed to the expected random variations in measurement value. These are often called *rogue data points* or *data outliers*. Sources of such abnormal error include sudden transient voltage surges on the main power supply and incorrect recording of data (e.g., writing down 146.1 when the actual measured value was 164.1). It is accepted practice in such cases to discard these rogue measurements, and a threshold level of a $\pm 3\sigma$ deviation is often used to determine what should be discarded. It is rare for measurement errors to exceed $\pm 3\sigma$ limits when only normal random effects are affecting the measured value.

While the aforementioned represents a reasonable theoretical approach to identifying and eliminating rogue data points, the practical implementation of such a procedure needs to be done with care. The main practical difficulty that exists in dealing with rogue data points is in establishing what the expected standard deviation of the measurements is. When a new set of measurements is being taken where the expected standard deviation is not known, the possibility exists that a rogue data point exists within the measurements. Simply applying a computer program to the measurements to calculate the standard deviation will produce an erroneous result because the calculated value will be biased by the rogue data point. The simplest way to overcome this difficulty is to plot a histogram of any new set of measurements and examine this manually to spot any data outliers. If no outliers are apparent, the standard deviation can be calculated and then used in a $\pm 3\sigma$ threshold against which to test all future measurements. However, if this initial data histogram shows up any outliers, these should be excluded from the calculation of the standard deviation.

It is interesting at this point to return to the problem of ensuring that there are no outliers in the set of data used to calculate the standard deviation of data and hence the threshold for rejecting outliers. We have suggested that a histogram of some initial measurements be drawn and examined for outliers. What would happen if the set of data given earlier in Example 3.13 was the initial data set that was examined for outliers by drawing a histogram? What would happen if we did not spot the outlier of 4.59? This question can be answered by looking at the effect on the calculated value of standard deviation if this rogue data point of 4.59 is included in the calculation. The standard deviation calculated over the 19 values, excluding the 4.59

measurement, is 0.052. The standard deviation calculated over the 20 values, including the 4.59 measurement, is 0.063 and the mean data value is changed to 4.42. This gives a $3\sigma$ threshold of 0.19, and the boundaries for the $\pm 3\sigma$ threshold operation are now 4.23 and 4.61. This does not exclude the data value of 4.59, which we identified previously as a being a rogue data point! This confirms the necessity of looking carefully at the initial set of data used to calculate the thresholds for rejection of the rogue data point to ensure that initial data do not contain any rogue data points. If drawing and examining a histogram do not clearly show that there are no rogue data points in the "reference" set of data, it is worth taking another set of measurements to see whether a reference set of data can be obtained that is more clearly free of rogue data points.

## Example 3.13

A set of measurements is made with a new pressure transducer. Inspection of a histogram of the first 20 measurements does not show any data outliers. The standard deviation of the measurements is calculated as 0.05 bar after this check for data outliers, and the mean value is calculated as 4.41. Following this, the following further set of measurements is obtained:

4.35 4.46 4.39 4.34 4.41 4.52 4.44 4.37 4.41 4.33 4.39 4.47 4.42 4.59 4.45 4.38 4.43 4.36 4.48 4.45

Use the $\pm 3\sigma$ threshold to determine whether there are any rogue data points in the measurement set.

■

## Solution

Because the calculated $\sigma$ value for a set of "good" measurements is given as 0.05, the $\pm 3\sigma$ threshold is $\pm 0.15$. With a mean data value of 4.41, the threshold for rogue data points is values below 4.26 (mean value minus $3\sigma$) or above 4.56 (mean value plue $3\sigma$). Looking at the set of measurements, we observe that the measurement of 4.59 is outside the $\pm 3\sigma$ threshold, indicating that this is a rogue data point.

■

### *3.6.12 Student t Distribution*

When the number of measurements of a quantity is particularly small (less than about 30 samples) and statistical analysis of the distribution of error values is required, the possible deviation of the mean of measurements from the true measurement value (the mean of the infinite population that the sample is part of) may be significantly greater than is suggested by analysis based on a $z$ distribution. In response to this, a statistician called William Gosset developed an alternative distribution function that gives a more accurate prediction of the error distribution when the number of samples is small. He published this under the pseudonym "Student" and the distribution is commonly called *student t distribution.* It should be noted that *t* distribution has the same requirement as the *z* distribution in terms of the necessity for data to belong to a Gaussian distribution.

The student *t* variable expresses the difference between the mean of a small sample ($x_{mean}$) and the population mean ($\mu$) in terms of the following ratio:

$$t = \frac{|\text{error in mean}|}{\text{standard error of the mean}} = \frac{|\mu - x_{mean}|}{\sigma/\sqrt{N}}. \tag{3.26}$$

Because we do not know the exact value of $\sigma$, we have to use the best approximation to $\sigma$ that we have, which is the standard deviation of the sample, $\sigma_x$. Substituting this value for $\sigma$ in Equation (3.26) gives

$$t = \frac{|\mu - x_{mean}|}{\sigma_x/\sqrt{N}}. \tag{3.27}$$

Note that the modulus operation ($|\cdots|$) on the error in the mean in Equations (3.26) and (3.27) means that *t* is always positive.

The shape of the probability distribution curve *F*(*t*) of the *t* variable varies according to the value of the number of degrees of freedom, k (= N − 1), with typical curves being shown in Figure 3.13. As $k \to \infty$, $F(t) \to F(z)$, that is, the distribution becomes a standard Gaussian one. For values of $k < \infty$, the curve of *F*(*t*) against *t* is both narrower and less high in the center than a standard Gaussian curve, but has the same properties of symmetry about $t = 0$ and a total area under the curve of unity.

In a similar way to *z* distribution, the probability that *t* will lie between two values $t_1$ and $t_2$ is given by the area under the *F*(*t*) curve between $t_1$ and $t_2$. The *t* distribution is published in the form of a standard table (see Table 3.3) that gives values of the area under the curve $\alpha$ for various values of k, where

$$\alpha = \int_{t_3}^{\infty} F(t)dt. \tag{3.28}$$

The area $\alpha$ is shown in Figure 3.14. $\alpha$ corresponds to the probability that *t* will have a value greater than $t_3$ to some specified confidence level. Because the total area under the *F*(*t*) curve is unity, there is also a probability of $(1 - \alpha)$ that *t* will have a value less than $t_3$. Thus, for a value $\alpha = 0.05$, there is a 95% probability (i.e., a 95% confidence level) that $t < t_3$.

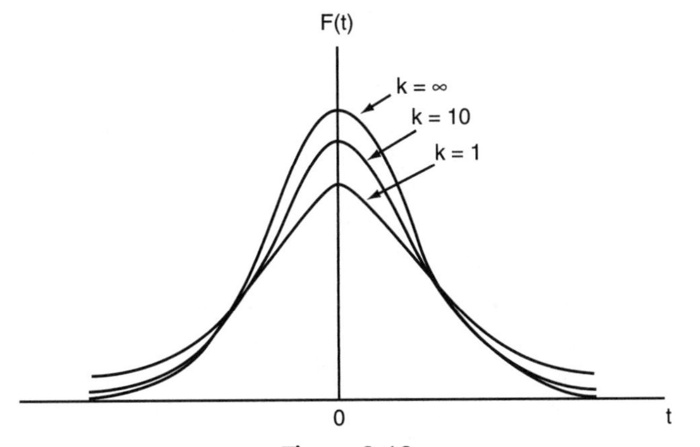


**Figure 3.13**
Typical *t*-distribution curves.

**Table 3.3 *t* Distribution**

| $k$ | $\alpha = 0.10$ | $\alpha = 0.05$ | $\alpha = 0.025$ | $\alpha = 0.01$ | $\alpha = 0.005$ | $\alpha = 0.001$ |
|---|---|---|---|---|---|---|
| 1 | 3.078 | 6.314 | 12.71 | 31.82 | 63.66 | 318.3 |
| 2 | 1.886 | 2.920 | 4.303 | 6.965 | 9.925 | 23.33 |
| 3 | 1.638 | 2.353 | 3.182 | 4.541 | 5.841 | 10.21 |
| 4 | 1.533 | 2.132 | 2.776 | 3.747 | 4.604 | 7.173 |
| 5 | 1.476 | 2.015 | 2.571 | 3.365 | 4.032 | 5.893 |
| 6 | 1.440 | 1.943 | 2.447 | 3.143 | 3.707 | 5.208 |
| 7 | 1.415 | 1.895 | 2.365 | 2.998 | 3.499 | 4.785 |
| 8 | 1.397 | 1.860 | 2.306 | 2.896 | 3.355 | 4.501 |
| 9 | 1.383 | 1.833 | 2.262 | 2.821 | 3.250 | 4.297 |
| 10 | 1.372 | 1.812 | 2.228 | 2.764 | 3.169 | 4.144 |
| 11 | 1.363 | 1.796 | 2.201 | 2.718 | 3.106 | 4.025 |
| 12 | 1.356 | 1.782 | 2.179 | 2.681 | 3.055 | 3.930 |
| 13 | 1.350 | 1.771 | 2.160 | 2.650 | 3.012 | 3.852 |
| 14 | 1.345 | 1.761 | 2.145 | 2.624 | 2.977 | 3.787 |
| 15 | 1.341 | 1.753 | 2.131 | 2.602 | 2.947 | 3.733 |
| 16 | 1.337 | 1.746 | 2.120 | 2.583 | 2.921 | 3.686 |
| 17 | 1.333 | 1.740 | 2.110 | 2.567 | 2.898 | 3.646 |
| 18 | 1.330 | 1.734 | 2.101 | 2.552 | 2.878 | 3.610 |
| 19 | 1.328 | 1.729 | 2.093 | 2.539 | 2.861 | 3.579 |
| 20 | 1.325 | 1.725 | 2.086 | 2.528 | 2.845 | 3.552 |
| 21 | 1.323 | 1.721 | 2.080 | 2.518 | 2.831 | 3.527 |
| 22 | 1.321 | 1.717 | 2.074 | 2.508 | 2.819 | 3.505 |
| 23 | 1.319 | 1.714 | 2.069 | 2.500 | 2.807 | 3.485 |
| 24 | 1.318 | 1.711 | 2.064 | 2.492 | 2.797 | 3.467 |
| 25 | 1.316 | 1.708 | 2.060 | 2.485 | 2.787 | 3.450 |
| 26 | 1.315 | 1.706 | 2.056 | 2.479 | 2.779 | 3.435 |
| 27 | 1.314 | 1.703 | 2.052 | 2.473 | 2.771 | 3.421 |
| 28 | 1.313 | 1.701 | 2.048 | 2.467 | 2.763 | 3.408 |
| 29 | 1.311 | 1.699 | 2.045 | 2.462 | 2.756 | 3.396 |
| 30 | 1.310 | 1.697 | 2.042 | 2.457 | 2.750 | 3.385 |

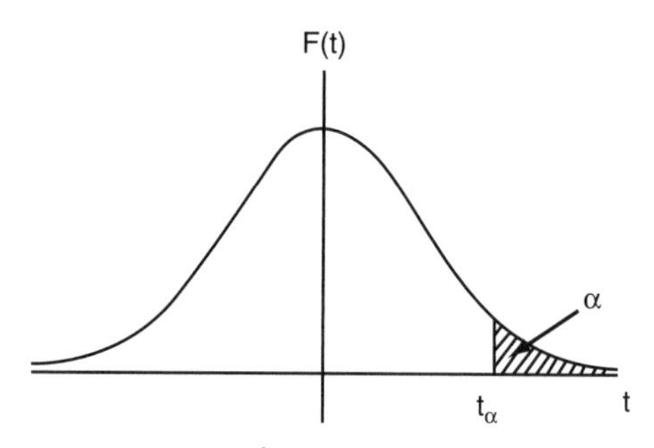


**Figure 3.14**
Meaning of area α for *t*-distribution curve.

Because of the symmetry of *t* distribution, α is also given by

$$\alpha = \int_{-\infty}^{-t_3} F(t)dt \tag{3.29}$$

as shown in Figure 3.15. Here, α corresponds to the probability that *t* will have a value less than $-t_3$, with a probability of $(1-\alpha)$ that *t* will have a value greater than $-t_3$.

Equations (3.28) and (3.29) can be combined to express the probability $(1-\alpha)$ that *t* lies between two values $-t_4$ and $+t_4$. In this case, α is the sum of two areas of α/2 as shown in Figure 3.16. These two areas can be represented mathematically as

$$\frac{\alpha}{2} = \int_{-\infty}^{-t_4} F(t)dt \text{ (left-hand area) and } \frac{\alpha}{2} = \int_{t_4}^{\infty} F(t)dt \text{ (right-hand area).}$$

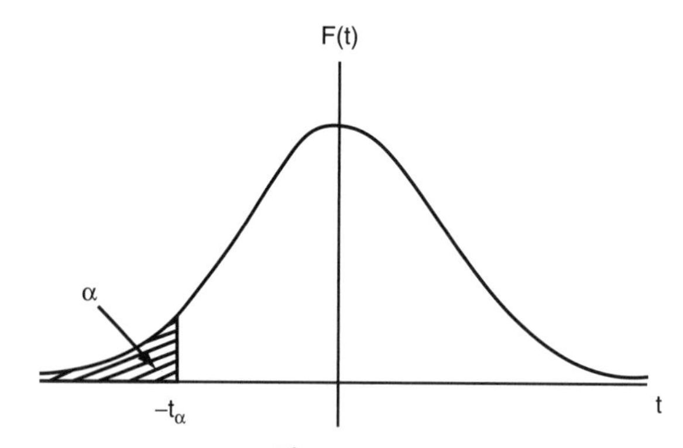


**Figure 3.15**
Alternative interpretation of area α for *t*-distribution curve.

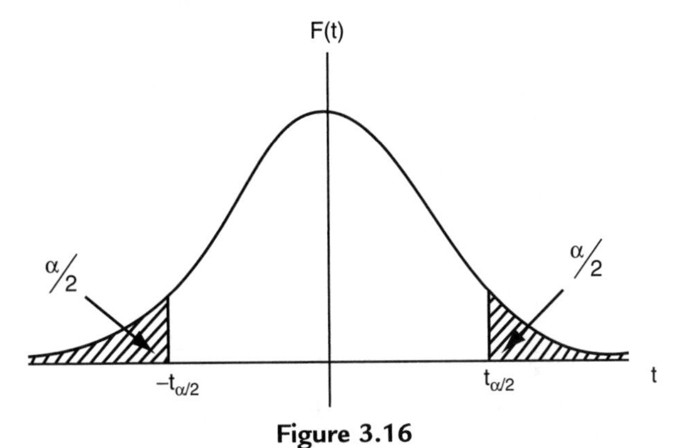


**Figure 3.16**
Area between $-\alpha/2$ and $+\alpha/2$ on *t*-distribution curve.

The values of $t_4$ can be found in any *t* distribution table, such as Table 3.3.

Referring back to Equation (3.27), this can be expressed in the form:

$$|\mu - x_{mean} = \frac{t\sigma_x}{\sqrt{N}}|.$$

Hence, upper and lower bounds on the expected value of the population mean μ (the true value of *x*) can be expressed as

$$-\frac{t_4\sigma_x}{\sqrt{N}} \leq \mu - x_{mean} \leq +\frac{t_4\sigma_x}{\sqrt{N}}$$

$$\text{or} \quad x_{mean} - \frac{t_4\sigma_x}{\sqrt{N}} \leq \mu \leq x_{mean} + \frac{t_4\sigma_x}{\sqrt{N}} \tag{3.30}$$

Out of interest, let us examine what would have happened if we had calculated the error bounds on μ using standard Gaussian (*z*-distribution) tables. For 95% confidence, the maximum error is given as $\pm 1.96\sigma/\sqrt{N}$, that is, ±0.96, which rounds to ±1.0 mm, meaning the mean internal diameter is given as 105.4 ± 1.0 mm. The effect of using *t* distribution instead of *z* distribution clearly expands the magnitude of the likely error in the mean value to compensate for the fact that our calculations are based on a relatively small number of measurements.

## Example 3.14

The internal diameter of a sample of hollow castings is measured by destructive testing of 15 samples taken randomly from a large batch of castings. If the sample mean is 105.4 mm with a standard deviation of 1.9 mm, express the upper and lower bounds to a confidence level of 95% on the range in which the mean value lies for internal diameter of the whole batch.

■

## Solution

For 15 samples ($N = 15$), the number of degrees of freedom ($k$) = 14.

For a confidence level of 95%, $\alpha = 1 - 0.95 = 0.05$. Looking up the value of $t$ in Table 3.3 for $k = 14$ and $\alpha/2 = 0.025$ gives $t = 2.145$. Thus, applying Equation (3.30):

$$105.4 - \frac{(2.145)(1.9)}{\sqrt{15}} \leq \mu \leq 105.4 + \frac{(2.145)(1.9)}{\sqrt{15}},$$

that is, $104.3 \leq \mu \leq 106.5$.

Thus, we would express the mean internal diameter of the whole batch of castings as $105.4 \pm 1.1$ mm.

■

## 3.7 Aggregation of Measurement System Errors

Errors in measurement systems often arise from two or more different sources, and these must be aggregated in the correct way in order to obtain a prediction of the total likely error in output readings from the measurement system. Two different forms of aggregation are required: (1) a single measurement component may have both systematic and random errors and (2) a measurement system may consist of several measurement components that each have separate errors.

### *3.7.1 Combined Effect of Systematic and Random Errors*

If a measurement is affected by both systematic and random errors that are quantified as $\pm x$ (systematic errors) and $\pm y$ (random errors), some means of expressing the combined effect of both types of errors is needed. One way of expressing the combined error would be to sum the two separate components of error, that is, to say that the total possible error is $e = \pm(x + y)$. However, a more usual course of action is to express the likely maximum error as

$$e = \sqrt{(x^2 + y^2)}. \tag{3.31}$$

It can be shown (ANSI/ASME, 1985) that this is the best expression for the error statistically, as it takes into account the reasonable assumption that the systematic and random errors are independent and so it is unlikely that both will be at their maximum or minimum value simultaneously.

### *3.7.2 Aggregation of Errors from Separate Measurement System Components*

A measurement system often consists of several separate components, each of which is subject to errors. Therefore, what remains to be investigated is how errors associated with each measurement system component combine together so that a total error calculation can be made for the complete measurement system. All four mathematical operations of addition, subtraction, multiplication, and division may be performed on measurements derived from different instruments/transducers in a measurement system. Appropriate techniques for the various situations that arise are covered later.

*Error in a sum*

If the two outputs $y$ and $z$ of separate measurement system components are to be added together, we can write the sum as $S = y + z$. If the maximum errors in $y$ and $z$ are $\pm ay$ and $\pm bz$, respectively, we can express the maximum and minimum possible values of $S$ as

$$S_{\max} = (y + ay) + (z + bz) \;;\; S_{\min} = (y - ay) + (z - bz) \;;\; \text{or } S = y + z \pm (ay + bz).$$

This relationship for $S$ is not convenient because in this form the error term cannot be expressed as a fraction or percentage of the calculated value for $S$. Fortunately, statistical analysis can be applied (see Topping, 1962) that expresses $S$ in an alternative form such that the most probable maximum error in $S$ is represented by a quantity $e$, where $e$ is calculated in terms of the *absolute* errors as

$$e = \sqrt{(ay)^2 + (bz)^2}. \tag{3.32}$$

Thus $S = (y + z) \pm e$. This can be expressed in the alternative form:

$$S = (y + z)(1 \pm f), \tag{3.33}$$

where $f = e/(y + z)$.

It should be noted that Equations (3.32) and (3.33) are only valid provided that the measurements are uncorrelated (i.e., each measurement is entirely independent of the others).

## Example 3.15

A circuit requirement for a resistance of 550 Ω is satisfied by connecting together two resistors of nominal values 220 and 330 Ω in series. If each resistor has a tolerance of ±2%, the error in the sum calculated according to Equations (3.32) and (3.33) is given by

$$e = \sqrt{(0.02 \times 220)^2 + (0.02 \times 330)^2} = 7.93 \;;\; f = 7.93/50 = 0.0144.$$

Thus the total resistance $S$ can be expressed as $S = 550\ \Omega \pm 7.93\ \Omega$ or $S = 550\ (1 \pm 0.0144)\ \Omega$, that is, $S = 550\ \Omega \pm 1.4\%$

■

*Error in a difference*

If the two outputs $y$ and $z$ of separate measurement systems are to be subtracted from one another, and the possible errors are $\pm ay$ and $\pm bz$, then the difference $S$ can be expressed (using statistical analysis as for calculating the error in a sum and assuming that the measurements are uncorrelated) as

$$S = (y - z) \pm e \quad \text{or} \quad S = (y - z)(1 \pm f),$$

where $e$ is calculated as in Equation (3.32) and $f = e/(y - z)$.

This example illustrates very poignantly the relatively large error that can arise when calculations are made based on the difference between two measurements.

## ■ Example 3.16

A fluid flow rate is calculated from the difference in pressure measured on both sides of an orifice plate. If the pressure measurements are 10.0 and 9.5 bar and the error in the pressure-measuring instruments is specified as $\pm 0.1\%$, then values for $e$ and $f$ can be calculated as

$$e = \sqrt{(0.001 \times 10)^2 + (0.001 \times 9.5)^2} = 0.0138 \;;\; f = 0.0138/0.5 = 0.0276$$

■

*Error in a product*

If outputs $y$ and $z$ of two measurement system components are multiplied together, the product can be written as $P = yz$. If the possible error in $y$ is $\pm ay$ and in $z$ is $\pm bz$, then the maximum and minimum values possible in $P$ can be written as

$$P_{max} = (y + ay)(z + bz) = yz + ayz + byz + aybz \;;\; P_{min} = (y - ay)(z - bz) = yz - ayz - byz + aybz.$$

For typical measurement system components with output errors of up to 1 or 2% in magnitude, both $a$ and $b$ are very much less than one in magnitude and thus terms in $aybz$ are negligible compared with other terms. Therefore, we have $P_{max} = yz(1 + a + b)$ ; $P_{min} = yz(1 - a - b)$. Thus the maximum error in product $P$ is $\pm(a + b)$. While this expresses the maximum possible error in $P$, it tends to overestimate the likely maximum error, as it is very unlikely that the errors in $y$ and $z$ will both be at the maximum or minimum value at the same time. A statistically better estimate of the likely maximum error $e$ in product $P$, provided that the measurements are uncorrelated, is given by Topping (1962):

$$e = \sqrt{a^2 + b^2}. \tag{3.34}$$

Note that in the case of multiplicative errors, $e$ is calculated in terms of *fractional* errors in $y$ and $z$ (as opposed to *absolute* error values used in calculating additive errors).

## Example 3.17

If the power in a circuit is calculated from measurements of voltage and current in which the calculated maximum errors are, respectively, $\pm 1$ and $\pm 2\%$, then the maximum likely error in the calculated power value, calculated using Equation (3.34) is $\pm\sqrt{0.01^2 + 0.02^2} = \pm 0.022$ or $\pm 2.2\%$.

■

### *Error in a quotient*

If the output measurement $y$ of one system component with possible error $\pm ay$ is divided by the output measurement $z$ of another system component with possible error $\pm bz$, then the maximum and minimum possible values for the quotient can be written as

$$Q_{max} = \frac{y + ay}{z - bz} = \frac{(y + ay)(z + bz)}{(z - bz)(z + bz)} = \frac{yz + ayz + byz + aybz}{z^2 - b^2z^2};$$

$$Q_{min} = \frac{y - ay}{z + bz} = \frac{(y - ay)(z - bz)}{(z + bz)(z - bz)} = \frac{yz - ayz - byz + aybz}{z^2 - b^2z^2}.$$

For $a << 1$ and $b << 1$, terms in $ab$ and $b^2$ are negligible compared with the other terms. Hence $Q_{max} = \frac{yz(1 + a + b)}{z^2}$; $Q_{min} = \frac{yz(1 - a - b)}{z^2}$; that is, $Q = \frac{y}{z} \pm \frac{y}{z}(a + b)$.

Thus the maximum error in the quotient is $\pm(a + b)$. However, using the same argument as made earlier for the product of measurements, a statistically better estimate (see Topping, 1962) of the likely maximum error in the quotient $Q$, provided that the measurements are uncorrelated, is that given in Equation (3.34).

## Example 3.18

If the density of a substance is calculated from measurements of its mass and volume where the respective errors are $\pm 2$ and $\pm 3\%$, then the maximum likely error in the density value using Equation (3.34) is $\pm\sqrt{0.02^2 + 0.003^2} = \pm 0.036$ or $\pm 3.6\%$.

■

### 3.7.3 Total Error When Combining Multiple Measurements

The final case to be covered is where the final measurement is calculated from several measurements that are combined together in a way that involves more than one type of arithmetic operation. For example, the density of a rectangular-sided solid block of material can be calculated from measurements of its mass divided by the product of measurements of its length, height, and width. Because errors involved in each stage of arithmetic are cumulative, the total measurement error can be calculated by adding together the two error values associated with the two multiplication stages involved in calculating the volume and then calculating the error in the final arithmetic operation when the mass is divided by the volume.

## Example 3.19

A rectangular-sided block has edges of lengths $a$, $b$, and $c$, and its mass is $m$. If the values and possible errors in quantities $a$, $b$, $c$, and $m$ are as shown, calculate the value of density and the possible error in this value.

$$a = 100 \text{ mm} \pm 1\%, \text{b} = 200 \text{ mm} \pm 1\%, \text{c} = 300 \text{ mm} \pm 1\%, \text{m} = 20 \text{ kg} \pm 0.5\%.$$

## Solution

Value of $ab = 0.02\ m^2 \pm 2\%$ [possible error $= 1\% + 1\% = 2\%$]
Value of $(ab)c = 0.006\ m^3 \pm 3\%$ [possible error $= 2\% + 1\% = 3\%$]
Value of $\frac{m}{abc} = \frac{20}{0.006} = 3330 \text{ kg/m}^3 \pm 3.5\%$ [possible error $= 3\% + 0.5\% = 3.5\%$]

## 3.8 Summary

This chapter introduced the subject of measurement uncertainty, and the length of the chapter gives testimony to the great importance attached to this subject. Measurement errors are a fact of life and, although we can do much to reduce the magnitude of errors, we can never reduce errors entirely to zero. We started the chapter off by noting that measurement uncertainty comes in two distinct forms, known respectively as *systematic error* and *random error*. We learned that the nature of systematic errors was such that the effect on a measurement reading was to make it either consistently greater than or consistently less than the true value of the measured quantity. Random errors, however, are entirely random in nature, such that successive measurements of a constant quantity are randomly both greater than and less than the true value of the measured quantity.

In our subsequent study of systematic measurement errors, we first examined all the sources of this kind of error. Following this, we looked at all the techniques available for reducing the magnitude of systematic errors arising from the various error sources identified. Finally, we examined ways of quantifying the remaining systemic measurement error after all reasonable means of reducing error magnitude had been applied.

Our study of random measurement errors also started off by studying typical sources of these. We observed that the nature of random errors means that we can get close to the correct value of the measured quantity by averaging over a number of measurements, although we noted that we could never actually get to the correct value unless we achieved the impossible task of having an infinite number of measurements to average over. We found that how close we get to the correct value depends on how many measurements we average over and how widely the measurements are spread. We then examined the two alternative ways of calculating an average in terms of the *mean* and *median* value of a set of measurements. Following this, we looked at ways of expressing the spread of measurements about the mean/median value. This led to the formal mathematical quantification of spread in terms of *standard deviation* and *variance*. We then started to look at graphical ways of expressing the spread. Initially, we considered representations of spread as a *histogram* and then went on to show how histograms expand into *frequency distributions* in the form of a smooth curve. We found that truly random data are described by a particular form of frequency distribution known as *Gaussian* (or *normal*). We introduced the *z variable* and saw how this can be used to estimate the number of measurements in a set of measurements that have an error magnitude between two specified values. Following this, we looked at the implications of the fact that we can only ever have a finite number of measurements. We saw that a variable called the *standard error of the mean* could be calculated that estimates the difference between the mean value of a finite set of measurements and the true value of the measured quantity (the mean of an infinite data set). We went on to look at how this was useful in estimating the likely error in a single measurement subject to random errors in the situation where it is not possible to average over a number of measurements. As an aside, we then went on to look at how the $z$ variable was useful in analyzing tolerances of manufactured components subject to random variations in a parallel way to the analysis of measurements subject to random variations. Following this, we went on to look at $\chi^2$ distribution. This can be used to quantify the variation in the variance of a finite set of measurements with respect to the variance of the infinite set that the finite set is part of. Up until this point in the chapter, all analysis of random errors assumed that the measurement set fitted a Gaussian distribution. However, this assumption must always be justified by applying *goodness of fit tests*, so these were explained in the following section, where we saw that a $\chi^2$ test is the most rigorous test available for goodness of fit. A particular problem that can affect the analysis of random errors adversely is the presence of rogue data points (data outliers) in measurement data. These were considered, and the conditions under which they can justifiably be excluded from the analyzed data set were explored. Finally, we saw that yet another problem that can affect the analysis of random errors is where the measurement set only has a small number of values. In this case, calculations based on $z$ distribution are inaccurate, and we explored the use of a better distribution called *t distribution*.

The chapter ended with looking at how the effects of different measurement errors are aggregated together to predict the total error in a measurement system. This process was considered in two parts. First, we looked at how systematic and random error magnitudes can be combined together in an optimal way that best predicts the likely total error in a particular measurement. Second, we looked at situations where two or more measurements of different quantities are combined together to give a composite measurement value and looked at the best way of dealing with each of the four arithmetic operations that can be carried out on different measurement components.

## 3.9 Problems

3.1. Explain the difference between systematic and random errors. What are the typical sources of these two types of errors?

3.2. In what ways can the act of measurement cause a disturbance in the system being measured?

3.3. In the circuit shown in Figure 3.17, the resistor values are given by $R_1 = 1000\ \Omega$ ; $R_2 = 1000\ \Omega$ ; $V = 20$ volts. The voltage across AB (i.e., across $R_2$) is measured by a voltmeter whose internal resistance is given by $R_m = 9500\ \Omega$.

   (a) What will be the reading on the voltmeter?

   (b) What would the voltage across AB be if the voltmeter was not loading the circuit (i.e., if $R_m$ = infinity)?

   (c) What is the measurement error due to the loading effect of the voltmeter?

3.4. Suppose that the components in the circuit shown in Figure 3.1a have the following values:

$$R_1 = 330\ \Omega;\ R_2 = 1000\ \Omega;\ R_3 = 1200\ \Omega;\ R_4 = 220\ \Omega; R_5 = 270\ \Omega.$$

If the instrument measuring the output voltage across $AB$ has a resistance of 5000 Ω, what is the measurement error caused by the loading effect of this instrument?

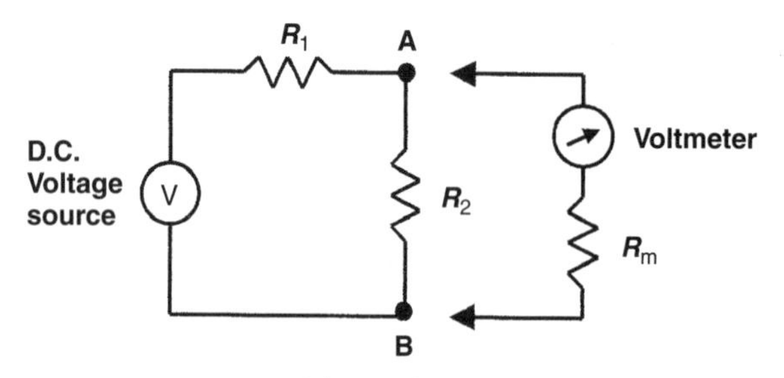


**Figure 3.17**
Circuit for Problem 3.3.

3.5. (a) Explain what is meant by the term "modifying inputs."
(b) Explain briefly what measures can be taken to reduce or eliminate the effect of modifying inputs.

3.6. Instruments are normally calibrated and their characteristics defined for particular standard ambient conditions. What procedures are normally taken to avoid measurement errors when using instruments subjected to changing ambient conditions?

3.7. The voltage across a resistance $R_5$ in the circuit of Figure 3.18 is to be measured by a voltmeter connected across it.
(a) If the voltmeter has an internal resistance ($R_m$) of 4750 Ω, what is the measurement error?
(b) What value would the voltmeter internal resistance need to be in order to reduce the measurement error to 1%?

3.8. In the circuit shown in Figure 3.19, the current flowing between A and B is measured by an ammeter whose internal resistance is 100 Ω. What is the measurement error caused by the resistance of the measuring instrument?

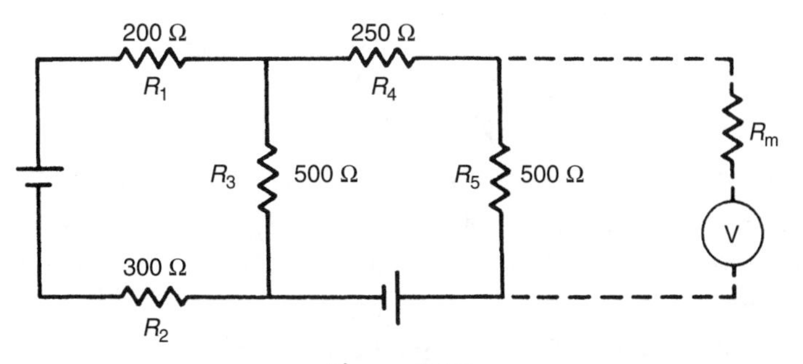


**Figure 3.18**
Circuit for Problem 3.7.

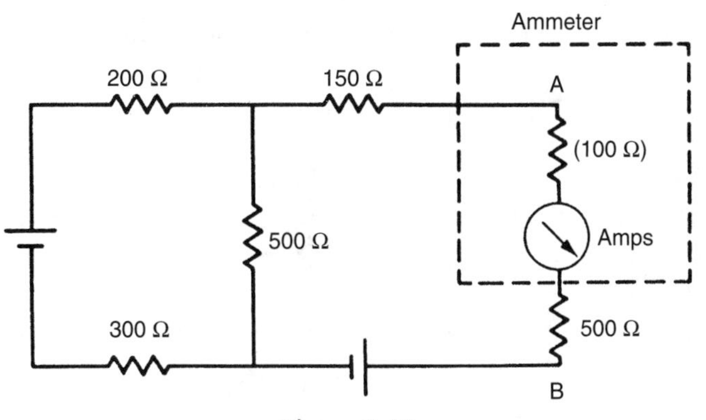


**Figure 3.19**
Circuit for Problem 3.8.

3.9. What steps can be taken to reduce the effect of environmental inputs in measurement systems?

3.10. (a) Explain why a voltmeter never reads exactly the correct value when it is applied to an electrical circuit to measure the voltage between two points.

(b) For the circuit shown in Figure 3.17, show that the voltage $E_m$ measured across points AB by the voltmeter is related to the true voltage $E_o$ by the following expression:

$$\frac{E_m}{E_o} = \frac{R_m(R_1 + R_2)}{R_1(R_2 + R_m) + R_2 R_m}$$

(c) If the parameters in Figure 3.17 have the following values, $R_1 = 500\ \Omega$; $R_2 = 500\ \Omega$; $R_m = 4750\ \Omega$, calculate the percentage error in the voltage value measured across points AB by the voltmeter.

3.11. The output of a potentiometer is measured by a voltmeter having resistance $R_m$, as shown in Figure 3.20. $R_t$ is the resistance of the total length $X_t$ of the potentiometer, and $R_i$ is the resistance between the wiper and common point $C$ for a general wiper position $X_i$. Show that the measurement error due to the resistance, $R_m$, of the measuring instrument is given by

$$Error = E \frac{R_i^2(R_t - R_i)}{R_t(R_i R_t + R_m R_t - R_i^2)}.$$

Hence show that the maximum error occurs when $X_i$ is approximately equal to $2X_t/3$. (Hint: differentiate the error expression with respect to $R_i$ and set to 0. Note that maximum error does not occur exactly at $X_i = 2X_t/3$, but this value is very close to the position where the maximum error occurs.)

3.12. In a survey of 15 owners of a certain model of car, the following values for average fuel consumption were reported:

25.5 30.3 31.1 29.6 32.4 39.4 28.9 30.0 33.3 31.4 29.5 30.5 31.7 33.0 29.2

Calculate mean value, median value, and standard deviation of the data set.

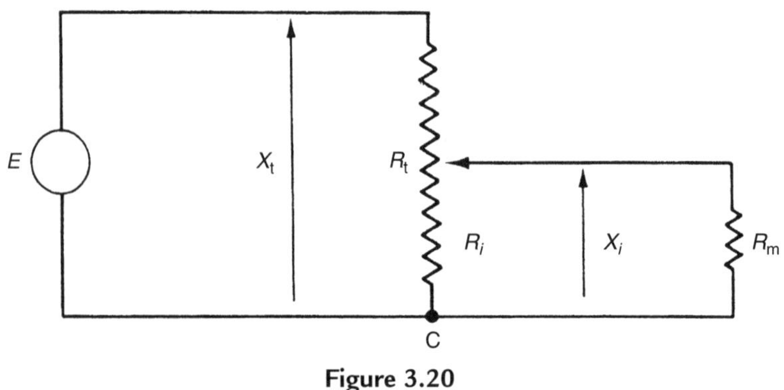


**Figure 3.20**
Circuit for Problem 3.11.

3.13. The following 10 measurements of the freezing point of aluminum were made using a platinum/rhodium thermocouple:

658.2 659.8 661.7 662.1 659.3 660.5 657.9 662.4 659.6 662.2

Find (a) median, (b) mean, (c) standard deviation, and (d) variance of the measurements.

3.14. The following 25 measurements were taken of the thickness of steel emerging from a rolling mill:

3.97 3.99 4.04 4.00 3.98 4.03 4.00 3.98 3.99 3.96 4.02 3.99 4.01

3.97 4.02 3.99 3.95 4.03 4.01 4.05 3.98 4.00 4.04 3.98 4.02

Find (a) median, (b) mean, (c) standard deviation, and (d) variance of the measurements.

3.15. The following 10 measurements were made of output voltage from a high-gain amplifier contaminated due to noise fluctuations:

1.53, 1.57, 1.54, 1.54 ,1.50, 1.51, 1.55, 1.54, 1.56, 1.53

Determine the mean value and standard deviation. Hence, estimate the accuracy to which the mean value is determined from these 10 measurements. If 1000 measurements were taken, instead of 10, but $\sigma$ remained the same, by how much would the accuracy of the calculated mean value be improved?

3.16. The following measurements were taken with an analogue meter of current flowing in a circuit (the circuit was in steady state and therefore, although measurements varied due to random errors, the current flowing was actually constant):

21.5, 22.1, 21.3, 21.7, 22.0, 22.2, 21.8, 21.4, 21.9, 22.1 mA

Calculate mean value, deviations from the mean, and standard deviation.

3.17. Using the measurement data given in Problem 3.14, draw a histogram of errors (use error bands 0.03 units wide, i.e., the center band will be from $-0.015$ to $+0.015$).

3.18. (a) What do you understand by the term *probability density function*?
(b) Write down an expression for a Gaussian probability density function of given mean value $\mu$ and standard deviation $\sigma$ and show how you would obtain the best estimates of these two quantities from a sample of population $n$.

3.19. Measurements in a data set are subject to random errors, but it is known that the data set fits a Gaussian distribution. Use standard Gaussian tables to determine the percentage of measurements that lie within the boundaries of $\pm 1.5\sigma$, where $\sigma$ is the standard deviation of the measurements.

3.20. Measurements in a data set are subject to random errors, but it is known that the data set fits a Gaussian distribution. Use error function tables to determine the value of $x$ required such that 95% of the measurements lie within the boundaries of $\pm x\sigma$, where $\sigma$ is the standard deviation of the measurements.

3.21. By applying error function tables for mean and standard deviation values calculated in Problem 3.14, estimate
(a) How many measurements are $<4.00$?
(b) How many measurements are $<3.95$?
(c) How many measurements are between 3.98 and 4.02?
Check your answers against real data.

3.22. The resolution of the instrument referred to in Problem 3.14 is clearly 0.01. Because of the way in which error tables are presented, estimations of the number of measurements in a particular error band are likely to be closer to the real number if boundaries of the error band are chosen to be between measurement values. In part c of Problem 3.21, values $>3.98$ are subtracted from values $>4.02$, thus excluding measurements equal to 3.98. Test this hypothesis out by estimating:
(a) How many measurements are $<3.995$?
(b) How many measurements are $<3.955$?
(c) How many measurements are between 3.975 and 4.025?
Check your answers against real data.

3.23. Measurements in a data set are subject to random errors, but it is known that the data set fits a Gaussian distribution. Use error function tables to determine the percentage of measurements that lie within the boundaries of $\pm 2\sigma$, where $\sigma$ is the standard deviation of the measurements.

3.24. A silicon-integrated circuit chip contains 5000 ostensibly identical transistors. Measurements are made of the current gain of each transistor. Measurements have a mean of 20.0 and a standard deviation of 1.5. The probability distribution of the measurements is Gaussian.
(a) Write down an expression for the number of transistors on the chip that have a current gain between 19.5 and 20.5.
(b) Show that this number of transistors with a current gain between 19.5 and 20.5 is approximately 1300.
(c) Calculate the number of transistors that have a current gain of 17 or more (this is the minimum current gain necessary for a transistor to be able to drive the succeeding stage of the circuit in which the chip is used).

3.25. In a particular manufacturing process, bricks are produced in batches of 10,000. Because of random variations in the manufacturing process, random errors occur in the target length of the bricks produced. If the bricks have a mean length of 200 mm with a standard deviation of 20 mm, show how the error function tables supplied can be used to calculate the following:
(a) number of bricks with a length between 198 and 202 mm.
(b) number of bricks with a length greater than 170 mm.

3.26. The temperature-controlled environment in a hospital intensive care unit is monitored by an intelligent instrument that measures temperature every minute and calculates the mean and standard deviation of the measurements. If the mean is 75°C and the standard deviation is 2.15,

(a) What percentage of the time is the temperature less than 70°C?

(b) What percentage of the time is the temperature between 73 and 77°C?

3.27. Calculate the standard error of the mean for measurements given in Problem 3.13. Hence, express the melting point of aluminum together with the possible error in the value expressed.

3.28. The thickness of a set of gaskets varies because of random manufacturing disturbances, but thickness values measured belong to a Gaussian distribution. If the mean thickness is 3 mm and the standard deviation is 0.25, calculate the percentage of gaskets that have a thickness greater than 2.5 mm.

3.29. If the measured variance of 25 samples of bread cakes taken from a large batch is 4.85 grams, calculate the true variance of the mass for a whole batch of bread cakes to a 95% significance level.

3.30. Calculate true standard deviation of the diameter of a large batch of tires to a confidence level of 99% if the measured standard deviation in the diameter for a sample of 30 tires is 0.63 cm.

3.31. One hundred fifty measurements are taken of the thickness of a coil of rolled steel sheet measured at approximately equidistant points along the center line of its length. Measurements have a mean value of 11.291 mm and a standard deviation of 0.263 mm. The smallest and largest measurements in the sample are 10.73 and 11.89 mm. Measurements are divided into eight data bins with boundaries at 10.695, 10.845, 10.995, 11.145, 11.295, 11.445, 11.595, 11.745, and 11.895. The first bin, containing measurements between 10,695 and 10.845, has eight measurements in it, and the count of measurements in the following successive bins is 12, 21, 34, 31, 25, 14, and 5. Apply the $\chi^2$ test to see whether the measurements fit a Gaussian distribution to a 95% confidence level.

3.32. The temperature in a furnace is regulated by a control system that aims to keep the temperature close to 800°C. The temperature is measured every minute over a 2-hour period, during which time the minimum and maximum temperatures measured are 782 and 819°C. Analysis of the 120 measurements shows a mean value of 800.3°C and a standard deviation of 7.58°C. Measurements are divided into eight data bins of 5°C width with boundaries at 780.5, 785.5, 790.5, 795.5, 800.5, 805.5, 810.5, 815.5, and 820.5. The measurement count in bin one from 780.5 to 785.5°C was 3, and the count in the other successive bins was 8, 21, 30, 28, 19, 9, and 2. Apply the $\chi^2$ test to see whether the measurements fit a Gaussian distribution to (a) a 90% confidence level and (b) a 95% confidence level (think carefully about whether the $\chi^2$ test will be reliable for the measurement counts observed and whether there needs to be any change in the number of data bins used for the $\chi^2$ test).

3.33. The volume contained in each sample of 10 bottles of expensive perfume is measured. If the mean volume of the sample measurements is 100.5 ml with a standard deviation of 0.64 ml, calculate the upper and lower bounds to a confidence level of 95% of the mean value of the whole batch of perfume from which the 10 samples were taken.

3.34. A 3-volt d.c. power source required for a circuit is obtained by connecting together two 1.5-volt batteries in series. If the error in the voltage output of each battery is specified as $\pm 1\%$, calculate the likely maximum error in the 3-volt power source that they make up.

3.35. A temperature measurement system consists of a thermocouple whose amplified output is measured by a voltmeter. The output relationship for the thermocouple is approximately linear over the temperature range of interest. The e.m.f./temp relationship of the thermocouple has a possible error of $\pm 1\%$, the amplifier gain value has a possible error of $\pm 0.5\%$, and the voltmeter has a possible error of $\pm 2\%$. What is the possible error in the measured temperature?

3.36. A pressure measurement system consists of a monolithic piezoresistive pressure transducer and a bridge circuit to measure the resistance change of the transducer. The resistance ($R$) of the transducer is related to pressure ($P$) according to $R = K_1 P$ and the output of the bridge circuit ($V$) is related to resistance ($R$) by $V = K_2 R$. Thus, the output voltage is related to pressure according to $V = K_1 K_2 P$. If the maximum error in $K_1$ is $\pm 2\%$, the maximum error in $K_2$ is $\pm 1.5\%$, and the voltmeter itself has a maximum measurement error of $\pm 1\%$, what is the likely maximum error in the pressure measurement?

3.37. A requirement for a resistance of 1220 Ω in a circuit is satisfied by connecting together resistances of 1000 and 220 Ω in series. If each resistance has a tolerance of $\pm 5\%$, what is the likely tolerance in the total resistance?

3.38. In order to calculate the heat loss through the wall of a building, it is necessary to know the temperature difference between inside and outside walls. Temperatures of 5 and 20°C are measured on each side of the wall by mercury-in-glass thermometers with a range of 0 to +50°C and a quoted inaccuracy of $\pm 1\%$ of full-scale reading.

(a) Calculate the likely maximum possible error in the calculated value for the temperature difference.

(b) Discuss briefly how using measuring instruments with a different measurement range may improve measurement accuracy.

3.39. A fluid flow rate is calculated from the difference in pressure measured across a venturi. Flow rate is given by $F = K(p_2 - p_1)$, where $p_1$ and $p_2$ are the pressures either side of the venturi and $K$ is a constant. The two pressure measurements are 15.2 and 14.7 bar.

(a) Calculate the possible error in flow measurement if pressure-measuring instruments have a quoted error of $\pm 0.2\%$ of their reading.

(b) Discuss briefly why using a differential pressure sensor rather than two separate pressure sensors would improve measurement accuracy.

3.40. The power dissipated in a car headlight is calculated by measuring the d.c. voltage drop across it and the current flowing through it ($P = V \times I$). If possible errors in the measured voltage and current values are $\pm 1$ and $\pm 2\%$, respectively, calculate the likely maximum possible error in the power value deduced.

3.41. The resistance of a carbon resistor is measured by applying a d.c. voltage across it and measuring the current flowing ($R = V/I$). If the voltage and current values are measured as $10 \pm 0.1$ V and $214 \pm 5$ mA, respectively, express the value of the carbon resistor.

3.42. The specific energy of a substance is calculated by measuring the energy content of a cubic meter volume of the substance. If the errors in energy measurement and volume measurement are $\pm 1$ and $\pm 2\%$, respectively, what is the possible error in the calculated value of specific energy (specific energy = energy per unit volume of material)?

3.43. In a particular measurement system, quantity $x$ is calculated by subtracting a measurement of a quantity $z$ from a measurement of a quantity $y$, that is, $x = y - z$. If the possible measurement errors in $y$ and $z$ are $\pm ay$ and $\pm bz$, respectively, show that the value of $x$ can be expressed as $x = y - z \pm (ay - bz)$.

(a) What is inconvenient about this expression for $x$, and what is the basis for the following expression for $x$ that is used more commonly?

$$x = (y - z) \pm e,$$

where $e = \sqrt{(ay)^2 + (bz)^2}$.

(b) In a different measurement system, quantity $p$ is calculated by multiplying together measurements of two quantities $q$ and $r$ such that $p = qr$. If the possible measurement errors in $q$ and $r$ are $\pm aq$ and $\pm br$, respectively, show that the value of $p$ can be expressed as $p = (qr)(1 \pm [a + b])$. The volume flow rate of a liquid in a pipe (the volume flowing in unit time) is measured by allowing the end of the pipe to discharge into a vertical-sided tank with a rectangular base (see Figure 3.21). The depth of the liquid in the tank is measured at the start as $h_1$ meters and 1 minute later it is measured as $h_2$ meters. If the length and width of the tank are $l$ and $w$ meters, respectively, write down an expression for the volume flow rate of the liquid in cubic meters per minute. Calculate the volume flow rate of the liquid if the measured parameters have the following values:

$$h_1 = 0.8 \text{ m} \; ; \; h_2 = 1.3 \text{ m} \; ; \; l = 4.2 \text{ m} \; ; \; w = 2.9 \text{ m}$$

If the possible errors in the measurements of $h_1, h_2, l$, and $w$ are $1, 1, 0.5$, and $0.5\%$, respectively, calculate the possible error in the calculated value of the flow rate.

3.44. The density of a material is calculated by measuring the mass of a rectangular-sided block of the material whose edges have lengths of $a$, $b$, and $c$. What is the possible error

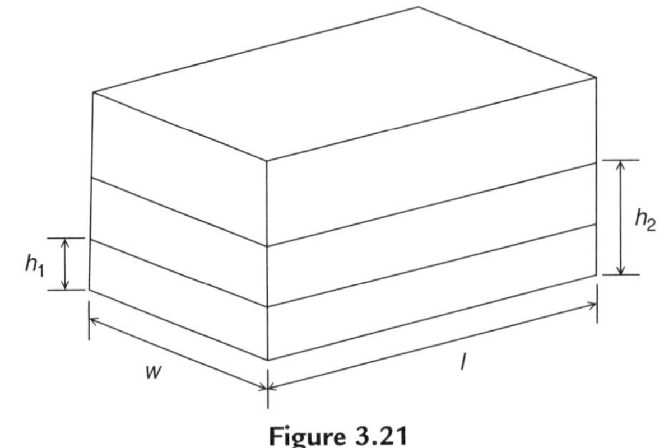


**Figure 3.21**
Diagram for Problem 3.44.

in the calculated density if the possible error in mass measurement is $\pm1.0\%$ and possible errors in length measurement are $\pm0.5\%$?

3.45. The density ($d$) of a liquid is calculated by measuring its depth ($c$) in a calibrated rectangular tank and then emptying it into a mass-measuring system. The length and width of the tank are ($a$) and ($b$), respectively, and thus the density is given by

$$d = m/(a \times b \times c),$$

where $m$ is the measured mass of the liquid emptied out. If the possible errors in the measurements of $a$, $b$, $c$, and $m$ are 1, 1, 2, and 0.5%, respectively, determine the likely maximum possible error in the calculated value of the density ($d$).

3.46. The volume flow rate of a liquid is calculated by allowing the liquid to flow into a cylindrical tank (stood on its flat end) and measuring the height of the liquid surface before and after the liquid has flowed for 10 minutes. The volume collected after 10 minutes is given by

$$Volume = (h_2 - h_1)\pi(d/2)^2,$$

where $h_1$ and $h_2$ are the starting and finishing surface heights and $d$ is the measured diameter of the tank.

(a) If $h_1 = 2$ m, $h_2 = 3$ m, and $d = 2$ m, calculate the volume flow rate in $m^3$/min.

(b) If the possible error in each measurement $h_1$, $h_2$, and $d$ is $\pm1\%$, determine the likely maximum possible error in the calculated value of volume flow rate (it is assumed that there is negligible error in the time measurement).

## References

ANSI/ASME Standards. (1985). *ASME performance test codes, supplement on instruments and apparatus, part 1: measurement uncertainty*. New York: American Society of Mechanical Engineers.

Topping, J. (1962). *Errors of observation and their treatment*. Chapman and Hall.

CHAPTER 4

# *Calibration of Measuring Sensors and Instruments*



## 4.1 Introduction

We just examined the various systematic and random measurement error sources in the last chapter. As far as systematic errors are concerned, we observed that recalibration at a suitable frequency was an important weapon in the quest to minimize errors due to drift in instrument characteristics. The use of proper and rigorous calibration procedures is essential in order to ensure that recalibration achieves its intended purpose; to reflect the importance of getting these procedures right, this whole chapter is dedicated to explaining the various facets of calibration.

We start in Section 4.2 by formally defining what calibration means, explaining how it is performed and considering how to calculate the frequency with which the calibration exercise should be repeated. We then go on to look at the calibration environment in Section 4.3, where we learn that proper control of the environment in which instruments are calibrated is an essential component in good calibration procedures. Section 4.4 then continues with a review of how the calibration of working instruments against reference instruments is linked by the calibration chain to national and international reference standards relating to the quantity that the instrument being calibrated is designed to measure. Finally, Section 4.5 emphasizes the importance of maintaining records of instrument calibrations and suggests appropriate formats for such records.

## 4.2 Principles of Calibration

Calibration consists of comparing the output of the instrument or sensor under test against the output of an instrument of known accuracy when the same input (the measured quantity) is applied to both instruments. This procedure is carried out for a range of inputs covering the whole measurement range of the instrument or sensor. Calibration ensures that the measuring accuracy of all instruments and sensors used in a measurement system is known over the whole measurement range, provided that the calibrated instruments and sensors are used in environmental conditions that are the same as those under which they were calibrated. For use of instruments and sensors under different environmental conditions, appropriate correction has to be made for the ensuing modifying inputs, as described in Chapter 3. Whether applied to instruments or sensors, calibration procedures are identical, and hence only the term instrument will be used for the rest of this chapter, with the understanding that whatever is said for instruments applies equally well to single measurement sensors.

Instruments used as a standard in calibration procedures are usually chosen to be of greater inherent accuracy than the process instruments that they are used to calibrate. Because such instruments are only used for calibration purposes, greater accuracy can often be achieved by specifying a type of instrument that would be unsuitable for normal process measurements. For instance, ruggedness is not a requirement, and freedom from this constraint opens up a much wider range of possible instruments. In practice, high-accuracy, null-type instruments are used very commonly for calibration duties, as the need for a human operator is not a problem in these circumstances.

Instrument calibration has to be repeated at prescribed intervals because the characteristics of any instrument change over a period. Changes in instrument characteristics are brought about by such factors as mechanical wear, and the effects of dirt, dust, fumes, chemicals, and temperature change in the operating environment. To a great extent, the magnitude of the drift in characteristics depends on the amount of use an instrument receives and hence on the amount of wear and the length of time that it is subjected to the operating environment. However, some drift also occurs even in storage as a result of aging effects in components within the instrument.

Determination of the frequency at which instruments should be calibrated is dependent on several factors that require specialist knowledge. If an instrument is required to measure some quantity and an inaccuracy of $\pm 2\%$ is acceptable, then a certain amount of performance degradation can be allowed if its inaccuracy immediately after recalibration is $\pm 1\%$. What is important is that the pattern of performance degradation be quantified, such that the instrument can be recalibrated before its accuracy has reduced to the limit defined by the application.

Susceptibility to the various factors that can cause changes in instrument characteristics varies according to the type of instrument involved. Possession of an in-depth knowledge of the mechanical construction and other features involved in the instrument is necessary in order to be able to quantify the effect of these quantities on the accuracy and other characteristics of an instrument. The type of instrument, its frequency of use, and the prevailing environmental conditions all strongly influence the calibration frequency necessary, and because so many factors are involved, it is difficult or even impossible to determine the required frequency of instrument recalibration from theoretical considerations. Instead, practical experimentation has to be applied to determine the rate of such changes. Once the maximum permissible measurement error has been defined, knowledge of the rate at which the characteristics of an instrument change allows a time interval to be calculated that represents the moment in time when an instrument will have reached the bounds of its acceptable performance level. The instrument must be recalibrated either at this time or earlier. This measurement error level that an instrument reaches just before recalibration is the error bound that must be quoted in the documented specifications for the instrument.

A proper course of action must be defined that describes the procedures to be followed when an instrument is found to be out of calibration, that is, when its output is different to that of the calibration instrument when the same input is applied. The required action depends very much on the nature of the discrepancy and the type of instrument involved. In many cases, deviations in the form of a simple output bias can be corrected by a small adjustment to the instrument (following which the adjustment screws must be sealed to prevent tampering). In other cases, the output scale of the instrument may have to be redrawn or scaling factors altered where the instrument output is part of some automatic control or inspection system. In extreme cases, where the calibration procedure shows signs of instrument damage, it may be necessary to send the instrument for repair or even scrap it.

Whatever system and frequency of calibration are established, it is important to review this from time to time to ensure that the system remains effective and efficient. It may happen that a less expensive (but equally effective) method of calibration becomes available with the passage of time, and such an alternative system must clearly be adopted in the interests of cost efficiency. However, the main item under scrutiny in this review is normally whether the calibration interval is still appropriate. Records of the calibration history of the instrument will be the primary basis on which this review is made. It may happen that an instrument starts to go out of calibration more quickly after a period of time, either because of aging factors within the instrument or because of changes in the operating environment. The conditions or mode of usage of the instrument may also be subject to change. As the environmental and usage conditions of an instrument may change beneficially as well as adversely, there is the possibility that the recommended calibration interval may decrease as well as increase.

## 4.3 Control of Calibration Environment

Any instrument used as a standard in calibration procedures must be kept solely for calibration duties and must never be used for other purposes. Most particularly, it must not be regarded as a spare instrument that can be used for process measurements if the instrument normally used for that purpose breaks down. Proper provision for process instrument failures must be made by keeping a spare set of process instruments. Standard calibration instruments must be totally separate.

To ensure that these conditions are met, the calibration function must be managed and executed in a professional manner. This will normally mean setting aside a particular place within the instrumentation department of a company where all calibration operations take place and where all instruments used for calibration are kept. As far as possible this should take the form of a separate room rather than a sectioned-off area in a room used for other purposes as well. This will enable better environmental control to be applied in the calibration area and will also offer better protection against unauthorized handling or use of calibration instruments. The level of environmental control required during calibration should be considered carefully with due regard to what level of accuracy is required in the calibration procedure, but should not be overspecified, as this will lead to unnecessary expense. Full air conditioning is not normally required for calibration at this level, as it is very expensive, but sensible precautions should be taken to guard the area from extremes of heat or cold; also, good standards of cleanliness should be maintained.

While it is desirable that all calibration functions are performed in this carefully controlled environment, it is not always practical to achieve this. Sometimes, it is not convenient or possible to remove instruments from a process plant, and in these cases, it is standard practice to calibrate them in situ. In these circumstances, appropriate corrections must be made for the deviation in the calibration environmental conditions away from those specified. This practice does not obviate the need to protect calibration instruments and maintain them in constant conditions in a calibration laboratory at all times other than when they are involved in such calibration duties on plant.

As far as management of calibration procedures is concerned, it is important that the performance of all calibration operations is assigned as the clear responsibility of just one person. That person should have total control over the calibration function and be able to limit access to the calibration laboratory to designated, approved personnel only. Only by giving this appointed person total control over the calibration function can the function be expected to operate efficiently and effectively. Lack of such definite management can only lead to unintentional neglect of the calibration system, resulting in the use of equipment in an out-of-date state of calibration and subsequent loss of traceability to reference standards. Professional management is essential so that the customer can be assured that an efficient calibration system is in operation and that the accuracy of measurements is guaranteed.

Calibration procedures that relate in any way to measurements used for quality control functions are controlled by the international standard ISO 9000 (this subsumes the old British quality standard BS 5750). One of the clauses in ISO 9000 requires that all persons using calibration equipment be adequately trained. The manager in charge of the calibration function is clearly responsible for ensuring that this condition is met. Training must be adequate and targeted at the particular needs of the calibration systems involved. People must understand what they need to know and especially why they must have this information. Successful completion of training courses should be marked by the award of qualification certificates. These attest to the proficiency of personnel involved in calibration duties and are a convenient way of demonstrating that the ISO 9000 training requirement has been satisfied.

## 4.4 Calibration Chain and Traceability

The calibration facilities provided within the instrumentation department of a company provide the first link in the calibration chain. Instruments used for calibration at this level are known as *working standards*. As such, working standard instruments are kept by the instrumentation department of a company solely for calibration duties, and for no other purpose, then it can be assumed that they will maintain their accuracy over a reasonable period of time because use-related deterioration in accuracy is largely eliminated. However, over the longer term, the characteristics of even such standard instruments will drift, mainly due to aging effects in components within them. Therefore, over this longer term, a program must be instituted for calibrating working standard instruments at appropriate intervals of time against instruments of yet higher accuracy. The instrument used for calibrating working standard instruments is known as a *secondary reference standard*. This must obviously be a very well-engineered instrument that gives high accuracy and is stabilized against drift in its performance with time. This implies that it will be an expensive instrument to buy. It also requires that the environmental conditions in which it is used be controlled carefully in respect of ambient temperature, humidity, and so on.

When the working standard instrument has been calibrated by an authorized standards laboratory, a calibration certificate will be issued. This will contain at least the following information:

- identification of the equipment calibrated
- calibration results obtained
- measurement uncertainty
- any use limitations on the equipment calibrated
- date of calibration
- authority under which the certificate is issued

The establishment of a company standards laboratory to provide a calibration facility of the required quality is economically viable only in the case of very large companies where large numbers of instruments need to be calibrated across several factories. In the case of small to medium size companies, the cost of buying and maintaining such equipment is not justified. Instead, they would normally use the calibration service provided by various companies that specialize in offering a standards laboratory. What these specialist calibration companies do effectively is to share out the high cost of providing this highly accurate but infrequently used calibration service over a large number of companies. Such standards laboratories are closely monitored by national standards organizations.

In the United States, the appropriate national standards organization for validating standards laboratories is the National Bureau of Standards, whereas in the United Kingdom it is the National Physical Laboratory. An international standard now exists (ISO/IEC 17025, 2005), which sets down criteria that must be satisfied in order for a standards laboratory to be validated. These criteria cover the management requirements necessary to ensure proper operation and effectiveness of a quality management system within the calibration or testing laboratory and also some technical requirements that relate to the competence of staff, specification, and maintenance of calibration/test equipment and practical calibration procedures used.

National standards organizations usually monitor both instrument calibration and mechanical testing laboratories. Although each different country has its own structure for the maintenance of standards, each of these different frameworks tends to be equivalent in its effect in ensuring that the requirements of ISO/IEC 17025 are met. This provides confidence that the goods and services that cross national boundaries from one country to another have been measured by properly calibrated instruments.

The national standards organizations lay down strict conditions that a standards laboratory has to meet before it is approved. These conditions control laboratory management, environment, equipment, and documentation. The person appointed as head of the laboratory must be suitably qualified, and independence of operation of the laboratory must be guaranteed. The management structure must be such that any pressure to rush or skip calibration procedures for production reasons can be resisted. As far as the laboratory environment is concerned, proper temperature and humidity control must be provided, and high standards of cleanliness and housekeeping must be maintained. All equipment used for calibration purposes must be maintained to reference standards and supported by calibration certificates that establish this traceability. Finally, full documentation must be maintained. This should describe all calibration procedures, maintain an index system for recalibration of equipment, and include a full inventory of apparatus and traceability schedules. Having met these conditions, a standards laboratory becomes an accredited laboratory for providing calibration services and issuing calibration certificates. This accreditation is reviewed at approximately 12 monthly intervals to ensure that the laboratory is continuing to satisfy the conditions for approval laid down.

*Primary reference standards*, as listed in Table 1.1, describe the highest level of accuracy achievable in the measurement of any particular physical quantity. All items of equipment used in standards laboratories as secondary reference standards have to be calibrated themselves against primary reference standards at appropriate intervals of time. This procedure is acknowledged by the issue of a calibration certificate in the standard way. National standards organizations maintain suitable facilities for this calibration. In the United States, this is the National Bureau of Standards, and in the United Kingdom it is the National Physical Laboratory. Similar national standards organizations exist in many other countries. In certain cases, such primary reference standards can be located outside national standards organizations. For instance, the primary reference standard for dimension measurement is defined by the wavelength of the orange-red line of krypton light, and it can therefore be realized in any laboratory equipped with an interferometer. In certain cases (e.g., the measurement of viscosity), such primary reference standards are not available and reference standards for calibration are achieved by collaboration between several national standards organizations who perform measurements on identical samples under controlled conditions [ISO 5725 (1994) and ISO 5725-2/Cor1 (2002)].

What has emerged from the foregoing discussion is that calibration has a chain-like structure in which every instrument in the chain is calibrated against a more accurate instrument immediately above it in the chain, as shown in Figure 4.1. All of the elements in the calibration chain must be known so that the calibration of process instruments at the bottom of the chain is traceable to the fundamental measurement standards. This knowledge of the full chain of instruments involved in the calibration procedure is known as *traceability* and is specified as a mandatory requirement in satisfying the ISO 9000 standard. Documentation must exist that shows that process instruments are calibrated by standard instruments linked by a chain of increasing accuracy back to national reference standards. There must be clear evidence to show that there is no break in this chain.

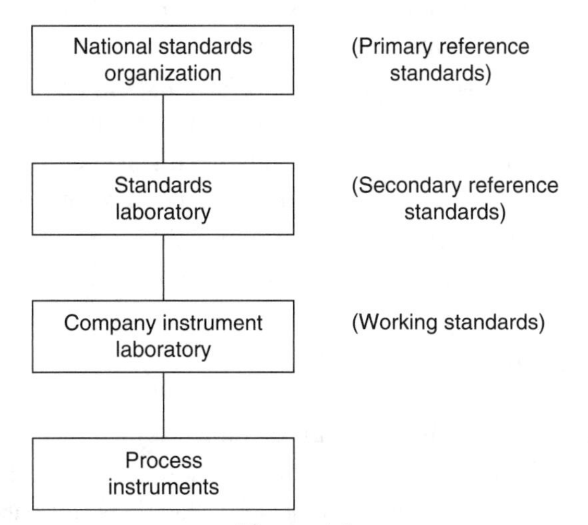


**Figure 4.1**
Instrument calibration chain.

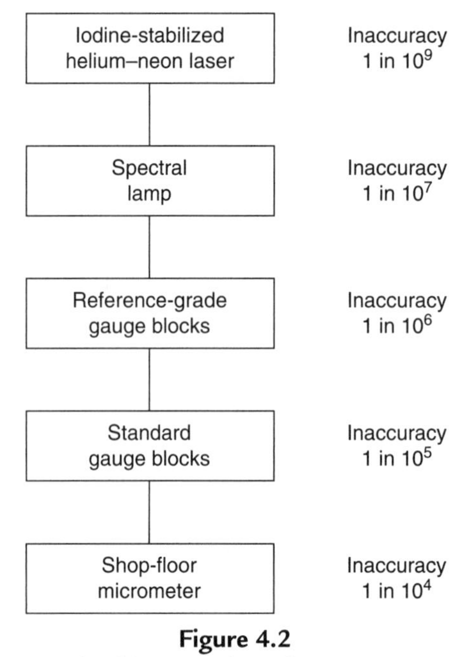


**Figure 4.2**
Typical calibration chain for micrometers.

To illustrate a typical calibration chain, consider the calibration of micrometers (Figure 4.2). A typical shop floor micrometer has an uncertainty (inaccuracy) of less than 1 in $10^4$. These would normally be calibrated in the instrumentation department or standards laboratory of a company against laboratory standard gauge blocks with a typical uncertainty of less than 1 in $10^5$. A specialist calibration service company would provide facilities for calibrating these laboratory standard gauge blocks against reference-grade gauge blocks with a typical uncertainty of less than 1 in $10^6$. More accurate calibration equipment still is provided by national standards organizations. The National Bureau of Standards and National Physical Laboratory maintain two sets of standards for this type of calibration, a working standard and a primary standard. Spectral lamps are used to provide a working reference standard with an uncertainty of less than 1 in $10^7$. The primary standard is provided by an iodine-stabilized helium–neon laser that has a specified uncertainty of less than 1 in $10^9$. All of the links in this calibration chain must be shown in any documentation that describes the use of micrometers in making quality-related measurements.

## 4.5 Calibration Records

An essential element in the maintenance of measurement systems and the operation of calibration procedures is the provision of full documentation. This must give a full description of the measurement requirements throughout the workplace, instruments used, and calibration system and procedures operated. Individual calibration records for each instrument must be included within this. This documentation is a necessary part of the quality manual, although it may exist physically as a separate volume if this is more convenient. An overriding constraint on the style in which the documentation is presented is that it should be simple and easy to read. This is often facilitated greatly by a copious use of appendices.

The starting point in the documentation must be a statement of what measurement limits have been defined for each measurement system documented. Such limits are established by balancing the costs of improved accuracy against customer requirements, and also with regard to what overall quality level has been specified in the quality manual. The technical procedures required for this, which involve assessing the type and magnitude of relevant measurement errors, are described in Chapter 3. It is customary to express the final measurement limit calculated as $\pm 2$ standard deviations, that is, within 95% confidence limits (for an explanation of these terms, see Chapter 3).

Instruments specified for each measurement situation must be listed next. This list must be accompanied by full instructions about the proper use of the instruments concerned. These instructions will include details about any environmental control or other special precautions that must be taken to ensure that the instruments provide measurements of sufficient accuracy to meet the measurement limits defined. The proper training courses appropriate to plant personnel who will use the instruments must be specified.

Having disposed of the question about what instruments are used, documentation must go on to cover the subject of calibration. Full calibration is not applied to every measuring instrument used in a workplace because ISO 9000 acknowledges that formal calibration procedures are not necessary for some equipment where it is uneconomic or technically unnecessary because the accuracy of the measurement involved has an insignificant effect on the overall quality target for a product. However, any equipment excluded from calibration procedures in this manner must be specified as such in the documentation. Identification of equipment that is in this category is a matter of informed judgment.

For instruments that are the subject of formal calibration, documentation must specify what standard instruments are to be used for the purpose and define a formal procedure of calibration. This procedure must include instructions for the storage and handling of standard calibration instruments and specify the required environmental conditions under which calibration is to be performed. Where a calibration procedure for a particular instrument uses published standard practices, it is sufficient to include reference to that standard procedure in the documentation rather than to reproduce the whole procedure. Whatever calibration system is established, a formal review procedure must be defined in the documentation that ensures its continued effectiveness at regular intervals. The results of each review must also be documented in a formal way.

A standard format for the recording of calibration results should be defined in the documentation. A separate record must be kept for every instrument present in the workplace, irrespective of whether the instrument is normally in use or is just kept as a spare. A form similar to that shown in Figure 4.3 should be used that includes details of the instrument's description, required calibration frequency, date of each calibration, and calibration results on each occasion. Where appropriate, documentation must also define the manner in which calibration results are to be recorded on the instruments themselves.

Documentation must specify procedures that are to be followed if an instrument is found to be outside the calibration limits. This may involve adjustment, redrawing its scale, or withdrawing an instrument, depending on the nature of the discrepancy and the type of instrument involved. Instruments withdrawn will either be repaired or be scrapped. In the case of withdrawn instruments, a formal procedure for marking them as such must be defined to prevent them being put back into use accidentally.

Two other items must also be covered by the calibration document. The traceability of the calibration system back to national reference standards must be defined and supported by calibration certificates (see Section 4.3). Training procedures must also be documented,

<table>
<tr><td colspan="2">Type of instrument:</td><td>Company serial number:</td></tr>
<tr><td colspan="2">Manufacturer's part number:</td><td>Manufacturer's serial number:</td></tr>
<tr><td colspan="2">Measurement limit:</td><td>Date introduced:</td></tr>
<tr><td colspan="3">Location:</td></tr>
<tr><td colspan="3">Instructions for use:</td></tr>
<tr><td colspan="2">Calibration frequency:</td><td>Signature of person responsible for calibration:</td></tr>
<tr><td colspan="3">CALIBRATION RECORD</td></tr>
<tr><td>Calibration date</td><td>Calibration results</td><td>Calibrated by</td></tr>
<tr><td></td><td></td><td></td></tr>
</table>

**Figure 4.3**
Typical format for instrument record sheets.

specifying the particular training courses to be attended by various personnel and what, if any, refresher courses are required.

All aspects of these documented calibration procedures will be given consideration as part of the periodic audit of the quality control system that calibration procedures are instigated to support. While the basic responsibility for choosing a suitable interval between calibration checks rests with the engineers responsible for the instruments concerned, the quality system auditor will need to see the results of tests that show that the calibration interval has been chosen correctly and that instruments are not going outside allowable measurement uncertainty limits between calibrations. Particularly important in such audits will be the existence of procedures instigated in response to instruments found to be out of calibration. Evidence that such procedures are effective in avoiding degradation in the quality assurance function will also be required.

## 4.6 Summary

Proper instrument calibration is an essential component in good measurement practice, and this chapter has been dedicated to explaining the various procedures that must be followed in order to perform calibration tasks efficiently and effectively. We have learned how working instruments are calibrated against a more accurate "reference" instrument that is maintained carefully and kept just for performing calibration tasks. We considered the importance of carefully designing and controlling the calibration environment in which calibration tasks are performed and observed that proper training of all personnel involved in carrying out calibration tasks had similar importance. We also learned that "first stage" calibration of a working instrument against a reference standard is part of a chain of calibrations that provides traceability of the working instrument calibration to national and international reference standards for the quantity being measured, with the latter representing the most accurate standards of measurement accuracy achievable. Finally, we looked at the importance of maintaining calibration records and suggested appropriate formats for these.

## 4.7 Problems

4.1. Explain the meaning of instrument calibration.
4.2. Explain why calibration is necessary.
4.3. Explain how the necessary calibration frequency is determined for a measuring instrument.
4.4. Explain the following terms:
(a) calibration chain
(b) traceability
(c) standards laboratory

4.5. Explain how the calibration procedure should be managed, particularly with regard to control of the calibration environment and choice of reference instruments.
4.6. Will a calibrated measuring instrument always be accurate? If not, explain why not and explain what procedures can be followed to ensure that accurate measurements are obtained when using calibrated instruments.
4.7. Why is there no fundamental reference standard for temperature calibration? How is this difficulty overcome when temperature sensors are calibrated?
4.8. Discuss the necessary procedures in calibrating temperature sensors.
4.9. Explain the construction and working characteristics of the following three kinds of instruments used as a reference standard in pressure sensor calibration: dead-weight gauge, U-tube manometer, and barometer.
4.10. Discuss the main procedures involved in calibrating pressure sensors.
4.11. Discuss the special equipment needed and procedures involved in calibrating instruments that measure the volume flow rate of liquids.
4.12. What kind of equipment is needed for calibrating instruments that measure the volume flow rate of gases? How is this equipment used?
4.13. Discuss the general procedures involved in calibrating level sensors.
4.14. What is the main item of equipment used in calibrating mass-measuring instruments? Sketch the following instruments and discuss briefly their mode of operation: beam balance, weigh beam, and pendulum scale.
4.15. Discuss how the following are calibrated: translational displacement transducers and linear-motion accelerometers.
4.16. Explain the general procedures involved in calibarting (a) vibration sensors and (b) shock sensors.
4.17. Discuss briefly the procedures involved in the following: rotational displacement sensors, rotational velocity sensors, and rotational acceleration sensors.
4.18. How are dimension-measuring instruments calibrated normally?
4.19. Discuss briefly the two main ways of calibrating angle-measuring instruments.
4.20. Discuss the equipment used and procedures involved in calibrating viscosity-measuring instruments.
4.21. Discuss briefly the calibration of moisture and humidity measurements.

## References

ISO/IEC 17025, (2005). *General requirements for the competence of testing and calibration laboratories.* Geneva: International Organisation for Standards, Geneva/International Electrotechnical Commission.

ISO 5725, (1994). *Precision of test methods: Determination of repeatability and reproducibility by inter-laboratory tests.* Geneva: International Organisation for Standards.

ISO 5725–2/Cor1, (2002). *Corrigdendum to part 2 of ISO 5725.* Geneva: International Organisation for Standards.

ISO 9000, (2008). *Quality management and quality assurance standards.* Geneva: International Organisation for Standards (individual parts published as ISO 9001, ISO 9002, ISO 9003, and ISO 9004).

CHAPTER 5

# Data Acquisition with LabVIEW



## 5.1 Introduction

This chapter is designed to introduce the reader to the concept of computer-based data acquisition and to LabVIEW, a software package developed by National Instruments. The main reason for focusing on LabVIEW is its prevalence in laboratory setting. To be sure there are other software tools that support laboratory data acquisition made by a range of vendors. These are reviewed briefly in the appendix due to their limited presence in the educational setting. We should also point out that Matlab and other software tools used to model and simulate dynamic systems are at times used in laboratory setting although their use is often limited to specialized applications such as real-time control. For this reason, these tools are not discussed in this chapter.

LabVIEW itself is as an extensive programming platform. It includes a multitude of functionalities ranging from basic algebraic operators to advanced signal processing components that can be integrated into rather sophisticated and complex programs. For pedagogical reasons we only introduce the main ideas from LabVIEW that are necessary for functioning in a typical undergraduate engineering laboratory environment. Advanced programming skills can be developed over time as the reader gains comfort with the basic functioning of LabVIEW and its external interfaces.

Specific topics discussed in this chapter and the associated learning objectives are as follows.

- Structure of personal computer (PC)-based data acquisition (DAQ) systems, the purpose of DAQ cards, and the role of LabVIEW in this context
- Development of simple virtual instruments (VIs) using basic functionalities of LabVIEW, namely arithmetic and logic operations
- Construction of functionally enhanced VIs using LabVIEW program flow control operations, such as the while loop and the case structure
- Development of VIs that allow for interaction with external hardware as, for instance, acquisition of external signals via DAQ card input channels and generation of functions using DAQ card output channels

These functionalities are essential to using LabVIEW in laboratory setting. Additional capabilities of LabVIEW are explored in the subsequent chapter on signal processing in LabVIEW.

## 5.2 Computer-Based Data Acquisition

In studying mechanical systems, it is often necessary to use electronic sensors to measure certain variables, such as temperature (using thermocouples or RTDs), pressure (using piezoelectric transducers), strain (using strain gauges), and so forth. Although it is possible to use oscilloscopes or multimeters to monitor these variables, it is often preferable to use a PC to view and record the data through the use of a DAQ card. One particular advantage of using computers in this respect is that data can be stored and converted to a format that can be used by spreadsheets (such as Microsoft Excel) or other software packages such as Matlab for more extensive analysis. Another advantage is that significant digital processing of data can be performed in real time via the same platform used to acquire the data. This can significantly improve the process of performing an experiment by making real-time data more useful for further processing.

### *5.2.1 Acquisition of Data*

One important step in the data acquisition process is the conversion of analogue signals received from sensing instruments to digital representations that can be processed by the computer. Because data must be stored in the computer's memory in the form of individual data points represented by binary numbers, incoming analogue data must be *sampled* at discrete time intervals and *quantized* to one of a set of predefined values. In most cases, this is accomplished using a digital-to-analogue (D/A) conversion component on the DAQ card inside the PC or interconnected to it via a Universal Serial Bus (USB) port. Note that both options are used commonly. However, laptop computers and/or low-profile PCs generally require the use of USB-based DAQ devices.

## 5.3 National Instruments LabVIEW

LabVIEW is a software package that provides the functional tools and a user interface for data acquisition. Figure 5.1 depicts a schematic of data flow in the data acquisition process. Note that the physical system may be a mechanical system, such as a beam subjected to stress, a chemical process such as a distillation column, a DC motor with both mechanical and electrical components, and so forth. The key issue here is that certain measurements are taken from the given physical system and are acquired and processed by the PC-based data acquisition system.

LabVIEW plays a pivotal role in the data acquisition process. Through the use of VIs, LabVIEW directs the real-time sampling of sensor data through the DAQ card (also known as the I/O card) and is capable of storing, processing, and displaying the collected data. In most cases, one or more sensors transmit analogue readings to the DAQ card in the computer. These analogue data are then converted to individual digital values by the DAQ card and are made available to LabVIEW, at which point they can be displayed to the user. Although LabVIEW is capable of some data analysis functions, it is often preferable to export the data to a spreadsheet for detailed analysis and graphical presentation.

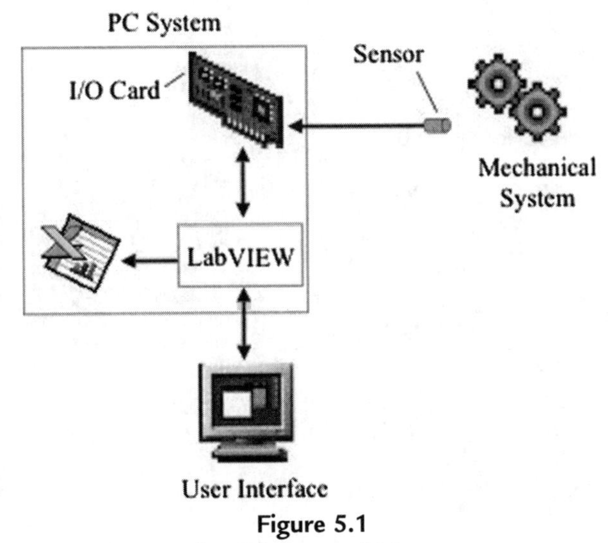


**Figure 5.1**
Schematic of data acquisition process.

### 5.3.1 Virtual Instruments

A VI is a program, created in the LabVIEW programming environment, that simulates physical or hard instruments such as oscilloscopes or function generators. A simple VI used to produce a waveform is depicted in Figure 5.2. The front panel (shown in Figure 5.2) acts as the *user interface*, while data acquisition (in this case the generation process) is performed by a combination of the PC and the DAQ card. Much like the front panel of a real instrument, the front panel window contains *controls* (i.e., knobs and switches) that allow the user to modify certain parameters during the experiment. These include a selector to choose the type of waveform and numerical controls to choose the frequency and amplitude of the generated waveform, as well as its phase, amplitude and offset.

The front panel of a VI typically also contains *indicators* that display data or other important information related to the experiment. In this case, a *graph* is used to depict the waveform. The block diagram (not shown but discussed later) is analogous to the wiring and internal components of a real instrument. The configuration of the VI's block diagram determines how front panel controls and indicators are related. It also incorporates functions such as communicating with the DAQ card and exporting data to disk files in spreadsheet format.

## 5.4 Introduction to Graphical Programming in LabVIEW

LabVIEW makes use of a graphical programming language that determines how each VI will work. This section discusses the inner workings of a simple LabVIEW VI used to add and subtract two numbers. While this VI is not particularly useful in a laboratory setting, it

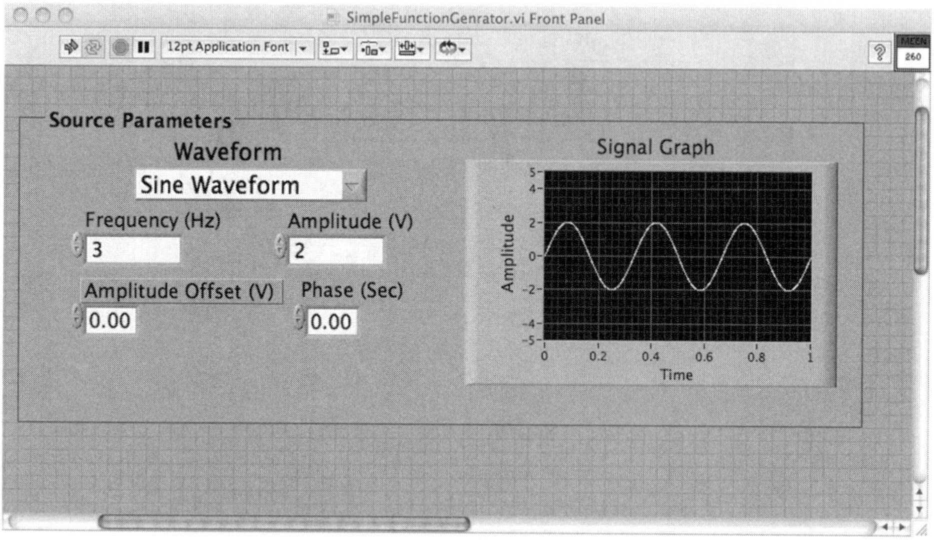

**Figure 5.2**
A simple function generator virtual instrument.

illustrates how basic LabVIEW components can be used to construct a VI and hence helps the reader move towards developing more sophisticated VIs. Figure 5.3 shows the front panel and block diagram of the VI, which accepts two numbers from the user (X and Y) via two simple *numeric controls* and produces the sum and difference (X + Y and X − Y) of the numbers displaced on the front panel via two simple *numeric indicators.*

The block diagram of the VI is a graphical, or more accurately a *data flow,* program that defines how the controls and indicators on the front panel are interrelated. Controls on the front panel of the VI show values that can be modified by the user while the VI is operating. Indicators display values that are output by the VI. Each control and indicator in the front panel window is associated with a *terminal* in the block diagram window. Wires in the block diagram window represent the flow of data within the VI. *Nodes* are additional programming elements that can perform operations on variables, perform input or output functions through the DAQ card, and serve a variety of other functions as well.

The two nodes in the VI shown in Figure 5.3 (add and subtract) have two inputs and one output, as for instance is depicted in Figure 5.4. Data can be passed to a node through its input terminals (usually on the left), and the results can be accessed through the node's output terminals, usually on the right.

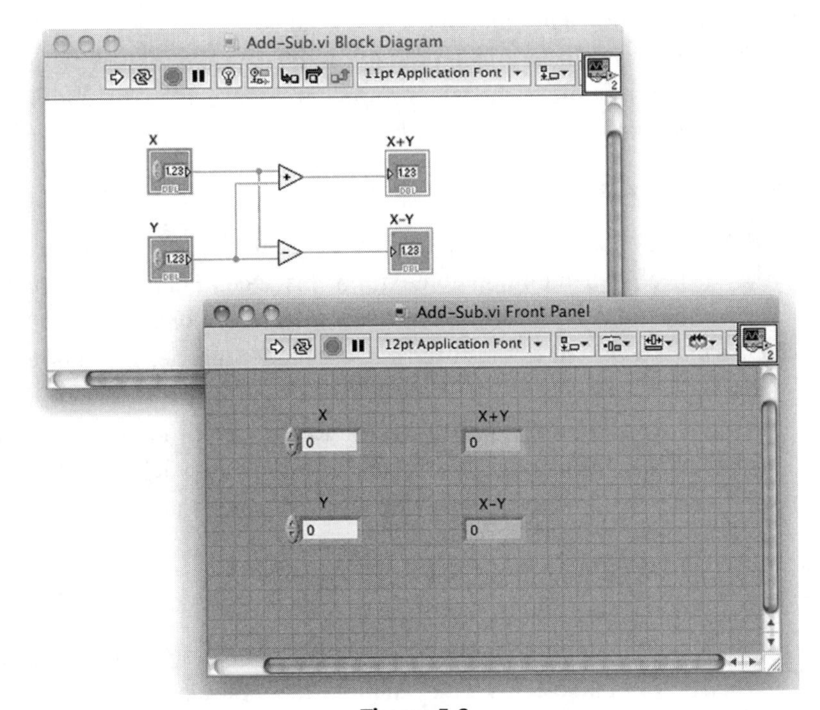


**Figure 5.3**
Addition and subtraction VI.

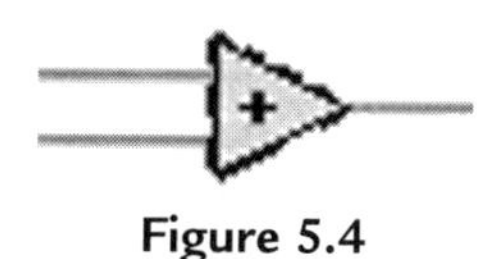

**Figure 5.4**
The add node.

Because LabVIEW diagrams are data flow driven, the sequence in which the various operations in the VI are executed is *not* determined by the *order* of a set of commands. Rather, a block diagram node executes when data are present at all of its input terminals. As a result, in the case of the block diagram in Figure 5.3, one does not know whether the add node or the subtract node will execute first. This issue has implications in more complex applications but is not particularly important in the present context. However, one cannot assume an order of execution merely on the basis of the position of the computational nodes (top to bottom or left to right). If a certain execution order is required or desired, one must explicitly build program flow control mechanisms into the VI, which in practice is not always possible nor is it in the spirit in which LabVIEW was originally designed.

### 5.4.1 Elements of the Tools Palette

The mouse pointer can perform a number of different functions in the LabVIEW environment depending on which pointer tool is selected. One can change the mouse pointer tool by selecting the desired tool from the tools palette shown in Figure 5.5. (If the tools palette does not appear on the screen, it can be displayed by selecting Tools on the View menu.)

Choices available in the tools palette are as follows:

- *Automatic tool selection.* Automatically selects the tool it assumes you need depending on context. For instance, it can be the *positioning tool*, the *wiring tool*, or the *text tool* as further noted later.
- *Operating tool.* This tool is used operate the front panel controls before or while the VI is running.
- *Positioning tool.* This tool is used to select, move, or resize objects in either the front panel or the diagram windows. For example, to delete a node in a diagram, one would first select the node with the positioning tool and then press the delete key.
- *Text tool.* This tool is used to add or change a label. The enter key is used to finalize the task.
- *Wiring tool.* This tool is used to wire objects together in the diagram window. When this tool is selected, you can move the mouse pointer over one of a node's input or output terminals to see the description of that terminal.

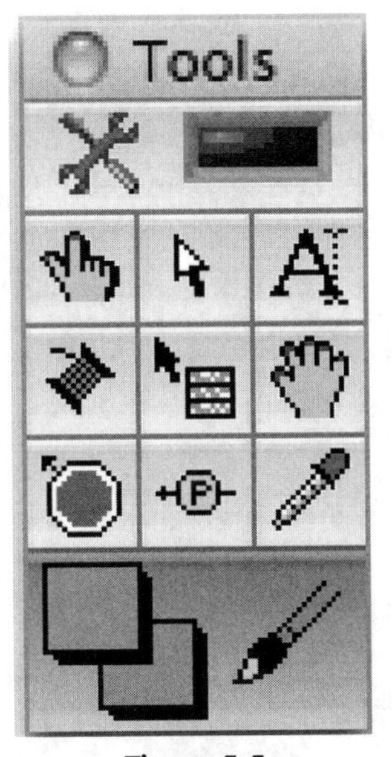


**Figure 5.5**
The tools palette.

One can add controls and indicators to the front panel of a VI by dragging and dropping them from the controls palette depicted in Figure 5.6 and which is visible *only* when the front panel window is *active.*

If, for some reason, the *controls palette* is not visible, one can access it by right clicking anywhere in the front panel window. Once a control or indicator is added to the front panel of a VI, the corresponding terminal is created automatically in the block diagram window. Adding additional nodes to the block diagram requires use of the *functions palette,* which is accessible once the block diagram window is visible. One can add arithmetic and logic elements to the block diagram window by dragging and dropping these elements from the functions palette. The functions to add, subtract, and so on can be found in the numeric subpalette of the programming section of the functions palette (Figure 5.7, the fourth icon from the left). One can also use constants from the numeric subpalette in a block diagram and wire those constants as inputs to various nodes.

## 5.5 Logic Operations in LabVIEW

In more complex VIs, one may encounter situations where the VI must react differently depending on the condition at hand. For example, in a data acquisition process, the VI may need to turn on a warning light when an input voltage exceeds a certain threshold and therefore it may

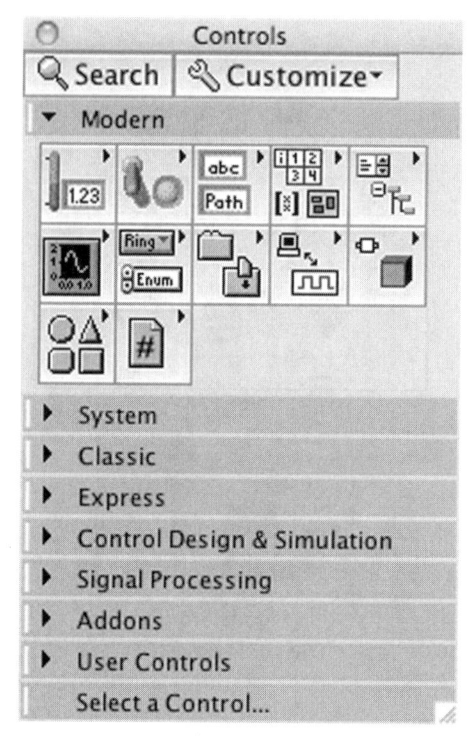


**Figure 5.6**
The controls palette.

be necessary to compare the input voltage with the threshold value. In addition, the VI needs to make a decision based on the result of the comparison (turn on a light on the screen or external to the DAQ system, produce an alarm sound, etc.). To allow for these types of applications, many logic operators are available in LabVIEW. These can be found in the comparison subpalette of the programming section of the functions palette. Note that as stated earlier, the functions palette is available when the diagram window is the top (focus) window on the screen. If one needs to identify the comparison subpalette, one can move the mouse pointer over the programming section of the palette to call out the different subpalettes.

Comparison nodes, as, for instance, depicted in Figure 5.8, are used to compare two numbers. Their output is either *true* or *false*. This value can be used to make decisions between two numbers as shown in Figure 5.8. Another important node in this respect is the *select* node, which is also available in the same subpalette. This node makes a decision based on the outcome of a *previous* comparison, such as depicted in Figure 5.8. If the select node's middle input is true, it selects its top input as its output. If its middle input is false, it selects its bottom input as its output. In this way, one can pair the comparison nodes with a select node to produce an appropriate action for subsequent processing.

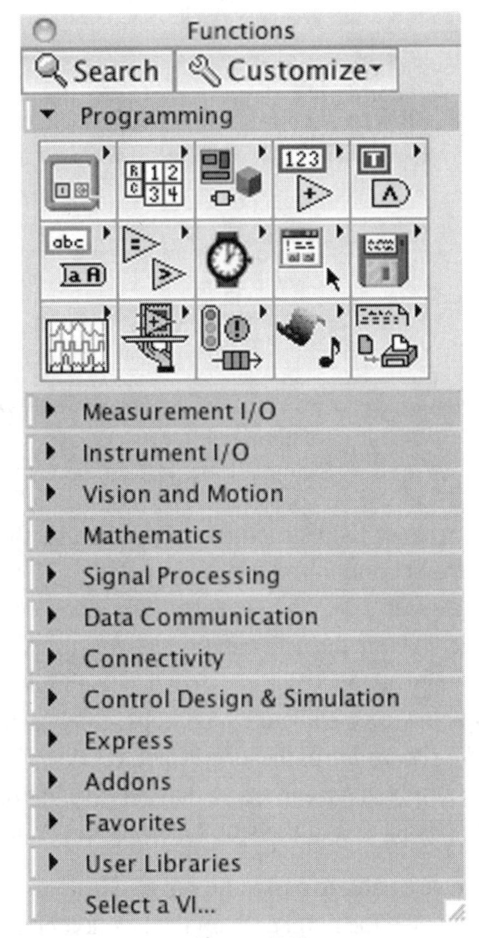


**Figure 5.7**
The functions palette.

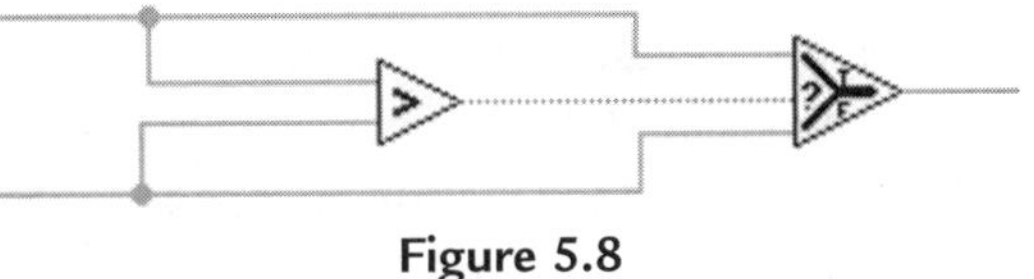

**Figure 5.8**
Logic example.

## 5.6 Loops in LabVIEW

In building more sophisticated VIs it will not be sufficient to simply perform an operation once. For example, if LabVIEW is being used to compute the factorial of an integer, $n$, the program will need to continue to multiply $n$ by $n - 1$, $n - 2$, and so forth. More pertinently, in data acquisition processes, one needs to acquire multiple samples of data and process the data. For this reason,

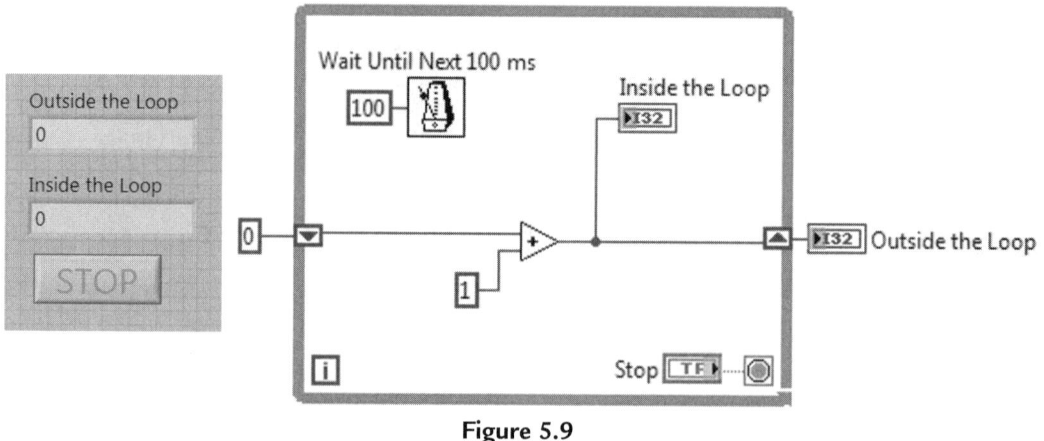


**Figure 5.9**

Example loop VI.

LabVIEW includes several types of loop structures. These can be found in the structures subpalette of the programming section in the functions palette. (Note that as with every one of LabVIEW's tools, one can use LabVIEW's help feature or its quick help feature to get more information on these constructs.) Here the user can find a *while* loop, a *for* loop, a *case statement*, and several other important loop structures. Figure 5.9 depicts a simple program that utilizes a loop. There are several important items to note about using loops. Everything contained inside the loop will be repeated until the *ending condition* of the loop is met. For a *while* loop, this will be some type of logical condition. In our example given earlier, we have used a button to *stop* the loop. The loop will stop when a value of *true* is passed to the stop button at the completion of the loop.

In addition, it is often important to be able to pass values from one iteration to the next. In the example given previously, we needed to take the number from the last iteration and add one to it. This is accomplished using a *shift register*. To add a shift register, one has to right click (or option click on a Mac) on the right or left side of the loop and use the menu that opens up to add the shift register. To use the shift register, one must wire the value to be passed to the next iteration to the right side as shown in Figure 5.9. To use the value from the previous iteration, one draws wire from the left side of the loop box to the terminal of one's choosing. In addition, elements initially wired together can be included in a loop simply by creating one around these elements. The wiring initially in place will be preserved. Finally, note that the metronome in the loop times the loop so that it runs every 100 ms. This element is available from the timing subpalette in the programming section of the functions patlette.

## 5.7 Case Structure in LabVIEW

The *case structure* is a programming construct that is useful in emulating a switch, similar to what appears on the front panel of a hard instrument to allow the user to select one of multiple available tasks. In terms of its appearance in LabVIEW, a case structure works much like a while loop as evident in Figure 5.10. However, significant differences exist between a case

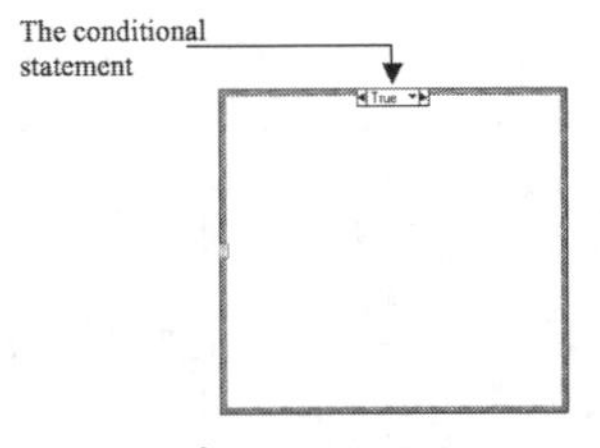


**Figure 5.10**
A case structure in LabVIEW.

structure and a while loop in that a case structure performs a *separate* operation for each case of the *conditional statement* that drives this structure. The conditional statement takes a value chosen by the user at runtime from among the set of values for which the given case structure is programmed to execute. This set can be {0,1}, as is the case initally when a case structure is added to a VI, and can be expanded during the programming stage by right clicking on the conditional statement and choosing "Add case" as necessary. For each case, the VI must include an appropriate execution plan that is placed within the bounding box of the case structure. The conditional statement associated with a case structure is typically driven by a *ring*, which is placed on the front panel of the given VI and appears outside the bounding box of the case structure in the block diagram panel of the VI.

This ring is connected to the question mark box on the right side of the case structure in the block diagram. This is illustrated in Figure 5.11 in conjunction with a four-function calculator implemented in LabVIEW. It is evident in Figure 5.11 that the *ring*, acting as an operation selector, drives the condition statement, whereas variables X and Y, implemented via numerical controls on the front panel, pass their value to the case structure, which embeds the actual mathematical operation for each of the four functions (add, subtract, multiply, and divide) in a dedicated panel. Figure 5.11 depicts the panel associated with the divide operation. Arrows in the condition statement of the case structure can be used during the programming stage to open each of the cases that the case structure is intended to implement. The ring outside the case structure must have as many elements as there are cases in the respective case structure. Here, it is important to make sure that the order of the selections in the ring and the order of the functions in the case structure correspond to each other. If the first option in the ring is "add," the first function of the case structure needs to implement the addition function. Exercise 5.6 deals with this issue at more length. Note that as is the case with other LabVIEW elements, right clicking on a given element allows the user to view a detailed description and examples associated with that element.

## 5.8 Data Acquisition Using LabVIEW

LabVIEW is primarily a data acquisition tool. The typical setup in the laboratory is shown in Figure 5.12. Aside from the hard instruments shown in Figure 5.12, which are also used to perform data acquisition and monitoring tasks, a terminal box is needed to obtain the input from sensors and to allow a way to produce output voltages from the data acquisition card.

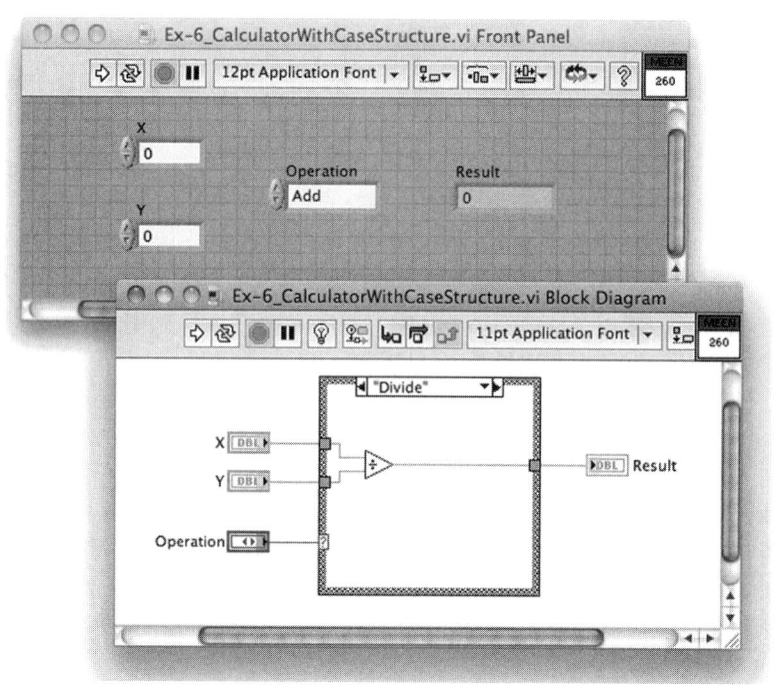


**Figure 5.11**
A menu ring used in conjunction with a case structure.

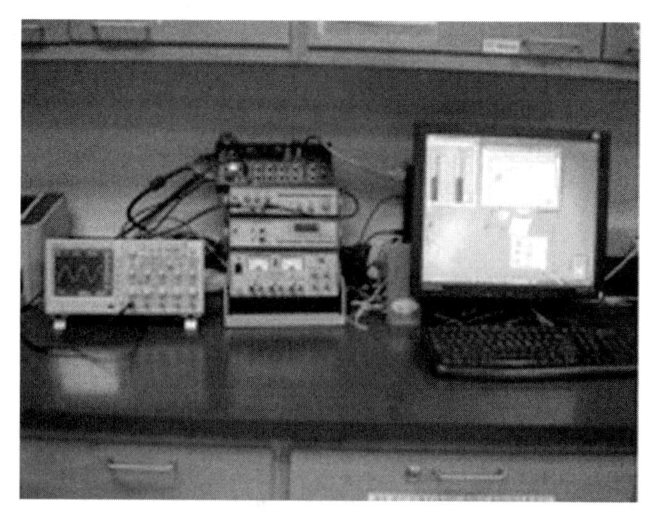

**Figure 5.12**
A typical setup for LabVIEW-based data acquisition.

**Figure 5.13**
A NI BNC-2120 connector block.

A typical connector block is shown in Figure 5.13. This is National Instrument BNC-2120. Pin numbers and the style of the terminal box will vary depending on the model being used. The experiments described later in this chapter use BNC-2120, but several will use different terminals. For the exercise discussed below the data acquisition card's differential mode is used, meaning that it reports the voltage difference across a channel (channel 0 in our case). The card can also operate in an absolute mode where it references the reading to the ground level. Finally, there is one last piece of equipment, which will be needed in the subsequent discussion: the function generator, which can be seen in Figure 5.13.

As the name suggests, this piece of equipment is used to generate a sine wave, a triangle wave, or a square wave. The amplitude of the wave can be adjusted by using the amplitude knob. The frequency of the wave can also be adjusted using the relevant buttons. The main use of this device in the present context is to produce an external input to a custom VI that is shown in Figure 5.14, which converts the voltage produced by the function generator into a temperature reading in Celsius and Fahrenheit units.

Functions related to communication with the DAQ card are handled by the "DAQ Assistant" function. The board ID and channel settings specify information about the DAQ card that tells the VI where to look for data. The algebraic operations that implement the conversion of data are more or less straightforward as depicted in Figure 5.14.

## 5.9 LabVIEW Function Generation

In this section we will create a VI that emulates a function generator, that is, the VI should produce a periodic signal with a user-selected shape, frequency, and amplitude to an output channel on the National Instruments DAQ card. In addition, the user should be able to select

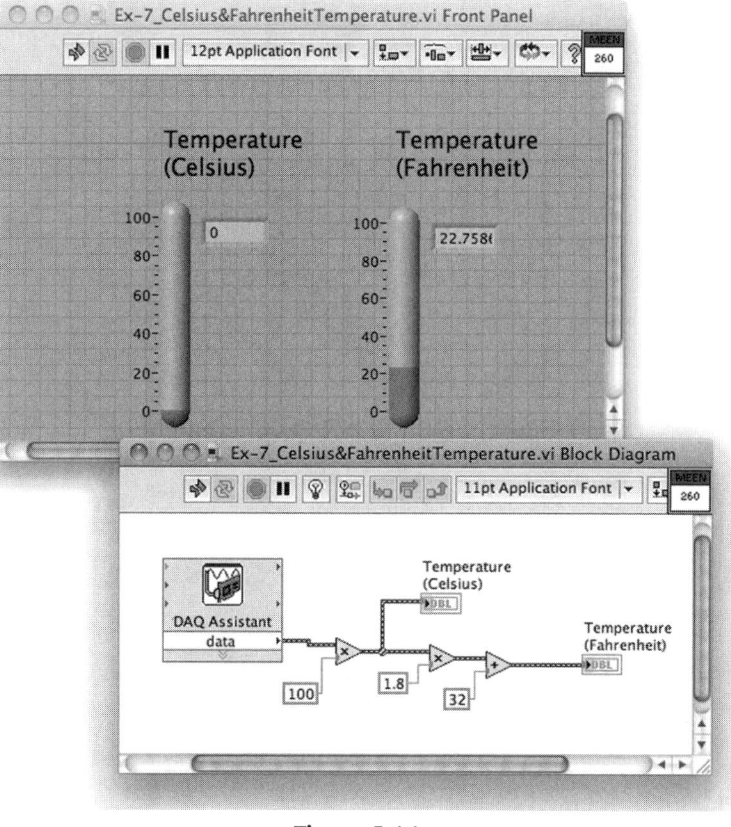


**Figure 5.14**
Thermometer VI.

the resolution of the waveform (i.e., the number of data points per period). The VI's front panel should also include a waveform graph that displays at least one period of the waveform selected as it appears in Figure 5.15.

The VI produces a sine wave, triangle wave, or square wave as selected by the user. This requires that a case structure be wrapped around the signal generation block and a ring block to communicate between the front panel and the case structure. This is discussed further in the LabVIEW exercises at the end of the chapter.

## 5.10 Summary

This chapter was meant to introduce the reader to the usage of LabVIEW in computer-based data acquisition. Basic LabVIEW programming concepts, namely controls, indicators, and simple nodes used to implement algebraic operations, as well as program flow control

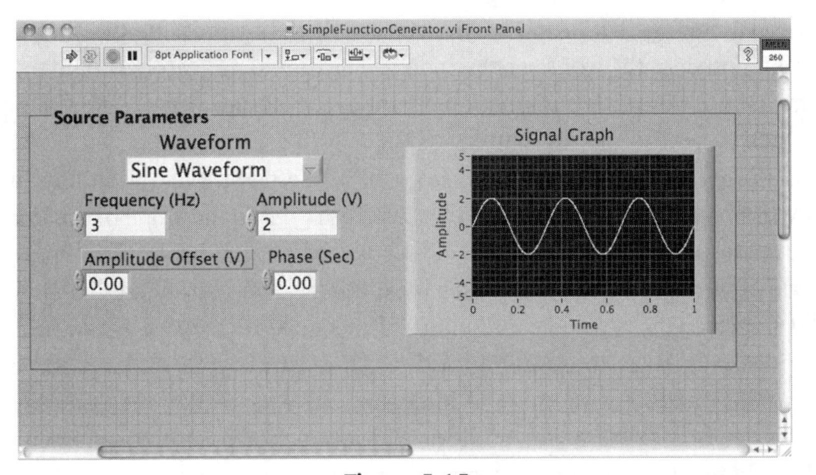


**Figure 5.15**
Function generator front panel.

constructs, that is, while loop and case structure, were introduced. These building blocks were used to implement increasingly more complex LabVIEW VIs. These VIs can be used as stand-alone LabVIEW programs or expanded to form sophisticated VIs in conjunction with additional LabVIEW elements. These include signal processing tools, which are discussed in the subsequent chapter. The exercises that follow enable the reader to practice constructing a number of VIs that are of value in a laboratory setting.

## 5.11 Problems

5.1. In order to demonstrate your understanding, build the VI found in Figure 5.3. To get started, click on the LabVIEW icon found on the desktop of your computer (assuming it is previously installed). This should pop up a window with several choices. Select "Blank VI" and click "OK." From here you should see the front panel (it is gray). To open the block diagram window go to the window and then the show block diagram option (Ctrl-E or Cmd-E also do the same). On the front panel, you will need to add the required numeric controls for X and Y and numeric indicators for X + Y and X − Y. These are available in the modern subpalette of the controls palette, which is available by right clicking in the front panel. You will then use the block diagram window and implement the *add* and *subtract* nodes (available in the numeric section of the programming subpalette of the functions palette, which itself can be viewed by right clicking in the block diagram window) and connect them to the appropriate controls and indicators. Note that you can change the labels of each entity by using the text tool from the tools palette. Changing the label of an indicator or control in the block diagram window changes its label in the front panel as well. Once you have inserted the necessary nodes and wired them together properly, you can run the VI by pressing the run button at the upper left corner of the front panel window. The digital indicators should then display the results of the addition and subtraction operations. You can also use continuous run button , which is next to the run button.

5.2. Create a program that will take the slope of a line passing between any two points. In other words, given two points, $(X_1,Y_1)$ and $(X_2,Y_2)$, find the slope of the line, $Y = mX + b$, that passes through these points. This program should have four digital inputs (as numeric controls) that allow the user to select the two coordinates. It should have one digital indicator that provides the value of the slope of the line between them.

5.3. Using ideas similar to those in Figure 5.8, demonstrate your understanding of the logic operations by adding an additional indicator that will display the greater of the two numbers determined from your addition and subtraction operations in Exercise 5.1. Is the number obtained by the addition operation always the greatest?

5.4. Demonstrate your understanding of logic operations by building a VI that takes as input three numbers and outputs the greatest and the smallest value of the three. This will involve very similar logic operations to the one performed in the earlier VI.

5.5. After you have had a brief exposure to the operation of loops and other structures in LabVIEW, practice by creating the program from Figure 5.9. When fully operational, the program should continue to increment by a counter 1 every 0.1 seconds until the stop button is pushed. Verify that this is true and that waiting longer to press the stop button results in a larger number. What is the difference between inside and outside loop displays? Why do you think this happens? If you let the program run and forgot to turn it off when you left the lab, would your program ever be able to reach infinity? Explain your answer.

5.6. The specific task that you will perform is to create a calculator that will add, subtract, multiply, or divide two numbers using a case structure as depicted earlier in Figure 5.11. This will require you to use two inputs, an output, and a ring on the front panel. You will need a case structure in the diagram window, which initially will have only two cases {0,1} but these can be extended by right clicking on the condition statement in the block diagram. The case structure itself can be found in the structures subpalette of the functions palette. The ring should be added to the front panel of the VI and can be found on the modern subpalette of the controls palette. The properties dialog box of the ring, which opens up by right clicking on the ring, allows you to add cases as you need (in this case, four cases). Make sure the order of the list on the ring corresponds to the case list.

5.7. Data acquisition is probably the most important aspect of LabVIEW. To be able to understand completely what takes place when you are taking data, it is important to know what tasks are performed by a VI that acquires data from an external source. The VI introduced earlier in Figure 5.14 takes a temperature measurement every time the run button is pushed. While this is useful for taking measurements where temperature is constant (e.g., room temperature), it is not useful in tracking temperature changes.

(The next exercise addresses this issue via a loop structure.) If a low-frequency square wave is used, however, it is possible to verify that the VI functions as intended. Connect a function generator to the DAQ card. Make sure that the function generator is set to produce a square wave and that the frequency is set to be about 0.1 Hz. You may wish to view the signal on an oscilloscope to ensure that the function generator is indeed producing the desired waveform. This can be done by connecting the output of the function generator via a BNC connector to Channel A (or Channel 1) on a typical two-channel scope. Be sure to set the trigger mode to external (EXT) and choose Channel A (or Channel 1) as the trigger source. You may have to initially use the ground function of the scope to ensure that the beam is zeroed at the midlevel of the screen and that the vertical and horizontal scales are set up properly. The same BNC connector can be used to connect the function generator to the BNC-2120, which acts as the interface between the DAQ card and the function generator.

5.8. In this exercise we need to add a loop to the VI from the previous exercise so that the VI can take readings continuously until a stop button on the front panel of the VI is pushed and the VI is stopped. The loop construct is in the structure subpalette of the programming section of the functions palette. You can also find the stop button in the Boolean subpalette of the classic section of the controls palette. Verify that this works and watch how the temperature reading changes with time. Change the frequency of the function generator and observe the effects of the output. What difficulties arise when the frequency is too fast relative to the timing capacity of the loop? If we had no digital indicator, only thermometers, would this be a problem?

5.9. In this exercise, we will add more functions to the VI from the previous exercise. If we only want to maintain some temperature (e.g., in a thermostat), we might only care about the highest value of the temperature. Use the *logic* operators to determine the highest temperature of all the measurements (hint: you will need a shift register). In LabVIEW, there are many different options for data types (scalar, dynamic data, waveform, etc.). For this part of the experiment, you will need to convert from the dynamic data to a single scalar. To do this, right click in the block diagram and select express, signal manipulation, from DDT. Then place the block, select single scalar, and then click OK.

This VI gives a little more insight into data acquisition with LabVIEW and some of the difficulties that may arise. With this VI, will you be able to tell how temperature changes as a function of time? What could we do to solve this problem (in general)? What are some of the problems with the various output styles that we have considered? What might be some more beneficial ideas?

5.10. Based on the function generator in Figure 5.15, random noise should be added to the signal by means of the "add noise" option of the signal simulation block, in the signal analysis toolbox. The output from the function generator should be added to the noise using an addition block, as it can add both signals and constants. Send this noisy signal to a new waveform graph. In order to extract the original signal from the noisy signal, a filter will be applied. The filter reduces the effect of portions of the signal that are above a specified frequency. This allows us to remove the high-frequency noise. The filter block can be found in the signal analysis toolbox. Configure the filter to remove noise from the signal. Your front panel window should look much like Figure 5.16.

Select three sets of operating conditions (i.e., amplitude and frequency) for the triangle wave pattern and measure the actual amplitude and frequency of the waveform shown on the oscilloscope. This will require you to connect the DAQ board to the scope inputs using BNC cables. Verify that your measured amplitudes and frequencies are the same as those you input to the VI. Vary the noise amplitude between 0 and twice the signal amplitude and take screen captures of the results.

## 5.12 Appendix: Software Tools for Laboratory Data Acquisition

### 5.12.1 Measurement Foundry

Measurement Foundry is produced by Data Translation and has significant capabilities for rapid application development in conjunction with DT-Open Laters for .NET class library and compliant hardware. It offers the ability to acquire, synchronize, and correlate data and to perform control loop operations. It also features automatic documentation of programs and links with Excel and Matlab.

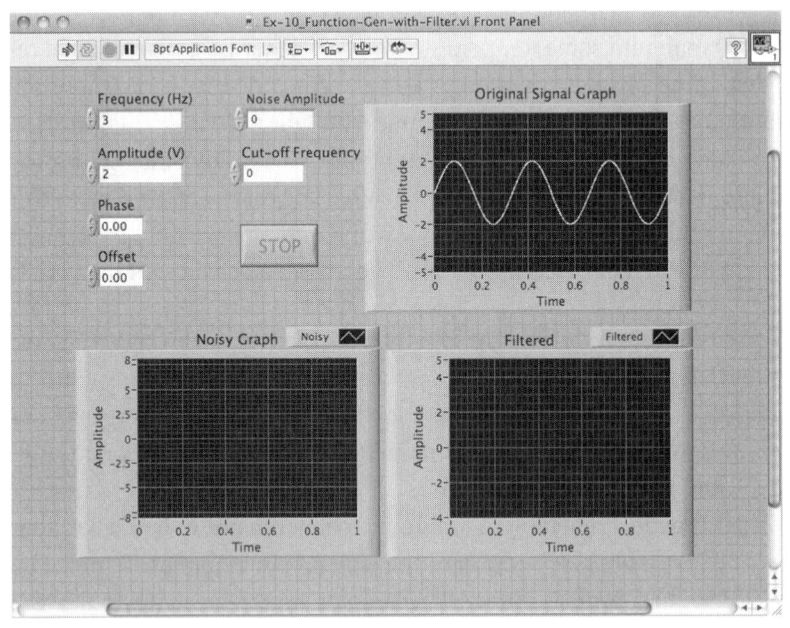


Figure 5.16
Function generator and filter front panel.

### 5.12.2 DasyLab

DasyLab from Measurement Computing Corporation (MCC) allows for the integration of graphical functions, real-time operations, including PID control, and real-time display via charts, meters, and graphs, and incorporates an extensive library of computational functions. It provides serial, OPC, ODBC, and network interface functions and supports data acquisition hardware from MCC, IOtech, and other vendors.

### 5.12.3 iNET-iWPLUS

This software, sold by Iomega, which is also a distributor of DAQ cards and sensors, can be used to generate analogue and digital output waveforms and run feedback/control loops, such as PID, and has capabilities similar in concept to LabVIEW, which allow for the creation of custom virtual instruments.

### 5.12.4 WinWedge

WinWedge is yet another data acquistion software that uses a Microsoft Windows Dynamic Data Exchange mechanism to provide data acquisition and instrument control in conjunction with Microsoft Excel, Access, and so on. It provides a menu-driven configuration program and is used mainly in conjunction with serial data streams.

CHAPTER 6

# Signal Processing with LabVIEW



## 6.1 Introduction

Analogue and digital filters are used extensively in sensor signal processing. In this chapter, both of these topics are discussed and examples using LabVIEW are presented. The main use of these signal-processing techniques is in pre- and postprocessing of sensor signals. In particular, analogue filters are often used to deal with the so-called aliasing phenomenon that is common in data acquisition systems. Digital filters are generally used to postprocess acquired signals and can be used in conjunction with sophisticated digital signal-processing techniques such as Fast Fourier Transform to perform spectral analysis of acquired signals. With this in mind, the chapter introduces the reader to the following topics:

- Analogue filters, their analysis, and synthesis using passive components (resistors and capacitors) as well as active components (operational amplifiers).

- Frequency response of a low-pass filter using Bode plots, illustrating the response of the filter with varying frequency of the input signal.
- The concept of a digital filter as counterpart to the analogue filter and built using software.
- Implementation of *moving average* (MA) and *autoregressive moving average* (ARMA) model-based digital filters and their implementation in LabVIEW.
- Using LabVIEW to develop virtual instruments (VIs) that implement digital filters and demonstrate their functioning using a graphical display of pre- and postfiltered signals.

These learning objectives are illustrated through detailed descriptions, examples, and exercises. The presentation refers to supporting hardware from National Instruments and other manufacturers (namely Analog Devices, a maker of micromechanical accelerometers and similar microelectromechanical components used as sensors in measurement and instrumentation).

The reader is expected to have access to a laboratory environment that allows for implementation of analogue filters via passive and active electronic components, test instruments such as oscilloscopes and function generators, and, more importantly, access to LabVIEW for implementation of the virtual instruments discussed in the sequel. Use of a small inexpensive fan, which can be the basis for studying the impact of rotor imbalance and vibration that is filtered with a combination of analogue and digital filters, is also useful. Other sources of real data can be used, however, so long as the measured signal manifests a combination of "good" and "bad" information, preferably at low and high frequencies, respectively, which can be used as the basis for the experimental component of the examples and exercises discussed in this chapter. It is possible, however, to produce virtual sensor data using LabVIEW itself, as it is done in several of the examples, to illustrate signal processing techniques, although it is always important to use real data at some point in the process of learning to apply signal-processing techniques.

## 6.2 Analogue Filters

Analogue filters are used primarily for two reasons: (i) to buffer and reduce the impedance of sensors for interface with data acquisition devices and (ii) to eliminate high-frequency noise from the original signal so as to prevent *aliasing* in analogue-to-digital conversion. Analogue filters can be constructed using *passive* components (namely resistors, capacitors, and, at times, inductors) or via a combination of passive and *active* components (transistors or, more commonly, operational amplifiers) leading to active filter designs. We consider both cases here.

### 6.2.1 Passive Filters

Passive filters are designed with a few simple electronic components (resistors and capacitors). The basic *low-pass* filter, depicted in Figure 6.1, can be used to remove (or attenuate) high-frequency noise in the original signal, in this case denoted by $v_i$. The underlying assumption is that any time function can be viewed as being a combination of sinusoidals. More exactly, any periodic function of time is approximated by an infinite series of sinusoids at frequencies that are multiples of the so-called fundamental frequency of the original signal (the so-called *spectral content* of the signal). Nonperiodic functions can be viewed in essentially the same way but we must allow for a continuous spectrum. For example, the original signal, $v_i$, may be approximated by

$$v_i = V_{i,1}\sin(\omega t) + V_{i,2}\sin(2\omega t) + V_{i,3}\sin(3\omega t) + \cdots, \tag{6.1}$$

where $V_{i,1}, V_{i,2}, V_{i,3}, \ldots$ are the amplitudes of the consecutively higher frequency components or *harmonics* of the original signal.

These higher frequency components may represent fluctuations (or, in many cases, electrical noise) that we may wish to attenuate to prevent aliasing (appearance of high-frequency components in the analogue signal as low-frequency aliases of these components) and generally to present a clean signal to the DAQ system. The filter in Figure 6.1 produces an output, $v_o$, which has the *same set* of components (in terms of the respective frequencies) as the original signal, $v_i$, but at reduced amplitudes:

$$v_o = V_{o,1}\sin(\omega t) + V_{o,2}\sin(2\omega t) + V_{o,3}\sin(3\,\omega t) + \cdots \tag{6.2}$$

where $V_{o,1}, V_{o,2}, V_{o,3}, \ldots$ are the amplitudes of the sinusoidal components in $v_o$. In general, $V_{o,1}$, $V_{o,2}, V_{o,3}, \ldots$ are *smaller* than their counterparts in the input signal, $V_{i,1}, V_{i,2}, V_{i,3}, \ldots$. For instance, depending on the values of $R$ and $C$ in the filter shown in Figure 6.1, $V_{o,1}$ may be very close to $V_{i,1}$, say 98% of this value, but $V_{o,2}$ may be about 70% of $V_{i,2}$ and so forth. The reason for this is that the given low-pass filter attenuates each signal according to its frequency; the higher the frequency, the larger the attenuation (hence the smaller the amplitude of the given component in the output signal). The precise amount of attenuation can be found from the so-called frequency response graph of the filer (commonly referred to as its *Bode* plot) as, for instance, depicted in Figure 6.2.

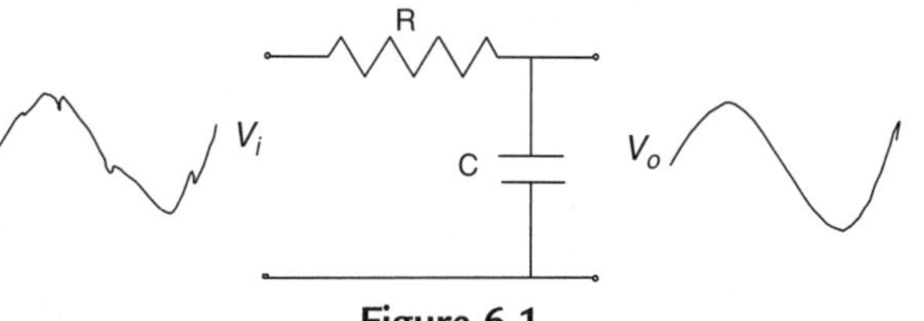


**Figure 6.1**
Passive low-pass filter.

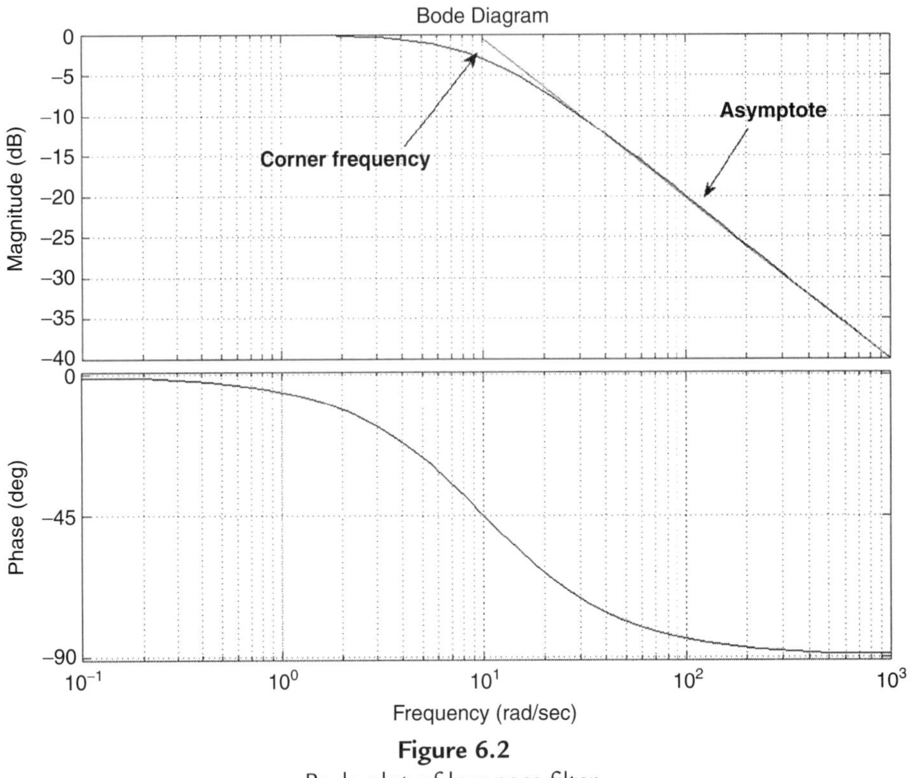


**Figure 6.2**
Bode plot of low-pass filter.

The specific equation for the filter from the application of Kirchoff's current law at the output node (assuming an open circuit or no load at the output) is given by

$$\frac{v_i - v_o}{R} + C\frac{d}{dt}(v_o) = 0. \tag{6.3}$$

This can be simplified into

$$RC\frac{dv_o}{dt} + v_o = v_i. \tag{6.4}$$

As a first-order differential equation, it can be transformed into transfer function form, using Laplace transform, as

$$\frac{\hat{v}_o}{\hat{v}_i} = \frac{1}{\tau s + 1}, \tag{6.5}$$

where $\tau = RC$ is the time constant of the filter, that is, essentially the time it takes for the filter to respond to a step input function by reaching 63% (almost two-thirds) of its steady-state output.

The inverse of $\tau$, that is, $\omega_c = 1/\tau$, is known as the *corner frequency* of the filter, that is, the frequency above which the filter starts to attenuate its input. These are discussed further below in the context of the so-called *Bode plot* of the filter. Assuming sinusoidal functions of the form $v_i = V_i(\omega t)$, whose Laplace transform is given by

$$\hat{v}_i = \frac{\omega V_i}{s^2 + \omega^2}, \tag{6.6}$$

and substituting in Equation (6.6), taking partial fractions and simplification (including dropping the transient response term as detailed in the appendix), the steady-state output would be $v_o = V_o \sin(\omega t + \phi)$, where

$$\frac{V_o}{V_i} = \frac{1}{\sqrt{(\tau\omega)^2 + 1}} \text{ and } \phi = -\tan^{-1}(\tau\omega). \tag{6.7}$$

In more standard form, we have

$$\left.\frac{V_o}{V_i}\right|_{dB} = 20\log\frac{1}{\sqrt{(\tau\omega)^2 + 1}} = 20\log 1 - 10\log\left((\tau\omega)^2 + 1\right) \tag{6.8}$$

$$\left.\frac{V_o}{V_i}\right|_{dB} = \begin{cases} 0 & \tau\omega << 1 \\ -20\log\tau\omega & 1 << \tau\omega \end{cases}. \tag{6.9}$$

The graph of the $V_o/V_i|_{dB}$ is precisely what was depicted earlier in Figure 6.2 where it is evident that higher frequencies, particularly those higher than the corner frequency of 10 r/s, are attenuated at an increasing rate, leading to the reduction of high-frequency components in the filtered signal.

### 6.2.2 Active Filters Using Op-amps

The passive filter shown previously is simple and can, in principle, be used to filter out undesirable components of a given signal. However, because it is made up of entirely passive components (resistors and capacitors), it has to draw current from the input and will, in addition, "load" the circuit connected to the output of the filter. Op-amps can eliminate this problem, as the current that is drawn from the input stage is very small (because op-amps have large internal resistances, of the order of 10 MΩ). Likewise, as active devices, op-amps supply current to drive their output and hence minimize the impact of the filter on any output circuit, such as the DAQ card, thereby less affecting the reading of the acquired signal. With this in mind, op-amps are often used in conjunction with resistors and capacitors to create an active filter. A sample configuration is given in Figure 6.3.

To be able to utilize this circuit, we must determine the relation between the input $v_i$ and the output $v_o$. The summation of currents at the inverting input is given by

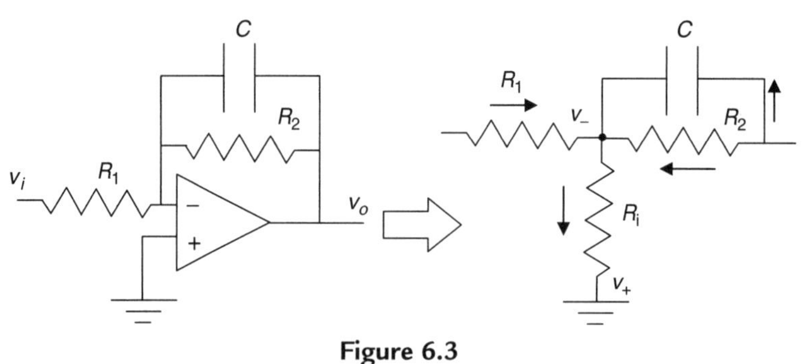


**Figure 6.3**
Active low-pass filter.

$$\frac{v_i - v_-}{R_1} + \frac{v_o - v_-}{R_2} + C\frac{d}{dt}(v_o - v_-) = 0.$$

Now, given that in a negative feedback configuration as shown earlier, $v_- = v_+ = 0$, we can simplify this to

$$\frac{v_i}{R_1} + \frac{v_o}{R_2} + C\frac{d}{dt}(v_o) = 0.$$

We can rewrite this as

$$v_o + R_2 C\frac{d}{dt}(v_o) = -\frac{R_2}{R_1}v_i.$$

This is very similar to a passive filter equation, with the exception of the gain on $v_i$. The negative sign means that the filtered signal will lag at least 180 degrees behind the unfiltered one. (This issue can be resolved via an inverting filter of gain of $-1$.) We can call this gain $k$, so $k = R_2/R_1$. In addition, we see that the time constant is given by $\tau = R_2 C$. So, we have two degrees of freedom in our filter configuration. We may adjust the gain constant, $k$, to amplify low-frequency signals. However, this will also raise the value of the crossover frequency and therefore allow noise to have a higher amplification. In addition, we can change the corner frequency ($\omega_c = 1/\tau$) by adjusting the relevant parameters. This will have effects on the gain as well, which must be taken into account. These can be better understood by examining the system in the frequency domain. In the frequency domain, the ratio of the output to the input voltage is given by

$$\left.\frac{V_o}{V_i}\right|_{dB} = 20\log\frac{k}{\sqrt{(\tau\omega)^2 + 1}} = 20\log k - 10\log\left((\tau\omega)^2 + 1\right)$$

$$\left.\frac{V_o}{V_i}\right|_{dB} = \begin{cases} 20\log k & \tau\omega << 1 \\ 20\log k - 20\log\tau\omega & 1 << \tau\omega \end{cases}.$$

We can understand better how the filter works by examining the Bode plot of the system, which is indeed similar to that in Figure 6.2, with the only exception being an upward shift of the graph by $20\log k$. As discussed earlier, the graph is meant to illustrate how the filter passes through signals of frequency up to its corner frequency and gradually attenuates those beyond this level as depicted in Figure 6.2.

### 6.2.3 Implementation on a Breadboard

In a laboratory setting, one can build and test various circuits, including low-pass filters using breadboards. Figure 6.4 depicts one such breadboard made by National Instruments (NI). Similar boards with more or less additional features are available from various manufacturers. This particular breadboard has a connector (lower left corner) that attaches readily to an NI DAQ card. Other breadboards are also available from NI and other manufacturers.

### 6.2.4 Building the Circuit

Assuming access to a breadboard similar to Figure 6.4, circuits can be built on the white section of the breadboard. The horizontal sections of holes in the wider strips of this section are connected together to allow placement of circuit components that may have

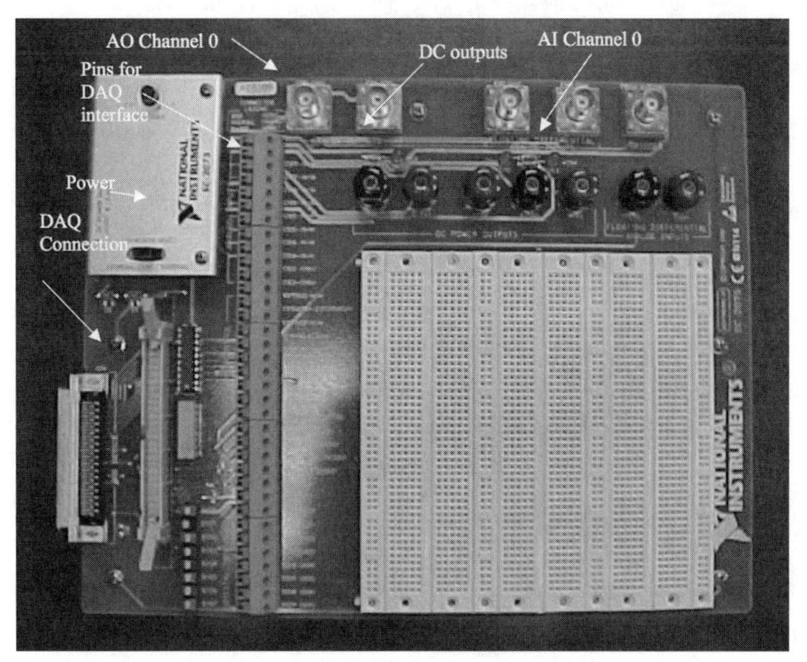


**Figure 6.4**
Breadboard for filter implementation.

different sizes, while the thinner sections provide a mechanism for access to power and ground (or common signals). In addition, as stated earlier, this particular board has the ability to interface with the data acquisition card. Inputs and outputs from the card can be taken from the pins shown in Figure 6.4 and connected either to the breadboard or to external devices.

### 6.2.5 Electronic Components

A variety of electronic components are used to build analogue filters with resistors, capacitors being the most common types. Resistors can be either fixed or variable. As the name implies, a fixed resistor has one resistance value and cannot be changed. A variable resistor, however, can be adjusted to have different values. Resistors are coded by color as shown in Figure 6.5.

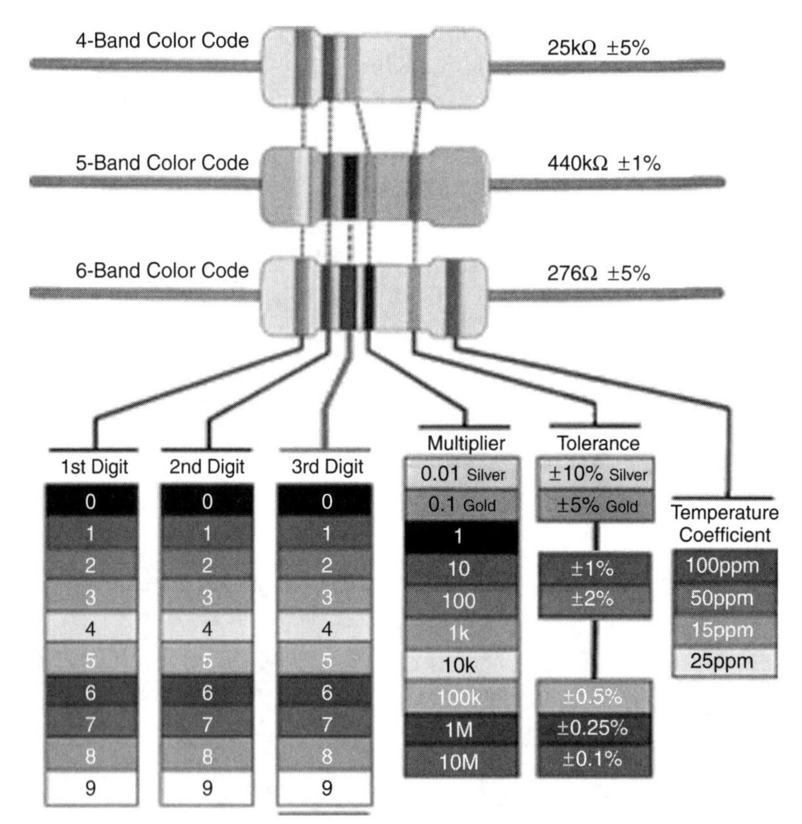


**Figure 6.5**
Resistor color-coding scheme.

**Table 6.1 Example Resistors**

| Band 1 | Band 2 | Band 3 | Multiplier | Tolerance | Value |
|---|---|---|---|---|---|
| Orange | Orange | N/A | Brown | Brown | 330 Ω ± 1% |
| Blue | Gray | Black | Green | Blue | 68 MΩ ± 0.25% |
| Black | Green | Green | Red | Red | 5500 Ω ± 2% |
| Red | White | Yellow | Gold | Violet | 29.4 Ω ± 0.1% |

Note that colors may not be evident but this same figure is generally available online at a number of sources.

Note that not all resistors have the same number of bands, so it is important to know the type of resistor at hand. If a resistor has only four bands, then it does not have the third digit. A few examples are illustrated in Table 6.1.

In addition to resistors, capacitors are used in building analogue filters. A picture of a ceramic plate capacitor is shown in Figure 6.6. Unlike resistors, capacitors have their value explicitly printed on them. Capacitance values can usually be read off in either a picofarad ($pF$) scale or a microfarad ($\mu F$) scale. Intermediate scales are used very seldom. In addition, it is important to note that a capacitor has some parasitic resistance as well, although this is usually very small.

Finally, op-amps are key components of analogue filters. A picture of a typical op-amp, LM 741, is depicted in Figure 6.7. It is important to be sure that the correct pins are used when connecting the op-amp. The data sheet for the device, generally available from the manufacturer, provides detailed specifications in this regard.

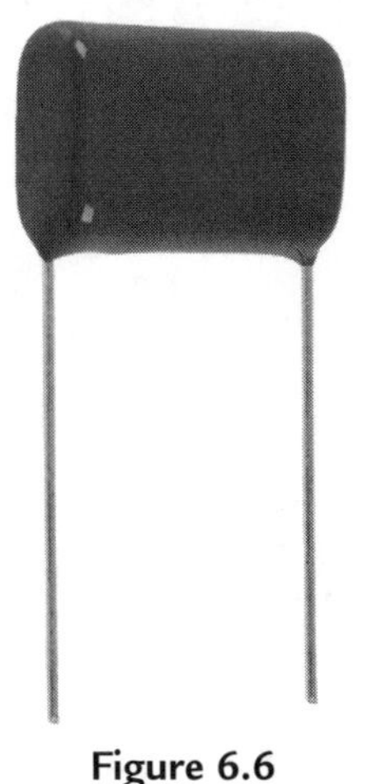

**Figure 6.6**
Ceramic plate capacitor.

**Figure 6.7**
Operational amplifier in DIP package.

### 6.2.6 Op-amps in Analogue Signal Processing

As stated earlier, op-amps can be used to perform simple or more complex tasks. For instance, as shown in Figure 6.8, to add two signals, $v_i$ and $v_r$, together, an op-amp and three resistors are used. This can be useful, for instance, in adjusting for a d.c. offset in an accelerometer signal or similar cases. If we were to derive the relationships between the inputs, we would arrive at the following:

$$v_o = -\frac{R_2}{R_1}(v_i + v_r).$$

Therefore, if we set $v_r$ to be the negative value of the DC offset from the accelerometer, we will be able to remove this offset successfully.

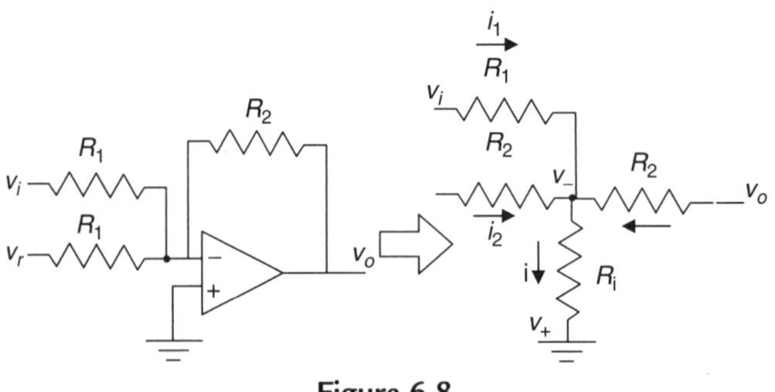


**Figure 6.8**
Amplifier schematic diagram.

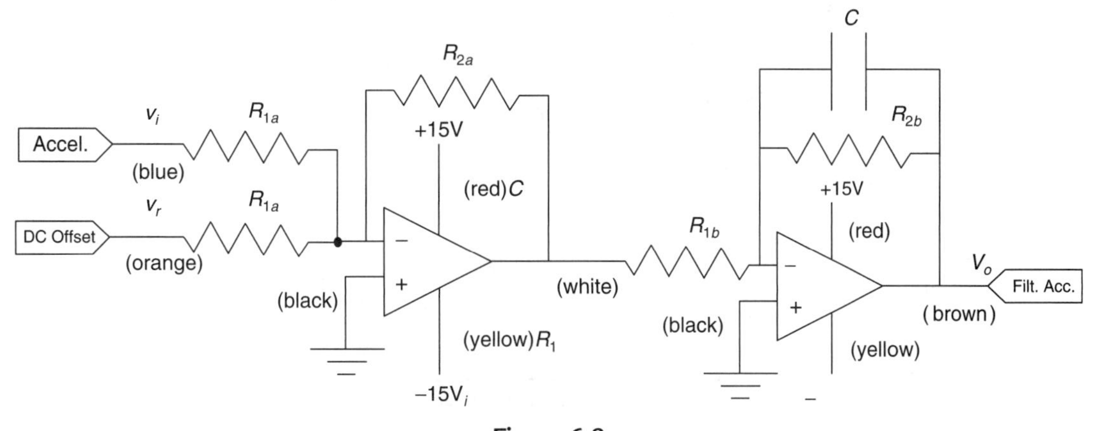


**Figure 6.9**
Offset removal and low-pass filtering.

More importantly, we will generally need to remove high-frequency noise in the sensor signal prior to sampling the signal with an A/D converter. For this reason, the combination of op-amps in Figure 6.9 is used.

In practical implementation, color coding the wiring is helpful in preventing confusion. Typically the coding scheme uses red, positive supply voltage; black, ground; and blue, signal. Other colors are used to supplement these and differentiate the different signals. Exercise 6.1 refers to this in more detail in the context of performing a simple acceleration measurement experiment.

## 6.3 Digital Filters

Digital filtering uses discrete data points sampled at regular intervals. These data points are usually sampled from an analogue device such as the output of a sensor (say an accelerometer used to measure vibration in a beam). Digital filters as shown schematically in Figure 6.10 rely not only on the current value of the measured variable, but also on its past values (either in raw or filtered form). This leads to two kinds of filters, which are examined in the sequel.

### *6.3.1 Input Averaging Filter*

In the input averaging filter, the previously unfiltered values of the given signal are used in the scheme. This filter takes the form of

$$y_k = \alpha u_k + (1 - \alpha)u_{k-1}. \tag{6.10}$$

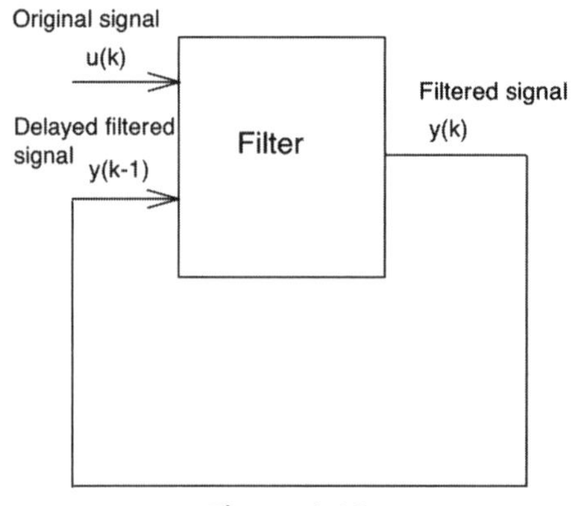


**Figure 6.10**
Diagram of a digital filter.

Note that we can write $\alpha_0 \equiv \alpha$ and $\alpha_1 \equiv 1 - \alpha$ and hence write the above as

$$y_k = \alpha_0 u_k + \alpha_1 u_{k-1} \tag{6.11}$$

and thus further extend this formulation to more complex filters such as

$$y_k = \alpha_0 u_k + \alpha_1 u_{k-1} + \alpha_2 u_{k-2} + \cdots + \alpha_n u_{k-n}. \tag{6.12}$$

This general form is called a *moving average* filter, as it in effect averages past values of the input signal, each with its respective weight. Selection of these weights is often an issue and can be formalized, although in the present context we will deal with simpler cases in which an intuitive approach to selecting these values can be used.

### 6.3.2 Filter with Memory

In a filter with memory, previously filtered values are used to adjust the new output. This filter takes the form

$$y_k = \alpha u_k + (1 - \alpha) y_{k-1},$$

where $\alpha$ is the weight on the current value of the unfiltered signal, $u_k$. The remainder is from the previous value of the filtered signal, $y_{k-1}$. Varying $\alpha$ will change the extent to which the input signal is filtered. In particular, a relatively large $\alpha$ weighs in the current value of the input signal, while a small $\alpha$ weighs in the past (filtered) signal, $y_k$. Normally, $\alpha \leq 1$. This is evident in the following example.

### 6.3.3 Example

A set of data points is measured from a continuous signal as given in Table 6.2.

**Table 6.2 Data for Digital-Filtering Example**

| $k$ | $u_k$ |
|---|---|
| 0 | 0.10 |
| 1 | 1.05 |
| 2 | 1.92 |
| 3 | 3.90 |
| 4 | 4.02 |
| 5 | 4.94 |

A simple input averaging filter with $\alpha$ values of 0.25, 0.5, 0.75, and 1.0 is used to filter these values as depicted later (Figure 6.11). The filters produce a continuum of response patterns indicating that no single value of $\alpha$ is best. However, one can argue, based on proximity to the general pattern of the input signal, that $\alpha = 0.5$ may be reasonable. In practice, one has to fine-tune $\alpha$ or similar parameters of a given filter to fit the application in mind. There are, as shown later, formal methods (based on digital signal processing) that allow the user to select the filter parameter to affect the frequency response of the filter (similar to the analogue case). The mathematical techniques underlying these tools are beyond the scope of this book although their application is discussed later.

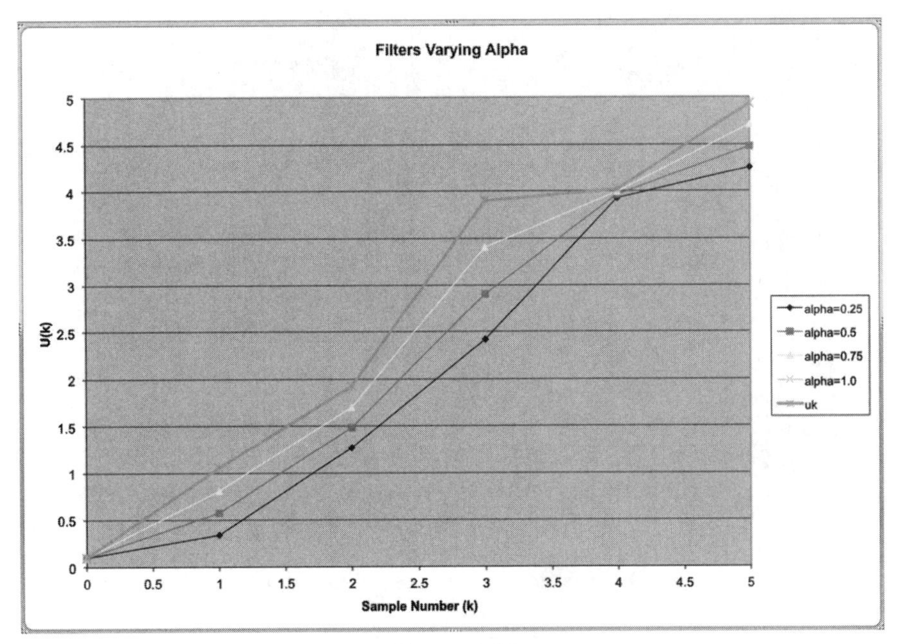


**Figure 6.11**
Simple averaging filter results.

### 6.3.4 LabVIEW Implementation

One can implement a digital filter in LabVIEW as shown in Figure 6.12. The block diagram of this filter appears in Figure 6.13. The filter parameters are set at 0.5 and 0.5 in the lower left corner of the front panel but can be changed if necessary. The block diagram depicts a sinusoidal signal generator, as well as a noise generator on the left side. The in-place operation allows the addition of individual data elements.

The in-place element structure allows a simple operation such as addition to occur on the corresponding elements of two dynamic arrays produced by the sinusoidal and noise signal generators. This node can be found in the structures subpalette of the programming section of the fuctions palette as depicted in Figure 6.14.

The filter itself is implemented as a *finite impulse response* (FIR) filter node (found in the advanced FIR filtering section of the filters subpalette of the signal-processing section of the functions palette*), which effectively implements a *moving average* filter type. The array of filter coefficients appears in the front panel of Figure 6.12. Note that in this case the source

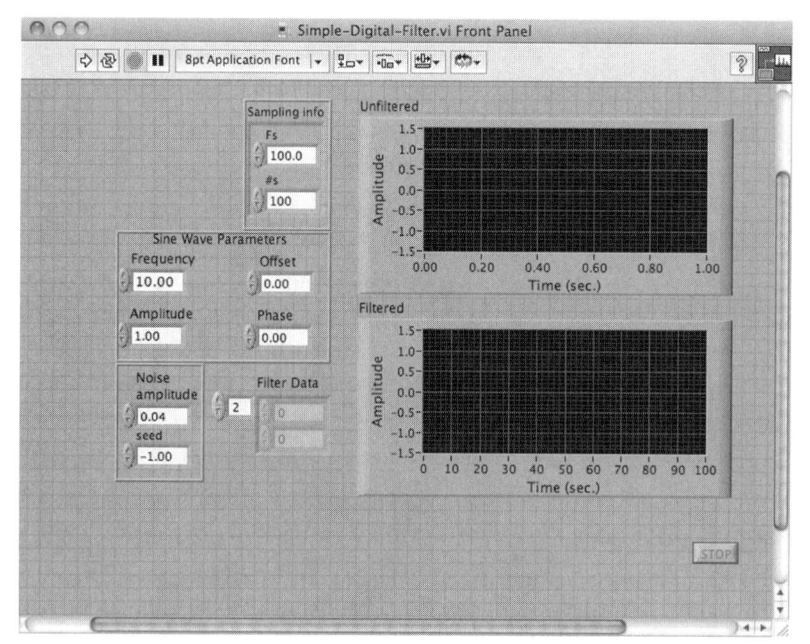


**Figure 6.12**
Front panel of a simple digital filter.

* The location of these nodes in the respective palette may change depending on the version of LabVIEW. It is, however, possible to search for a specific type of node by name and locate it irrespective of its location.

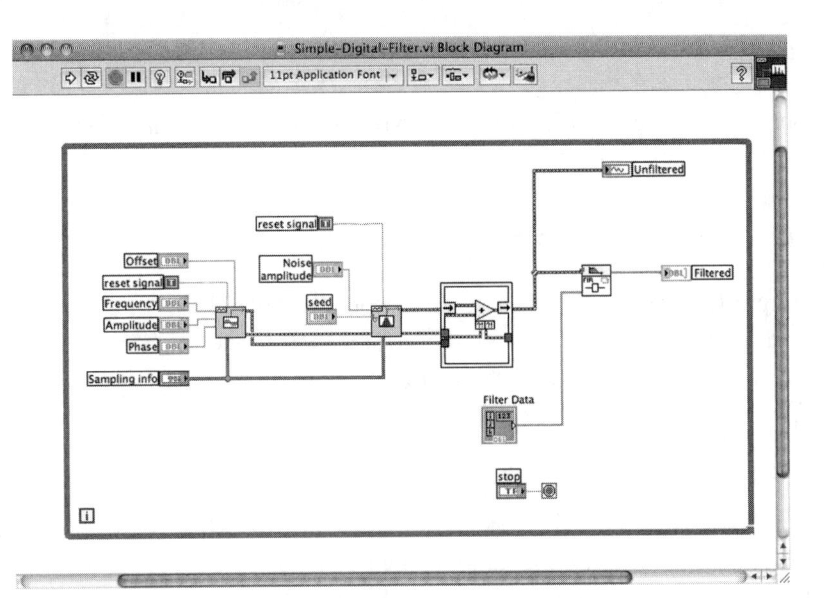


**Figure 6.13**
Diagram of a simple digital filter.

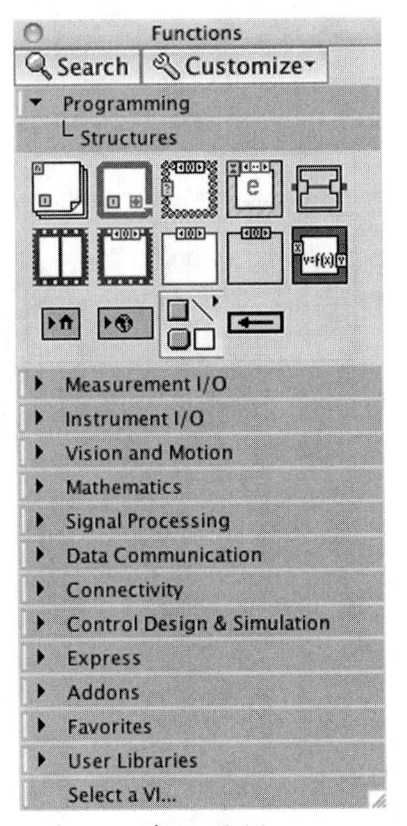


**Figure 6.14**
Structures subpalette of the functions palette.

signal is generated internally but it is possible to do the same task using an external input signal, as, for instance, generated by a function generator.

### 6.3.5 Higher Order Digital Filters

While the simple filter in the previous section works reasonably well, one can build more effective filters using a combination of autoregressive terms and moving average terms, via an ARMA model, also referred to as an infinite impulse response (IIR) filter. The general equation for such a filter is given as

$$y_k = -a_1 y_{k-1} - a_2 y_{k-2} - a_3 y_{k-3} + \cdots + b_0 u_k + b_1 u_{k-1} + b_2 u_{k-2} + b_3 u_{k-3} + \cdots \quad (6.13)$$

Note that $a_i$ and $b_i$ must be chosen properly for stable and effective performance. This is not a trivial task and requires advanced techniques that extend beyond the scope of this text. The essential idea is to place the so-called poles and zeros (the roots of the denominator and numerator of the corresponding discrete time transfer function) in reasonable locations in the complex plane. However, there are well-known design strategies that have performed well, including Butterworth, Chebyshev, and Bessel, that are programmed in LabVIEW. It is also possible to produce the filter coefficients in Matlab or a similar tool and use a similar technique as in the previous section (albeit using an IIR filter node) to implement the given filter. Implementation using LabVIEW built-in functions is depicted in Figure 6.15. Filter parameters

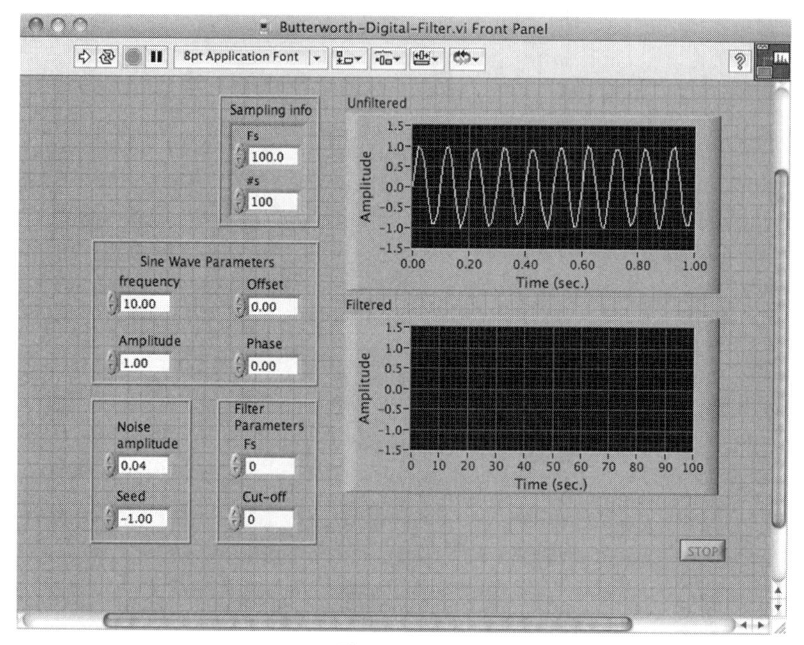


**Figure 6.15**
Front panel of Butterworth filter design.

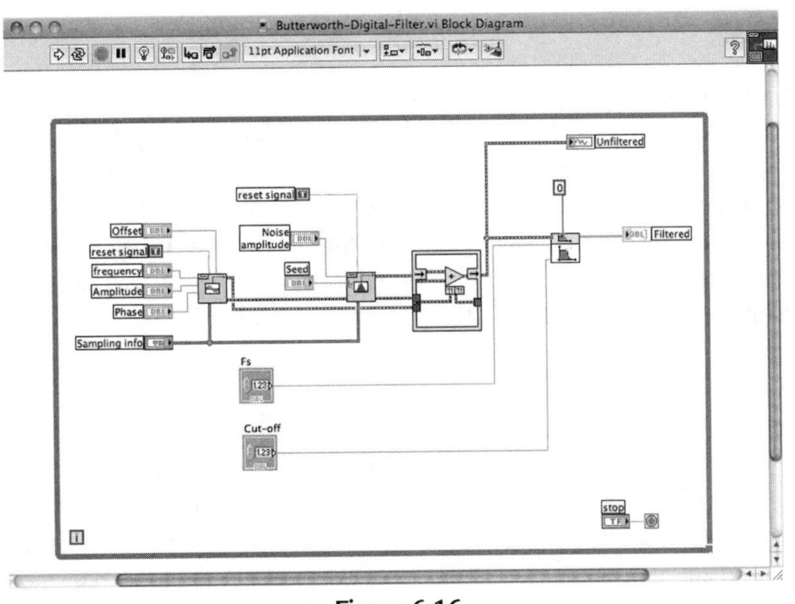


**Figure 6.16**
Block diagram of Butterworth filter design.

include the sampling frequency, which in this case is the same frequency used to generate the signal in the first place. The cutoff frequency is also another required parameter, which is chosen to correspond to the frequency at which the noise components start to dominate the real signal. The block diagram is depicted in Figure 6.16. The filter node (from the signal processing section of the function palette) requires filter type (set to zero for a low-pass filter) as well as the two parameters mentioned earlier. Note that the unconnected terminal of the filter node (the high-frequency cutoff) is not required for a low-pass or high-pass filter but is required for a bandpass filter.

The response of the filter depicted in the front panel diagram in Figure 6.15 typifies the low-pass filtering effect of a Butterworth filter, which indeed performs better than the simple filter we designed earlier.

## 6.4 Conclusions

This chapter considered signal-processing techniques used commonly in conjunction with computer-based data acquisition systems. As a prelude to digital signal processing via LabVIEW, we considered analogue signal processing using operational amplifiers (op-amps). This process is often essential to successful computer-based data acquisition, as the original signal may be contaminated with noise, which leads to aliasing in the digitize signal (a process akin to the appearance of ghost images on television screens). We further discussed digital filtering using moving average and autoregressive moving average models.

## 6.5 Problems

6.1. Implement the analogue filter design described in Figure 6.10 using a pair of LM-741 op-amps and appropriate resistors and capacitors. You will need to choose a reasonable capacitor value for $C$, say $0.1pF$, and then choose an appropriate resistor value for $R_{2b}$ to achieve a corner frequency of 10 Hz (effectively 60 rad/s). For this exercise, design the gain of your filter to be unity (1) by selecting $R_{2a} = R_{2b}$. Likewise, choose $R_{1a} = R_{2a}$ because you will use the first stage only to invert the signal produced by the function generator (since it will be inverted again by the second stage filter). In building the circuit, pay close attention to the color coding marked on the figure. Do *not* power up the board or connect power to the circuit without first ensuring that connections are made properly.

6.2. Perform preliminary evaluation of the filter in the previous exercise by generating a signal using a function generator capable of producing a sinusoidal function of varying frequency (similar to that shown in the previous chapter). You can connect the function generator output to $v_i$ input in the filter and ground $v_r$ input since it is not used in this exercise. (It is always good practice to ground unused inputs to prevent the circuit from picking up and introducing noise in the actual signal.) You will need an oscilloscope to evaluate the functioning of the filter. Vary the frequency of the signal from 1 to 100 Hz and make an estimate of the filter performance (in terms of attenuation of the signal amplitude) by recording (reading off the scope) the amplitude of the signal as a function of frequency. Compare your graph with Figure 6.2. Note that the figure is drawn on a logarithmic scale. Be mindful of this fact in your comparison. Bear in mind the corner frequency of the filter, which is expected to be at 10 Hz.

6.3. Create the VI shown in Figure 6.17. This VI can be used to evaluate the analogue filter discussed in the previous exercises. The signal generated by the VI is filtered with the analogue filter implemented in the previous exercises. The spectrum of both unfiltered and filtered signals is shown in the respective panels of the front panel of the VI. Set the wave type to be a sine wave, the frequency to be 10 Hz, the amplitude to be between 2 and 5 V, and the noise amplitude to be 10% of the signal amplitude. Use an oscilloscope to verify that you are producing the correct signal. Note that the VI should be implemented such that the Analogue Output Channel 0 on the National Instruments interface board is used to produce the generated signal. In addition, examine the spectrum of the signal given in the VI. Later you will want to filter the noise, not the signal. Having an idea of where the spectrum of the signal dominates enables one to filter the signal properly.

6.4. You can use the signal generated by the VI (via Analogue Output Channel 0) of Figure 6.17 with the analogue filter from the previous exercises. The output of the filter should be used as Analogue Input Channel 0 so you can evaluate the performance of

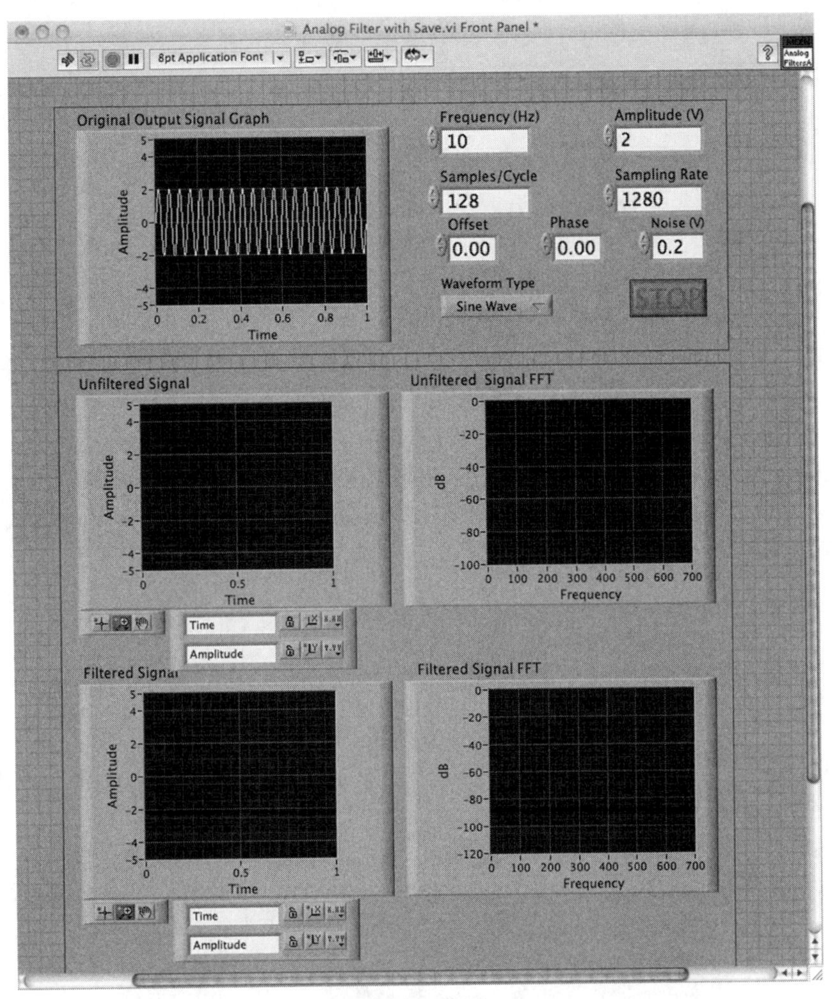


**Figure 6.17**
Front panel of analogue filter test VI.

the analogue filter using the VI. Set the noise amplitude to zero in your VI. Starting with 1 Hz, adjust the frequency in steps of 2, 5, and 10, up to 1000 Hz. Determine the frequency at which the filtered wave amplitude is 1/10th the original signal amplitude. Also, review the spectrum of the signal as it appears in the respective panel of the VI.

6.5. In this step, you will evaluate the performance of the filter as a function of noise amplitude. You will first set the signal frequency at 10 Hz, and set the amplitude to be 2 V. Adjust the noise amplitude from 0 to 2 V. Measure the amplitude of the filtered signal and fill in Table 6.3. In addition, note the impact of increasing the noise amplitude on the frequency spectrum of the filtered and unfiltered signals.

**Table 6.3 Noise Attenuation as a Function of Amplitude**

| Noise amp (V) | Filtered amp (V) |
|---|---|
| 0.1 | |
| 0.2 | |
| 0.5 | |
| 0.8 | |
| 1 | |
| 1.5 | |
| 2 | |
| 3 | |
| 5 | |

6.6. You can demonstrate further how the filter may be used in a real system. To do this you can use a small fan with a small weight attached to one of the blades with sturdy tape that prevents the added weight to fly off as the fan rotates. A semiconductor-based (MEMS) accelerometer similar to that shown in Figure 6.18 may be attached to the base of the fan to record the acceleration due to an unbalanced rotor. We wish to filter out all the noise that occurs above the fan's rotational frequency. In order to do this, we must first know what the frequency is. Determine the frequency from the fan specifications, which should be available from manufacturer data sheets.

6.7. Create the VI for use with the accelerometer as shown in Figure 6.19. The VI must be able to read the accelerometer signal and produce filtered and unfiltered displays of this signal along with the spectrum of the signal.

6.8. Connect the accelerometer to the amplifier in Figure 6.9. The accelerometer, in the packaging shown in Figure 6.18, is manufactured by Analog Devices (and is connected

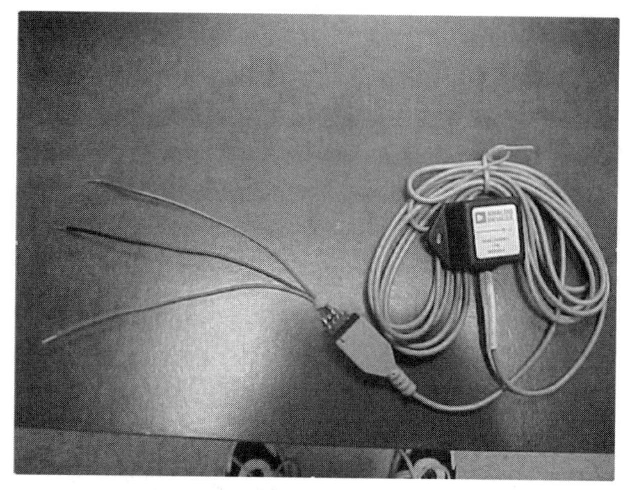

**Figure 6.18**
MEMS accelerometer (Analog Devices).

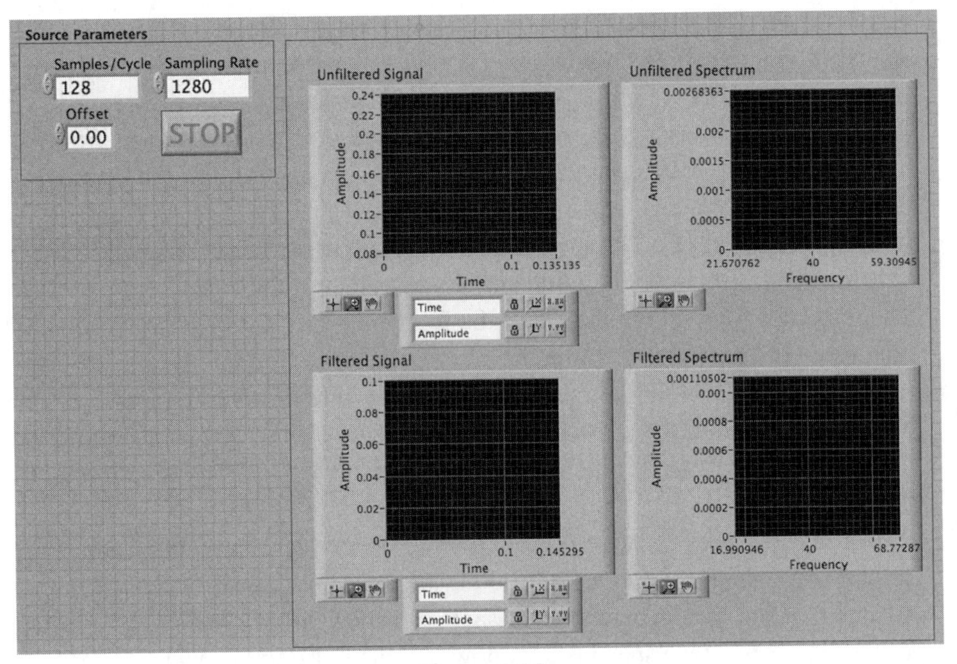


**Figure 6.19**
VI for use with accelerometer.

readily with the filter). Examine the output of the filter on an oscilloscope to ensure that the signal from the accelerometer makes sense. Assuming that the accelerometer is attached to the base of the fan (in normally operating condition, i.e., without any added mass to create rotor unbalance) you should see the fan frequency on the oscilloscope. The accelerometer may have an offset that can be removed in the next step.

6.9. Using the VI developed earlier, produce the offset necessary to eliminate the accelerometer offset.

6.10. Given the frequency of the fan at its high setting, calculate the corner frequency that would be needed to filter the rotational speed of the fan. You may use the same filter that you used previously or, if the corner frequency is very different, you may change your resistances to achieve a better design. In addition, you will need to increase the gain of your amplifier to amplify the accelerometer voltage and subtract off the DC offset.

6.11. Examine the effects of adding an imbalance to the fan. Power down the board, disconnect the balanced fan, and connect the unbalanced fan following the same procedure.

6.12. Modify the filter in Figure 6.12 so that it allows a dial gauge to represent $\alpha$, the filter parameter. This is a simple modification but allows you to better understand the filter structure.

6.13. Create a case diagram to allow multiple types of noise (uniform, etc.) to be added to the signal in Figure 6.12. Does changing the wave type seem to have any effect on the quality of the filtered output? Does the filter reproduce each wave exactly?

6.14. What effect does changing the frequency have on the spectrum of filtered and unfiltered signals? What happens to the frequency spectrum of the filtered signal at high frequencies?

6.15. What is the effect of noise amplitude? What happens when it is small? What is the result when it gets to be the same amplitude as that of the signal, or larger? How did changing the noise amplitude affect the frequency spectrum? What information does this give you?

6.16. In using unbalanced fan vibration as the source of your measurements, compare the filtered responses at low and high speeds with unfiltered responses. For which fan speed is the filtered response cleaner? How does this relate to your cutoff frequency?

6.17. For the experiments discussed earlier the fan speeds were fairly close, so designing a filter that would be acceptable for both speeds is not very difficult. What are some of the problems with trying to use the same filter design if the two speeds are drastically different? (Hint: If the speed is low, how will the noise frequencies change?)

6.18. What are some potential problems of using filters in real systems? (Hint: In the earlier exercises, we knew *exactly* what we were looking for. What if we do not?)

6.19. Why does high-frequency noise get reduced in low-pass digital filters?

6.20. Would using more sample points in the filter be beneficial in eliminating high-frequency noise? What would happen if too many points are used, that is, close to the total number of data points collected?

6.21. What are some potential problems of using filters in real systems? (Hint: Think if you need data in real time and a higher order filter is needed.)

## 6.6 Appendix

### 6.6.1 Simple Filter Solution

We start with

$$\frac{\hat{v}_o}{\hat{v}_i} = \frac{1}{\tau s + 1}. \tag{6.14}$$

Assuming sinusoidal functions of the form

$$v_i = V_i(\omega t), \text{ or } \hat{v}_i = \frac{\omega V_i}{s^2 + \omega^2}.$$

We have the following partial fraction exapansion

$$\hat{v}_o = \frac{1}{\tau s + 1}\frac{\omega V_i}{s^2 + \omega^2} = \frac{A}{s + 1/\tau} + \frac{B}{s + j\omega} + \frac{C}{s - j\omega}, \tag{6.15}$$

where

$$A = \lim_{s \to -1/\tau} (s + 1/\tau)\hat{v}_o(s) = \frac{\omega V_i}{(-1/\tau)^2 + \omega^2} = \frac{\omega \tau^2 V_i}{1 + \tau^2 \omega^2} \tag{6.16}$$

$$B = \lim_{s \to -j\omega} (s + j\omega)\hat{v}_o(s) = \frac{\omega V_i}{(-j\tau\omega + 1)(-2j\omega)} \tag{6.17}$$

$$C = \lim_{s \to j\omega} (s - j\omega)\hat{v}_o(s) = \frac{\omega V_i}{(j\tau\omega + 1)(2j\omega)} \tag{6.18}$$

Now noting that

$$v_o(t) = Ae^{-t/\tau} + Be^{-j\omega t} + Ce^{j\omega t} \tag{6.19}$$

and that the first term dies out with time, we look at the steady state value of $v_o\,(t)$ as

$$v_o(t) = \frac{\omega V_i}{(-j\tau\omega + 1)(-2j\omega)}e^{-j\omega t} + \frac{\omega V_i}{(j\tau\omega + 1)(2j\omega)}e^{j\omega t} \tag{6.20}$$

and further into

$$v_o(t) = \frac{(j\tau\omega + 1)V_i}{(\tau^2\omega^2 + 1)(-2j)}e^{-j\omega t} + \frac{(-j\tau\omega + 1)V_i}{(\tau^2\omega^2 + 1)(2j)}e^{j\omega t} \tag{6.21}$$

and further into

$$v_0(t) = \frac{V_i}{(\tau^2\omega^2 + 1)(-2j)}((j\tau\omega + 1)e^{-j\omega t} - (-j\tau\omega + 1)e^{j\omega t}) \tag{6.22}$$

and

$$v_0(t) = \frac{V_i}{(\tau^2\omega^2 + 1)(-2j)}(+j\tau\omega(e^{j\omega t} + e^{-j\omega t}) - (e^{j\omega t} - e^{-j\omega t})) \tag{6.23}$$

and

$$v_0(t) = \frac{V_i}{(\tau^2\omega^2 + 1)(-2j)}(2j\tau\omega\cos(\omega t) - 2j\sin(\omega t)) \tag{6.24}$$

and

$$v_0(t) = \frac{V_i}{\tau^2\omega^2 + 1}(-\tau\omega\cos(\omega t) + \sin(\omega t)) \tag{6.25}$$

and

$$v_0(t) = V_i(-\sin(\phi)\cos(\omega t) + \cos(\phi)\sin(\omega t)) \tag{6.26}$$

where $\tan(\phi) = \tau\omega$. We thus have

$$v_0(t) = V_i \sin(\omega t - \varphi). \tag{6.27}$$

### 6.6.2 Matlab Solution to the Butterworth Filter Design

As you observed simple first-order filters may do well for eliminating random noise, but they do not do well at attenuating signals with a certain frequency. For this reason, there are many other digital-filtering approaches. Many revolve around designing an analogue filter and approximating it as a digital filter. This is the case for the Butterworth filter. Luckily, this design process can be automated nicely using Matlab. Matlab uses a command called butter to generate the coefficients for a filter with a certain order and cutoff frequency. A sample command is given here.

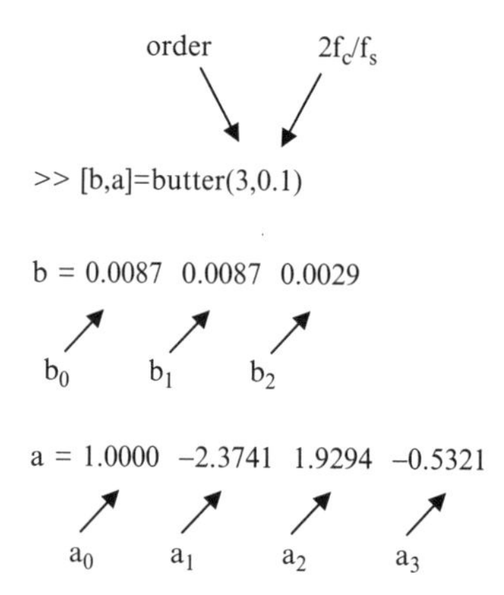


The command just given generates a third-order Butterworth filter. The first term in the command (3 in this example) is the order of the filter. By setting this term, you can control how many past data points the filter uses. The second term in the argument is the ratio of the cutoff frequency to the Nyquist frequency. The Nyquist frequency is one-half of the sampling rate, so this ratio, r, is given by

$$r = \frac{f_c}{f_s/2}.$$

These frequencies are both given in Hz, which makes $r$ a unitless quantity. It is important to note that digital filters only filter frequencies relative to the sampling frequency. If the cutoff frequency is increased and the sampling frequency is increased by the same amount, the same result is achieved. In addition, the maximum value of $r$ is 1.0 (why do you think this is?). However, this would correspond to an unfiltered response.

Recall that the equation for the filter is given by

$$y_k = \frac{1}{a_0}(-a_1 y_{k-1} - a_2 y_{k-2} - a_3 y_{k-3} + b_0 u_k + b_1 u_{k-1} + b_2 u_{k-2} + b_3 u_{k-3}).$$

The $b$ vector contains the $b_i$ coefficients. The leftmost number corresponds to $b_0$. The next number corresponds to $b_1$ and so on. The coefficients of $a$ are given by the next vector. Notice that our Butterworth filter has an $a_0$ term. This isn't present in the main part of the lab. The reason for this is that the term is always set to be 1, as seen by the values of $a$ given earlier. So, realizing that $a_0$ is always one allows us to simplify this expression into the following:

$$y_k = -a_1 y_{k-1} - a_2 y_{k-2} - a_3 y_{k-3} + b_0 u_k + b_1 u_{k-1} + b_2 u_{k-2} + b_3 u_{k-3}.$$

Here are a few additional examples of Butterworth filter designs:

```
>> [b, a] = butter (3, 0.05)
b = 0.0004 0.0012 0.0012 0.0004
a = 1.0000 −2.6862 2.4197 −0.7302
>> [b, a] = butter (3, 0.2)
b = 0.0181 0.0543 0.0543 0.0181
a = 1.0000 −1.7600 1.1829 −0.2781
```

For more information on this subject, see *Digital Signal Processing* by Proakis and Manolakis.

CHAPTER 7

# Electrical Indicating and Test Instruments



## 7.1 Introduction

The mode of operation of most measuring instruments is to convert the measured quantity into an electrical signal. Usually, this output quantity is in the form of voltage, although other forms of output, such as signal frequency or phase changes, are sometimes found.

We shall learn in this chapter that the magnitude of voltage signals can be measured by various electrical indicating and test instruments. These can be divided broadly into electrical


Measurement and Instrumentation: Theory and Application

meters (in both analogue and digital forms) and various types of oscilloscopes. As well as signal level voltages, many of these instruments can also measure higher magnitude voltages, and this is indicated where appropriate. The oscilloscope is particularly useful for interpreting instrument outputs that exist in the form of a varying phase or frequency of an electrical signal.

Electrical meters exist in both digital and analogue forms, although use of an analogue form now tends to be restricted to panel meters, where the analogue form of the output display means that abnormal conditions of monitored systems are identified more readily than is the case with the numeric form of output given by digital meters. The various forms of digital and analogue meters found commonly are presented in Sections 7.2 and 7.3.

The oscilloscope is a very versatile measuring instrument widely used for signal measurement, despite the measurement accuracy provided being inferior to that of most meters. Although existing in both analogue and digital forms, most instruments used professionally are now digital, with analogue versions being limited to inexpensive, low-specification instruments intended for use in educational establishments. Although of little use to professional users, the features of analogue instruments are covered in this chapter because students are quite likely to meet these when doing practical work associated with their course. As far as digital oscilloscopes are concerned, the basic type of instrument used is known as a digital storage oscilloscope. More recently, digital phosphor oscilloscopes have been introduced, which have a capability of detecting and recording rapid transients in voltage signals. A third type is the digital sampling oscilloscope, which is able to measure very high-frequency signals. A fourth and final type is a personal computer (PC)-based oscilloscope, which is effectively an add-on unit to a standard PC. All of these different types of oscilloscopes are discussed in Section 7.4.

## 7.2 Digital Meters

All types of digital meters are basically modified forms of the *digital voltmeter* (DVM), irrespective of the quantity that they are designed to measure. Digital meters designed to measure quantities other than voltage are, in fact, digital voltmeters that contain appropriate electrical circuits to convert current or resistance measurement signals into voltage signals. *Digital multimeters* are also essentially digital voltmeters that contain several conversion circuits, thus allowing the measurement of voltage, current, and resistance within one instrument.

Digital meters have been developed to satisfy a need for higher measurement accuracies and a faster speed of response to voltage changes than can be achieved with analogue instruments. They are technically superior to analogue meters in almost every respect. The binary nature of the output reading from a digital instrument can be applied readily to a display that is in the form of discrete numerals. Where human operators are required to measure and record signal voltage levels, this form of output makes an important contribution to measurement reliability and accuracy, as the problem of analogue meter parallax error is eliminated and the possibility of gross error through misreading the meter output is reduced greatly. The availability in many instruments of a direct output in digital form is also very useful in the rapidly expanding range of computer control applications. Quoted inaccuracy values are between ±0.005% (measuring d.c. voltages) and ±2%. Digital meters also have very high input impedance (10 MΩ compared with 1–20 KΩ for analogue meters), which avoids the measurement system loading problem (see Chapter 3) that occurs frequently when analogue meters are used. Additional advantages of digital meters are their ability to measure signals of frequency up to 1 MHz and the common inclusion of features such as automatic ranging, which prevents overload and reverse polarity connection, etc.

The major part of a digital voltmeter is the circuitry that converts the analogue voltage being measured into a digital quantity. As the instrument only measures d.c. quantities in its basic mode, another necessary component within it is one that performs a.c.–d.c. conversion and thereby gives it the capacity to measure a.c. signals. After conversion, the voltage value is displayed by means of indicating tubes or a set of solid-state light-emitting diodes. Four-, five-, or even six-figure output displays are used commonly, and although the instrument itself may not be inherently more accurate than some analogue types, this form of display enables measurements to be recorded with much greater accuracy than that obtainable by reading an analogue meter scale.

Digital voltmeters differ mainly in the technique used to affect the analogue-to-digital conversion between the measured analogue voltage and the output digital reading. As a general rule, the more expensive and complicated conversion methods achieve a faster conversion speed. Some common types of DVM are discussed here.

### 7.2.1 Voltage-to-Time Conversion Digital Voltmeter

This is the simplest form of DVM and is a ramp type of instrument. When an unknown voltage signal is applied to input terminals of the instrument, a negative slope ramp waveform is generated internally and compared with the input signal. When the two are equal, a pulse is generated that opens a gate, and at a later point in time a second pulse closes the gate when the negative ramp voltage reaches zero. The length of time between the gate opening and closing is monitored by an electronic counter, which produces a digital display according to the level of the input voltage signal. Its main drawbacks are nonlinearities in the shape of the ramp waveform used and lack of noise rejection; these problems lead to a typical inaccuracy of ±0.05%. It is relatively inexpensive, however.

### 7.2.2 Potentiometric Digital Voltmeter

This uses a servo principle, in which the error between the unknown input voltage level and a reference voltage is applied to a servo-driven potentiometer that adjusts the reference voltage until it balances the unknown voltage. The output reading is produced by a mechanical drum-type digital display driven by the potentiometer. This is also a relatively inexpensive form of DVM that gives excellent performance for its price.

### 7.2.3 Dual-Slope Integration Digital Voltmeter

This is another relatively simple form of DVM that has better noise-rejection capabilities than many other types and gives correspondingly better measurement accuracy (inaccuracy as low as ±0.005%). Unfortunately, it is quite expensive. The unknown voltage is applied to an integrator for a fixed time, $T_1$, following which a reference voltage of opposite sign is applied to the integrator, which discharges down to a zero output in an interval, $T_2$, measured by a counter. The output–time relationship for the integrator is shown in Figure 7.1, from which the unknown voltage, $V_i$, can be calculated geometrically from the triangle as

$$V_i = V_{ref}(T_1/T_2). \tag{7.1}$$

### 7.2.4 Voltage-to-Frequency Conversion Digital Voltmeter

In this instrument, the unknown voltage signal is fed via a range switch and an amplifier into a converter circuit whose output is in the form of a train of voltage pulses at a frequency proportional to the magnitude of the input signal. The main advantage of this type of DVM is its ability to reject a.c. noise.

### 7.2.5 Digital Multimeter

This is an extension of the DVM. It can measure both a.c. and d.c. voltages over a number of ranges through inclusion within it of a set of switchable amplifiers and attenuators. It is used widely in circuit test applications as an alternative to the analogue multimeter and includes protection circuits that prevent damage if high voltages are applied to the wrong range.

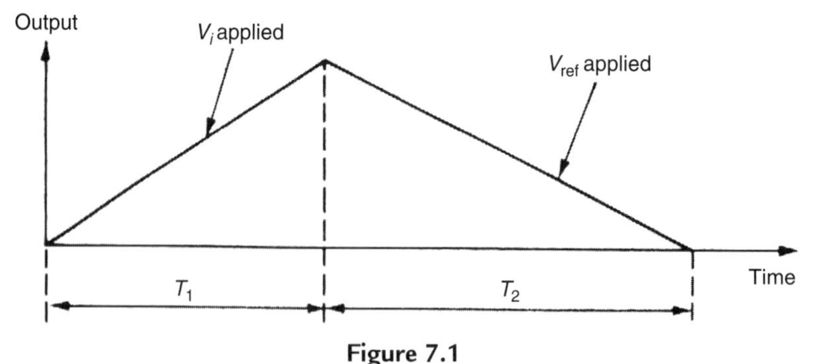


**Figure 7.1**
Output-time relationship for an integrator in a dual-slope digital voltmeter.

## 7.3 Analogue Meters

Despite the technical superiority of digital meters, particularly in terms of their greater accuracy and much higher input impedance, analogue meters continue to be used in a significant number of applications. First, they are often preferred as indicators in system control panels. This is because deviations of controlled parameters away from the normal expected range are spotted more easily by a pointer moving against a scale in an analogue meter rather than by variations in the numeric output display of a digital meter. A typical, commercially available analogue panel meter is shown in Figure 7.2. Analogue instruments also tend to suffer less from noise and isolation problems, which favor their use in some applications. In addition, because analogue instruments are usually passive instruments that do not need a power supply, this is often very useful in measurement applications where a suitable main power supply is not readily available. Many examples of analogue meters also remain in use for historical reasons.

Analogue meters are electromechanical devices that drive a pointer against a scale. They are prone to measurement errors from a number of sources that include inaccurate scale marking during manufacture, bearing friction, bent pointers, and ambient temperature variations. Further human errors are introduced through parallax error (not reading the scale from directly above) and mistakes in interpolating between scale markings. Quoted inaccuracy values are between $\pm 0.1$ and $\pm 3\%$. Various types of analogue meters are used as discussed here.

### 7.3.1 Moving Coil Meter

A moving coil meter is a very commonly used form of analogue voltmeter because of its sensitivity, accuracy, and linear scale, although it only responds to d.c. signals. As shown schematically in Figure 7.3, it consists of a rectangular coil wound round a soft iron core that is suspended in the field of a permanent magnet. The signal being measured is applied to the coil, which produces a radial magnetic field. Interaction between this induced field and the field

**Figure 7.2**
Eltime analogue panel meter *(reproduced by permission of Eltime Controls)*.

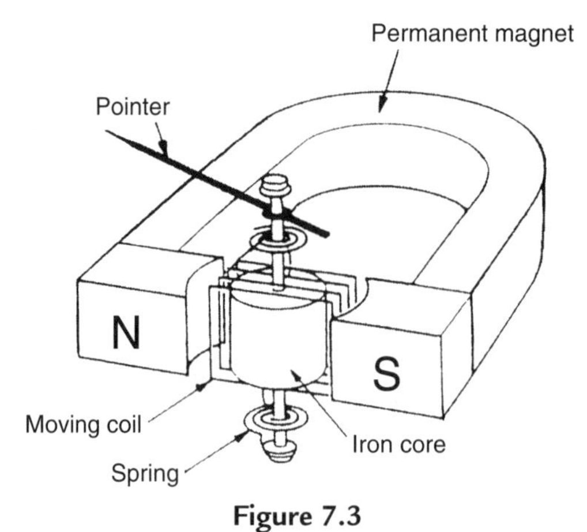


**Figure 7.3**
Mechanism of a moving coil meter.

produced by the permanent magnet causes torque, which results in rotation of the coil. The amount of rotation of the coil is measured by attaching a pointer to it that moves past a graduated scale. The theoretical torque produced is given by

$$T = B\,I\,h\,w\,N, \tag{7.2}$$

where $B$ is the flux density of the radial field, $I$ is the current flowing in the coil, $h$ is the height of the coil, $w$ is the width of the coil, and $N$ is the number of turns in the coil. If the iron core is cylindrical and the air gap between the coil and pole faces of the permanent magnet is uniform, then the flux density $B$ is constant and Equation (7.2) can be rewritten as

$$T = K\,I, \tag{7.3}$$

that is, torque is proportional to the coil current and the instrument scale is linear.

As the basic instrument operates at low current levels of one milliamp or so, it is only suitable for measuring voltages up to around 2 volts. If there is a requirement to measure higher voltages, the measuring range of the instrument can be increased by placing a resistance in series with the coil, such that only a known proportion of the applied voltage is measured by the meter. In this situation the added resistance is known as a *shunting resistor*.

While Figure 7.3 shows the traditional moving coil instrument with a long U-shaped permanent magnet, many newer instruments employ much shorter magnets made from recently developed magnetic materials such as Alnico and Alcomax. These materials produce a substantially greater flux density, which, in addition to allowing the magnet to be smaller, has additional advantages in allowing reductions to be made in the size of the coil and in increasing the usable range of deflection of the coil to about 120°. Some versions of the instrument also have either a specially shaped core or specially shaped magnet pole faces to cater for special situations where a nonlinear scale, such as a logarithmic one, is required.

### 7.3.2 Moving Iron Meter

As well as measuring d.c. signals, the moving iron meter can also measure a.c. signals at frequencies up to 125 Hz. It is the least expensive form of meter available and, consequently, this type of meter is also used commonly for measuring voltage signals. The signal to be measured is applied to a stationary coil, and the associated field produced is often amplified by the presence of an iron structure associated with the fixed coil. The moving element in the instrument consists of an iron vane suspended within the field of the fixed coil. When the fixed coil is excited, the iron vane turns in a direction that increases the flux through it.

The majority of moving-iron instruments are either of the attraction type or of the repulsion type. A few instruments belong to a third combination type. The attraction type, where the iron vane is drawn into the field of the coil as the current is increased, is shown schematically in Figure 7.4a. The alternative repulsion type is sketched in Figure 7.4b. For an excitation current, $I$, the torque produced that causes the vane to turn is given by

$$T = \frac{I^2 dM}{2d\theta},$$

where $M$ is the mutual inductance and $\theta$ is the angular deflection. Rotation is opposed by a spring that produces a backwards torque given by

$$T_s = K\theta.$$

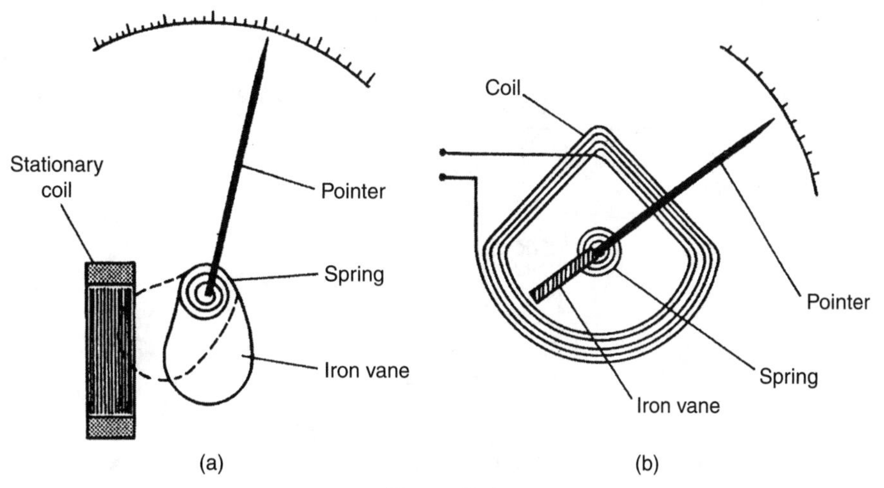


**Figure 7.4**
Mechanisms of moving iron meters: (a) attraction type and (b) repulsion type.

At equilibrium, $T = T_s$, and $\theta$ is therefore given by

$$\theta = \frac{I^2 dM}{2K d\theta}. \tag{7.4}$$

The instrument thus has a square-law response where the deflection is proportional to the square of the signal being measured, that is, the output reading is a root-mean-squared (r.m.s.) quantity.

The instrument can typically measure voltages in the range of 0 to 30 volts. However, it can be modified to measure higher voltages by placing a resistance in series with it, as in the case of moving coil meters. A series resistance is particularly beneficial in a.c. signal measurements because it compensates for the effect of coil inductance by reducing the total resistance/inductance ratio, and hence measurement accuracy is improved. A switchable series resistance is often provided within the casing of the instrument to facilitate range extension. However, when the voltage measured exceeds about 300 volts, it becomes impractical to use a series resistance within the case of the instrument because of heat-dissipation problems, and an external resistance is used instead.

### *7.3.3 Clamp-on Meters*

These are used for measuring circuit currents and voltages in a noninvasive manner that avoids having to break the circuit being measured. The meter clamps onto a current-carrying conductor, and the output reading is obtained by transformer action. The principle of operation is illustrated in Figure 7.5, where it can be seen that the clamp-on jaws of the instrument act as a transformer core and the current-carrying conductor acts as a primary winding. Current induced in the secondary winding is rectified and applied to a moving coil meter. Although it is

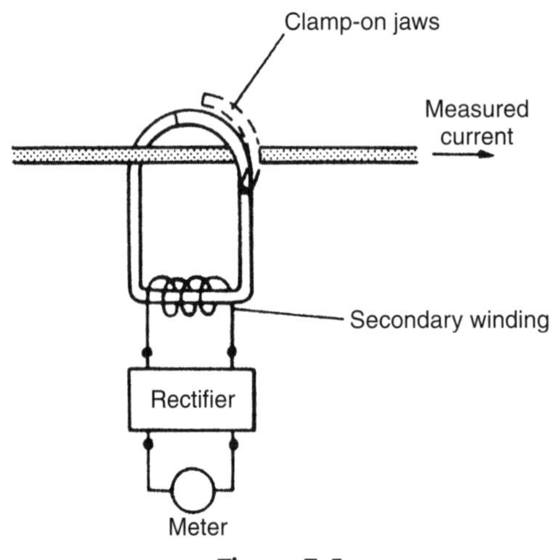


**Figure 7.5**
Schematic drawing of a clamp-on meter.

a very convenient instrument to use, the clamp-on meter has low sensitivity and the minimum current measurable is usually about 1 amp.

### *7.3.4 Analogue Multimeter*

The analogue multimeter is now less common than its counterpart, the digital multimeter, but is still widely available. It is a multifunction instrument that can measure current and resistance, as well as d.c. and a.c. voltage signals. Basically, the instrument consists of a moving coil analogue meter with a switchable bridge rectifier to allow it to measure a.c. signals, as shown in Figure 7.6. A set of rotary switches allows the selection of various series and shunt resistors, which make the instrument capable of measuring both voltage and current over a number of ranges. An internal power source is also provided to allow it to measure resistances as well. While this instrument is very useful for giving an indication of voltage levels, the compromises in its design that enable it to measure so many different quantities necessarily mean that its accuracy is not as good as instruments that are purposely designed to measure just one quantity over a single measuring range.

### *7.3.5 Measuring High-Frequency Signals with Analogue Meters*

One major limitation in using analogue meters for a.c. voltage measurement is that the maximum frequency measurable directly is low—2 kHz for the dynamometer voltmeter and only 100 Hz in the case of the moving iron instrument. A partial solution to this limitation is to rectify the voltage signal and then apply it to a moving coil meter, as shown in Figure 7.7. This extends the upper measurable frequency limit to 20 kHz. However, inclusion of the bridge rectifier makes the measurement system particularly sensitive to environmental temperature

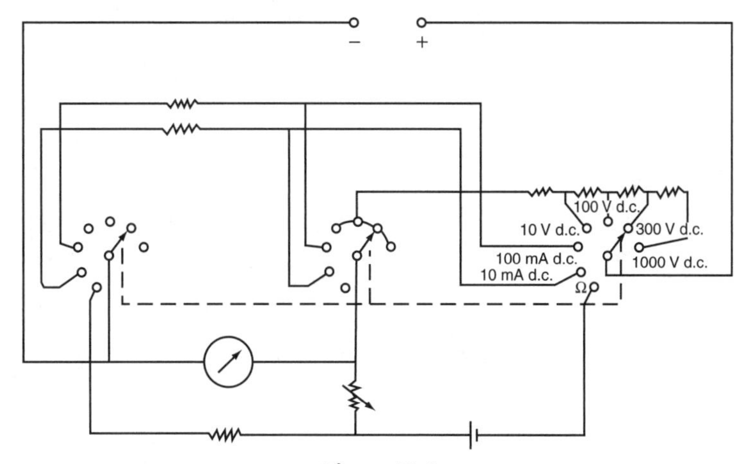


**Figure 7.6**
Circuitry of an analogue multimeter.

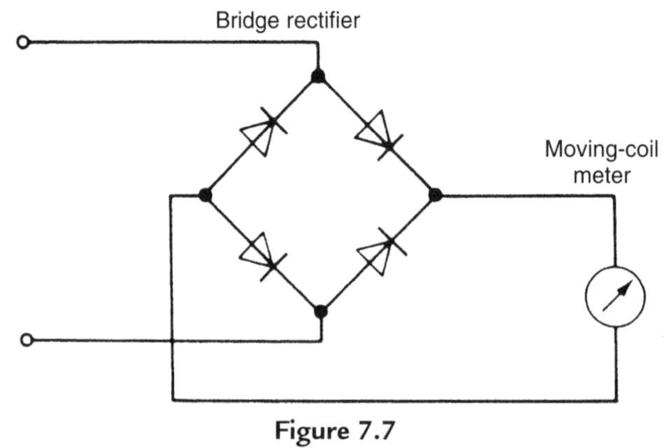


**Figure 7.7**
Measurement of high-frequency voltage signals.

changes, and nonlinearities significantly affect measurement accuracy for voltages that are small relative to the full-scale value.

### 7.3.6 Calculation of Meter Outputs for Nonstandard Waveforms

The two examples given here provide an exercise in calculating the output reading from various types of analogue voltmeters. These examples also serve as a useful reminder of the mode of operation of each type of meter and the form that the output takes.

## Example 7.1

Calculate the reading that would be observed on a moving coil ammeter when it is measuring the current in the circuit shown in Figure 7.8.

## Solution

A moving coil meter measures mean current.

$$I_{mean} = \frac{1}{2\pi}\left(\int_0^{\pi} \frac{5\omega t}{\pi} d\omega t + \int_{\pi}^{2\pi} 5\sin(\omega t)d\omega t\right) = \frac{1}{2\pi}\left(\left[\frac{5(\omega t)^2}{2\pi}\right]_0^{\pi} + 5[-\cos(\omega t)]_{\pi}^{2\pi}\right)$$

$$= \frac{1}{2\pi}\left(\frac{5\pi^2}{2\pi} - 0 - 5 - 5\right) = \frac{1}{2\pi}\left(\frac{5\pi}{2} - 10\right) = \frac{5}{2\pi}\left(\frac{\pi}{2} - 2\right) = -0.342 \text{ amps}$$

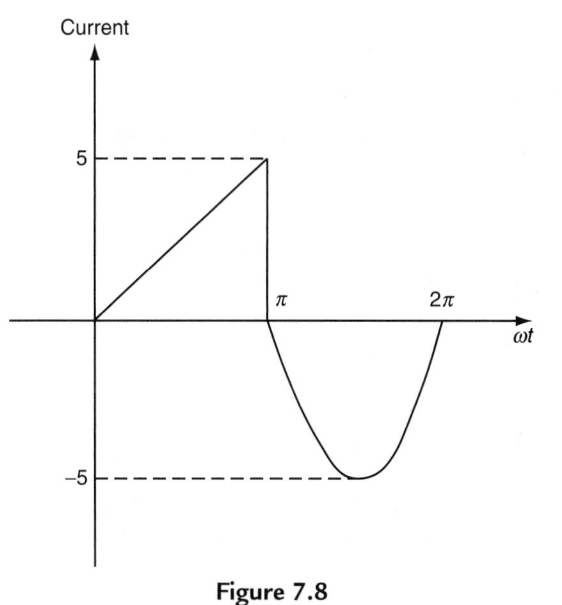


**Figure 7.8**
Circuit for Examples 7.1 and 7.2.

## Example 7.2

Calculate the reading that would be observed on a moving iron ammeter when it is measuring the current in the circuit shown in Figure 7.8.

## Solution

A moving iron meter measures r.m.s. current.

$$I^2_{rms} = \frac{1}{2\pi}\left(\int_0^{\pi}\frac{25(\omega t)^2}{\pi^2}d\omega t + \int_{\pi}^{2\pi} 25\sin^2(\omega t)d\omega t\right) = \frac{1}{2\pi}\left(\int_0^{\pi}\frac{25(\omega t)^2}{\pi^2}d\omega t + \int_{\pi}^{2\pi}\frac{25(1-\cos 2\omega t)}{2}d\omega t\right)$$

$$= \frac{25}{2\pi}\left(\left[\frac{(\omega t)^3}{3\pi^2}\right]_0^{\pi} + \left[\frac{\omega t}{2} - \frac{\sin 2\omega t}{4}\right]_{\pi}^{2\pi}\right) = \frac{25}{2\pi}\left(\frac{\pi}{3} + \frac{2\pi}{2} - \frac{\pi}{2}\right)$$

$$= \frac{25}{2\pi}\left(\frac{\pi}{3} + \frac{\pi}{2}\right) = \frac{25}{2}\left(\frac{1}{3} + \frac{1}{2}\right) = 10.416$$

Thus, $I_{rms} = \sqrt{(I^2_{rms})} = 3.23$ amp

## 7.4 Oscilloscopes

The oscilloscope is probably the most versatile and useful instrument available for signal measurement. While oscilloscopes still exist in both analogue and digital forms, analogue models tend to be low specification, low-cost instruments produced for educational use in schools, colleges, and universities. Almost all oscilloscopes used for professional work now tend to be digital models. These can be divided into digital storage oscilloscopes, digital phosphor oscilloscopes, and digital sampling oscilloscopes.

The basic function of an oscilloscope is to draw a graph of an electrical signal. In the most common arrangement, the $y$ axis (vertical) of the display represents the voltage of a measured signal and the $x$ axis (horizontal) represents time. Thus, the basic output display is a graph of the variation of the magnitude of the measured voltage with time.

The oscilloscope is able to measure a very wide range of both a.c. and d.c. voltage signals and is used particularly as an item of test equipment for circuit fault finding. In addition to measuring voltage levels, it can also measure other quantities, such as the frequency and phase of a signal. It can also indicate the nature and magnitude of noise that may be corrupting the measurement signal. The most expensive models can measure signals at frequencies up to 25 GHz, while the least expensive models can only measure signals up to 10 MHz. One particularly strong merit of the oscilloscope is its high input impedance, typically 1 MΩ, which means that the instrument has a negligible loading effect in most measurement situations. As a test instrument, it is often required to measure voltages whose frequency and magnitude are totally unknown. The set of rotary switches that alter its time base so easily, and the circuitry that protects it from damage when high voltages are applied to it on the wrong range, make it ideally suited for such applications. However, it is not a particularly accurate instrument and is best used where only an approximate measurement is required. In the best instruments, inaccuracy can be limited to $\pm 1\%$ of the reading, but inaccuracy can approach $\pm 5\%$ in the least expensive instruments.

The most important aspects in the specification of an oscilloscope are its bandwidth, rise time, and accuracy. Bandwidth is defined as the range of frequencies over which the oscilloscope amplifier gain is within 3 dB* of its peak value, as illustrated in Figure 7.9. The −3-dB point is where the gain is 0.707 times its maximum value. In most oscilloscopes, the amplifier is direct coupled, which means that it amplifies d.c. voltages by the same factor as low-frequency a.c. ones. For such instruments, the minimum frequency measurable is zero and the bandwidth can be interpreted as the maximum frequency where the sensitivity (deflection/volt) is within 3 dB of the peak value. In all measurement situations, the oscilloscope chosen for use must be such

* The decibel, commonly written dB, is used to express the ratio between two quantities. For two voltage levels, $V_1$ and $V_2$, the difference between the two levels is expressed in decibels as $20\log_{10}(V_1/V_2)$. It follows from this that $20\log_{10}(0.7071) = -3\,dB$.

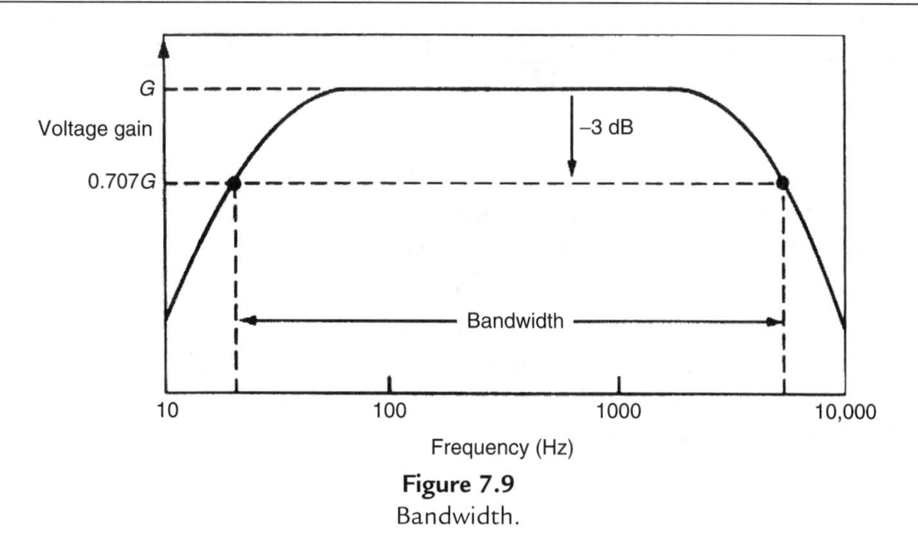


**Figure 7.9**
Bandwidth.

that the maximum frequency to be measured is well within the bandwidth. The −3-dB specification means that an oscilloscope with a specified inaccuracy of ±2% and a bandwidth of 100 MHz will have an inaccuracy of ±5% when measuring 30-MHz signals; this inaccuracy will increase still further at higher frequencies. Thus, when applied to signal-amplitude measurement, the oscilloscope is only usable at frequencies up to about 0.3 times its specified bandwidth.

Rise time is the transit time between 10 and 90% levels of the response when a step input is applied to the oscilloscope. Oscilloscopes are normally designed such that

$$\text{bandwidth} \times \text{rise time} = 0.35.$$

Thus, for a bandwidth of 100 MHz, rise time = 0.35/100,000,000 = 3.5 ns.

All oscilloscopes are relatively complicated instruments constructed from a number of subsystems, and it is necessary to consider each of these in turn in order to understand how the complete instrument functions. To achieve this, it is useful to start with an explanation of an analogue oscilloscope, as this was the original form in which oscilloscopes were made and many of the terms used to describe the function of oscilloscopes emanate from analogue forms.

### *7.4.1 Analogue Oscilloscope (Cathode Ray Oscilloscope)*

Analogue oscilloscopes were originally called cathode ray oscilloscopes because a fundamental component within them is a cathode ray tube. In recent times, digital oscilloscopes have almost entirely replaced analogue versions in professional use. However, some very inexpensive versions of analogue oscilloscopes still exist that find educational uses in schools, colleges, and universities. The low cost of basic analogue models is their only merit, as their inclusion of a cathode ray tube makes them very fragile, and the technical performance of digital equivalents is greatly superior.

The *cathode ray tube* within an analogue oscilloscope is shown schematically in Figure 7.10. The cathode consists of a barium and strontium oxide-coated, thin, heated filament from which a stream of electrons is emitted. The stream of electrons is focused onto a well-defined spot on a fluorescent screen by an electrostatic focusing system that consists of a series of metal discs and cylinders charged at various potentials. Adjustment of this focusing mechanism is provided by a *focus* control on the front panel of an oscilloscope. An *intensity* control varies the cathode heater current and therefore the rate of emission of electrons, and thus adjusts the intensity of the display on the screen. These and other typical controls are shown in the illustration of the front panel of a simple oscilloscope given in Figure 7.11. It should be noted that the layout shown is only one example. Every model of oscilloscope has a different layout of control knobs, but the functions provided remain similar irrespective of the layout of the controls with respect to each other.

Application of potentials to two sets of deflector plates mounted at right angles to one another within the tube provide for deflection of the stream of electrons, such that the spot where the electrons are focused on the screen is moved. The two sets of deflector plates are normally known as horizontal and vertical deflection plates, according to the respective motion caused to the spot on the screen. The magnitude of any signal applied to the deflector plates can be calculated by measuring the deflection of the spot against a cross-wires graticule etched on the screen.

### Channel

One channel describes the basic subsystem of an electron source, focusing system, and deflector plates. This subsystem is often duplicated one or more times within the cathode ray tube to provide a capability of displaying two or more signals at the same time on the screen.

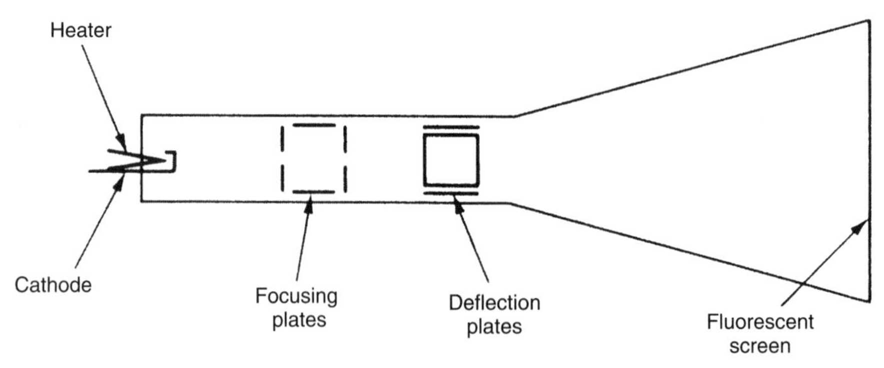


**Figure 7.10**
Cathode ray tube.

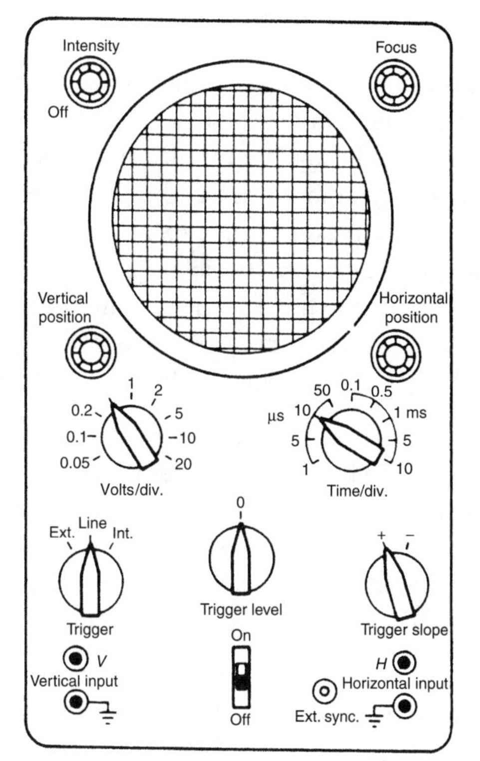


**Figure 7.11**
Controls of a simple oscilloscope.

The common oscilloscope configuration with two channels can therefore display two separate signals simultaneously.

*Single-ended input*

This type of input only has one input terminal plus a ground terminal per oscilloscope channel and, consequently, only allows signal voltages to be measured relative to ground. It is normally only used in simple oscilloscopes.

*Differential input*

This type of input is provided on more expensive oscilloscopes. Two input terminals plus a ground terminal are provided for each channel, which allows the potentials at two nongrounded points in a circuit to be compared. This type of input can also be used in single-ended mode to measure a signal relative to ground by using just one of the input terminals plus ground.

### *Time base circuit*

The purpose of a time base is to apply a voltage to the horizontal deflector plates such that the horizontal position of the spot is proportional to time. This voltage, in the form of a ramp known as a sweep waveform, must be applied repetitively, such that the motion of the spot across the screen appears as a straight line when a d.c. level is applied to the input channel. Furthermore, this time base voltage must be synchronized with the input signal in the general case of a time-varying signal, such that a steady picture is obtained on the oscilloscope screen. The length of time taken for the spot to traverse the screen is controlled by a *time/div* switch, which sets the length of time taken by the spot to travel between two marked divisions on the screen, thereby allowing signals at a wide range of frequencies to be measured.

Each cycle of the sweep waveform is initiated by a pulse from a pulse generator. The input to the pulse generator is a sinusoidal signal known as a triggering signal, with a pulse being generated every time the triggering signal crosses a preselected slope and voltage level condition. This condition is defined by *trigger level* and *trigger slope* switches. The former selects the voltage level on the trigger signal, commonly zero, at which a pulse is generated, while the latter selects whether pulsing occurs on a positive or negative going part of the triggering waveform.

Synchronization of the sweep waveform with the measured signal is achieved most easily by deriving the trigger signal from the measured signal, a procedure known as *internal triggering*. Alternatively, *external triggering* can be applied if the frequencies of the triggering signal and measured signals are related by an integer constant such that the display is stationary. External triggering is necessary when the amplitude of the measured signal is too small to drive the pulse generator; it is also used in applications where there is a requirement to measure the phase difference between two sinusoidal signals of the same frequency. It is very convenient to use 50-Hz line voltage for external triggering when measuring signals at mains frequency; this is often given the name *line triggering*.

### *Vertical sensitivity control*

This consists of a series of attenuators and preamplifiers at the input to the oscilloscope. These condition the measured signal to the optimum magnitude for input to the main amplifier and vertical deflection plates, thus enabling the instrument to measure a very wide range of different signal magnitudes. Selection of the appropriate input amplifier/attenuator is made by setting a *volts/div* control associated with each oscilloscope channel. This defines the magnitude of the input signal that will cause a deflection of one division on the screen.

### *Display position control*

This allows the position at which a signal is displayed on the screen to be controlled in two ways. The horizontal position is adjusted by a *horizontal position* knob on the oscilloscope front panel, and similarly a *vertical position* knob controls the vertical position. These controls adjust the position of the display by biasing the measured signal with d.c. voltage levels.

### *7.4.2 Digital Storage Oscilloscopes*

Digital storage oscilloscopes are the most basic form of digital oscilloscopes but even these usually have the ability to perform extensive waveform processing and provide permanent storage of measured signals. When first created, a digital storage oscilloscope consisted of a conventional analogue cathode ray oscilloscope with the added facility that the measured analogue signal could be converted to digital format and stored in computer memory within the instrument. These stored data could then be reconverted to analogue form at the frequency necessary to refresh the analogue display on the screen, producing a nonfading display of the signal on the screen.

While examples of such early digital oscilloscopes might still be found in some workplaces, modern digital storage oscilloscopes no longer use cathode ray tubes and are entirely digital in construction and operation. The front panel of any digital oscilloscope has a similar basic layout to that shown for an analogue oscilloscope in Figure 7.11, except that the controls for "focusing" and "intensity" are not needed in a digital instrument. The block diagram in Figure 7.12 shows typical components used in the digital storage oscilloscope. A typical commercial instrument was also shown earlier in Figure 5.1. The first component (as in an analogue oscilloscope) is an amplifier/attenuator unit that allows adjustment of the magnitude of the input voltage signal to an appropriate level. This is followed by an analogue-to-digital converter that samples the input signal at discrete points in time. The sampled signal values are stored in the acquisition memory component before passing into a microprocessor. This carries out signal processing functions, manages the front panel control settings, and prepares the output display. Following this, the output signal is stored in a display memory module before being output to the display itself. This consists of either a monochrome or a multicolor liquid crystal display (see Chapter 8). The signal displayed is actually a sequence of individual dots rather than a continuous line as displayed by an analogue oscilloscope. However, as the density of dots increases, the display becomes closer and closer to a continuous line. The density of the

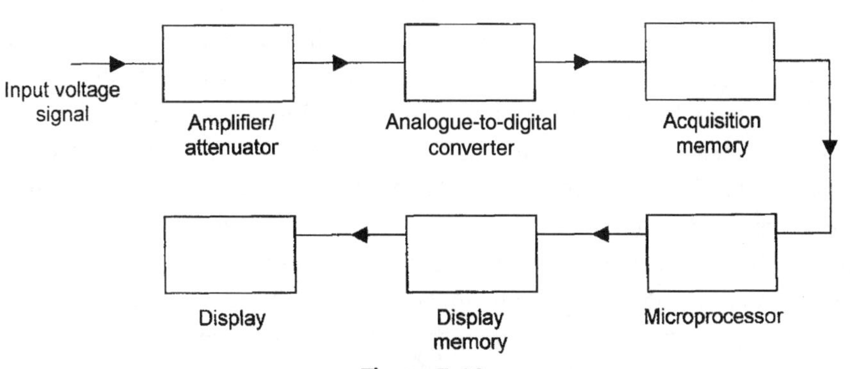


**Figure 7.12**
Components of a digital storage oscilloscope.

dots is entirely dependent on the sampling rate at which the analogue signal is digitized and the rate at which the memory contents are read to reconstruct the original signal. As the speed of sampling and signal processing is a function of instrument cost, more expensive instruments give better performance in terms of dot density and the accuracy with which the analogue signal is recorded and represented. Nevertheless, the cost of computing power is now sufficiently low to mean that all but the least expensive instruments now have a display that looks very much like a continuous trace.

In addition to their ability to display the magnitude of voltage signals and other parameters, such as signal phase and frequency, most digital oscilloscopes can also carry out analysis of the measured waveform and compute signal parameters such as maximum and minimum signal levels, peak-peak values, mean values, r.m.s. values, rise time, and fall time. These additional functions are controlled by extra knobs and push buttons on the front panel. They are also ideally suited to capturing transient signals when set to single-sweep mode. This avoids the problem of the very careful synchronization that is necessary to capture such signals on an analogue oscilloscope. In addition, digital oscilloscopes often have facilities to output analogue signals to devices such as chart recorders and output digital signals in a form compatible with standard interfaces such as IEEE488 and RS232.

The principal limitation of a digital storage oscilloscope is that the only signal information captured is the status of the signal at each sampling instant. Thereafter, no new signal information is captured during the time that the previous sample is being processed. This means that any signal changes occurring between sampling instants, such as fast transients, are not detected. This problem is overcome in the digital phosphor oscilloscope.

### *7.4.3 Digital Phosphor Oscilloscope*

This newer type of oscilloscope, first introduced in 1998, uses a parallel-processing architecture instead of the serial-processing architecture found in digital storage oscilloscopes. The components of the instrument are shown schematically in Figure 7.13. The amplifier/attenuator and analogue-to-digital converter are the same as in a digital storage oscilloscope. However, the signal processing mechanism is substantially different. Output from the analogue-to-digital converter passes into a digital phosphor memory unit, which is, in fact, entirely electronic and not composed of chemical phosphor as its name might imply. Thereafter, data follow two parallel paths. First, a microprocessor processes data acquired at each sampling instant according to the settings on the control panel and sends the processed signal to the instrument display unit. In addition to this, a snapshot of the input signal is sent directly to the display unit at a rate of 30 images per second. This enhanced processing capability enables the instrument to have a higher waveform capture rate and to detect very fast signal transients missed by digital storage oscilloscopes.

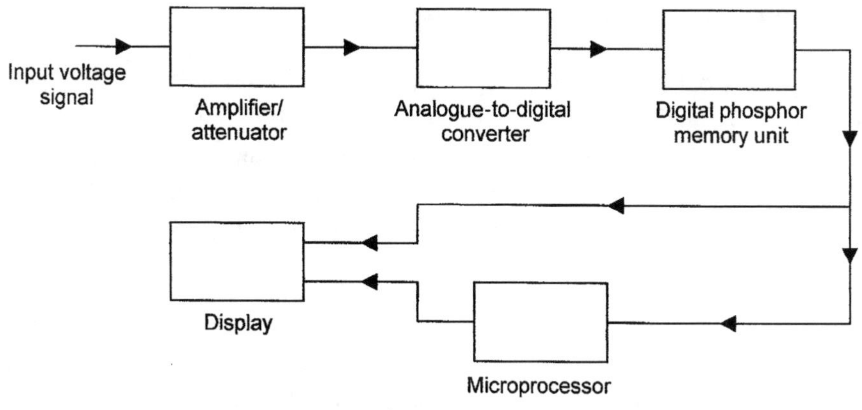


**Figure 7.13**
Components of a digital phosphor oscilloscope.

### *7.4.4 Digital Sampling Oscilloscope*

The digital sampling oscilloscope has a bandwidth of up to 25 GHz, which is about 10 times better than that achieved by other types of oscilloscopes. This increased bandwidth is achieved by reversing the positions of the analogue-to-digital converter and the amplifier, as shown in the block diagram in Figure 7.14. This reversal means that the sampled signal applied to the amplifier has a much lower frequency than the original signal, allowing use of a low bandwidth amplifier. However, the fact that the input signal is applied directly to the analogue-to-digital converter without any scaling means that the instrument can only be used to measure signals whose peak magnitude is within a relatively small range of typically 1 volt peak-peak. In contrast, both digital storage and digital phosphor oscilloscopes can typically deal with inputs up to 500 volts.

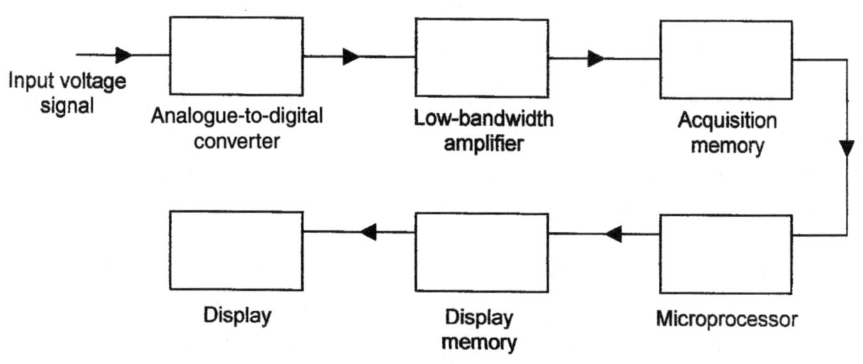


**Figure 7.14**
Components of a digital sampling oscilloscope.

### 7.4.5 Personal Computer-Based Oscilloscope

A PC-based oscilloscope consists of a hardware unit that connects to a standard PC via either a USB or a parallel port. The hardware unit provides signal scaling, analogue-to-digital conversion, and buffer memory functions found in a conventional oscilloscope. More expensive PC-based oscilloscopes also provide some high-speed digital signal processing functions within the hardware unit. The host PC itself provides the control interface and display facilities.

The primary advantage of a PC-based oscilloscope over other types is one of cost; the cost saving is achieved because use of the PC obviates the need for a display unit and front control panel found in other forms of oscilloscopes. The larger size of a PC display compared with a conventional oscilloscope often makes the output display easier to read. A further advantage is one of portability, as a laptop plus add-on hardware unit is usually smaller and lighter than a conventional oscilloscope. PC-based oscilloscopes also facilitate the transfer of output data into standard PC software such as spreadsheets and word processors.

Although PC-based oscilloscopes have a number of advantages over converntional oscilloscopes, they also have disadvantages. First, electromagnetic noise originating in PC circuits requires the hardware unit to be well shielded in order to avoid corruption of the measured signal. Second, signal sampling rates can be limited by the mode of connection of the hardware unit into the PC.

## 7.5 Summary

This chapter looked at the various ways of measuring electrical signals that form the output of most types of measuring instruments. We noted that these signals were usually in the form of varying voltages, although a few instruments have an output where either the phase or the frequency of an electrical signal changes. We observed that varying voltages could be measured either by electrical meters or by one of several forms of oscilloscopes. We also learned that the latter are also able to interpret frequency and phase changes in signals.

Our discussion started with electrical meters, which we found now mainly existed in digital form, but we noted that analogue forms also exist, which are mainly used as meters in control panels. We looked first of all at the various forms of digital meters and followed this with a presentation on the types of analogue meters still in use.

Our discussion on oscilloscopes also revealed that both analogue and digital forms exist, but we observed that analogue instruments are now predominantly limited to less expensive versions used in education markets. However, because the students at which this book is aimed are quite likely to meet analogue oscilloscopes for practical work during their course, we started off by looking at the features of such instruments. We then went on to look at the four alternative forms of digital oscilloscope that form the basis for almost all oscilloscopes used professionally. We learned that the basic form is known as a digital storage oscilloscope and that even this is superior in most respects to an analogue oscilloscope. Where better performance is needed, particularly if the observed signal has fast transients, we saw that a new type known as a digital phosphor oscilloscope is used. A third kind, known as a digital sampling oscilloscope, is designed especially for measuring very high-frequency signals. However, we noted that this could also measure voltage signals that were up to 1 volt peak-to-peak in magnitude. Finally, we looked at the merits of PC-based oscilloscopes. In addition to offering oscilloscope facilities at a lower cost than other forms of oscilloscopes, we learned that these had several other advantages but also some disadvantages.

## 7.6 Problems

7.1. Summarize the advantages of digital meters over their analogue counterparts.

7.2. Explain the four main alternative mechanisms used for affecting analogue-to-digital conversion in a digital voltmeter.

7.3. What sort of applications are analogue meters still commonly found in?

7.4. Explain the mode of operation of a moving coil meter.

7.5. Explain the mode of operation of a moving iron meter.

7.6. How does an oscilloscope work?

7.7. What are the main differences between analogue and digital oscilloscopes?

7.8. Explain the following terms: (a) bandwidth and (b) rise time. In designing oscilloscopes, what relationship is sort between bandwidth and rise time?

7.9. Explain the following terms in relation to an oscilloscope: (a) channel, (b) single-ended input, (c) differential input, (d) time base, (e) vertical sensitivity, and (f) display position control.

7.10. Sketch a block diagram showing the main components in a digital storage oscilloscope and explain the mode of operation of the instrument.

7.11. Draw a block diagram showing the main components in a digital phosphor oscilloscope. What advantages does a digital phosphor oscilloscope have over a digital storage one?

7.12. Illustrate the main components in a digital sampling oscilloscope by sketching a block diagram of them. What performance advantages does a digital sampling oscilloscope have over a digital storage one?

7.13. What is a PC-based oscilloscope? Discuss its advantages and disadvantages compared with a digital oscilloscope.

7.14. What are the main differences among a digital storage oscilloscope, a digital phosphor oscilloscope, and a digital sampling oscilloscope? How do these differences affect their performance and typical usage?

CHAPTER 8

# Display, Recording, and Presentation of Measurement Data



## 8.1 Introduction

Earlier chapters in this book have been essentially concerned with describing ways of producing high-quality, error-free data at the output of a measurement system. Having gotten that data, the next consideration is how to present it in a form where it can be readily used and analyzed. This chapter therefore starts by covering the techniques available to either display measurement data for current use or record it for future use. Following this, standards of good practice for presenting data in either graphical or tabular form are covered, using either paper or a computer monitor screen as the display medium. This leads to a discussion of mathematical regression techniques for fitting best lines through data points on a graph. Confidence tests to assess the correctness of the line fitted are also described. Finally, correlation tests are described that determine the degree of association between two sets of data when both are subject to random fluctuations.

## 8.2 Display of Measurement Signals

Measurement signals in the form of a varying electrical voltage can be displayed either by an *oscilloscope* or by any of the *electrical meters* described earlier in Chapter 7. However, if signals are converted to digital form, other display options apart from meters become possible, such as electronic output displays or use of a computer monitor.

### 8.2.1 Electronic Output Displays

Electronic displays enable a parameter value to be read immediately, thus allowing for any necessary response to be made immediately. The main requirement for displays is that they should be clear and unambiguous. Two common types of character formats used in displays, seven-segment and 7 × 5 dot matrix, are shown in Figure 8.1. Both types of displays have the advantage of being able to display alphabetic as well as numeric information, although the seven-segment format can only display a limited 9-letter subset of the full 26-letter alphabet. This allows added meaning to be given to the number displayed by including a word or letter code. It also allows a single display unit to send information about several parameter values, cycling through each in turn and including alphabetic information to indicate the nature of the variable currently displayed.

Electronic output units usually consist of a number of side-by-side cells, where each cell displays one character. Generally, these accept either serial or parallel digital input signals, and

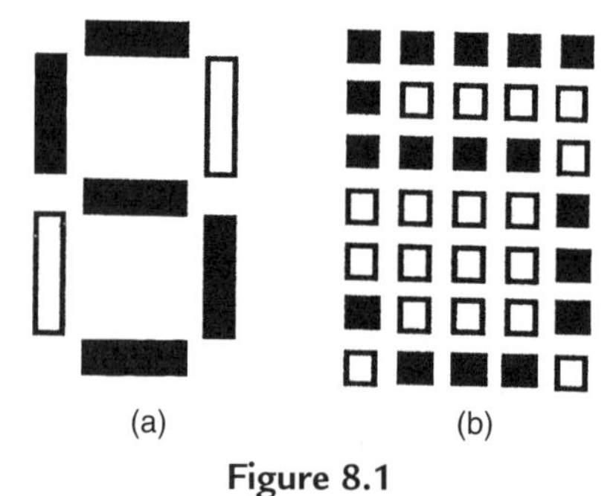


**Figure 8.1**
Character formats used in electronic displays: (a) seven segment and (b) 7 × 5 dot matrix.

the input format can be either binary-coded decimal or ACSII. Technologies used for the individual elements in the display are either light-emitting diodes or liquid-crystal elements.

### 8.2.2 Computer Monitor Displays

Now that computers are part of the furniture in most homes, the ability of computers to display information is widely understood and appreciated. Computers are now both inexpensive and highly reliable and provide an excellent mechanism for both displaying and storing information. As well as alphanumeric displays of industrial plant variable and status data, for which the plant operator can vary the size of font used to display the information at will, it is also relatively easy to display other information, such as plant layout diagrams and process flow layouts. This allows not only the value of parameters that go outside control limits to be displayed, but also their location on a schematic map of the plant. Graphical displays of the behavior of a measured variable are also possible. However, this poses difficulty when there is a requirement to display the variable's behavior over a long period of time, as the length of the time axis is constrained by the size of the monitor's screen. To overcome this, the display resolution has to decrease as the time period of the display increases.

*Touch screens* have the ability to display the same sort of information as a conventional computer monitor, but also provide a command-input facility in which the operator simply has to touch the screen at points where images of keys or boxes are displayed. A full "qwerty" keyboard is often provided as part of the display. The sensing elements behind the screen are protected by glass and continue to function even if the glass gets scratched. Touch screens are usually totally sealed, thus providing intrinsically safe operation in hazardous environments.

## 8.3 Recording of Measurement Data

As well as displaying the current values of measured parameters, there is often a need to make continuous recordings of measurements for later analysis. Such records are particularly useful when faults develop in systems, as analysis of the changes in measured parameters in the time before the fault is discovered can often quickly indicate the reason for the fault. Options for recording data include chart recorders, digital oscilloscopes, digital data recorders, and hard-copy devices such as inkjet and laser printers. The various types of recorders used are discussed here.

### 8.3.1 Chart Recorders

Chart recorders have particular advantages in providing a non-corruptible record that has the merit of instant "viewability." This means that all but paperless forms of chart recorders satisfy regulations set for many industries that require variables to be monitored and recorded continuously with hard-copy output. ISO 9000 quality assurance procedures and ISO 14000 environmental protection systems set similar requirements, and special regulations in the defense industry go even further by requiring hard-copy output to be kept for 10 years. Hence, while many people have been predicting the demise of chart recorders, the reality of the situation is that they are likely to be needed in many industries for many years to come.

Originally, all chart recorders were electromechanical in operation and worked on the same principle as a galvanometric moving coil meter (see analogue meters in Chapter 7) except that the moving coil to which the measured signal was applied carried a pen, as shown in Figure 8.2, rather than carrying a pointer moving against a scale as it would do in a meter. The pen drew an ink trace on a strip of ruled chart paper that was moved past the pen at constant speed by an electrical motor. The resultant trace on chart paper showed variations with time in the magnitude of the measured signal. Even early recorders commonly had two or more pens of different colors so that several measured parameters could be recorded simultaneously.

The first improvement to this basic recording arrangement was to replace the galvanometric mechanism with a servo system, as shown in Figure 8.3, in which the pen is driven by a servomotor, and a sensor on the pen feeds back a signal proportional to pen position. In this form, the instrument is known as a *potentiometric recorder*. The servo system reduces the typical inaccuracy of the recorded signal to $\pm 0.1\%$, compared to $\pm 2\%$ in a galvanometeric mechanism recorder. Typically, the measurement resolution is around 0.2% of the full-scale reading. Originally, the servo motor was a standard d.c. motor, but brushless servo motors are now invariably used to avoid the commutator problems that occur with d.c. motors. The position signal is measured by a potentiometer in less expensive models, but more expensive models achieve better performance and reliability using a noncontacting ultrasonic sensor to provide feedback on pen position. The difference between the pen position and

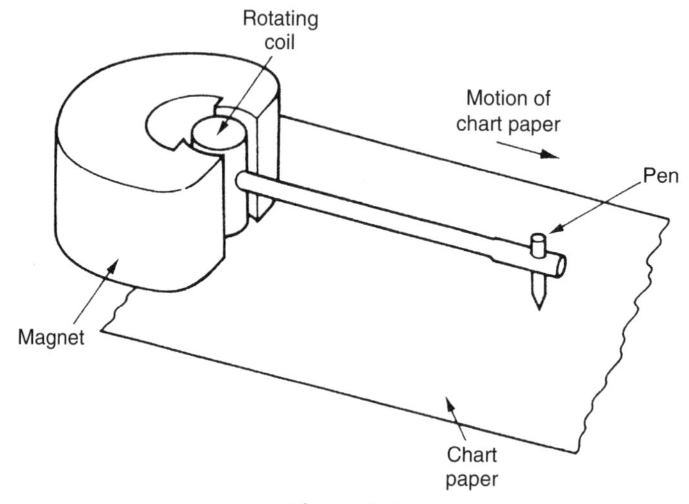


**Figure 8.2**
Original form of galvanometric chart recorder.

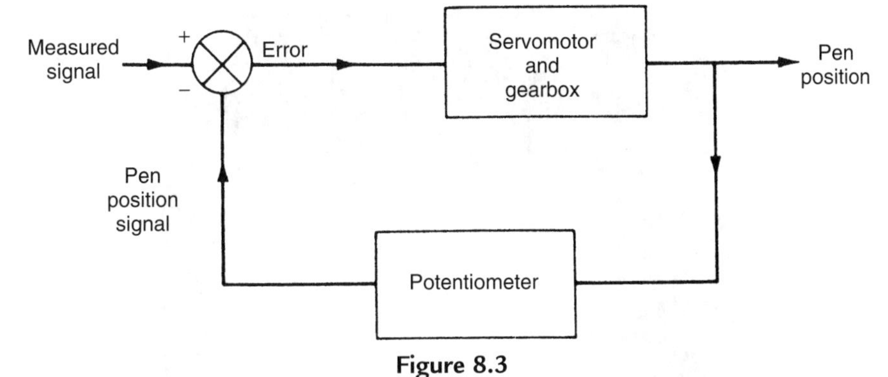


**Figure 8.3**
Servo system of a potentiometric chart recorder.

the measured signal is applied as an error signal that drives the motor. One consequence of this electromechanical balancing mechanism is that the instrument has a slow response time, in the range of 0.2–2.0 seconds, which means that electromechanical potentiometric recorders are only suitable for measuring d.c. and slowly time-varying signals.

All current potentiometric chart recorders contain a microprocessor controller, where the functions vary according to the particular chart recorder. Common functions are selection of range and chart speed, along with specification of alarm modes and levels to detect when measured variables go outside acceptable limits. Basic recorders can record up to three different signals using three different colored pens. However, multipoint recorders can have 24 or more inputs and plot six or more different colored traces simultaneously. As an alternative to pens, which can run out of ink at inconvenient times, recorders using a heated stylus recording signals on heat-sensitive paper are available. Another variation is the circular chart recorder, in which the chart paper is circular in shape and is rotated rather than moving translationally. Finally, paperless forms of recorder exist where the output display is generated entirely electronically. These various forms are discussed in more detail later.

### *Pen strip chart recorder*

A pen strip chart recorder refers to the basic form of the electromechanical potentiometric chart recorder mentioned earlier. It is also called a *hybrid chart recorder* by some manufacturers. The word "hybrid" was used originally to differentiate chart recorders that had a microprocessor controller from those that did not. However, because all chart recorders now contain a microprocessor, the term hybrid has become superfluous.

Strip chart recorders typically have up to three inputs and up to three pens in different colors, allowing up to three different signals to be recorded. A typical commercially available model is shown in Figure 8.4. Chart paper comes in either roll or fan-fold form. The drive mechanism

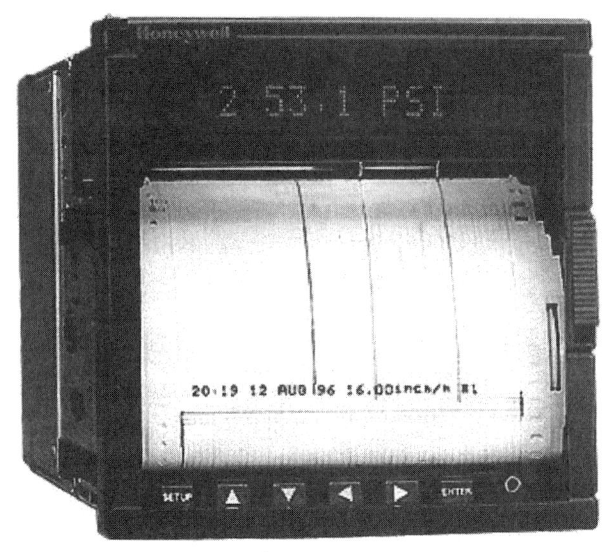


**Figure 8.4**
Honeywell DPR100 strip chart recorder *(reproduced by permission of Honeywell International, Inc.).*

can be adjusted to move the chart paper at different speeds. The fastest speed is typically 6000 mm/hour and the slowest is typically 1 mm/hour.

As well as recording signals as a continuous trace, many models also allow for the printing of alphanumeric data on the chart to record date, time, and other process information. Some models also have a digital numeric display to provide information on the current values of recorded variables.

#### *Multipoint strip chart recorder*

A multipoint strip chart recorder is a modification of the pen strip chart recorder that uses a dot matrix print head striking against an ink ribbon instead of pens. A typical model might allow up to 24 different signal inputs to be recorded simultaneously using a six-color ink ribbon. Certain models of such recorders also have the same enhancements as pen strip chart recorders in terms of printing alphanumeric information on the chart and providing a digital numeric output display.

#### *Heated-stylus chart recorder*

A heated-stylus chart recorder is another variant that records the input signal by applying a heated stylus to heat-sensitive chart paper. The main purpose of this alternative printing mechanism is to avoid the problem experienced in other forms of paper-based chart recorders of pen cartridges or printer ribbons running out of ink at inconvenient times.

### Circular chart recorder

A circular chart recorder consists of a servo-driven pen assembly that records the measured signal on a rotating circular paper chart, as shown in Figure 8.5. The rotational speed of the chart can be typically adjusted between one revolution in 1 hour to one revolution in 31 days. Recorded charts are replaced and stored after each revolution, which means replacement intervals that vary between hourly and monthly according to the chart speed. The major advantage of a circular chart recorder over other forms is compactness. Some models have up to four different colored pen assemblies, allowing up to four different parameters to be recorded simultaneously.

### Paperless chart recorder

A paperless chart recorder, sometimes alternatively called a *virtual chart recorder* or a *digital chart recorder*, displays the time history of measured signals electronically using a color-matrix liquid crystal display. This avoids the chore of periodically replacing chart paper and ink cartridges associated with other forms of chart recorders. Reliability is also enhanced compared with electromechanical recorders. As well as displaying the most recent time history of measured signals on its screen, the instrument also stores a much larger past history. This stored data can be recalled in batches and redisplayed on the screen as required. The only downside compared with other forms of chart recorders is this limitation of only displaying one screen full of information at a time. Of course, conventional recorders allow the whole past history of signals to be viewed at the same time on hard-copy, paper recordings. Otherwise, specifications are very similar to other forms of chart recorders, with vertical motion of the screen display

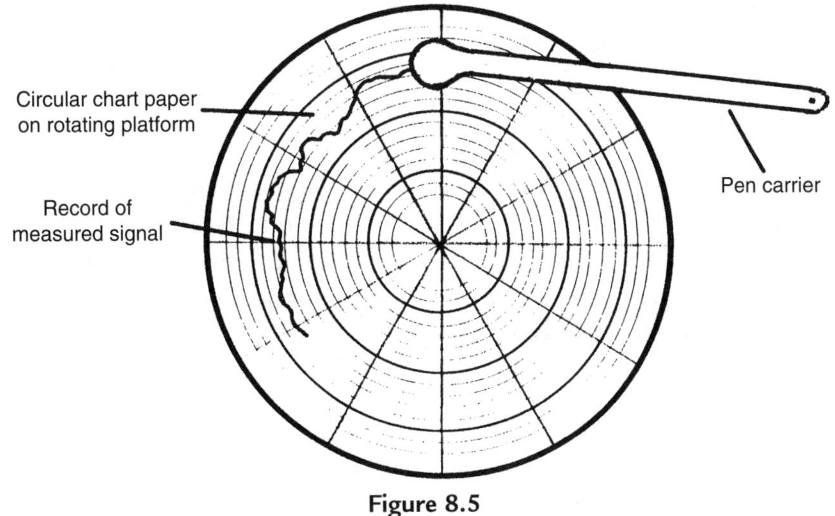


**Figure 8.5**
Circular chart recorder.

varying between 1 and 6000 mm/hour, typical inaccuracy less than ±0.1%, and capability of recording multiple signals simultaneously in different colors.

*Videographic recorder*

A videographic recorder provides exactly the same facilities as a paperless chart recorder but has additional display modes, such as bar graphs (histograms) and digital numbers. However, it should be noted that the distinction is becoming blurred between the various forms of paperless recorders described earlier and videographic recorders as manufacturers enhance the facilities of their instruments. For historical reasons, many manufacturers retain the names that they have traditionally used for their recording instruments but there is now much overlap between their respective capabilities as the functions provided are extended.

### 8.3.2 Ink-Jet and Laser Printers

Standard computer output devices in the form of ink-jet and laser printers are now widely used as an alternative means of storing measurement system output in paper form. Because a computer is a routine part of many data acquisition and processing operations, it often makes sense to output data in a suitable form to a computer printer rather than a chart recorder. This saves the cost of a separate recorder and is facilitated by the ready availability of software that can output measurement data in a graphical format.

### 8.3.3 Other Recording Instruments

Many of the devices mentioned in Chapters 5 and 7 have facilities for storing measurement data digitally. These include data logging acquisition devices and digital storage oscilloscopes. These data can then be converted into hard-copy form as required by transferring it to either a chart recorder or a computer and printer.

### 8.3.4 Digital Data Recorders

Digital data recorders, also known as *data loggers*, have already been introduced in Chapter 5 in the context of data acquisition. They provide a further alternative way of recording measurement data in a digital format. Data so recorded can then be transferred at a future time to a computer for further analysis, to any of the forms of measurement display devices discussed in Section 8.2, or to one of the hard-copy output devices described in Section 8.3.

Features contained within a data recorder/data logger obviously vary according to the particular manufacturer/model under discussion. However, most recorders have facilities to handle measurements in the form of both analogue and digital signals. Common analogue input signals allowed include d.c. voltages, d.c. currents, a.c. voltages, and a.c. currents. Digital inputs can usually be either in the form of data from digital measuring instruments or discrete data representing events such as switch closures or relay operations. Some models also provide alarm facilities to alert operators to abnormal conditions during data recording operations.

Many data recorders provide special input facilities optimized for particular kinds of measurement sensors, such as accelerometers, thermocouples, thermistors, resistance thermometers, strain gauges (including strain gauge bridges), linear variable differential transformers, and rotational differential transformers. Some instruments also have special facilities for dealing with inputs from less common devices such as encoders, counters, timers, tachometers, and clocks. A few recorders also incorporate integral sensors when they are designed to measure a particular type of physical variable.

The quality of data recorded by a digital recorder is a function of the cost of the instrument. Paying more usually means getting more memory to provide a greater data storage capacity, greater resolution in the analogue-to-digital converter to give better recording accuracy, and faster data processing to allow greater data sampling frequency.

## 8.4 Presentation of Data

The two formats available for presenting data on paper are tabular and graphical, and the relative merits of these are compared later. In some circumstances, it is clearly best to use only one or the other of these two alternatives alone. However, in many data collection exercises, part of the measurements and calculations are expressed in tabular form and part graphically, making best use of the merits of each technique. Very similar arguments apply to the relative merits of graphical and tabular presentations if a computer screen is used for presentation instead of paper.

### *8.4.1 Tabular Data Presentation*

A tabular presentation allows data values to be recorded in a precise way that exactly maintains the accuracy to which the data values were measured. In other words, the data values are written down exactly as measured. In addition to recording raw data values as measured, tables often also contain further values calculated from raw data. An example of a tabular data presentation is given in Table 8.1. This records results of an experiment to determine the strain induced in a bar of material subjected to a range of stresses. Data were obtained by applying a sequence of forces to the end of the bar and using an extensometer to measure the change in length. Values of the stress and strain in the bar are calculated from these measurements and are also included in Table 8.1. The final row, which is of crucial importance in any tabular presentation, is the estimate of possible error in each calculated result.

**Table 8.1 Table of Measured Applied Forces and Extensometer Readings and Calculations of Stress and Strain**

| | Force Applied (KN) | Extensometer Reading (Divisions) | Stress ($N/m^2$) | Strain |
|---|---|---|---|---|
| | 0 | 0 | 0 | 0 |
| | 2 | 4.0 | 15.5 | $19.8 \times 10^{-5}$ |
| | 4 | 5.8 | 31.0 | $28.6 \times 10^{-5}$ |
| | 6 | 7.4 | 46.5 | $36.6 \times 10^{-5}$ |
| | 8 | 9.0 | 62.0 | $44.4 \times 10^{-5}$ |
| | 10 | 10.6 | 77.5 | $52.4 \times 10^{-5}$ |
| | 12 | 12.2 | 93.0 | $60.2 \times 10^{-5}$ |
| | 14 | 13.7 | 108.5 | $67.6 \times 10^{-5}$ |
| Possible error in measurements (%) | ±0.2 | ±0.2 | ±1.5 | ±1.0 |

A table of measurements and calculations should conform to several rules as illustrated in Table 8.1:

- The table should have a title that explains what data are being presented within the table.
- Each column of figures in the table should refer to the measurements or calculations associated with one quantity only.
- Each column of figures should be headed by a title that identifies the data values contained in the column.
- Units in which quantities in each column are measured should be stated at the top of the column.
- All headings and columns should be separated by bold horizontal (and sometimes vertical) lines.
- Errors associated with each data value quoted in the table should be given. The form shown in Table 8.1 is a suitable way to do this when the error level is the same for all data values in a particular column. However, if error levels vary, then it is preferable to write the error boundaries alongside each entry in the table.

### *8.4.2 Graphical Presentation of Data*

Presentation of data in graphical form involves some compromise in the accuracy to which data are recorded, as the exact values of measurements are lost. However, graphical presentation has important advantages over tabular presentation.

- Graphs provide a pictorial representation of results that is comprehended more readily than a set of tabular results.
- Graphs are particularly useful for expressing the quantitative significance of results and showing whether a linear relationship exists between two variables. Figure 8.6 shows a

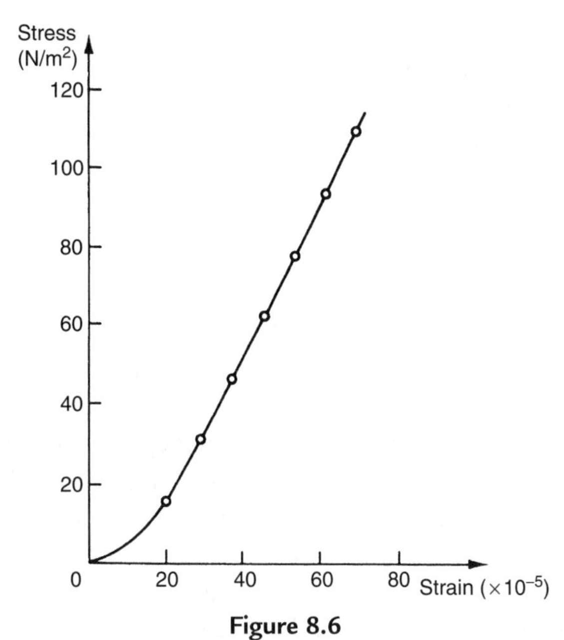


**Figure 8.6**
Sample graphical presentation of data: graph of stress against strain.

graph drawn from the stress and strain values given in Table 8.1. Construction of the graph involves first of all marking the points corresponding to the stress and strain values. The next step is to draw some line through these data points that best represents the relationship between the two variables. This line will normally be either a straight one or a smooth curve. Data points will not usually lie exactly on this line but instead will lie on either side of it. The magnitude of the excursions of the data points from the line drawn will depend on the magnitude of the random measurement errors associated with data.

- Graphs can sometimes show up on a data point that is clearly outside the straight line or curve that seems to fit the rest of the data points. Such a data point is probably due either to a human mistake in reading an instrument or to a momentary malfunction in the measuring instrument itself. If the graph shows such a data point where a human mistake or instrument malfunction is suspected, the proper course of action is to repeat that particular measurement and then discard the original data point if the mistake or malfunction is confirmed.

Like tables, the proper representation of data in graphical form has to conform to certain rules:

- The graph should have a title or caption that explains what data are being presented in the graph.
- Both axes of the graph should be labeled to express clearly what variable is associated with each axis and to define the units in which the variables are expressed.

- The number of points marked along each axis should be kept reasonably small—about five divisions is often a suitable number.
- No attempt should be made to draw the graph outside the boundaries corresponding to the maximum and minimum data values measured, that is, in Figure 8.6, the graph stops at a point corresponding to the highest measured stress value of 108.5.

*Fitting curves to data points on a graph*

The procedure of drawing a straight line or smooth curve as appropriate that passes close to all data points on a graph, rather than joining data points by a jagged line that passes through each data point, is justified on account of the random errors known to affect measurements. Any line between data points is mathematically acceptable as a graphical representation of data if the maximum deviation of any data point from the line is within the boundaries of the identified level of possible measurement errors. However, within the range of possible lines that could be drawn, only one will be the optimum one. This optimum line is where the sum of negative errors in data points on one side of the line is balanced by the sum of positive errors in data points on the other side of the line. The nature of data points is often such that a perfectly acceptable approximation to the optimum can be obtained by drawing a line through the data points by eye. In other cases, however, it is necessary to fit a line mathematically, using regression techniques.

*Regression techniques*

Regression techniques consist of finding a mathematical relationship between measurements of two variables, $y$ and $x$, such that the value of variable $y$ can be predicted from a measurement of the other variable, $x$. However, regression techniques should not be regarded as a magic formula that can fit a good relationship to measurement data in all circumstances, as the characteristics of data must satisfy certain conditions. In determining the suitability of measurement data for the application of regression techniques, it is recommended practice to draw an approximate graph of the measured data points, as this is often the best means of detecting aspects of data that make it unsuitable for regression analysis. Drawing a graph of data will indicate, for example, whether any data points appear to be erroneous. This may indicate that human mistakes or instrument malfunctions have affected the erroneous data points, and it is assumed that any such data points will be checked for correctness.

Regression techniques cannot be applied successfully if the deviation of any particular data point from the line to be fitted is greater than the maximum possible error calculated for the measured variable (i.e., the predicted sum of all systematic and random errors). The nature of some measurement data sets is such that this criterion cannot be satisfied, and any attempt to apply regression techniques is doomed to failure. In that event, the only valid course of action is to express the measurements in tabular form. This can then be used as an $x-y$ look-up table, from which values of the variable $y$ corresponding to particular values of $x$ can be read off. In many cases, this problem of large errors in some data points only becomes apparent during the process of attempting to fit a relationship by regression.

A further check that must be made before attempting to fit a line or curve to measurements of two variables, $x$ and $y$, is to examine data and look for any evidence that both variables are subject to random errors. It is a clear condition for the validity of regression techniques that only one of the measured variables is subject to random errors, with no error in the other variable. If random errors do exist in both measured variables, regression techniques cannot be applied and recourse must be made instead to correlation analysis (covered later in this chapter). Simple examples of a situation where both variables in a measurement data set are subject to random errors are measurements of human height and weight, and no attempt should be made to fit a relationship between them by regression.

Having determined that the technique is valid, the regression procedure is simplest if a straight-line relationship exists between the variables, which allows a relationship of the form $y = a + bx$ to be estimated by linear least-squares regression. Unfortunately, in many cases, a straight-line relationship between points does not exist, which is shown readily by plotting raw data points on a graph. However, knowledge of physical laws governing data can often suggest a suitable alternative form of relationship between the two sets of variable measurements, such as a quadratic relationship or a higher order polynomial relationship. Also, in some cases, the measured variables can be transformed into a form where a linear relationship exists. For example, suppose that two variables, $y$ and $x$, are related according to $y = ax^c$. A linear relationship from this can be derived, using a logarithmic transformation, as $\log(y) = \log(a) + c\log(x)$.

Thus, if a graph is constructed of $\log(y)$ plotted against $\log(x)$, the parameters of a straight-line relationship can be estimated by linear least-squares regression.

All quadratic and higher order relationships relating one variable, $y$, to another variable, $x$, can be represented by a power series of the form:

$$y = a_0 + a_1 x + a_2 x^2 + \cdots + a_p x^p.$$

Estimation of the parameters $a_0 \ldots a_p$ is very difficult if $p$ has a large value. Fortunately, a relationship where $p$ only has a small value can be fitted to most data sets. Quadratic least-squares regression is used to estimate parameters where $p$ has a value of two; for larger values of $p$, polynomial least-squares regression is used for parameter estimation.

Where the appropriate form of relationship between variables in measurement data sets is not obvious either from visual inspection or from consideration of physical laws, a method that is effectively a trial and error one has to be applied. This consists of estimating the parameters of successively higher order relationships between $y$ and $x$ until a curve is found that fits data sufficiently closely. What level of closeness is acceptable is considered later in the section on confidence tests.

*Linear least-squares regression*

If a linear relationship between $y$ and $x$ exists for a set of $n$ measurements, $y_1 \ldots y_n, x_1 \ldots x_n$, then this relationship can be expressed as $y = a + bx$, where coefficients $a$ and $b$ are constants. The purpose of least-squares regression is to select optimum values for $a$ and $b$ such that the line gives the best fit to the measurement data.

The deviation of each point $(x_i, y_i)$ from the line can be expressed as $d_i$, where $d_i = y_i - (a + bx_i)$.

The best-fit line is obtained when the sum of squared deviations, $S$, is a minimum, that is, when

$$S = \sum_{i=1}^{n} (d_i^2) = \sum_{i=1}^{n} (y_i - a - bx_i)^2$$

is a minimum.

The minimum can be found by setting partial derivatives $\partial S/\partial a$ and $\partial S/\partial b$ to zero and solving the resulting two simultaneous (normal) equations:

$$\partial S/\partial a = \sum 2(y_i - a - bx_i)(-1) = 0 \tag{8.1}$$

$$\partial S/\partial b = \sum 2(y_i - a - bx_i)(-x_i) = 0 \tag{8.2}$$

Values of the coefficients $a$ and $b$ at the minimum point can be represented by $\hat{a}$ and $\hat{b}$, which are known as the least-squares estimates of $a$ and $b$. These can be calculated as follows.

From Equation (8.1),

$$\sum y_i = \sum \hat{a} + \hat{b} \sum x_i = n\hat{a} + \hat{b} \sum x_i$$

and thus,

$$\hat{a} = \frac{\sum y_i - \hat{b} \sum x_i}{n}. \tag{8.3}$$

From Equation (8.2),

$$\sum (x_i y_i) = \hat{a} \sum x_i + \hat{b} \sum x_i^2. \tag{8.4}$$

Now substitute for $\hat{a}$ in Equation (8.4) using Equation (8.3):

$$\sum (x_i y_i) = \frac{\left(\sum y_i - \hat{b} \sum x_i\right)}{n} \sum x_i + \hat{b} \sum x_i^2.$$

Collecting terms in $\hat{b}$,

$$\hat{b}\left[\sum x_i^2 - \frac{\left(\sum x_i\right)^2}{n}\right] = \sum (x_i y_i) - \frac{\sum x_i \sum y_i}{n}.$$

Rearranging gives

$$\hat{b}\left[\sum x_i^2 - n\left\{\left(\sum x_i/n\right)\right\}^2\right] = \sum (x_i y_i) - n\sum (x_i/n)\sum (y_i/n),$$

which can be expressed as

$$\hat{b}\left[\sum x_i^2 - nx_m^2\right] = \sum (x_i y_i) - nx_m y_m,$$

where $x_m$ and $y_m$ are the mean values of $x$ and $y$. Thus,

$$\hat{b} = \frac{\sum (x_i y_i) - nx_m y_m}{\sum x_i^2 - nx_m^2}. \quad (8.5)$$

And, from Equation (8.3):

$$\hat{a} = y_m - \hat{b}x_m. \quad (8.6)$$

## Example 8.1

In an experiment to determine the characteristics of a displacement sensor with a voltage output, the following output voltage values were recorded when a set of standard displacements was measured:

| Displacement (cm) | 1.0 | 2.0 | 3.0 | 4.0 | 5.0 | 6.0 | 7.0 | 8.0 | 9.0 | 10.0 |
|---|---|---|---|---|---|---|---|---|---|---|
| Voltage | 2.1 | 4.3 | 6.2 | 8.5 | 10.7 | 12.6 | 14.5 | 16.3 | 18.3 | 21.2 |

Fit a straight line to this set of data using least-squares regression and estimate the output voltage when a displacement of 4.5 cm is measured.

■

## Solution

Let $y$ represent the output voltage and $x$ represent the displacement. Then a suitable straight line is given by $y=a+bx$. We can now proceed to calculate estimates for the coefficients $a$ and $b$ using Equations (8.5) and (8.6). The first step is to calculate the mean values of $x$ and $y$. These are found to be $x_m = 5.5$ and $y_m = 11.47$. Next, we need to tabulate $x_i y_i$ and $x_i^2$ for each pair of data values:

| $x_i$ | $y_i$ | $x_iy_i$ | $x_i^2$ |
|---|---|---|---|
| 1.0 | 2.1 | 2.1 | 1 |
| 2.0 | 4.3 | 8.6 | 4 |
| 3.0 | 6.2 | 18.6 | 9 |
| ⋮ | ⋮ | ⋮ | ⋮ |
| ⋮ | ⋮ | ⋮ | ⋮ |
| 10.0 | 21.2 | 212.0 | 100 |

Now calculate the values needed from this table—$n = 10$; $\sum(x_iy_i) = 801.0$; $\sum(x_i^2) = 385$—and enter these values into Equations (8.5) and (8.6).

$$\hat{b} = \frac{801.0 - (10 \times 5.5 \times 11.47)}{385 - (10 \times 5.5^2)} = 2.067; \hat{a} = 11.47 - (2.067 \times 5.5) = 0.1033;$$

that is, $y = 0.1033 + 2.067x$.

Hence, for $x = 4.5$, $y = 0.1033 + (2.067 \times 4.5) = 9.40$ volts. Note that in this solution we have only specified the answer to an accuracy of three figures, which is the same accuracy as the measurements. Any greater number of figures in the answer would be meaningless.

■

Least-squares regression is often appropriate for situations where a straight-line relationship is not immediately obvious, for example, where $y \propto x^2$ or $y \propto \exp(x)$.

## Example 8.2

From theoretical considerations, it is known that the voltage ($V$) across a charged capacitor decays with time ($t$) according to the relationship $V = K \exp(-t/\tau)$. Estimate values for $K$ and $\tau$ if the following values of $V$ and $t$ are measured.

| $V$ | 8.67 | 6.55 | 4.53 | 3.29 | 2.56 | 1.95 | 1.43 | 1.04 | 0.76 |
|---|---|---|---|---|---|---|---|---|---|
| $t$ | 0 | 1 | 2 | 3 | 4 | 5 | 6 | 7 | 8 |

■

## Solution

If $V = K \exp(-T/\tau)$, then $\log_e(V) = \log_e(K) - t/\tau$. Now let $y = \log_e(V)$, $a = \log(K)$, $b = -1/\tau$, and $x = t$. Hence, $y = a + bx$, which is the equation of a straight line whose coefficients can be estimated by applying Equations (8.5) and (8.6). Therefore, proceed in the same way as Example 8.1 and tabulate the values required:

| $V$ | $\log_e(V)$ $(y_i)$ | $t$ $(x_i)$ | $(x_i y_i)$ | $(x_i^2)$ |
| --- | --- | --- | --- | --- |
| 8.67 | 2.16 | 0 | 0 | 0 |
| 6.55 | 1.88 | 1 | 1.88 | 1 |
| 4.53 | 1.51 | 2 | 3.02 | 4 |
| ⋮ | ⋮ | ⋮ | ⋮ | ⋮ |
| ⋮ | ⋮ | ⋮ | ⋮ | ⋮ |
| 0.76 | −0.27 | 8 | −2.16 | 64 |

Now calculate the values needed from this table—n $= 9$; $\sum(x_i y_i) = 15.86$; $\sum(x_i^2) = 204$; $x_m = 4.0$; $y_m = 0.9422$—and enter these values into Equations (8.5) and (8.6).

$$\hat{b} = \frac{15.86 - (9 \times 4.0 \times 0.9422)}{204 - (9 \times 4.0^2)} = -0.301; \; \hat{a} = 0.9422 + (0.301 \times 4.0) = 2.15$$

$$K = \exp(a) = \exp(2.15) = 8.58; \; \tau = -1/b = -1/(-0.301) = 3.32$$

■

*Quadratic least-squares regression*

Quadratic least-squares regression is used to estimate the parameters of a relationship, $y = a + bx + cx^2$, between two sets of measurements, $y_1 \ldots y_n, x_1 \ldots x_n$.

The deviation of each point $(x_i, y_i)$ from the line can be expressed as $d_i$, where $d_i = y_i - (a + bx_i + cx_i^2)$.

The best-fit line is obtained when the sum of the squared deviations, $S$, is a minimum, that is, when

$$S = \sum_{i=1}^{n} (d_i^2) = \sum_{i=1}^{n} (y_i - a - bx_i + cx_i^2)^2$$

is a minimum.

The minimum can be found by setting the partial derivatives $\partial S/\partial a$, $\partial S/\partial b$, and $\partial S/\partial c$ to zero and solving the resulting simultaneous equations, as for the linear least-squares regression case given earlier. Standard computer programs to estimate the parameters $a$, $b$, and $c$ by numerical methods are widely available and therefore a detailed solution is not presented here.

*Polynomial least-squares regression*

Polynomial least-squares regression is used to estimate the parameters of the $p$th order relationship $y = a_0 + a_1 x + a_2 x^2 + \cdots + a_p x^p$ between two sets of measurements, $y_1 \ldots y_n, x_1 \ldots x_n$.

The deviation of each point $(x_i, y_i)$ from the line can be expressed as $d_i$, where

$$d_i = y_i - \left(a_0 + a_1 x_i + a_2 x_i^2 + \cdots + a_p x_i^p\right).$$

The best-fit line is obtained when the sum of squared deviations given by

$$S = \sum_{i=1}^{n} \left(d_i^2\right)$$

is a minimum.

The minimum can be found as before by setting $p$ partial derivatives $\partial S/\partial a_0 \ldots \partial S/\partial a_p$ to zero and solving the resulting simultaneous equations. Again, as for the quadratic least-squares regression case, standard computer programs to estimate the parameters $a_0 \ldots a_p$ by numerical methods are widely available and therefore a detailed solution is not presented here.

### *Confidence tests in curve fitting by least-squares regression*

Once data have been collected and a mathematical relationship that fits the data points has been determined by regression, the level of confidence that the mathematical relationship fitted is correct must be expressed in some way. The first check that must be made is whether the fundamental requirement for the validity of regression techniques is satisfied, that is, whether the deviations of data points from the fitted line are all less than the maximum error level predicted for the measured variable. If this condition is violated by any data point that a line or curve has been fitted to, then use of the fitted relationship is unsafe and recourse must be made to tabular data presentation, as described earlier.

The second check concerns whether random errors affect both measured variables. If attempts are made to fit relationships by regression to data where both measured variables contain random errors, any relationship fitted will only be approximate and it is likely that one or more data points will have a deviation from the fitted line or curve greater than the maximum error level predicted for the measured variable. This will show up when the appropriate checks are made.

Having carried out the aforementioned checks to show that there are no aspects of data that suggest that regression analysis is not appropriate, the next step is to apply least-squares regression to estimate the parameters of the chosen relationship (linear, quadratic, etc.). After this, some form of follow-up procedure is clearly required to assess how well the estimated relationship fits the data points. A simple curve-fitting confidence test is to calculate the sum of squared deviations $S$ for the chosen $y/x$ relationship and compare it with the value of $S$ calculated for the next higher order regression curve that could be fitted to data. Thus if a straight-line relationship is chosen, the value of $S$ calculated should be of a similar magnitude to that obtained by fitting a quadratic relationship. If the value of $S$ were substantially lower for a quadratic relationship, this would indicate that a quadratic relationship was a better fit to data than a straight-line one and further tests would be needed to examine whether a cubic or higher order relationship was a better fit still.

Other more sophisticated confidence tests exist, such as the *F-ratio test*. However, these are outside the scope of this book.

### *Correlation tests*

Where both variables in a measurement data set are subject to random fluctuations, correlation analysis is applied to determine the degree of association between the variables. For example, in the case already quoted of a data set containing measurements of human height and weight, we certainly expect some relationship between the variables of height and weight because a tall person is heavier *on average* than a short person. Correlation tests determine the strength of the relationship (or interdependence) between the measured variables, which is expressed in the form of a correlation coefficient.

For two sets of measurements $y_1 \ldots y_n$ , $x_1 \ldots x_n$ with means $x_m$ and $y_m$, the correlation coefficient $\Phi$ is given by

$$\Phi = \frac{\sum (x_i - x_m)(y_i - y_m)}{\sqrt{\left[\sum (x_i - x_m)^2\right]\left[\sum (y_i - y_m)^2\right]}}.$$

The value of $|\Phi|$ always lies between 0 and 1, with 0 representing the case where the variables are completely independent of one another and 1 is the case where they are totally related to one another. For $0 < |\Phi| < 1$, linear least-squares regression can be applied to find relationships between the variables, which allows $x$ to be predicted from a measurement of $y$, and $y$ to be predicted from a measurement of $x$. This involves finding two separate regression lines of the form:

$$y = a + bx \quad \text{and} \quad x = c + dy.$$

These two lines are not normally coincident, as shown in Figure 8.7. Both lines pass through the centroid of the data points but their slopes are different.

As $|\Phi| \rightarrow 1$, the lines tend to coincidence, representing the case where the two variables are totally dependent on one another.

As $|\Phi| \rightarrow 0$, the lines tend to orthogonal ones parallel to the $x$ and $y$ axes. In this case, the two sets of variables are totally independent. The best estimate of $x$ given any measurement of $y$ is $x_m$, and the best estimate of $y$ given any measurement of $x$ is $y_m$.

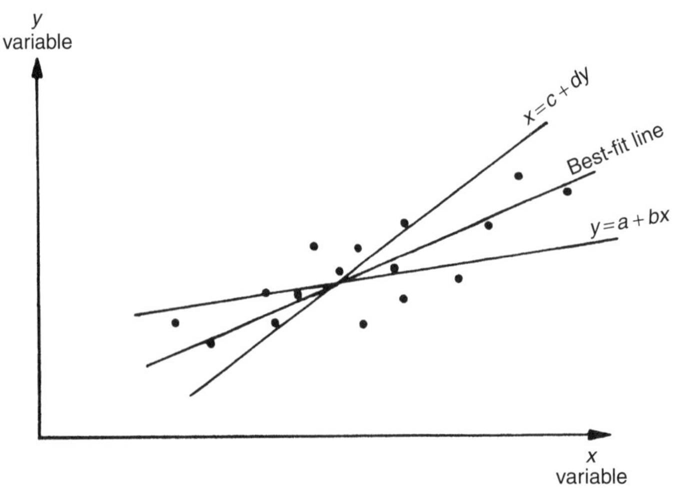


**Figure 8.7**
Relationship between two variables with random fluctuations.

For the general case, the best fit to data is the line that bisects the angle between the lines on Figure 8.7.

## 8.5 Summary

This chapter began by looking at the various ways that measurement data can be displayed, either using electronic display devices or using a computer monitor. We then went on to consider how measurement data could be recorded in a way that allows future analysis. We noted that this facility was particularly useful when faults develop in systems, as analysis of the changes in measured parameters in the time before the fault is discovered can often quickly indicate the reason for the fault. Options available for recording data are numerous and include chart recorders, digital oscilloscopes, digital data recorders, and hard-copy devices such as ink-jet and laser printers. We gave consideration to each of these and indicated some of the circumstances in which each alternative recording device might be used.

The next subject of study in the chapter was recommendations for good practice in the presentation of data. We looked at both graphical and tabular forms of presentation using either paper or a computer monitor screen as the display medium. We then went on to consider the best way of fitting lines through data points on a graph. This led to a discussion of mathematical regression techniques and the associated confidence tests necessary to assess the correctness of the line fitted using regression. Finally, we looked at correlation tests. These are used to determine the degree of association between two sets of data when they are both subject to random fluctuations.

## 8.6 Problems

8.1. What are the main ways available for displaying parameter values to human operators responsible for controlling industrial manufacturing systems? (Discussion on electronic displays and computer monitors is expected.)

8.2. Discuss the range of instruments and techniques available for recording measurement signals, mentioning particularly the frequency response characteristics of each instrument or technique and the upper frequency limit for signals in each case.

8.3. Discuss the features of the main types of chart recorders available for recording measurement signals.

8.4. What is a digital data recorder and how does it work?

8.5. (a) Explain the derivation of the expression

$$\ddot{\theta} + \frac{K_i^2\dot{\theta}}{JR} + \frac{K_s\theta}{J} = \frac{K_i V_t}{JR}$$

describing the dynamic response of a galvanometric chart recorder following a step change in the electrical voltage output of a transducer connected to its input. Explain also what all the terms in the expression stand for. (Assume that impedances of both the transducer and the recorder have a resistive component only and that there is negligible friction in the system.)

(b) Derive expressions for the measuring system natural frequency, $\omega_n$, the damping factor, $\xi$, and the steady-state sensitivity.

(c) Explain simple ways of increasing and decreasing the damping factor and describe the corresponding effect on measurement sensitivity.

(d) What damping factor gives the best system bandwidth?

(e) What aspects of the design of a chart recorder would you modify in order to improve the system bandwidth? What is the maximum bandwidth typically attainable in chart recorders, and if such a maximum-bandwidth instrument is available, what is the highest frequency signal that such an instrument would be generally regarded as being suitable for measuring if the accuracy of the signal amplitude measurement is important?

8.6. Discuss the relative merits of tabular and graphical methods of recording measurement data.

8.7. What would you regard as good practice in recording measurement data in graphical form?

8.8. What would you regard as good practice in recording measurement data in tabular form?

8.9. Explain the technique of linear least-squares regression for finding a relationship between two sets of measurement data.

8.10. Explain the techniques of (a) quadratic least-squares regression and (b) polynomial least-squares regression. How would you determine whether either quadratic or polynomial least-squares regression provides a better fit to a set of measurement data than linear least-squares regression?

8.11. During calibration of a platinum resistance thermometer, the following temperature and resistance values were measured:

| Resistance (Ω) | 212.8 | 218.6 | 225.3 | 233.6 | 240.8 | 246.6 |
|---|---|---|---|---|---|---|
| Temperature (°C) | 300 | 320 | 340 | 360 | 380 | 400 |

The temperature measurements were made using a primary reference standard instrument for which the measurement errors can be assumed to be zero. The resistance measurements were subject to random errors but it can be assumed that there are no systematic errors in them.

(a) Determine the sensitivity of measurement in Ω/°C in as accurate a manner as possible.

(b) Write down the temperature range that this sensitivity value is valid for.

(c) Explain the steps that you would take to test the validity of the type of mathematical relationship that you have used for data.

8.12. Theoretical considerations show that quantities $x$ and $y$ are related in a linear fashion such that $y=ax+b$. Show that the best estimate of the constants a and b are given by

$$\hat{a} = \frac{\sum(x_i - x_m)(y_i - y_m)}{\sum(x_i - x_m)^2} \quad ; \quad \hat{b} = y_m - \hat{a}x_m$$

Explain carefully the meaning of all the terms in the aforementioned two equations.

8.13. The characteristics of a chromel–constantan thermocouple is known to be approximately linear over the temperature range of 300–800°C. The output e.m.f. was measured practically at a range of temperatures, and the following table of results was obtained. Using least-squares regression, calculate coefficients $a$ and $b$ for the relationship $T=a+bE$ that best describes the temperature–e.m.f. characteristic.

| Temp (°C) | 300 | 325 | 350 | 375 | 400 | 425 | 450 | 475 | 500 | 525 | 550 |
|---|---|---|---|---|---|---|---|---|---|---|---|
| e.m.f. (mV) | 21.0 | 23.2 | 25.0 | 26.9 | 28.6 | 31.3 | 32.8 | 35.0 | 37.2 | 38.5 | 40.7 |
| Temp (°C) | 575 | 600 | 625 | 650 | 675 | 700 | 725 | 750 | 775 | 800 | |
| e.m.f. (mV) | 43.0 | 45.2 | 47.6 | 49.5 | 51.1 | 53.0 | 55.5 | 57.2 | 59.0 | 61.0 | |

8.14. Measurements of the current ($I$) flowing through a resistor and the corresponding voltage drop ($V$) are shown:

| $I$ (amps) | 1 | 2 | 3 | 4 | 5 |
|---|---|---|---|---|---|
| $V$ (volts) | 10.8 | 20.4 | 30.7 | 40.5 | 50.0 |

Instruments used to measure voltage and current were accurate in all respects except that they each had a zero error that the observer failed to take account of or to correct at the time of measurement. Determine the value of the resistor from data measured.

8.15. A measured quantity $y$ is known from theoretical considerations to depend on variable $x$ according to the relationship $y = a + bx^2$. For the following set of measurements of $x$ and $y$, use linear least-squares regression to determine the estimates of parameters $a$ and $b$ that fit data best.

| $x$ | 0 | 1 | 2 | 3 | 4 | 5 |
|---|---|---|---|---|---|---|
| $y$ | 0.9 | 9.2 | 33.4 | 72.5 | 130.1 | 200.8 |

8.16. The mean time to failure ($MTTF$) of an integrated circuit is known to obey a law of the following form: $MTTF = C \exp T_0/T$, where $T$ is the operating temperature and $C$ and $T_o$ are constants. The following values of $MTTF$ at various temperatures were obtained from accelerated life tests.

| *MTTF* (hours) | 54 | 105 | 206 | 411 | 941 | 2145 |
|---|---|---|---|---|---|---|
| Temperature (°K) | 600 | 580 | 560 | 540 | 520 | 500 |

(a) Estimate the values of C and $T_o$. [Hint: $\log_e(MTTF) = \log_e(C) + T_0/T$. This equation is now a straight-line relationship between log($MTTF$) and $1/T$, where log($C$) and $T_o$ are constants.]

(b) For an $MTTF$ of 10 years, calculate the maximum allowable temperature.

CHAPTER 9

# Variable Conversion Elements




**Measurement and Instrumentation: Theory and Application**

## 9.1 Introduction

We have already observed that outputs from measurement sensors often take the form of voltage signals. These can be measured using the voltage indicating and test instruments discussed in chapter 7. However, we have also discovered that sensor output does not take the form of an electrical voltage in many cases. Examples of these other forms of sensor output include translational displacements and changes in various electrical parameters such as resistance, inductance, capacitance, and current. In some cases, the output may alternatively take the form of variations in the phase or frequency of an a.c. electrical signal.

We therefore need to have a means of converting sensor outputs that are initially in some nonvoltage form into a more convenient form. This can be achieved by putting various types of variable conversion elements into the measurement system. We consider these in this chapter. First, we will see that bridge circuits are a particularly important type of variable conversion element; these are covered in some detail. Following this, we look at various alternative techniques for transducing the outputs of a measurement sensor into a form that is measured more readily.

## 9.2 Bridge Circuits

Bridge circuits are used very commonly as a variable conversion element in measurement systems and produce an output in the form of a voltage level that changes as the measured physical quantity changes. They provide an accurate method of measuring resistance, inductance, and capacitance values and enable the detection of very small changes in these quantities about a nominal value. They are of immense importance in measurement system technology because so many transducers measuring physical quantities have an output that is expressed as a change in resistance, inductance, or capacitance. A displacement-measuring strain gauge, which has a varying resistance output, is but one example of this class of transducers. Normally, excitation of the bridge is by a d.c. voltage for resistance measurement and by an a.c. voltage for inductance or capacitance measurement. Both null and deflection types of bridges exist, and, in a like manner to instruments in general, null types are employed mainly for calibration purposes and deflection types are used within closed loop automatic control schemes.

### *9.2.1 Null-Type d.c. Bridge (Wheatstone Bridge)*

A null-type bridge with d.c. excitation, known commonly as a Wheatstone bridge, has the form shown in Figure 9.1. The four arms of the bridge consist of the unknown resistance $R_u$, two equal value resistors $R_2$ and $R_3$, and variable resistor $R_v$ (usually a decade resistance box). A d.c. voltage $V_i$ is applied across the points AC, and resistance $R_v$ is varied until the voltage measured across points BD is zero. This null point is usually measured with a high sensitivity galvanometer.

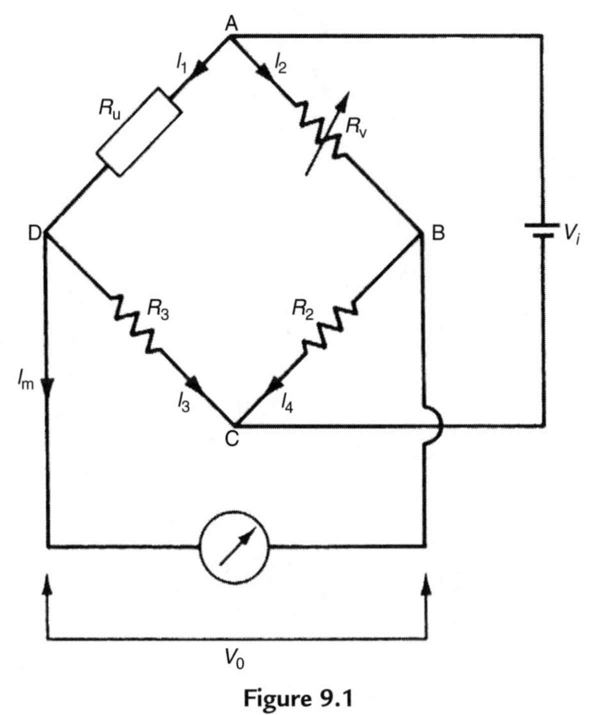


**Figure 9.1**
Wheatstone bridge.

To analyze the Whetstone bridge, define the current flowing in each arm to be $I_1 \ldots I_4$ as shown in Figure 9.1. Normally, if a high impedance voltage-measuring instrument is used, current $I_m$ drawn by the measuring instrument will be very small and can be approximated to zero. If this assumption is made, then, for $I_m = 0$: $I_1 = I_3$ and $I_2 = I_4$.

Looking at path ADC, we have voltage $V_i$ applied across resistance $R_u + R_3$ and by Ohm's law:

$$I_1 = \frac{V_i}{R_u + R_3}.$$

Similarly, for path ABC,

$$I_2 = \frac{V_i}{R_v + R_2}.$$

Now we can calculate the voltage drop across AD and AB:

$$V_{AD} = I_1 R_v = \frac{V_i R_u}{R_u + R_3} \quad ; \quad V_{AB} = I_2 R_v = \frac{V_i R_v}{R_v + R_2}.$$

By the principle of superposition, $V_o = V_{BD} = V_{BA} + V_{AD} = -V_{AB} + V_{AD}$.

Thus,

$$V_o = -\frac{V_i R_v}{R_v + R_2} + \frac{V_i R_u}{R_u + R_3}. \tag{9.1}$$

At the null point $V_o = 0$, so

$$\frac{R_u}{R_u + R_3} = \frac{R_v}{R_v + R_2}.$$

Inverting both sides,

$$\frac{R_u + R_3}{R_u} = \frac{R_v + R_2}{R_v},$$

that is,

$$\frac{R_3}{R_u} = \frac{R_2}{R_v}$$

or

$$R_u = \frac{R_3 R_v}{R_2}. \tag{9.2}$$

Thus, if $R_2 = R_3$, then $R_u = R_v$. As $R_v$ is an accurately known value because it is derived from a variable decade resistance box, this means that $R_u$ is also accurately known.

A null-type bridge is somewhat tedious to use as careful adjustment of variable resistance is needed to get exactly to the null point. However, it provides a highly accurate measurement of resistance, leading to this being the preferred type when sensors are being calibrated.

### 9.2.2 Deflection-Type d.c. Bridge

A deflection type bridge with d.c. excitation is shown in Figure 9.2. This differs from the Wheatstone bridge mainly in that variable resistance $R_v$ is replaced by fixed resistance $R_1$ of the same value as the nominal value of unknown resistance $R_u$. As resistance $R_u$ changes, so output voltage $V_0$ varies, and this relationship between $V_0$ and $R_u$ must be calculated.

This relationship is simplified if we again assume that a high impedance voltage-measuring instrument is used and the current drawn by it, $I_m$, can be approximated to zero. (The case when this assumption does not hold is covered later in this section.) The analysis is then exactly the same as for the preceding example of the Wheatstone bridge, except that $R_v$ is replaced by $R_1$. Thus, from Equation (9.1), we have

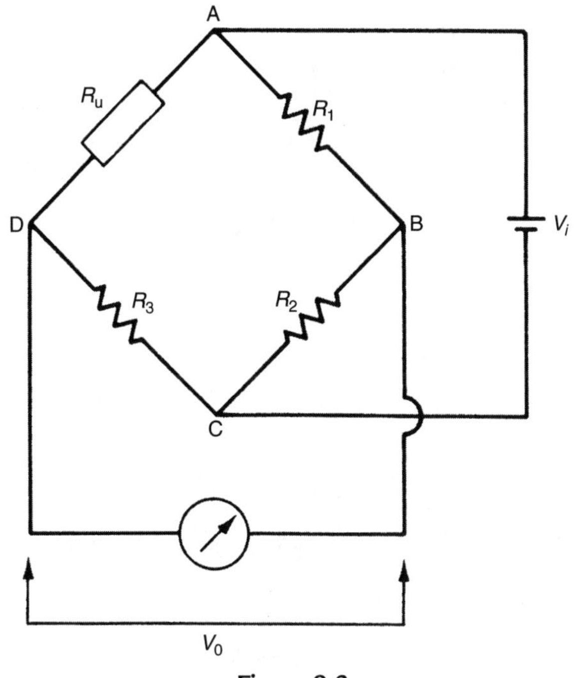


**Figure 9.2**
Deflection-type d.c. bridge.

$$V_0 = V_i \left( \frac{R_u}{R_u + R_3} - \frac{R_1}{R_1 + R_2} \right). \tag{9.3}$$

When $R_u$ is at its nominal value, that is, for $R_u = R_1$, it is clear that $V_0 = 0$ (since $R_2 = R_3$). For other values of $R_u$, $V_0$ has negative and positive values that vary in a nonlinear way with $R_u$.

The deflection-type bridge is somewhat easier to use than a null-type bridge because the output measurement is given directly in the form of a voltage measurement. However, its measurement accuracy is not as good as that of a null-type bridge. Despite its inferior accuracy, ease of use means that it is the preferred form of bridge in most general measurement situations unless the greater accuracy of a null-type bridge is absolutely necessary.

## Example 9.1

A certain type of pressure transducer, designed to measure pressures in the range 0–10 bar, consists of a diaphragm with a strain gauge cemented to it to detect diaphragm deflections. The strain gauge has a nominal resistance of 120 Ω and forms one arm of a Wheatstone bridge circuit, with the other three arms each having a resistance of 120 Ω. Bridge output is measured by an instrument whose input impedance can be assumed infinite. If, in order to limit heating effects, the maximum permissible gauge current is 30 mA, calculate the maximum permissible bridge excitation voltage. If the sensitivity of the strain gauge is 338 mΩ/bar and the maximum bridge excitation voltage is used, calculate the bridge output voltage when measuring a pressure of 10 bar.

■

## ■ Solution

This is the type of bridge circuit shown in Figure 9.2 in which the components have the following values:

$$R_1 = R_2 = R_3 = 120\ \Omega$$

Defining $I_1$ to be the current flowing in path ADC of the bridge, we can write

$$V_i = I_1(R_u + R_3).$$

At balance, $R_u = 120$ and the maximum value allowable for $I_1$ is 0.03A. Hence, $V_i = 0.03$ $(120 + 120) = 7.2$ V. Thus, the maximum bridge excitation voltage allowable is 7.2 volts.

For a pressure of 10 bar applied, the resistance change is 3.38 Ω, that is, $R_u$ is then equal to 123.38 Ω. Applying Equation (9.3), we can write

$$V_0 = V_i\left(\frac{R_u}{R_u + R_3} - \frac{R_1}{R_1 + R_2}\right) = 7.2\left(\frac{123.38}{243.38} - \frac{120}{240}\right) = 50mV.$$

Thus, if maximum permissible bridge excitation voltage is used, the output voltage is 50 mV when a pressure of 10 bar is measured.

■

The nonlinear relationship between output reading and measured quantity exhibited by Equation (9.3) is inconvenient and does not conform with the normal requirement for a linear input–output relationship. The method of coping with this nonlinearity varies according to the form of primary transducer involved in the measurement system.

One special case is where the change in unknown resistance $R_u$ is typically small compared with the nominal value of $R_u$. If we calculate the new voltage $V_0'$ when the resistance $R_u$ in Equation (9.3) changes by an amount $\delta R_u$, we have

$$V_0' = V_i\left(\frac{R_u + \delta R_u}{R_u + \delta R_u + R_3} - \frac{R_1}{R_1 + R_2}\right). \tag{9.4}$$

The change of voltage output is therefore given by

$$\delta V_0 = V_0{}' - V_0 = \frac{V_i \delta R_u}{R_u + \delta R_u + R_3}.$$

If $\delta R_u << R_u$, then the following linear relationship is obtained:

$$\frac{\delta V_0}{\delta R_u} = \frac{V_i}{R_u + R_3}. \tag{9.5}$$

This expression describes the measurement sensitivity of the bridge. Such an approximation to make the relationship linear is valid for transducers such as strain gauges where the typical changes of resistance with strain are very small compared with nominal gauge resistance.

However, many instruments that are inherently linear themselves, at least over a limited measurement range, such as resistance thermometers, exhibit large changes in output as the input quantity changes, and the approximation of Equation (9.5) cannot be applied. In such cases, specific action must be taken to improve linearity in the relationship between bridge output voltage and measured quantity. One common solution to this problem is to make the values of the resistances $R_2$ and $R_3$ at least 10 times those of $R_1$ and $R_u$ (nominal). The effect of this is best observed by looking at a numerical example.

Consider a platinum resistance thermometer with a range of 0–50°C whose resistance at 0°C is 500 Ω and whose resistance varies with temperature at the rate of 4 Ω/°C. Over this range of measurement, the output characteristic of the thermometer itself is nearly perfectly linear. (Note that the subject of resistance thermometers is discussed further in Chapter 14.)

Taking first the case where $R_1 = R_2 = R_3 = 500\ \Omega$ and $V_i = 10$ V, and applying Equation (9.3):

At 0°C, $V_0 = 0$

At 25°C, $R_u = 600\ \Omega$ and $V_0 = 10\left(\frac{600}{1100} - \frac{500}{1000}\right) = 0.455$ V

At 50°C, $R_u = 700\ \Omega$ and $V_0 = 10\left(\frac{700}{1200} - \frac{500}{1000}\right) = 0.833$ V

This relationship between $V_0$ and $R_u$ is plotted as curve A in Figure 9.3 and nonlinearity is apparent. Inspection of the manner in which output voltage $V_0$ above changes for equal steps of temperature change also clearly demonstrates nonlinearity.

For the temperature change from 0 to 25°C, the change in $V_0$ is (0.455 − 0) = 0.455 V
For the temperature change from 25 to 50°C, the change in $V_0$ is (0.833 − 0.455) = 0.378 V

If the relationship was linear, the change in $V_0$ for the 25–50°C temperature step would also be 0.455 V, giving a value for $V_0$ of 0.910 V at 50°C.

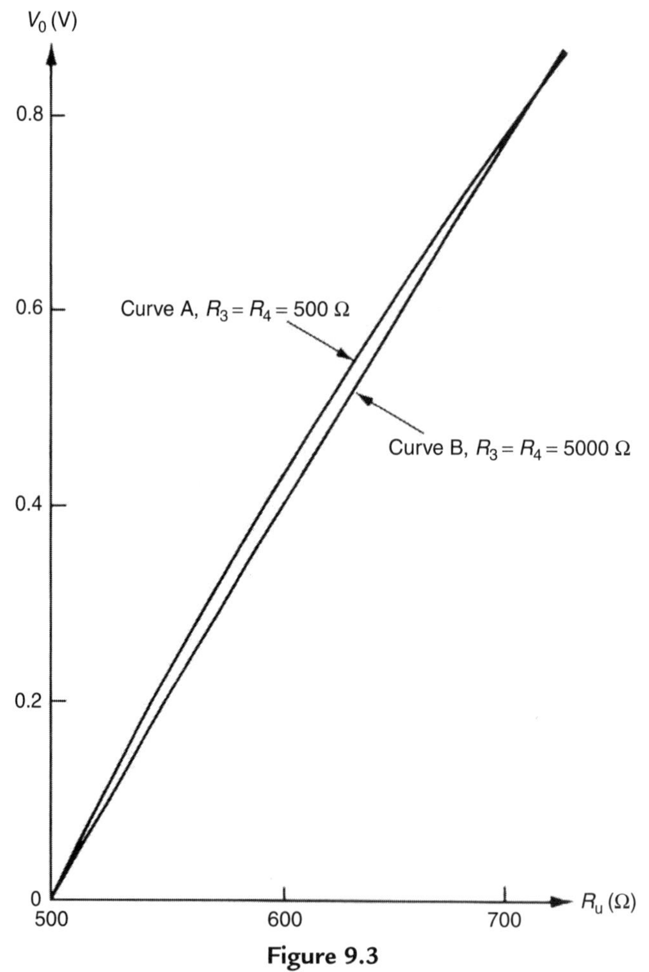


**Figure 9.3**
Linearization of bridge circuit characteristic.

Now take the case where $R_1 = 500\ \Omega$ but $R_2 = R_3 = 5000\ \Omega$ and let $V_i = 26.1$ V:

$$\text{At } 0^\circ\text{C}, \quad V_0 = 0$$

$$\text{At } 25^\circ\text{C}, \quad R_u = 600\ \Omega \quad \text{and} \quad V_0 = 26.1\left(\frac{600}{5600} - \frac{500}{5500}\right) = 0.424\ \text{V}$$

$$\text{At } 50^\circ\text{C}, \quad R_u = 700\ \Omega \quad \text{and} \quad V_0 = 26.1\left(\frac{700}{5700} - \frac{500}{5500}\right) = 0.833\ \text{V}$$

This relationship is shown as curve B in Figure 9.3 and a considerable improvement in linearity is achieved. This is more apparent if differences in values for $V_0$ over the two temperature steps are inspected.

From 0 to 25°C, the change in $V_0$ is 0.424 V
From 25 to 50°C, the change in $V_0$ is 0.409 V

Changes in $V_0$ over the two temperature steps are much closer to being equal than before, demonstrating the improvement in linearity. However, in increasing the values of $R_2$ and $R_3$, it was also necessary to increase the excitation voltage from 10 to 26.1 V to obtain the same output levels. In practical applications, $V_i$ would normally be set at the maximum level consistent with limitation of the effect of circuit heating in order to maximize the measurement sensitivity ($V_0/\delta R_u$ relationship). It would therefore not be possible to increase $V_i$ further if $R_2$ and $R_3$ were increased, and the general effect of such an increase in $R_2$ and $R_3$ is thus a decrease in the sensitivity of the measurement system.

The importance of this inherent nonlinearity in the bridge output relationship is diminished greatly if the primary transducer and bridge circuit are incorporated as elements within an intelligent instrument. In that case, digital computation is applied to produce an output in terms of the measured quantity that automatically compensates for nonlinearity in the bridge circuit.

#### *Case where current drawn by measuring instrument is not negligible*

For various reasons, it is not always possible to meet the condition that impedance of the instrument measuring the bridge output voltage is sufficiently large for the current drawn by it to be negligible. Wherever the measurement current is not negligible, an alternative relationship between bridge input and bridge output must be derived that takes the current drawn by the measuring instrument into account.

Thévenin's theorem is again a useful tool for this purpose. Replacing voltage source $V_i$ in Figure 9.4a by a zero internal resistance produces the circuit shown in Figure 9.4b, or the equivalent representation shown in Figure 9.4c. It is apparent from Figure 9.4c that equivalent circuit resistance consists of a pair of parallel resistors, $R_u$ and $R_3$, in series with the parallel resistor pair $R_1$ and $R_2$. Thus, $R_{DB}$ is given by

$$R_{DB} = -\frac{R_1 R_2}{R_1 + R_2} + \frac{R_u R_3}{R_u + R_3}. \tag{9.6}$$

The equivalent circuit derived via Thévenin's theorem with resistance $R_m$ of the measuring instrument connected across the output is shown in Figure 9.4d. The open-circuit voltage across DB, $E_0$, is the output voltage calculated earlier [Equation (9.3)] for the case of $R_m = 0$:

$$E_0 = V_i\left(\frac{R_u}{R_u + R_3} - \frac{R_1}{R_1 + R_2}\right). \tag{9.7}$$

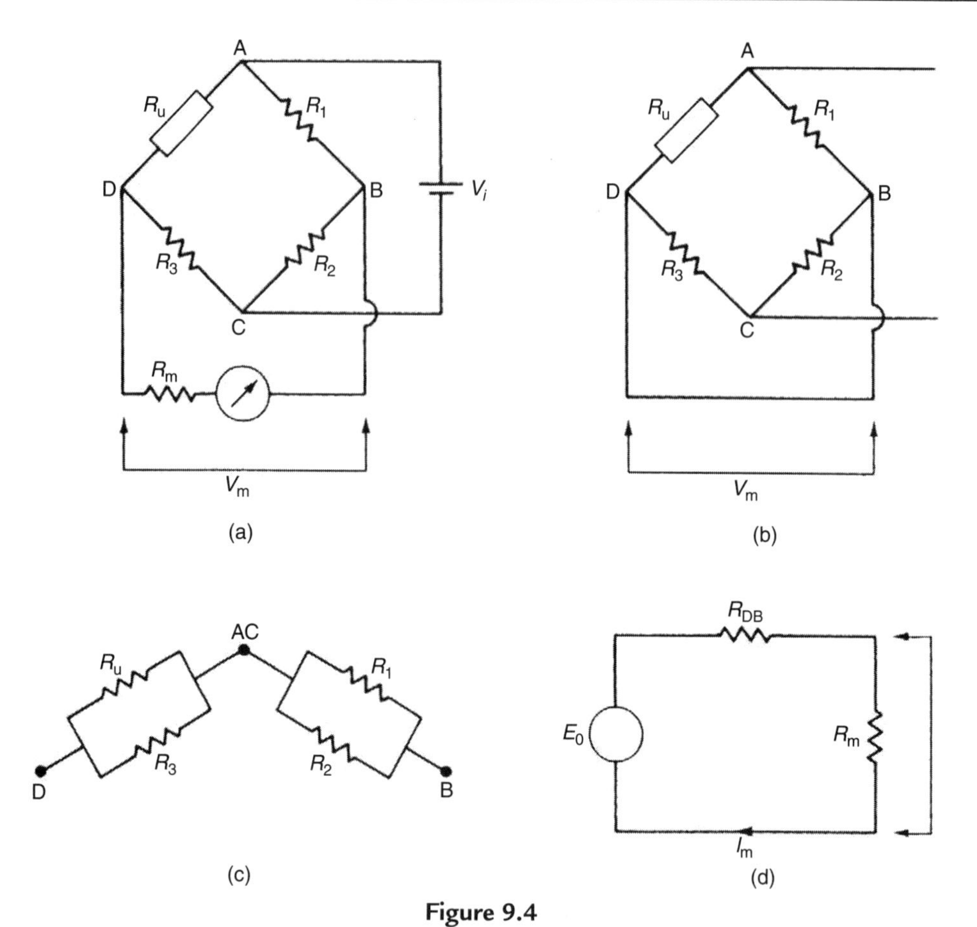


**Figure 9.4**
(a) A bridge circuit. (b) Equivalent circuit by Thévenin's theorem. (c) Alternative representation. (d) Equivalent circuit for alternative representation.

If the current flowing is $I_m$ when the measuring instrument of resistance $R_m$ is connected across DB, then, by Ohm's law, $I_m$ is given by

$$I_m = \frac{E_0}{R_{DB} + R_m}. \tag{9.8}$$

If $V_m$ is the voltage measured across $R_m$, then, again by Ohm's law:

$$V_m = I_m R_m = \frac{E_0 R_m}{R_{DB} + R_m}. \tag{9.9}$$

Substituting for $E_0$ and $R_{DB}$ in Equation (9.9), using the relationships developed in Equations (9.6) and (9.7), we obtain

$$V_m = \frac{V_i[R_u/(R_u + R_3) - R_1/(R_1 + R_2)]R_m}{R_1R_2/(R_1 + R_2) + R_uR_3/(R_u + R_3) + R_m}.$$

Simplifying:

$$V_m = \frac{V_i R_m(R_u R_2 - R_1 R_3)}{R_1R_2(R_u + R_3) + R_uR_3(R_1 + R_2) + R_m(R_1 + R_2)(R_u + R_3)}. \quad (9.10)$$

## Example 9.2

A bridge circuit, as shown in Figure 9.5, is used to measure the value of the unknown resistance $R_u$ of a strain gauge of nominal value 500 Ω. The output voltage measured across points DB in the bridge is measured by a voltmeter. Calculate the measurement sensitivity in volts per ohm change in $R_u$ if

(a) resistance $R_m$ of the measuring instrument is neglected
(b) account is taken of the value of $R_m$

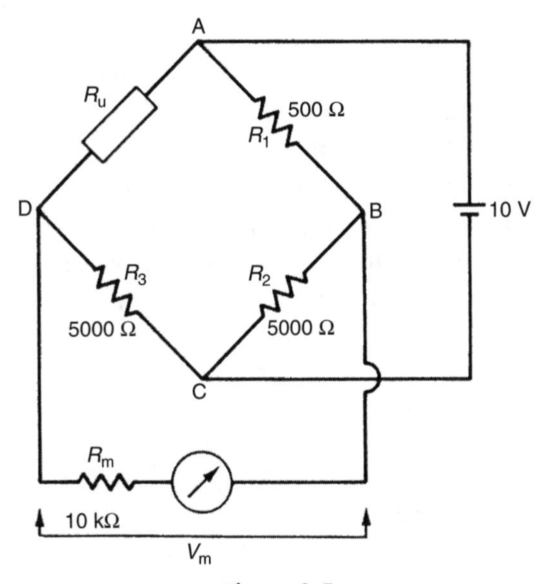


**Figure 9.5**
Bridge circuit for Example 9.2.

## Solution

For $R_u = 500\ \Omega$, $V_m = 0$. To determine sensitivity, calculate $V_m$ for $R_u = 501\ \Omega$.

(a) Applying Equation (9.3):

$$V_m = V_i\left(\frac{R_u}{R_u + R_3} - \frac{R_1}{R_1 + R_2}\right).$$

Substituting in values:

$$V_m = 10\left(\frac{501}{1001} - \frac{500}{1000}\right) = 5.00\ \text{m}V.$$

Thus, if resistance of the measuring circuit is neglected, the measurement sensitivity is 5.00 mV per ohm change in $R_u$.

(b) Applying Equation (9.10) and substituting in values:

$$V_m = \frac{10 \times 10^4 \times 500(501 - 500)}{500^2(1001) + 500 \times 501(1000) + 10^4 \times 1000 \times 1001} = 4.76\ \text{m}V.$$

Thus, if proper account is taken of the 10-KΩ value of the resistance of $R_m$, the true measurement sensitivity is shown to be 4.76 mV per ohm change in $R_u$.

### 9.2.3 *Error Analysis*

In the application of bridge circuits, the contribution of component value tolerances to total measurement system accuracy limits must be clearly understood. The analysis given here applies to a null-type (Wheatstone) bridge, but similar principles can be applied for a deflection-type bridge. The maximum measurement error is determined by first finding the value of $R_u$ in Equation (9.2) with each parameter in the equation set at that limit of its tolerance that produces the maximum value of $R_u$. Similarly, the minimum possible value of $R_u$ is calculated, and the required error band is then the span between these maximum and minimum values.

## Example 9.3

In the Wheatstone bridge circuit of Figure 9.1, $R_v$ is a decade resistance box with a specified inaccuracy $\pm 0.2\%$ and $R_2 = R_3 = 500\ \Omega \pm 0.1\%$. If the value of $R_v$ at the null position is 520.4 Ω, determine the error band for $R_u$ expressed as a percentage of its nominal value.

## Solution

Applying Equation (9.2) with $R_v = 520.4\ \Omega + 0.2\% = 521.44\ \Omega$, $R_3 = 5000\ \Omega + 0.1\% = 5005\ \Omega$, $R_2 = 5000\ \Omega - 0.1\% = 4995\ \Omega$, we get

$$R_v = \frac{521.44 \times 5005}{4995} = 522.48\ \Omega\ (= +0.4\%).$$

Applying Equation (9.2) with $R_v = 520.4\ \Omega - 0.2\% = 519.36\ \Omega$, $R_3 = 5000\ \Omega - 0.1\% = 4995\ \Omega$, $R_2 = 5000\ \Omega + 0.1\% = 5005\ \Omega$, we get

$$R_v = \frac{519.36 \times 4995}{5005} = 518.32\ \Omega\ (= -0.4\%).$$

Thus, the error band for $R_u$ is $\pm 0.4\%$.

The cumulative effect of errors in individual bridge circuit components is clearly seen. Although the maximum error in any one component is $\pm 0.2\%$, the possible error in the measured value of $R_u$ is $\pm 0.4\%$. Such a magnitude of error is often not acceptable, and special measures are taken to overcome the introduction of error by component value tolerances. One such practical measure is the introduction of apex balancing. This is one of many methods of bridge balancing that all produce a similar result. ■

*Apex balancing*

One form of apex balancing consists of placing an additional variable resistor $R_5$ at junction $C$ between resistances $R_2$ and $R_3$ and applying excitation voltage $V_i$ to the wiper of this variable resistance, as shown in Figure 9.6.

For calibration purposes, $R_u$ and $R_v$ are replaced by two equal resistances whose values are known accurately, and $R_5$ is varied until output voltage $V_0$ is zero. At this point, if the portions of resistance on either side of the wiper on $R_5$ are $R_6$ and $R_7$ (such that $R_5 = R_6 + R_7$), we can write:

$$R_3 + R_6 = R_2 + R_7.$$

We have thus eliminated any source of error due to tolerance in the value of $R_2$ and $R_3$, and the error in the measured value of $R_u$ depends only on the accuracy of one component, the decade resistance box $R_v$.

## Example 9.4

A potentiometer $R_5$ is put into the apex of the bridge shown in Figure 9.6 to balance the circuit. The bridge components have the following values: $R_u = 500\ \Omega$, $R_v = 500\ \Omega$, $R_2 = 515\ \Omega$, $R_3 = 480\ \Omega$, $R_5 = 100\ \Omega$.

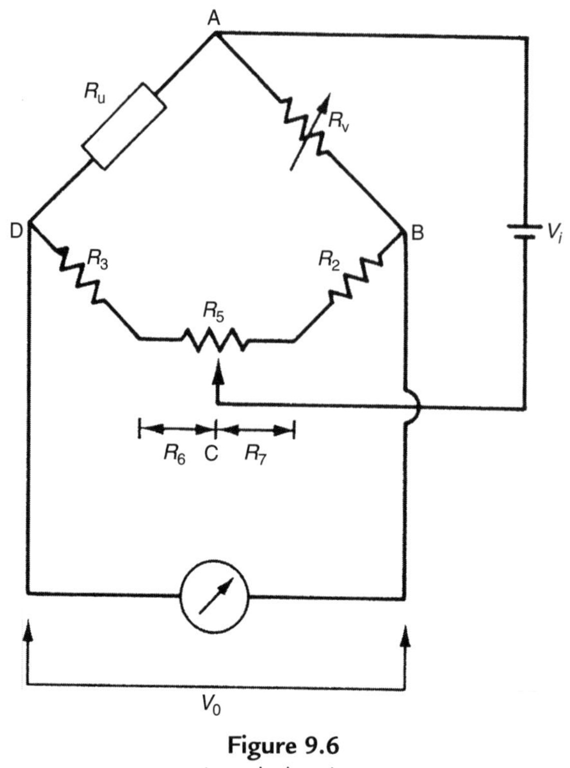


**Figure 9.6**
Apex balancing.

Determine the required value of resistances $R_6$ and $R_7$ of the parts of the potentiometer track either side of the slider in order to balance the bridge and compensate for the unequal values of $R_2$ and $R_3$.

■

## ■ Solution

For balance, $R_2 + R_7 = R_3 + R_6$ ; hence, $515 + R_7 = 480 + R_6$. Also, because $R_6$ and $R_7$ are the two parts of potentiometer track $R_5$ whose resistance is 100 Ω: $R_6 + R_7 = 100$; thus $515 + R_7 = 480 + (100 - R_7)$; that is, $2R_7 = 580 - 515 = 65$. Thus, $R_7 = 32.5$; hence, $R_6 = 100 - 32.5 = 67.5$ Ω.

■

### 9.2.4 a.c. Bridges

Bridges with a.c. excitation are used to measure unknown impedances (capacitances and inductances). Both null and deflection types exist. As for d.c. bridges, null types are more accurate but also more tedious to use. Therefore, null types are normally reserved for use in

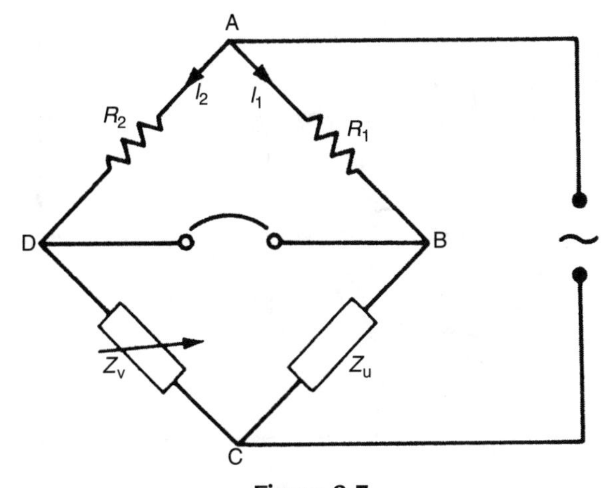


**Figure 9.7**
Null-type impedance bridge.

calibration duties and any other application where very high measurement accuracy is required. Otherwise, in all other general applications, deflection types are preferred.

*Null-type impedance bridge*

A typical null-type impedance bridge is shown in Figure 9.7. The null point can be detected conveniently by monitoring the output with a pair of headphones connected via an operational amplifier across points BD. This is a much less expensive method of null detection than application of an expensive galvanometer required for a d.c. Wheatstone bridge.

Referring to Figure 9.7, at the null point, $I_1R_1 = I_2R_2$ ; $I_1Z_u = I_2Z_v$.

Thus,

$$Z_u = \frac{Z_v R_1}{R_2} \tag{9.11}$$

If $Z_u$ is capacitive, that is, $Z_u = 1/j\omega C_u$, then $Z_v$ must consist of a variable capacitance box, which is readily available. If $Z_u$ is inductive, then $Z_u = R_u + j\omega L_u$.

Note that the expression for $Z_u$ as an inductive impedance has a resistive term in it because it is impossible to realize a pure inductor. An inductor coil always has a resistive component, although this is made as small as possible by designing the coil to have a high $Q$ factor ($Q$ factor is the ratio inductance/resistance). Therefore, $Z_v$ must consist of a variable resistance box and a variable inductance box. However, the latter is not readily available because it is difficult and hence expensive to manufacture a set of fixed value inductors to make up a variable inductance box. For this reason, an alternative kind of null-type bridge circuit, known as the *Maxwell bridge*, is used commonly to measure unknown inductances.

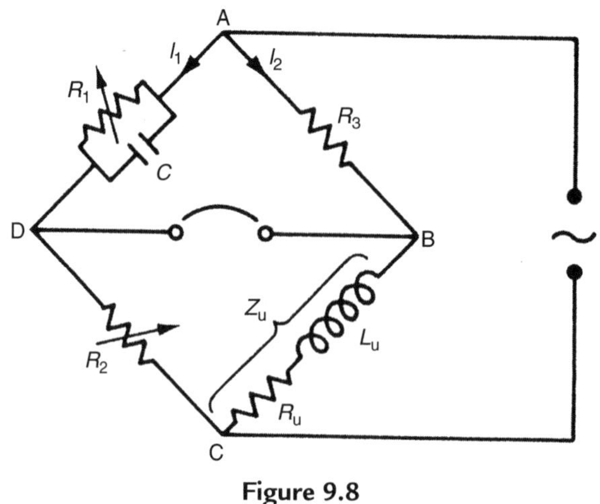


**Figure 9.8**
Maxwell bridge.

*Maxwell bridge*

A Maxwell bridge is shown in Figure 9.8. The requirement for a variable inductance box is avoided by introducing instead a second variable resistance. The circuit requires one standard fixed-value capacitor, two variable resistance boxes, and one standard fixed-value resistor, all of which are components that are readily available and inexpensive. Referring to Figure 9.8, we have at the null output point:

$$I_1 Z_{AD} = I_2 Z_{AB} \quad ; \quad I_1 Z_{DC} = I_2 Z_{BC}$$

Thus,

$$\frac{Z_{BC}}{Z_{AB}} = \frac{Z_{DC}}{Z_{AD}}.$$

or

$$Z_{BC} = \frac{Z_{DC} Z_{AB}}{Z_{AD}} \tag{9.12}$$

The quantities in Equation (9.12) have the following values:

$$\frac{1}{Z_{AD}} = \frac{1}{R_1} + j\omega C \quad \text{or} \quad Z_{AD} = \frac{R_1}{1 + j\omega C R_1}$$

$$Z_{AB} = R_3 \quad ; \quad Z_{BC} = R_u + j\omega L_u \quad ; \quad Z_{DC} = R_2$$

Substituting the values into Equation (9.12),

$$R_u + j\omega L_u = \frac{R_2 R_3 (1 + j\omega C R_1)}{R_1}.$$

Taking real and imaginary parts:

$$R_u = \frac{R_2 R_3}{R_1} \quad ; \quad L_u = R_2 R_3 C. \tag{9.13}$$

This expression [Equation (9.13)] can be used to calculate the quality factor ($Q$ value) of the coil:

$$Q = \frac{\omega L_u}{R_u} = \frac{\omega R_2 R_3 C R_1}{R_2 R_3} = \omega C R_1.$$

If a constant frequency ω is used, $Q \approx R_1$.

Thus, the Maxwell bridge can be used to measure the $Q$ value of a coil directly using this relationship.

## Example 9.5

In the Maxwell bridge shown in Figure 9.8, let the fixed-value bridge components have the following values: $R_3 = 5\ \Omega$ ; $C = 1$ mF. Calculate the value of the unknown impedance ($L_u$, $R_u$) if $R_1 = 159\ \Omega$ and $R_2 = 10\ \Omega$ at balance.

■

## Solution

Substituting values into the relations developed in Equation (9.13),

$$R_u = \frac{R_2 R_3}{R_1} = \frac{10 \times 5}{159} = 0.3145\ \Omega \quad ; \quad L_u = R_2 R_3 C = \frac{10 \times 5}{1000} = 50 \text{ mH}$$

■

## Example 9.6

Calculate the $Q$ factor for the unknown impedance in Example 9.5 at a supply frequency of 50 Hz.

■

## Solution

$$Q = \frac{\omega L_u}{R_u} = \frac{2\pi 50(0.05)}{0.3145} = 49.9$$

■

*Deflection-type a.c. bridge*

A common deflection type of a.c. bridge circuit is shown in Figure 9.9.

For capacitance measurement:

$$Z_u = 1/j\omega C_u \quad ; \quad Z_1 = 1/j\omega C_1.$$

For inductance measurement (making the simplification that the resistive component of the inductor is small and approximates to zero):

$$Z_u = j\omega L_u \quad ; \quad Z_1 = j\omega L_1.$$

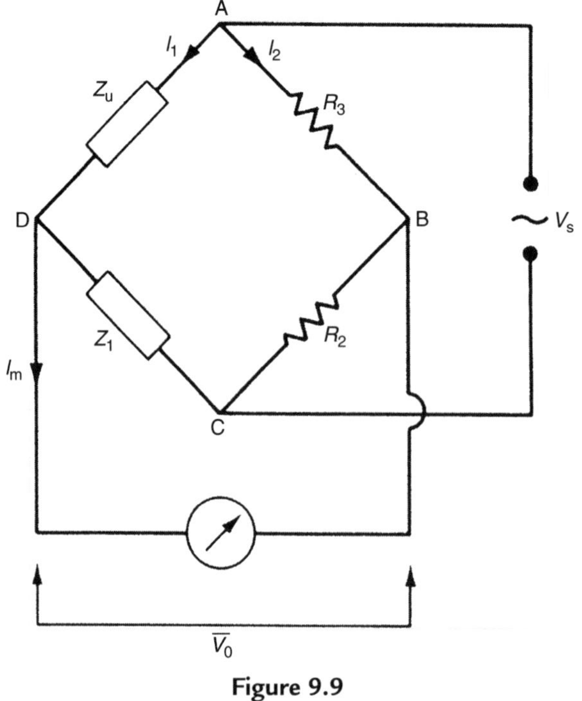


**Figure 9.9**
Common deflection-type a.c. bridge.

Analysis of the circuit to find the relationship between $V_0$ and $Z_u$ is simplified greatly if one assumes that $I_m$ is negligible. This is valid provided that the instrument measuring $V_0$ has high impedance. For $I_m=0$, currents in the two branches of the bridge, as defined in Figure 9.9, are given by

$$I_1 = \frac{V_s}{Z_1 + Z_u} \quad ; \quad I_2 = \frac{V_s}{R_2 + R_3}.$$

Also $V_{AD}=I_1Z_u$ and $V_{AB}=I_2R_3$.

Hence,

$$V_0 = V_{BD} = V_{AD} - V_{AB} = V_s\left(\frac{Z_u}{Z_1 + Z_u} - \frac{R_3}{R_2 + R_3}\right).$$

Thus, for capacitances,

$$V_0 = V_s\left(\frac{1/C_u}{1/C_1 + 1/C_u} - \frac{R_3}{R_2 + R_3}\right) = V_s\left(\frac{C_1}{C_1 + C_u} - \frac{R_3}{R_2 + R_3}\right) \tag{9.14}$$

and, for inductances,

$$V_0 = V_s\left(\frac{L_u}{L_1 + L_u} - \frac{R_3}{R_2 + R_3}\right). \tag{9.15}$$

This latter relationship [Equation (9.15)] is, in practice, only approximate, as inductive impedances are never pure inductances as assumed but always contain a finite resistance (i.e., $Z_u=j\omega L_u+R$). However, the approximation is valid in many circumstances.

## Example 9.7

A deflection bridge as shown in Figure 9.9 is used to measure an unknown capacitance, $C_u$. The components in the bridge have the following values:

$$V_s = 20\ V_{rms}, C_1 = 100\ \mu\text{F}, R_2 = 60\ \Omega, R_3 = 40\ \Omega$$

If $C_u = 100\ \mu\text{F}$, calculate the output voltage, $V_0$.

■

## Solution

From Equation (9.14),

$$V_0 = V_s\left(\frac{C_1}{C_1 + C_u} - \frac{R_3}{R_2 + R_3}\right) = 20(0.5 - 0.4) = 2\ V_{rms}.$$

■

## Example 9.8

An unknown inductance, $L_u$, is measured using a deflection type of bridge as shown in Figure 9.9. Components in the bridge have the following values:

$$V_s = 10\ V_{rms}, L_1 = 20\ mH, R_2 = 100\ \Omega, R_3 = 100\ \Omega$$

If the output voltage $V_0$ is $1V_{rms}$, calculate the value of $L_u$.

## Solution

From Equation (9.15),

$$\frac{L_u}{L_1 + L_u} = \frac{V_0}{V_s} + \frac{R_3}{R_2 + R_3} = 0.1 + 0.5 = 0.6.$$

Thus,

$$L_u = 0.6(L_1 + L_u)\ ;\ 0.4\ L_u = 0.6\ L_1\ ;\ L_u = \frac{0.6 L_1}{0.4} = 30\ mH.$$

### 9.2.5 *Commercial Bridges*

Ready-built bridges are available commercially, although these are substantially more expensive than a "homemade" bridge made up from discrete components and a voltmeter to measure the output voltage.

## 9.3 Resistance Measurement

Devices that convert the measured quantity into a change in resistance include a resistance thermometer, thermistor, wire-coil pressure gauge, and strain gauge. Standard devices and methods available for measuring change in resistance, which is measured in units of *ohms* (Ω), include a d.c. bridge circuit, voltmeter–ammeter method, resistance–substitution method, digital voltmeter, and ohmmeter. Apart from the ohmmeter, these instruments are normally only used to measure medium values of resistance in the range of 1 Ω to 1 MΩ, but this range is entirely adequate for all current sensors that convert the measured quantity into a change in resistance.

### 9.3.1 *d.c. Bridge Circuit*

d.c. bridge circuits, as discussed earlier, provide the most commonly used method of measuring medium value resistance values. The best measurement accuracy is provided by the null output-type Wheatstone bridge, and inaccuracy values of less than ±0.02% are achievable with commercially available instruments. Deflection-type bridge circuits are simpler to use in practice than the null output type, but their measurement accuracy is inferior and the nonlinear output relationship is an additional difficulty. Bridge circuits are particularly useful in converting resistance changes into voltage signals that can be input directly into automatic control systems.

### 9.3.2 Voltmeter–Ammeter Method

The voltmeter–ammeter method consists of applying a measured d.c. voltage across the unknown resistance and measuring the current flowing. Two alternatives exist for connecting the two meters, as shown in Figure 9.10. In Figure 9.10a, the ammeter measures current flowing in both the voltmeter and the resistance. The error due to this is minimized when the measured resistance is small relative to voltmeter resistance. In the alternative form of connection, Figure 9.10b, the voltmeter measures voltage drop across the unknown resistance and the ammeter. Here, the measurement error is minimized when the unknown resistance is large with respect to ammeter resistance. Thus, method (a) is best for measurement of small resistances and method (b) for large ones.

Having thus measured the voltage and current, the value of resistance is then calculated very simply by Ohm's law. This is a suitable method wherever the measurement inaccuracy of up to $\pm 1\%$ that it gives is acceptable.

### 9.3.3 Resistance–Substitution Method

In the voltmeter–ammeter method just given, either the voltmeter is measuring the voltage across the ammeter, as well as across the resistance, or the ammeter is measuring the current flow through the voltmeter, as well as through the resistance. The measurement error caused by this is avoided in the resistance–substitution technique. In this method, the unknown resistance in a circuit is replaced temporarily by a variable resistance. The variable resistance is adjusted until the measured circuit voltage and current are the same as existed with the unknown resistance in place. The variable resistance at this point is equal in value to the unknown resistance.

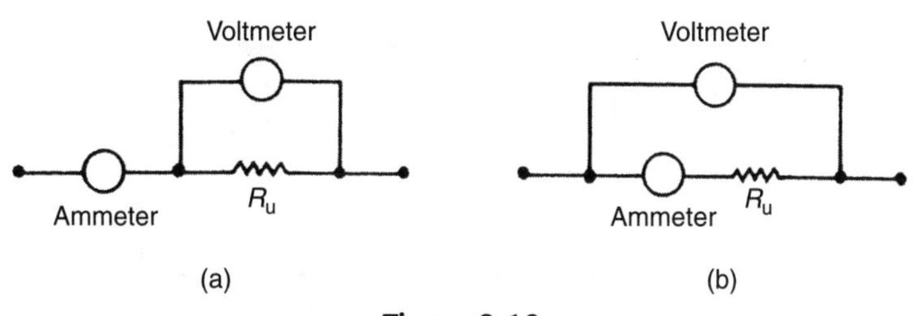


**Figure 9.10**
Voltmeter-ammeter method of measuring resistance.

### 9.3.4 Use of Digital Voltmeter to Measure Resistance

A digital voltmeter can also be used for measuring resistance if an accurate current source is included within it that passes current through the resistance. This can give a measurement inaccuracy as small as $\pm 0.1\%$.

### 9.3.5 Ohmmeter

Ohmmeters are used to measure resistances over a wide range from a few milliohms up to 50 MΩ. The first generation of ohmmeters contained a battery that applied a known voltage across a combination of the unknown resistance and a known resistance in series, as shown in Figure 9.11. Measurement of voltage, $V_m$, across known resistance, $R$, allows unknown resistance, $R_u$, to be calculated from

$$R_u = \frac{R(V_b - V_m)}{V_m},$$

where $V_b$ is the battery voltage. Unfortunately, this mode of resistance measurement gives a typical inaccuracy of $\pm 2\%$, which is only acceptable in a very limited number of applications. Because of this, first-generation ohmmeters have been mostly replaced by a new type of electronic ohmmeter.

The electronic ohmmeter contains two circuits. The first circuit generates a constant current ($I$) that is passed through the unknown resistance. The second circuit measures the voltage ($V$) across the resistance. The resistance is then given by Ohm's law as $R = V/I$. Electronic ohmmeters can achieve measurement inaccuracy as low as $\pm 0.02\%$.

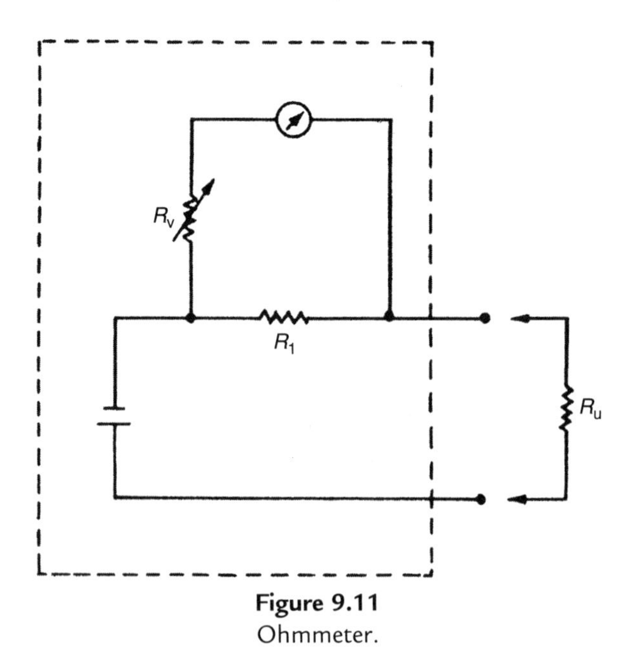


**Figure 9.11**
Ohmmeter.

Most *digital and analogue multimeters* contain circuitry of the same form as in an ohmmeter, and hence can be used similarly to obtain measurements of resistance.

## 9.4 Inductance Measurement

The main device that has an output in the form of a change in inductance is the inductive displacement sensor. Inductance is measured in *Henry* (H). It can only be measured accurately by an a.c. bridge circuit, and various commercial inductance bridges are available. However, when such a commercial inductance bridge is not available immediately, the following method can be applied to give an approximate measurement of inductance.

This approximate method consists of connecting the unknown inductance in series with a variable resistance, in a circuit excited with a sinusoidal voltage, as shown in Figure 9.12. Variable resistance is adjusted until the voltage measured across the resistance is equal to that measured across the inductance. The two impedances are then equal, and value of the inductance, $L$, can be calculated from

$$L = \frac{\sqrt{(R^2 - r^2)}}{2\pi f},$$

where $R$ is the value of the variable resistance, $r$ is the value of the inductor resistance, and $f$ is the excitation frequency.

## 9.5 Capacitance Measurement

Devices that have an output in the form of a change in capacitance include a capacitive level gauge, capacitive displacement sensor, capacitive moisture meter, and capacitive hygrometer. Capacitance is measured in units of *farads* (F). Like inductance, capacitance can only be measured accurately by an a.c. bridge circuit, and various types of capacitance bridges are

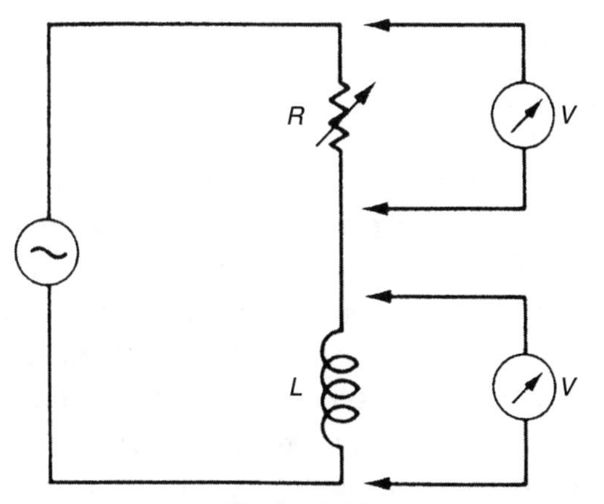


**Figure 9.12**
Approximate method of measuring inductance.

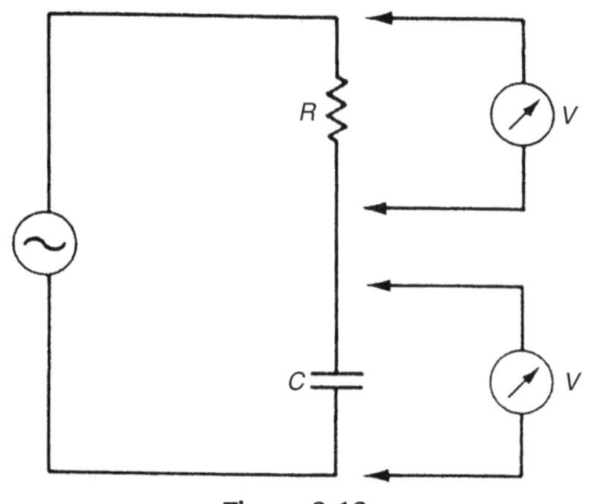


**Figure 9.13**
Approximate method of measuring capacitance.

available commercially. In circumstances where a proper capacitance bridge is not available immediately, and if an approximate measurement of capacitance is acceptable, one of the following two methods can be considered.

The first of these, shown in Figure 9.13, consists of connecting the unknown capacitor in series with a known resistance in a circuit excited at a known frequency. An a.c. voltmeter is used to measure the voltage drop across both the resistor and the capacitor. The capacitance value is then given by

$$C = \frac{V_r}{2\pi f R V_c},$$

where $V_r$ and $V_c$ are the voltages measured across the resistance and capacitance, respectively, $f$ is the excitation frequency, and $R$ is the known resistance.

An alternative approximate method of measurement is to measure the time constant of the capacitor connected in an RC circuit.

## 9.6 Current Measurement

Current measurement is needed for devices such as the thermocouple-gauge pressure sensor and the ionization gauge that have an output in the form of a varying electrical current. It is often also needed in signal transmission systems that convert the measured signal into a varying current. Any of the digital and analogue voltmeters discussed in chapter 7 can measure current if the meter is placed in series with the current-carrying circuit and the same frequency limits apply for the measured signal as they do for voltage measurement. The upper frequency limit for a.c. current measurement can be raised by rectifying the current prior to measurement. To minimize the loading effect on the measured system, any current-measuring instrument must have a small resistance. This is opposite of the case for voltage measurement where the instrument is required to have a high resistance for minimal circuit loading.

In addition to the requirement to measure signal level currents, many measurement applications also require higher magnitude electrical currents to be measured. Hence, the following discussion covers the measurement of currents at both the signal level and higher magnitudes.

Analogue meters are useful in applications where there is a need to display the measured value on a control panel. Moving coil instruments are used as panel meters to measure d.c. current in the milliamp range up to one ampere. Moving iron meters can measure both d.c. and a.c. up to several hundred amps directly. To measure larger currents with electromechanical meters, it is necessary to insert a shunt resistance into the circuit and measure the voltage drop across it. Apart from the obvious disturbance of the measured system, one particular difficulty that results from this technique is the large power dissipation in the shunt. In the case of a.c. current measurement, care must also be taken to match the resistance and reactance of the shunt to that of the measuring instrument so that frequency and waveform distortion in the measured signal are avoided.

*Current transformers* provide an alternative method of measuring high-magnitude currents, which avoids the difficulty of designing a suitable shunt. Different versions of these exist for transforming both d.c. and a.c. currents. A d.c. current transformer is shown in Figure 9.14. The central d.c. conductor in the instrument is threaded through two magnetic cores that carry two high impedance windings connected in series opposition. It can be shown that current flowing in the windings when excited with an a.c. voltage is proportional to the d.c. current in the central conductor. This output current is commonly rectified and then measured by a d.c. voltmeter.

An a.c. current transformer typically has a primary winding consisting of only a few copper turns wound on a rectangular or ring-shaped core. Secondary winding, however, would normally have several hundred turns according to the current step-down ratio required. The

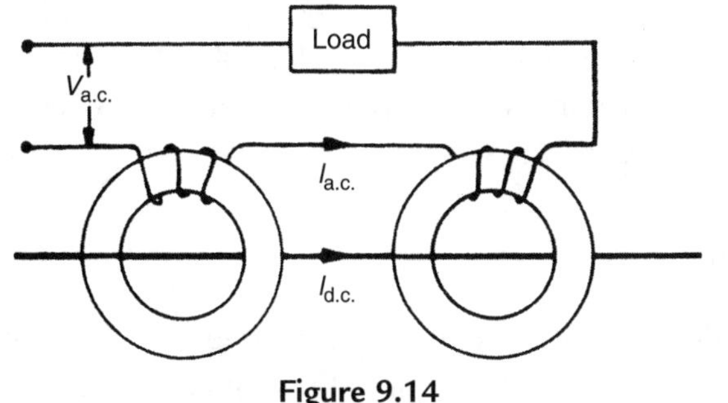


**Figure 9.14**
Current transformer.

output of secondary winding is measured by any suitable current-measuring instrument. The design of current transformers is substantially different from that of voltage transformers. The rigidity of its mechanical construction has to be sufficient to withstand the large forces arising from short-circuit currents, and special attention has to be paid to the insulation between its windings for similar reasons. A low-loss core material is used, and flux densities are kept as small as possible to reduce losses. In the case of very high currents, primary winding often consists of a single copper bar that behaves as a single-turn winding. The clamp-on meter, described in chapter 7, is a good example of this.

All of the other instruments for measuring voltage discussed in Chapter 7 can be applied to current measurement by using them to measure the voltage drop across a known resistance placed in series with the current-carrying circuit. The digital voltmeter is widely applied for measuring currents accurately by this method, and the oscilloscope is used frequently to obtain approximate measurements in circuit test applications. Finally, mention must also be made of the use of digital and analogue multimeters for current measurement, particularly in circuit test applications. These instruments include a set of switchable-dropping resistors and so can measure currents over a wide range. Protective circuitry within such instruments prevents damage when high currents are applied on the wrong input range.

## 9.7 Frequency Measurement

Frequency measurement is required as part of those devices that convert the measured physical quantity into a frequency change, such as a variable reluctance velocity transducer, stroboscopes, vibrating-wire force sensor, resonant wire pressure sensor, turbine flowmeter, Doppler-shift ultrasonic flowmeter, transit-time ultrasonic flowmeter, vibrating level sensor, quartz moisture meter, and quartz thermometer. In addition, the output relationship in some forms of a.c. bridge circuits used for measuring inductance and capacitance requires accurate measurement of the bridge excitation frequency.

Frequency is measured in units of *hertz* (Hz). A digital counter/timer is the most common instrument for measuring frequency. Alternatively, a phase-locked loop can be used. An oscilloscope is also used commonly, especially in circuit test and fault diagnosis applications. Finally, for measurements within the audio frequency range, a Wien bridge is another instrument that is sometimes used.

### 9.7.1 Digital Counter/Timer

A digital counter/timer is the most accurate and flexible instrument available for measuring frequency. Inaccuracy can be reduced down to 1 part in $10^8$, and all frequencies between d.c. and several gigahertz can be measured. The essential component within a counter/timer

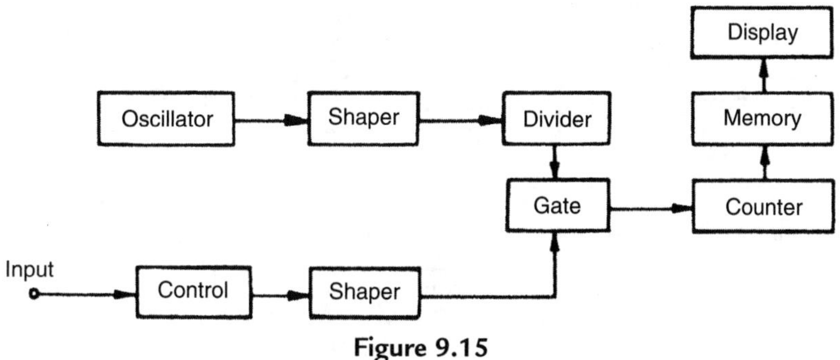


**Figure 9.15**
Digital counter/timer system.

instrument is an oscillator, which provides a very accurately known and stable reference frequency, which is typically either 100 kHz or 1 MHz. This is often maintained in a temperature-regulated environment within the instrument to guarantee its accuracy. The oscillator output is transformed by a pulse shaper circuit into a train of pulses and applied to an electronic gate, as shown in Figure 9.15. Successive pulses at the reference frequency alternately open and close the gate. The input signal of unknown frequency is similarly transformed into a train of pulses and applied to the gate. The number of these pulses that get through the gate during the time that it is open during each gate cycle is proportional to the frequency of the unknown signal.

The accuracy of measurement obviously depends on how far the unknown frequency is above the reference frequency. As it stands, therefore, the instrument can only accurately measure frequencies that are substantially above 1 MHz. To enable the instrument to measure much lower frequencies, a series of decade frequency dividers are provided within it. These increase the time between the reference frequency pulses by factors of 10, and a typical instrument can have gate pulses separated in time by between 1 μs and 1 second.

Improvement in the accuracy of low-frequency measurement can be obtained by modifying the gating arrangements such that the signal of unknown frequency is made to control the opening and closing of the gate. The number of pulses at the reference frequency that pass through the gate during the open period is then a measure of the frequency of the unknown signal.

### *9.7.2 Phase-Locked Loop*

A phase-locked loop is a circuit consisting of a phase-sensitive detector, a voltage-controlled oscillator (VCO), and amplifiers, connected in a closed loop system as shown in Figure 9.16. In a VCO, the oscillation frequency is proportional to the applied voltage. Operation of a phase-locked loop is as follows. The phase-sensitive detector compares the phase of the

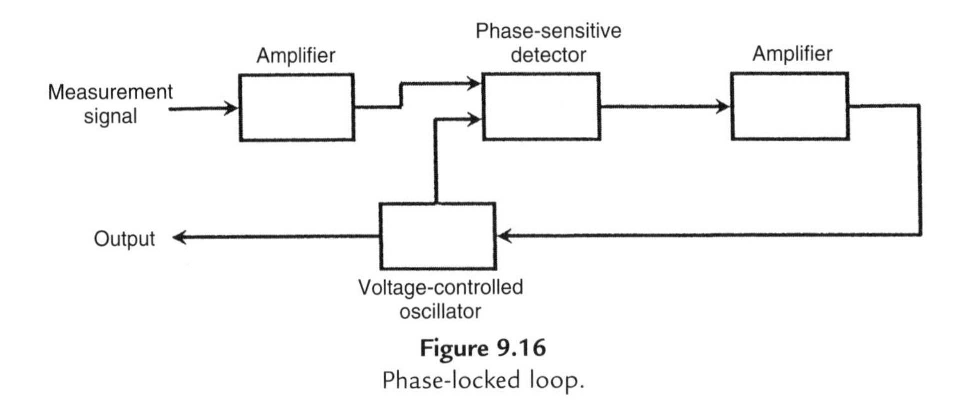


**Figure 9.16**
Phase-locked loop.

amplified input signal with the phase of the VCO output. Any phase difference generates an error signal, which is amplified and fed back to the VCO. This adjusts the frequency of the VCO until the error signal goes to zero, and thus the VCO becomes locked to the frequency of the input signal. The d.c. output from the VCO is then proportional to the input signal frequency.

### 9.7.3 Oscilloscope

Many digital oscilloscopes (particularly the more expensive ones) have a push button on the front panel that causes the instrument to automatically compute and display the frequency of the input signal as a numeric value.

Where this direct facility is not available (in some digital oscilloscopes and all analogue ones), two alternative ways of using the instrument to measure frequency are available. First, the internal time base can be adjusted until the distance between two successive cycles of the measured signal can be read against the calibrated graticule on the screen. Measurement accuracy by this method is limited, but can be optimized by measuring between points in the cycle where the slope of the waveform is steep, generally where it is crossing through from the negative to the positive part of the cycle. Calculation of the unknown frequency from this measured time interval is relatively simple. For example, suppose that the distance between two cycles is 2.5 divisions when the internal time base is set at 10 ms/div. The cycle time is therefore 25 ms and hence the frequency is 1000/25, that is, 40 Hz. Measurement accuracy is dependent on how accurately the distance between two cycles is read, and it is very difficult to reduce the error level below $\pm 5\%$ of the reading.

An alternative way of using an oscilloscope to measure frequency is to generate *Lisajous patterns*. These are produced by applying a known reference frequency sine wave to the $y$ input (vertical deflection plates) of the oscilloscope and the unknown frequency sinusoidal signal to the $x$ input (horizontal deflection plates). A pattern is produced on the screen according to the frequency ratio between the two signals, and if the numerator and denominator in the

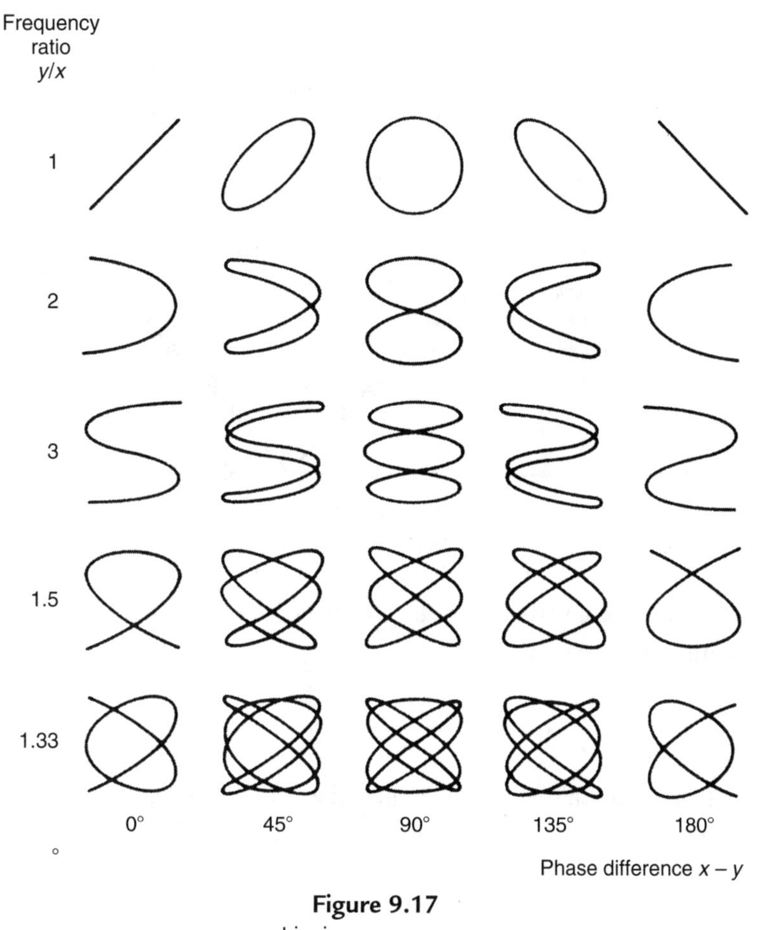


**Figure 9.17**
Lisajous patterns.

ratio of the two signals both represent an integral number of cycles, the pattern is stationary. Examples of these patterns are shown in Figure 9.17, which also shows that phase difference between the waveforms has an effect on the shape. Frequency measurement proceeds by adjusting the reference frequency until a steady pattern is obtained on the screen and then calculating the unknown frequency according to the frequency ratio that the pattern obtained represents.

### *9.7.4 Wien Bridge*

The Wien bridge, shown in Figure 9.18, is a special form of a.c. bridge circuit that can be used to measure frequencies in the audio range. An alternative use of the instrument is as a source of audio frequency signals of accurately known frequency. A simple set of headphones is often used to detect the null-output balance condition. Other suitable instruments for this

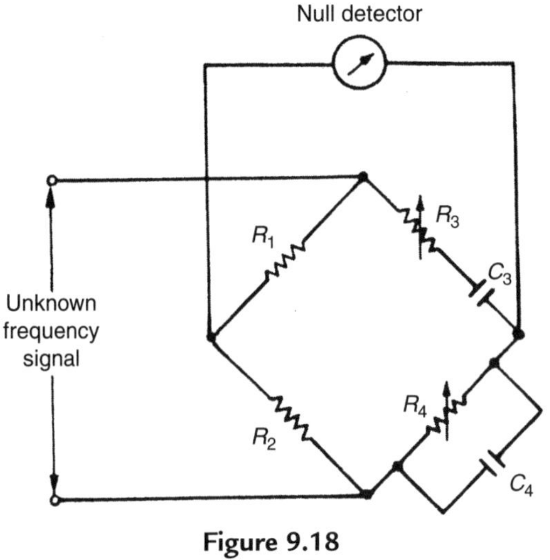


**Figure 9.18**
Wien bridge.

purpose are the oscilloscope and the electronic voltmeter. At balance, the unknown frequency is calculated according to

$$f = \frac{1}{2\pi R_3 C_3}.$$

The instrument is very accurate at audio frequencies, but at higher frequencies, errors due to losses in the capacitors and stray capacitance effects become significant.

## 9.8 Phase Measurement

Instruments that convert the measured variable into a phase change in a sinusoidal electrical signal include a transit time ultrasonic flowmeter, radar level sensor, LVDT, and resolver. The most accurate instrument for measuring the phase difference between two signals is the electronic counter/timer. However, other methods also exist. These include plotting the signals on an *X–Y* plotter using an oscilloscope and a phase-sensitive detector.

### *9.8.1 Electronic Counter/Timer*

In principle, the phase difference between two sinusoidal signals can be determined by measuring the time that elapses between the two signals crossing the time axis. However, in practice, this is inaccurate because zero crossings are susceptible to noise contamination. The normal solution to this problem is to amplify/attenuate the two signals so that they have the same amplitude and then measure the time that elapses between the two signals crossing some nonzero threshold value.

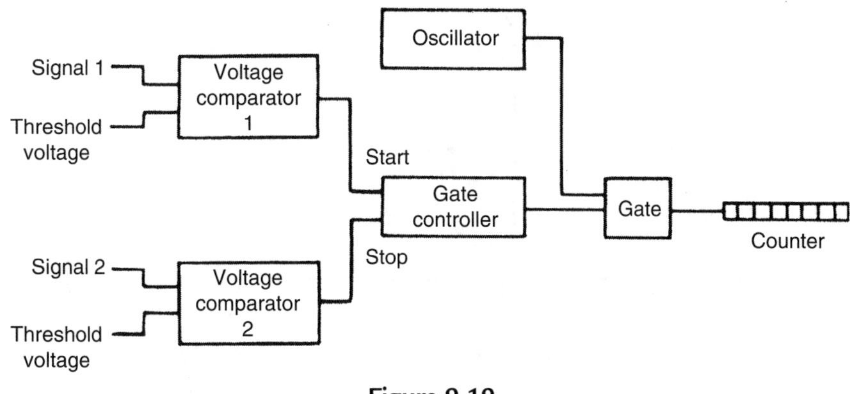


**Figure 9.19**
Phase measurement with digital counter/timer.

The basis of this method of phase measurement is a digital counter/timer with a quartz-controlled oscillator providing a frequency standard, which is typically 10 MHz. The crossing points of the two signals through the reference threshold voltage level are applied to a gate that starts and then stops pulses from the oscillator into an electronic counter, as shown in Figure 9.19. The elapsed time, and hence phase difference, between the two input signals is then measured in terms of the counter display.

### *9.8.2 X–Y Plotter*

This is a useful technique for approximate phase measurement but is limited to low frequencies because of the very limited bandwidth of an *X–Y* plotter. If two input signals of equal magnitude are applied to the *X* and *Y* inputs of a plotter, the plot obtained is an ellipse, as shown in Figure 9.20. If the *X* and *Y* inputs are given by

$$V_X = V\sin(\omega t) \quad ; \quad V_Y = V\sin(\omega t + \phi).$$

At $t=0$, $V_X=0$ and $V_Y=V\sin\phi$. Thus, from Figure 9.20, for $V_X = 0$, $V_Y = \pm h$:

$$\sin\phi = \pm \mathrm{h}/\mathrm{V}. \tag{9.16}$$

Solution of Equation (9.4) gives four possible values for $\phi$ but the ambiguity about which quadrant $\phi$ is in can usually be solved by observing the two signals plotted against time on a dual-beam oscilloscope.

### *9.8.3 Oscilloscope*

As for the case of frequency measurement, many digital oscilloscopes (particularly the more expensive ones) have a push button on the front panel that causes the instrument to automatically compute and display the phase of the input signal as a numeric value.

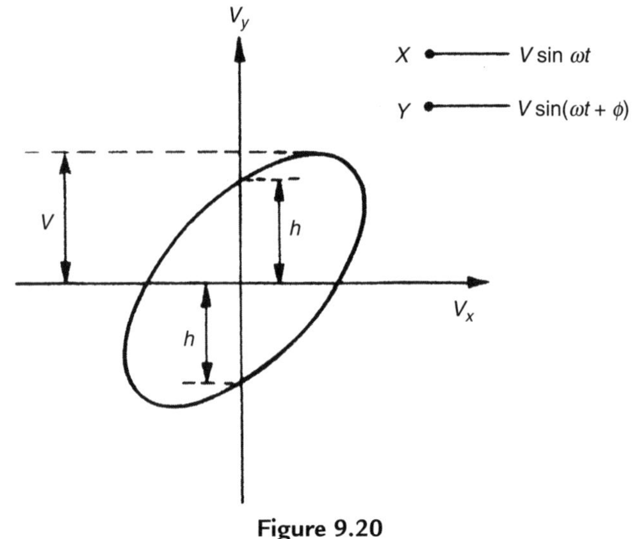


**Figure 9.20**
Phase measurement using $X$–$Y$ plotter.

Where this direct facility is not available (in some digital oscilloscopes and all analogue ones), approximate measurement of the phase difference between signals can be made using any dual-beam oscilloscope. The two signals are applied to the two oscilloscope inputs and a suitable time base is chosen such that the time between the crossing points of the two signals can be measured. The phase difference of both low- and high-frequency signals can be measured by this method, with the upper frequency limit measurable being dictated by the bandwidth of the oscilloscope (which is normally very high).

### *9.8.4 Phase-Sensitive Detector*

A phase-sensitive detector can be used to measure the phase difference between two signals that have an identical frequency. This can be exploited in measurement devices such as the varying-phase output resolver (see Chapter 20).

## 9.9 Summary

This chapter has been concerned with looking at ways of dealing with outputs from a measurement sensor that are not in the form of a readily measureable voltage signal. We started off by identifying the various alternative forms of output that we might have to deal with. Our list included translational displacement change outputs; changes in various electrical parameters, such as resistance, inductance, capacitance, and current; and changes in the phase or frequency of an a.c. electrical signal.

This led us to realize that we needed mechanisms for converting these sensor outputs that are initially in some nonvoltage form into a more convenient form. Such mechanisms are collectively called variable conversion elements. Because mechanisms for measuring translational displacements are needed for other purposes as well, we have deferred consideration of these until Chapter 19, where the subject of translational measurement is considered in detail. The rest of this chapter has therefore only been concerned with looking at the ways of dealing with nonvoltage electrical parameter outputs and outputs in the form of a frequency of phase change in an electrical signal.

We learned first of all in this study that bridge circuits are a particularly important type of variable conversion element, and we therefore went on to cover these in some detail. One particularly important thing that we learned was that bridge circuits exist in two forms, null type and deflection type. Of these, null types are more tedious to use but provide better measurement accuracy, leading to these being the preferred form when sensors are being calibrated. We noted also that both d.c. and a.c. bridges exist, with the former being used to interpret the output of sensors that exhibit a change in resistance and the latter for sensors that convert the measured quantity into a change in either inductance or capacitance.

We then went on to look at the ways of dealing with sensor outputs in other forms. In turn, we covered resistance measurement (alternative ways to using a d.c. bridge circuit), inductance measurement, capacitance measurement, and current measurement. Finally, we looked at ways of interpreting the output of sensors that is in the form of a change in either the frequency or the phase of an electrical signal.

## 9.10 Problems

9.1. Explain what a d.c. bridge circuit is and why it is so useful in measurement systems. List a few measurement sensors for which you would commonly use a d.c. bridge circuit to convert the sensor output into a change in output voltage of the bridge.

9.2. If elements in the d.c. bridge circuit shown in Figure 9.2 have the following values—$R_u = 110\,\Omega, R_1 = 100\,\Omega, R_2 = 1000\,\Omega, R_3 = 1000\,\Omega, V_i = 10$ V—calculate output voltage $V_0$ if the impedance of the voltage-measuring instrument is assumed to be infinite. [Hint: Apply Equation (9.3).]

9.3. Suppose that the resistive components in the d.c. bridge shown in Figure 9.2 have the following nominal values: $R_u = 3$ KΩ ; $R_1 = 6$ KΩ ; $R_2 = 8$ KΩ ; $R_3 = 4$ KΩ. The actual value of each resistance is related to the nominal value according to $R_{\text{actual}} = R_{\text{nominal}} + \partial R$, where ∂R has the following values: $\partial R_u = 30\,\Omega$ ; $\partial R_1 = -20\,\Omega$ ; $\partial R_2 = 40\,\Omega$ ; $\partial R_3 = -50\,\Omega$. Calculate the open circuit bridge output voltage if bridge supply voltage $V_i$ is 50 V.

9.4. (a) Suppose that the unknown resistance $R_u$ in Figure 9.2 is a resistance thermometer whose resistance at 100°C is 500 Ω and whose resistance varies with temperature at the rate of 0.5 Ω/°C for small temperature changes around 100°C. Calculate the sensitivity of the total measurement system for small changes in temperature around 100°C, given the following resistance and voltage values measured at 15°C by instruments calibrated at 15°C: $R_1 = 500\ \Omega$ ; $R_2 = R_3 = 5000\ \Omega$ ; $V_i = 10$ V.

(b) If the resistance thermometer is measuring a fluid whose true temperature is 104°C, calculate the error in the indicated temperature if the ambient temperature around the bridge circuit is 20°C instead of the calibration temperature of 15°C, given the following additional information:

Voltage-measuring instrument zero drift coefficient = +1.3 mV/°C
Voltage-measuring instrument sensitivity drift coefficient = 0
Resistances $R_1$, $R_2$, and $R_3$ have a positive temperature coefficient of +0.2% of nominal value/°C
Voltage source $V_i$ is unaffected by temperature changes.

9.5. Four strain gauges of resistance 120 Ω each are arranged into a d.c. bridge configuration such that each of the four arms in the bridge has one strain gauge in it. The maximum permissible current in each strain gauge is 100 mA. What is the maximum bridge supply voltage allowable, and what power is dissipated in each strain gauge with that supply voltage?

9.6. (a) Suppose that the variables shown in Figure 9.2 have the following values: $R_1 = 100\ \Omega$, $R_2 = 100\ \Omega$, $R_3 = 100\ \Omega$; $V_i = 12$ V. $R_u$ is a resistance thermometer with a resistance of 100 Ω at 100°C and a temperature coefficient of +0.3 Ω/°C over the temperature range from 50 to 150°C (i.e., the resistance increases as the temperature goes up). Draw a graph of bridge output voltage $V_0$ for 10-degree steps in temperature between 100 and 150°C [calculating $V_0$ according to Equation (9.3)].

(b) Discuss briefly whether you expect the graph that you have just drawn to be a straight line.

(c) Draw a graph of $V_0$ for similar temperature values if $R_2 = R_3 = 1000\ \Omega$ and all other components have the same values as given in part (a). Note that the line through the data points is straighter than that drawn in part (a) but the output voltage is much less at each temperature point.

(d) Discuss briefly the change in linearity of the graph drawn for part (c) and the change in measurement sensitivity compared with the graph drawn for part (a).

9.7. The unknown resistance $R_u$ in a d.c. bridge circuit, connected as shown in Figure 9.4a, is a resistance thermometer. The thermometer has a resistance of 350 Ω at 50°C and its temperature coefficient is +1 Ω/°C (the resistance increases as the temperature rises). The components of the system have the following values: $R_1 = 350\ \Omega$, $R_2 = R_3 = 2\ \mathrm{K}\Omega$, $R_m = 20\ \mathrm{K}\Omega$, $V_i = 5$ V. What is the output voltage reading when the temperature is 100°C? [Hint: Use Equation (9.10).]

9.8. The active element in a load cell is a strain gauge with a nominal resistance of 500 Ω in its unstressed state. The cell has a sensitivity of +0.5 Ω per Newton of applied force and is connected in a d.c. bridge circuit where the other three arms of the bridge each have a resistance of 500 Ω.

(a) If the bridge excitation voltage is 20 V, what is the measurement sensitivity of the system in volts per Newton for small applied forces?

(b) What is the bridge output voltage when measuring an applied force of 500 Newtons?

9.9. Suppose that the unknown resistance $Ru$ in Figure 9.2 is a resistance thermometer whose resistance at 100°C is 600 Ω and whose resistance varies with temperature at the rate of +0.4 Ω/°C for small temperature changes around 100°C. Calculate the sensitivity of the total measurement system for small changes in temperature around 100°C, given the following resistance and voltage values:

$$R_1 = 600\ \Omega;\ R_2 = R_3 = 6000\ \Omega;\ V_i = 20\ \text{V}$$

Assume that the ambient temperature around the bridge circuit was the same as that at which the voltage-measuring instrument and all bridge component values were calibrated.

9.10. The unknown resistance $R_u$ of a resistance thermometer is measured by a deflection-type bridge circuit of the form shown in Figure 9.2, where the parameters have the following values:

$$R_1 = 100\ \Omega;\ R_2 = R_3 = 1000\ \Omega;\ V_i = 20\ \text{V}$$

The thermometer has a resistance of 100 Ω at 0°C and the resistance varies with temperature at the rate of 0.4 Ω/°C for small temperature changes around 0°C.

(a) Calculate the bridge sensitivity in units of volts per ohm.

(b) Calculate the sensitivity of the total measurement system in units of volts/°C for small temperature changes around 0°C.

9.11. The unknown resistance $R_u$ of a strain gauge is to be measured by a bridge circuit of the form shown in Figure 9.5 but where the bridge components and the excitation voltage are different to the values shown in Figure 9.5, having instead the following values:

Nominal strain gauge resistance = 100 Ω; $R_m$ = 10 KΩ (unchanged)

$R_1 = 100\ \Omega$, $R_2 = 1000\ \Omega$, $R_3 = 1000\ \Omega$, $V_i = 10$ V

(a) Using Thévenin's theorem, derive an expression for the sensitivity of the bridge in terms of the change in output voltage $V_m$ that occurs when there is a small change in the resistance of the strain gauge.

(b) If the strain gauge is part of a pressure transducer with a sensitivity of 400 mΩ/bar, calculate the pressure measurement sensitivity in bridge output volts ($V_m$) per bar.

9.12. The unknown resistance $R_u$ of a thermistor is to be measured by a bridge circuit of the form shown in Figure 9.5 but where the bridge components and the excitation voltage are different to the values shown in Figure 9.5, having instead the following values:

$$R_1 = 1000\ \Omega,\ R_2 = 1000\ \Omega,\ R_3 = 1000\ \Omega,\ V_i = 10\ \text{V},\ R_m = 20\ \text{K}\Omega$$

The resistance ($R_u$) of the thermistor is related to the measured temperature (T) in degrees kelvin (°K) according to the following expression:

$$R_u = 1000 \exp\left[3675\left(\frac{1}{T} - 0.003354\right)\right]$$

Draw a graph of the bridge output in volts for values of the measured temperature in steps of 5°C between 0 and 50°C.

9.13. In the d.c. bridge circuit shown in Figure 9.21, the resistive components have the following values: $R_1 = R_2 = 120\ \Omega$ ; $R_3 = 117\ \Omega$ ; $R_4 = 123\ \Omega$ ; $R_a = R_p = 1000\ \Omega$.
(a) What are the resistance values of the parts of the potentiometer track either side of the slider when the potentiometer is adjusted to balance the bridge?
(b) What then is the effective resistance of each of the two left-hand arms of the bridge when the bridge is balanced?

9.14. List a few measurement transducers and sensors for which you would commonly use an a.c. bridge circuit to convert the sensor output into a change in output voltage of the bridge.

9.15. A Maxwell bridge, designed to measure the unknown impedance ($R_u$, $L_u$) of a coil, is shown in Figure 9.8.

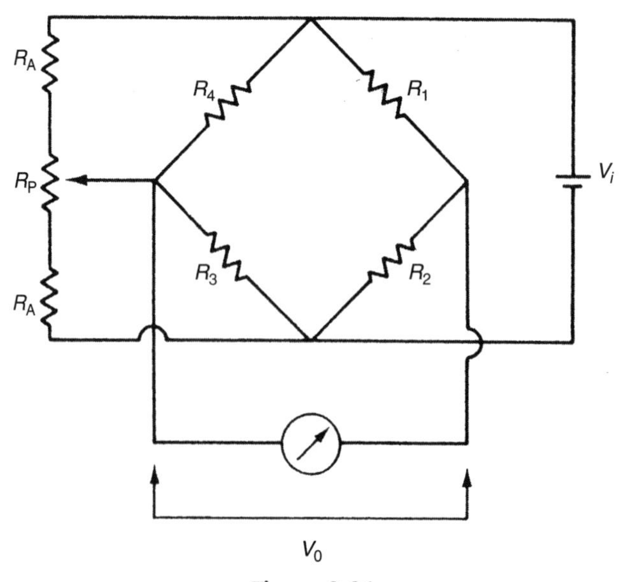


**Figure 9.21**
A d.c. bridge with apex balancing.

(a) Derive an expression for $R_u$ and $L_u$ under balance conditions.
(b) If the fixed bridge component values are $R_3 = 100\ \Omega$ and $C = 20\ \mu F$, calculate the value of the unknown impedance if $R_1 = 3183\ \Omega$ and $R_2 = 50\ \Omega$ at balance.
(c) Calculate the $Q$ factor for the coil if the supply frequency is 50 Hz.

9.16. A deflection bridge as shown in Figure 9.9 is used to measure an unknown inductance, $L_u$. The components in the bridge have the following values: $V_s = 30\ V_{rms}$, $L_1 = 80$ mH, $R_2 = 70\ \Omega$, $R_3 = 30\ \Omega$. If $L_u = 50$ mH, calculate the output voltage, $V_0$.

9.17. An unknown capacitance, $C_u$, is measured using a deflection bridge as shown in Figure 9.9. The components of the bridge have the following values: $V_s = 10\ V_{rms}$, $C_1 = 50\ \mu F$, $R_2 = 80\ \Omega$, $R_3 = 20\ \Omega$. If the output voltage is 3 $V_{rms}$, calculate the value of $C_u$.

9.18. A Hays bridge is often used for measuring the inductance of high $Q$ coils and has the configuration shown in Figure 9.22.
(a) Obtain the bridge balance conditions.
(b) Show that if the $Q$ value of an unknown inductor is high, the expression for the inductance value when the bridge is balanced is independent of frequency.
(c) If the $Q$ value is high, calculate the value of the inductor if the bridge component values at balance are as follow: $R_2 = R_3 = 1000\ \Omega$ ; $C = 0.02\ \mu F$.

9.19. Discuss alternative methods of measuring the frequency of an electrical signal and indicate the likely measurement accuracy obtained with each method.

9.20. Using the Lisajous figure method of measuring frequency, a reference frequency signal of 1 kHz is applied to the Y channel of an oscilloscope and the unknown frequency is applied to the X channel. Determine the unknown frequency for each of the oscilloscope displays shown in Figure 9.23. Also indicate the approximate phase difference of the unknown signal with respect to the reference signal.

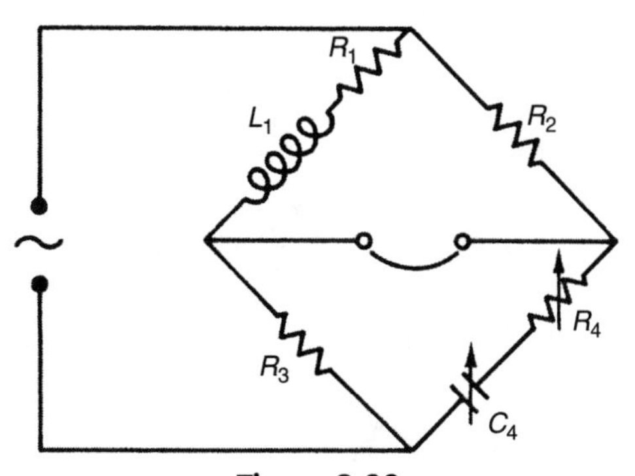


**Figure 9.22**
Hays bridge.

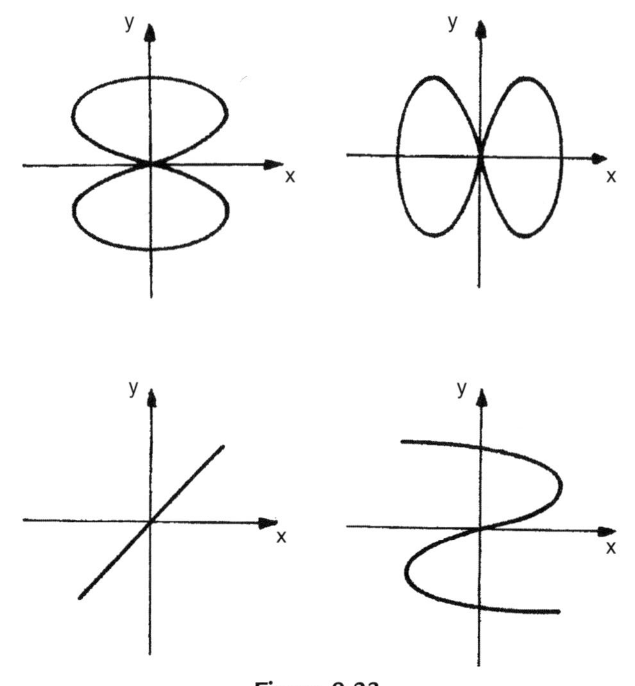


**Figure 9.23**
Oscilloscope displays for Problem 9.20.

CHAPTER 10

# Measurement Signal Transmission



## 10.1 Introduction

There is often a necessity in many measurement systems to transmit measurement signals over quite large distances from the point of measurement to the place where the signals are recorded and/or used in a process control system. The need to separate the processing/recording parts of a measurement system from the point of measurement can arise for several reasons. One major reason for the separation is the environment around the point of measurement, which is often hostile toward one or more components in the rest of the measurement system. Extremes of temperature, humidity, weather, or fumes are typical examples of environments at the point of measurement that are too hostile for other measurement system components. Remoteness of the point of measurement can be another reason for transmitting measured signals to another point. We often see this problem in environmental and weather monitoring systems. One example is water quality measurement in rivers, where sensors may actually be anchored in the river. In this sort of situation, no mains-powered electricity supply is available. While it is possible to use battery or solar power for the sensors themselves, mains power is normally needed for the necessary signal processing. Furthermore, there are usually no buildings available to protect the signal processing elements from the environment. For these reasons, transmission of the measurements to another point is necessary.


Measurement and Instrumentation: Theory and Application

We therefore devote this chapter to a study of the various ways in which measurement data can be transmitted. We will see that the need to transmit measurement signals over what can sometimes be large distances creates several problems that we will investigate. As discovered in the following pages, of the many difficulties associated with long distance signal transmission, contamination of the measurement signal by noise is the most serious. Many sources of noise exist in industrial environments, such as radiated electromagnetic fields from electrical machinery and power cables, induced electromagnetic fields through wiring loops, and spikes (large transient voltages) that sometimes occur on the mains a.c. power supply. Our investigation into signal transmission techniques shows us that signals can be transmitted electrically, pneumatically, optically, or by radiotelemetry in either analogue or digital format. We also discover that optical data transmission can be further divided into fiber-optic transmission and optical wireless transmission, according to whether a fiber-optic cable or just a plain air path is used as the transmission medium.

## 10.2 Electrical Transmission

The simplest method of electrical transmission is to transmit the measurement signal as a varying analogue voltage. However, this mode of transmission often causes the measurement signal to become corrupted by noise. To avoid such corruption, the signal can be transmitted as a varying current instead of as a varying voltage. An alternative solution is to transmit the signal by superimposing it on an a.c. carrier system. All of these methods are discussed here.

### *10.2.1 Transmission as Varying Voltages*

As most signals already exist in an electrical form as varying analogue voltages, the simplest mode of transmission is to maintain the signals in the same form. However, electrical transmission suffers problems of signal attenuation and also exposes signals to corruption through induced noise. Therefore, special measures have to be taken to overcome these problems.

Because output signal levels from many types of measurement transducers are very low, *signal amplification* prior to transmission is essential if a reasonable signal-to-noise ratio is to be obtained after transmission. Amplification at the input to the transmission system is also required to compensate for the attenuation of the signal that results from the resistance of the signal wires.

It is also usually necessary to provide *shielding* for the signal wires. Shielding consists of surrounding the signal wires in a cable with a metal shield connected to the earth. This provides a high degree of noise protection, especially against capacitive-induced noise due to the proximity of signal wires to high-current power conductors.

### 10.2.2 Current Loop Transmission

The signal-attenuation effect of conductor resistances can be minimized if varying voltage signals are transmitted as varying current signals. This technique, which also provides high immunity to induced noise, is known as current loop transmission and uses currents in the range between 4 and 20 mA* to represent the voltage level of the analogue signal. It requires a voltage-to-current converter of the form shown in Figure 10.1, which is commonly known as a *4- to 20-mA current loop interface*. Two voltage-controlled current sources are used, one providing a constant 4-mA output used as the power supply current and the other providing a variable 0- to 16-mA output scaled and proportional to the input voltage level. The net output current therefore varies between 4 and 20 mA, corresponding to analogue signal levels between zero and the maximum value. Use of a positive, nonzero current level to represent a zero value of the transmitted signal enables transmission faults to be identified readily. If the transmitted current is zero, this automatically indicates the presence of a transmission fault, as the minimum value of current that represents a proper signal is 4 mA.

Current to voltage conversion is usually required at termination of the transmission line to change the transmitted currents back to voltages. An operational amplifier, connected as shown in Figure 10.2, is suitable for this purpose. Output voltage $V$ is simply related to the input current $I$ by $V = IR$.

The advent of intelligent devices has also led to the development of a modified current loop interface known as the *extended 4- to 20-mA current interface protocol*. This provides for the transmission of command/status information and the device power supply in analogue form on

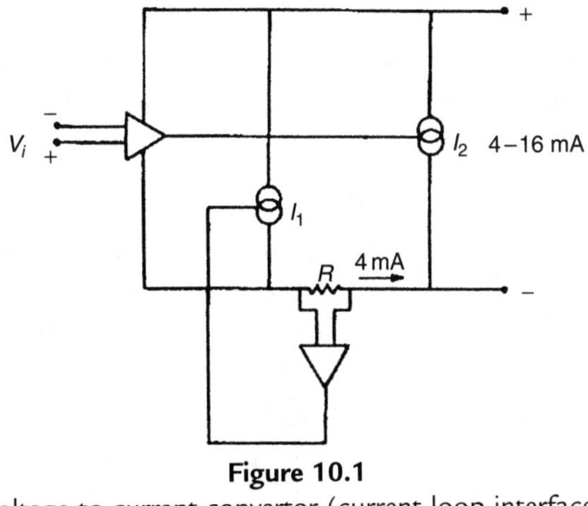


**Figure 10.1**
Voltage-to-current convertor (current loop interface).

* The 4- to 20-mA standard was agreed in 1972, prior to which a variety of different current ranges were used for signal transmission.

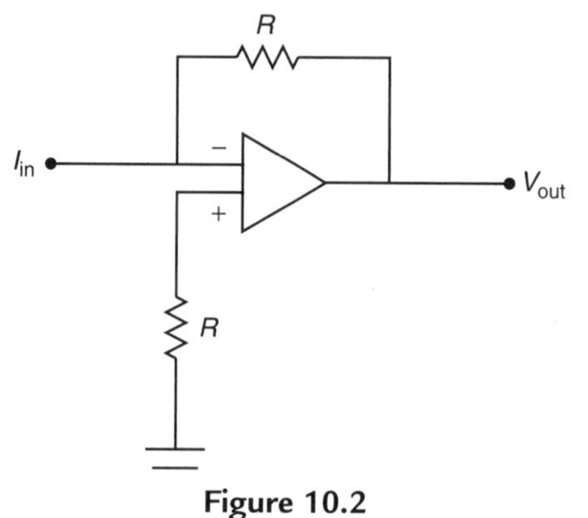


**Figure 10.2**
Current-to-voltage convertor.

signal wires. In this extended protocol, signals in the range of 3.8 to 20.5 mA are regarded as "normal" measurement signals, thus allowing for under- and overrange from the 4- to 20-mA measurement signal standard. The current bands immediately outside this in the range of 3.6 to 3.8 mA and 20.5 to 21.0 mA are used for the conveyance of commands to the sensor/transmitter and the receipt of status information from it. This means that, if signal wires are also used to carry the power supply to the sensor/transmitter, the power supply current must be limited to 3.5 mA or less to avoid the possibility of it being interpreted as a measurement signal or fault indicator. Signals greater than 21 mA (and less than 3.6 mA if signal wires are not carrying a power supply) are normally taken to indicate either a short circuit or an open circuit in the signal wiring.

### *10.2.3 Transmission Using an a.c. Carrier*

Another solution to the problem of noise corruption in low-level d.c. voltage signals is to transfer the signal onto an a.c. carrier system before transmission and extract it from the carrier at the end of the transmission line. Both amplitude modulation (AM) and frequency modulation (FM) can be used for this.

*Amplitude modulation* consists of translating the varying voltage signal into variations in the amplitude of a carrier sine wave at a frequency of several kilohertz. An a.c. bridge circuit is used commonly for this as part of the system for transducing the outputs of sensors that have a varying resistance (R), capacitance (C), or inductance (L) form of output. Referring back to Equations (9.14) and (9.15) in Chapter 9, for a sinusoidal bridge excitation voltage of $V_s = V\sin(\omega t)$, the output can be represented by $V_o = FV\sin(\omega t)$. $V_o$ is a sinusoidal voltage at the same frequency as the bridge excitation frequency, and its amplitude, $FV$, represents the magnitude of the sensor input (R, C, or L) to the bridge. For example, in the case of Equation (9.15):

$$FV = \left(\frac{L_u}{L_1 + L_u} - \frac{R_3}{R_2 + R_3}\right)V.$$

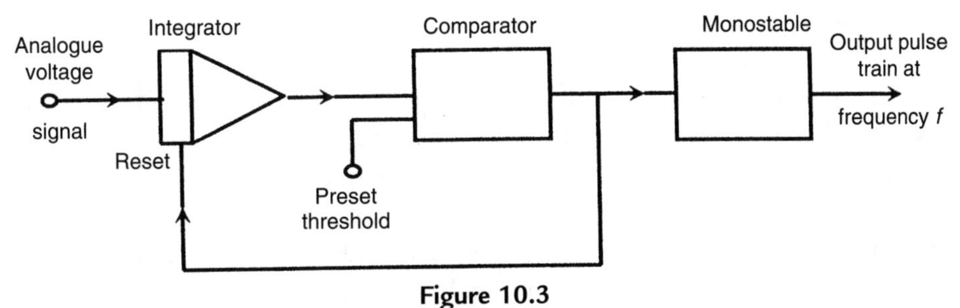


**Figure 10.3**
Voltage-to-frequency convertor.

After shifting the d.c. signal onto a high-frequency a.c. carrier, a high-pass filter can be applied to the AM signal. This successfully rejects noise in the form of low-frequency drift voltages and mains interference. At the end of the transmission line, demodulation is carried out to extract the measurement signal from the carrier.

*Frequency modulation* achieves even better noise rejection than AM and involves translating variations in an analogue voltage signal into frequency variations in a high-frequency carrier signal. A suitable voltage-to-frequency conversion circuit is shown in Figure 10.3 in which the analogue voltage signal input is integrated and applied to the input of a comparator preset to a certain threshold voltage level. When this threshold level is reached, the comparator generates an output pulse that resets the integrator and is also applied to a monostable. This causes the frequency, $f$, of the output pulse train to be proportional to the amplitude of the input analogue voltage.

At the end of the transmission line, the FM signal is usually converted back to an analogue voltage by a frequency-to-voltage converter. A suitable conversion circuit is shown in Figure 10.4 in which the input pulse train is applied to an integrator that charges up for a specified time. The charge on the integrator decays through a leakage resistor, and a balance voltage is established between the input charge on the integrator and the decaying charge at the output. This output balance voltage is proportional to the input pulse train at frequency $f$.

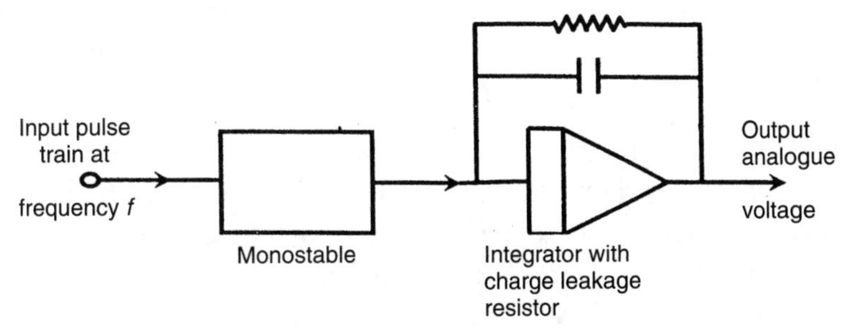


**Figure 10.4**
Frequency-to-voltage convertor.

## 10.3 Pneumatic Transmission

In recent years, pneumatic transmission tends to have been replaced by other alternatives in most new implementations of instrumentation systems, although many examples can still be found in operation in process industries. Pneumatic transmission consists of transmitting analogue signals as a varying pneumatic pressure level that is usually in the range of 3–15 p.s.i. (Imperial units are still used commonly in process industries, although the equivalent range in SI units is 207–1034 mbar, which is often rounded to 200–1000 mbar in metric systems.) A few systems also use alternative ranges of 3–27 or 6–48 p.s.i. Frequently, the initial signal is in the form of a varying voltage level that is converted into a corresponding pneumatic pressure. However, in some examples of pneumatic transmission, the signal is in varying current form to start with, and a current-to-pressure converter is used to convert the 4- to 20-mA current signals into pneumatic signals prior to transmission. Pneumatic transmission has the advantage of being intrinsically safe, and it provides similar levels of noise immunity to current loop transmission. However, one disadvantage of using air as the transmission medium is that the transmission speed is much slower than electrical or optical transmission. A further potential source of error would arise if there were a pressure gradient along the transmission tube. This would introduce a measurement error because air pressure changes with temperature.

Pneumatic transmission is found particularly in pneumatic control systems where sensors, actuators, or both are pneumatic. Typical pneumatic sensors are the pressure thermometer (see Chapter 14) and the motion-sensing nozzle flapper (see Chapter 19); a typical actuator is a pneumatic cylinder that converts pressure into linear motion. A pneumatic amplifier is often used to amplify the pneumatic signal to a suitable level for transmission.

## 10.4 Fiber-Optic Transmission

Light has a number of advantages over electricity as a medium for transmitting information. For example, it is intrinsically safe, and noise corruption of signals by neighboring electromagnetic fields is almost eliminated. The most common form of optical transmission consists of transmitting light along a fiber-optic cable, although wireless transmission also exists as described in Section 10.5.

Apart from noise reduction, optical signal attenuation along a fiber-optic link is much less than electric signal attenuation along an equivalent length of metal conductor. However, there is an associated cost penalty because of the higher cost of a fiber-optic system compared with the cost of metal conductors. In short fiber-optic links, cost is dominated by the terminating transducers needed to transform electrical signals into optical ones and vice versa. However, as the length of the link increases, the cost of the fiber-optic cable itself becomes more significant.

Fiber-optic cables are used for signal transmission in three distinct ways. First, relatively short fiber-optic cables are used as part of various instruments to transmit light from conventional sensors to a more convenient location for processing, often in situations where space is very short at the point of measurement. Second, longer fiber-optic cables are used to connect remote instruments to controllers in instrumentation networks. Third, even longer links are used for data transmission systems in telephone and computer networks. These three application classes have different requirements and tend to use different types of fiber-optic cable.

Signals are normally transmitted along a fiber-optic cable in digital format, although analogue transmission is sometimes used. If there is a requirement to transmit more than one signal, it is more economical to multiplex the signals onto a single cable rather than transmit the signals separately on multiple cables. *Multiplexing* involves switching the analogue signals in turn, in a synchronized sequential manner, into an analogue-to-digital converter that outputs onto the transmission line. At the other end of the transmission line, a digital-to-analogue converter transforms the digital signal back into analogue form and it is then switched in turn onto separate analogue signal lines.

### 10.4.1 Principles of Fiber Optics

The central part of a fiber-optic system is a light-transmitting cable containing one or more fibers made from either glass or plastic. This is terminated at each end by a transducer, as shown in Figure 10.5. At the input end, the transducer converts the signal from the electrical form in which most signals originate into light. At the output end, the transducer converts the transmitted light back into an electrical form suitable for use by data recording, manipulation, and display systems. These two transducers are often known as the transmitter and receiver, respectively.

*Fiber-optic cable* consists of an inner cylindrical core surrounded by a cylindrical cladding and a protective jacket, as shown in Figure 10.6. The jacket consists of several layers of polymer material and serves to protect the core from mechanical shocks that might affect its optical or physical properties. The refractive index of the inner core is greater than that of the surrounding cladding material, and the relationship between the two refractive indices affects the transmission characteristics of light along the cable. The amount of attenuation of light as it travels along the cable varies with the wavelength of the light transmitted. This characteristic is very nonlinear, and a graph of attenuation against wavelength shows a number of peaks

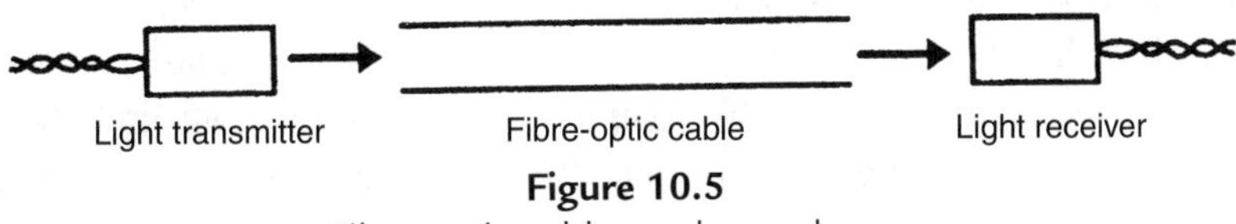


**Figure 10.5**
Fiber-optic cables and transducers.

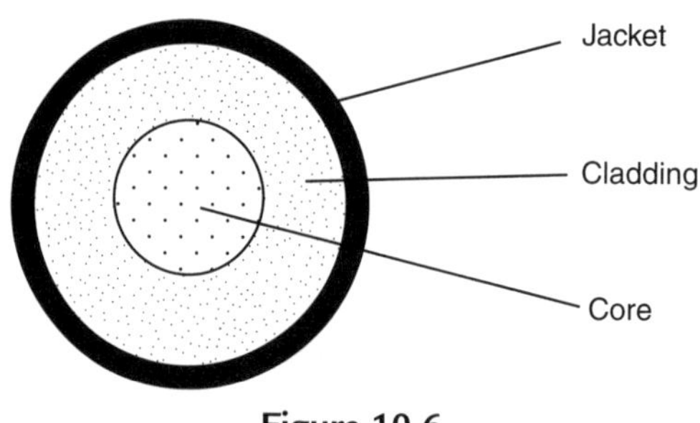


**Figure 10.6**
Cross section through fiber-optic cable.

and troughs. The position of these peaks and troughs varies according to the material used for the fibers. It should be noted that fiber manufacturers rarely mention these nonlinear attenuation characteristics and quote the value of attenuation that occurs at the most favorable wavelength.

Two forms of cables exist, known as single mode and multimode. Single mode cables (sometimes known as monomode cables) have a small diameter core, typically 6 μm, whereas multimode cables have a much larger core, typically between 50 and 200 μm in diameter. Both glass and plastic in different combinations are used in various forms of cable. One option is to use different types of glass fiber for both the core and the cladding. A second, and less expensive, option is to have a glass fiber core and a plastic cladding. This has the additional advantage of being less brittle than the all-glass version. Finally, all-plastic cables also exist, where two types of plastic fiber with different refractive indices are used. This is the least expensive form of all, but has the disadvantage of having high attenuation characteristics, making it unsuitable for transmission of light over medium to large distances.

Protection is normally given to the cable by enclosing it in the same types of insulating and armoring materials used for copper cables. This protects the cable against various hostile operating environments and also against mechanical damage. When suitably protected, fiber-optic cables can even withstand being engulfed in flames.

The *fiber-optic transmitter* is responsible for converting the electric signal from a measurement sensor into light and transferring this into the fiber-optic cable. It is theoretically possible to encode the measurement signal by modulating the intensity, frequency, phase, or polarization of the light injected into the cable, but light intensity modulation has now achieved a dominant position in most fiber-optic transmission systems. Either laser diodes or light-emitting diodes (LED) can be used as the light source in the transmitter. Laser diodes generate coherent light of a higher power than the incoherent light produced by LED. However, laser diodes are more complex, more expensive, and less reliable than LED. Also the relationship between the input current and the light output is more linear for light-emitting diodes. Hence, the latter are preferred in most applications.

The characteristics of the light source chosen for the transmitter must closely match the attenuation characteristics of the light path through the cable and the spectral response of the receiving transducer. This is because the proportion of the power from the light source coupled into the fiber-optic cable is more important than the absolute output power of the emitted light. This proportion is maximized by making purpose-designed transmitters that have a spherical lens incorporated into the chip during manufacture. This produces an approximately parallel beam of light into the cable with a typical diameter of 400 μm.

The proportion of light entering the fiber-optic cable is also governed by the quality of the end face of the cable and the way it is bonded to the transmitter. A good end face can be produced by either polishing or cleaving. Polishing involves grinding the fiber end down with progressively finer polishing compounds until a surface of the required quality is obtained. Attachment to the transmitter is then normally achieved by gluing. This is a time-consuming process but uses inexpensive materials. Cleaving makes use of special kits that nick the fiber, break it very cleanly by applying mechanical force, and then attach it to the transmitter by crimping. This is a much faster method but cleaving kits are quite expensive. Both methods produce good results.

A further factor that affects the proportion of light transmitted into the optic fibers in the cable is transmitter alignment. It is very important to achieve proper alignment of the transmitter with the center of the cable. The effect of misalignment depends on the relative diameters of the beam and the core of the cable. Figure 10.7 shows the effect on the proportion of power transmitted into the cable for cases of (a) cable core diameter > beam diameter, (b) cable core diameter = beam diameter, and (c) cable core diameter < beam diameter. This shows that some degree of misalignment can be tolerated except where the beam and cable core diameters are equal. The cost of producing exact alignment of the transmitter and cable is very high, as it requires the light source to be aligned exactly in its housing, the fiber to be aligned exactly in its connector, and the housing to be aligned exactly with the connector. Therefore, great cost savings can be achieved wherever some misalignment can be tolerated in the specification for the cable.

The *fiber-optic receiver* is the device that converts the optical signal back into electrical form. It is usually either a PIN diode or phototransistor. Phototransistors have good sensitivity but only have a low bandwidth. However, PIN diodes have a much higher bandwidth but a lower sensitivity. If both high bandwidth and high sensitivity are required, then special avalanche photodiodes are used, but at a severe cost penalty. The same considerations about losses at the interface between the cable and the receiver apply as for the transmitter, and both polishing and cleaving are used to prepare the fiber ends.

Output voltages from the receiver are very small and amplification is always necessary. The system is very prone to noise corruption at this point. However, the development of receivers that incorporate an amplifier is finding great success in reducing the scale of this noise problem.

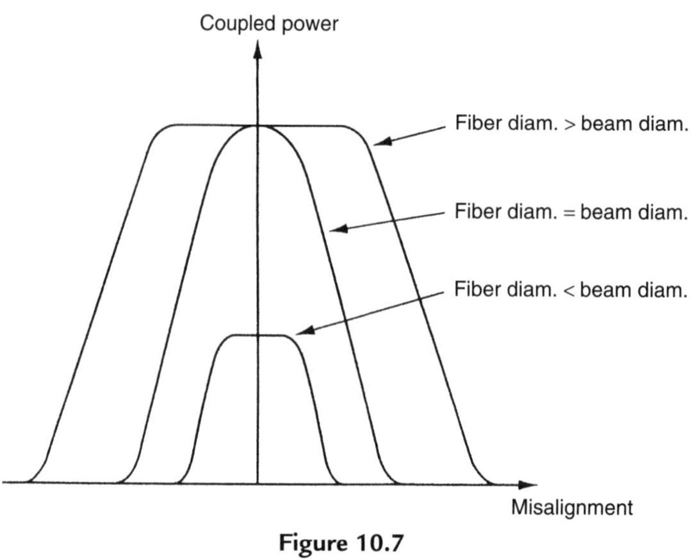


**Figure 10.7**
Effect of transmitter alignment on light power transmitted.

### *10.4.2 Transmission Characteristics*

Single mode cables have very simple transmission characteristics because the core has a very small diameter and light can only travel in a straight line down it. However, multimode cables have quite complicated transmission characteristics because of the relatively large diameter of the core.

While the transmitter is designed to maximize the amount of light that enters the cable in a direction parallel to its length, some light will inevitably enter multimode cables at other angles. Light that enters a multimode cable at any angle other than normal to the end face will be refracted in the core. It will then travel in a straight line until it meets the boundary between the core and cladding materials. At this boundary, some of the light will be reflected back into the core and some will be refracted in the cladding.

For materials of refractive indices $n_1$ and $n_2$ as shown in Figure 10.8, light entering from the external medium with refractive index $n_0$ at angle $\alpha_0$ will be refracted at angle $\alpha_1$ in the core and, when it meets the core-cladding boundary, part will be reflected at angle $\beta_1$ back into the core and part will be refracted at angle $\beta_2$ in the cladding. $\alpha_1$ and $\alpha_0$ are related by Snell's law according to

$$n_0 \sin\alpha_0 = n_1 \sin\alpha_1. \tag{10.1}$$

Similarly, $\beta_1$ and $\beta_2$ are related by

$$n_1 \sin\beta_1 = n_2 \sin\beta_2. \tag{10.2}$$

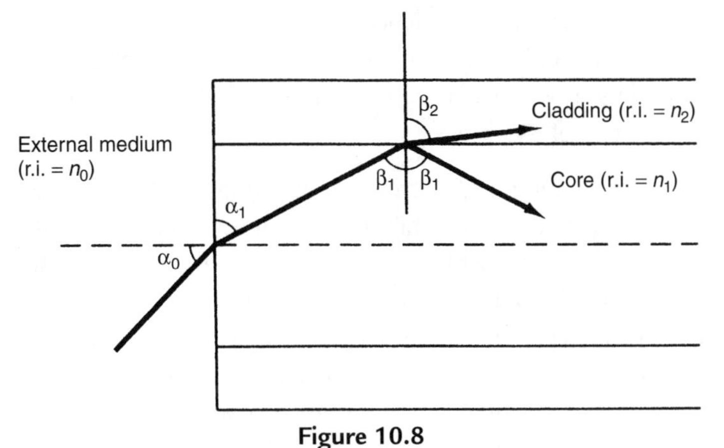


**Figure 10.8**
Transmission of light through cable.

Light that enters the cladding is lost and contributes to attenuation of the transmitted signal in the cable. However, observation of Equation (10.1) shows how this loss can be prevented. If $\beta_2 = 90°$, then the refracted ray will travel along the boundary between the core and cladding, and if $\beta_2 > 90°$, all of the beam will be reflected back into the core. The case where $\beta_2 = 90°$, corresponding to incident light at angle $\alpha_c$, is therefore the critical angle for total internal reflection to occur at the core/cladding boundary. The condition for this is that $\sin\beta_2 = 1$.

Setting $\sin\beta_2 = 1$ in Equation (10.1):

$$\frac{n_1 \sin\beta_1}{n_2} = 1.$$

Thus

$$\sin\beta_1 = {}^{n_2}/_{n_1}.$$

Inspection of Figure 10.8 shows that $\cos\alpha_1 = \sin\beta_1$. Hence,

$$\sin\alpha_1 = \sqrt{1 - \cos^2\alpha_1} = \sqrt{1 - \sin^2\beta_1} = \sqrt{1 - (n_2/n_1)^2}.$$

From Equation (10.1),

$$\sin\alpha_c = \sin\alpha_0 = \frac{n_1}{n_0} \sin\alpha_1.$$

Thus

$$\sin\alpha_c = \frac{n_1}{n_0}\sqrt{1 - \left(\frac{n_2}{n_1}\right)^2}.$$

Therefore, provided that the angle of incidence of light into the cable is greater than the critical angle given by $\theta = \sin^{-1}\alpha_c$, all of the light will be internally reflected at the core/cladding boundary. Further reflections will occur as light passes down the fibers, and it will thus travel in a zigzag fashion to the end of the cable.

While attenuation has been minimized, there is a remaining problem that the transmission time of the parts of the beam that travel in this zigzag manner will be greater than light that enters the fiber at 90° to the face and so travels in a straight line to the other end. In practice, incident light rays to the cable will be spread over the range given by $\sin^{-1}\alpha_c < \theta < 90°$, and so the transmission times of these separate parts of the beam will be distributed over a corresponding range. These differential delay characteristics of the light beam are known as modal dispersion. The practical effect is that a step change in light intensity at the input end of the cable will be received over a finite period of time at the output. Multimode cables where this happens are known as *step index* cables.

It is possible to largely overcome this latter problem in multimode cables by using cables made solely from glass fibers in which the refractive index changes gradually over the cross section of the core rather than abruptly at the core/cladding interface as in the step index cable discussed so far. This special type of cable is known as a *graded index* cable and it progressively bends light incident at less than 90° to its end face rather than reflecting it off the core/cladding boundary. Although parts of the beam away from the center of the cable travel further, they also travel faster than the beam passing straight down the center of the cable because the refractive index is lower away from the center. Hence, all parts of the beam are subject to approximately the same propagation delay. In consequence, a step change in light intensity at the input produces an approximate step change of light intensity at the output. The alternative solution is to use a single mode cable. This usually propagates light in a single mode only, which means that time dispersion of the signal is almost eliminated.

### 10.4.3 Multiplexing Schemes

Various types of multiplexing schemes are available. It is outside the scope of this text to discuss these in detail, and interested readers are recommended to consult a specialist text in fiber-optic transmission. It is sufficient to note here that wavelength division multiplexing is used predominantly in fiber-optic transmission systems. This uses a multiplexer in the transmitter to merge the different input signals together and a demultiplexer in the receiver to separate out the separate signals again. A different modulated frequency is used to transmit each signal. Because a single optic fiber is capable of propagating in excess of 100 different wavelengths without cross-interference, multiplexing allows more than 100 separate distributed sensors to be addressed. Wavelength division multiplexing systems normally use single mode cable of 9 μm diameter, although there are also examples of usage of 50- or 62.5-μm-diameter multimode cable.

## 10.5 Optical Wireless Telemetry

Wireless telemetry allows signal transmission to take place without laying down a physical link in the form of electrical or fiber-optic cable. This can be achieved using either radio or light waves to carry the transmitted signal across a plain air path between a transmitter and a receiver.

Optical wireless transmission was first developed in the early 1980s. It consists of a light source (usually infrared) transmitting encoded data information across an open, unprotected air path to a light detector. Three distinct modes of optical telemetry are possible, known as point to point, directed, and diffuse.

- *Point-to-point telemetry* uses a narrowly focused, fine beam of light, which is used commonly for transmission between adjacent buildings. A data transmission speed of 5 Mbit/s is possible at the maximum transmission distance of 1000 m. However, if the transmission distance is limited to 200 m, a transmission speed of 20 Mbit/s is possible. Point-to-point telemetry is used commonly to connect electrical or fiber-optic Ethernet networks in adjacent buildings.
- *Directed telemetry* transmits a slightly divergent beam of light that is directed toward reflective surfaces, such as walls and ceilings in a room. This produces a wide area of coverage and means that the transmitted signal can be received at a number of points. However, the maximum transmission rate possible is only 1 Mbit/s at the maximum transmission distance of 70 m. If the transmission distance is limited to 20 m, a transmission speed of 10 Mbit/s is possible.
- *Diffuse telemetry* is similar to directed telemetry but the beam is even more divergent. This increases the area of coverage but reduced transmission speed and range. At the maximum range of 20 m, the maximum speed of transmission is 500 kbit/s, although this increases to 2 MBit/s at a reduced range of 10 m.

In practice, implementations of optical wireless telemetry are relatively uncommon because the transmission of data across an open, unprotected air path is susceptible to random interruption. In cases where immunity to electromagnetic noise is particularly important, open path optical transmission is sometimes used because it provides immunity of the transmitted signal to electromagnetic noise at low cost. However, other ways of providing immunity to the transmitted signal against electromagnetic noise are often preferred, despite their higher cost.

The usual alternative to optical transmission for solving electromagnetic noise problems is to use fiber-optic transmission. In cases where laying a physical fiber-optic cable link is difficult, radio transmission is used commonly. This is preferred over optical transmission because it is much less prone to interference than optical transmission over an open air path, as radio waves can pass through most materials. However, there are some situations where radio transmission is subject to interference from neighboring radio-frequency systems operating at a similar wavelength and, in such circumstances, optical transmission is sometimes a better option.

## 10.6 Radiotelemetry (Radio Wireless Transmission)

Radiotelemetry is normally used over transmission distances up to 400 miles, although this can be extended by special techniques to provide communication through space over millions of miles. However, radiotelemetry is also used commonly over quite short distances to transmit signals where physical electrical or fiber-optic links are difficult to install or maintain. This occurs particularly when the source of the signals is mobile. The great advantage that radiotelemetry has over optical wireless transmission through an air medium is that radio waves are attenuated much less by most types of obstacles between the energy transmitter and receiver. Hence, as noted earlier, radiotelemetry usually performs better than optical wireless telemetry and is therefore used much more commonly.

In radiotelemetry, data are usually transmitted in a frequency-modulated format. A typical scheme is shown in Figure 10.9, although other arrangements also exist. In this particular scheme shown, 18 data channels are provided over the frequency range from 0.4 to 70 kHz, as given in the table that follows. Each channel is known as a subcarrier frequency and can be used to transmit data for a different physical variable. Thus, the system can transmit information from 18 different sensors simultaneously.

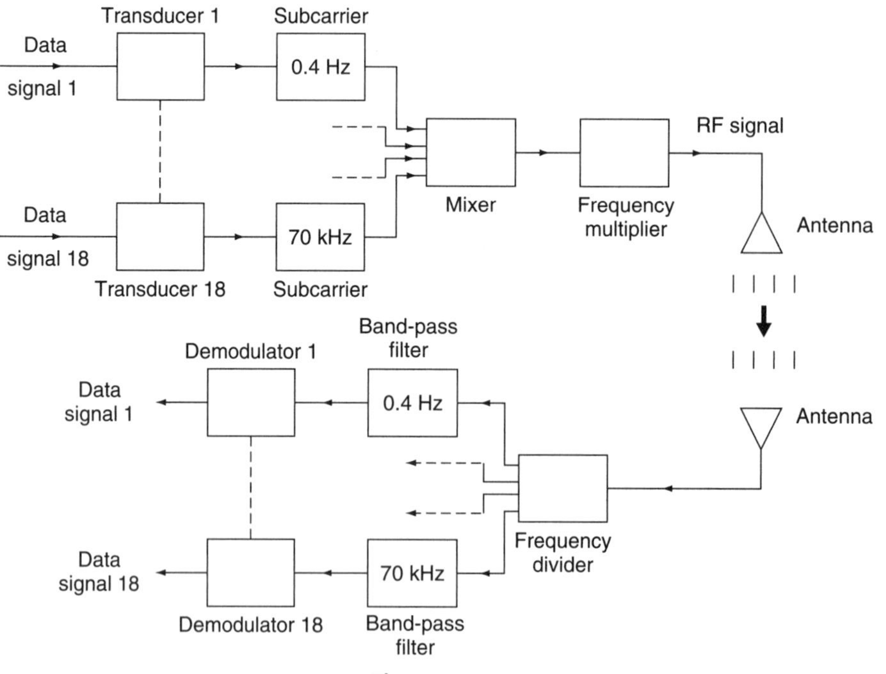


**Figure 10.9**
Radio transmission using FM/FM system.

| Band | 1 | 2 | 3 | 4 | 5 | 6 | 7 | 8 | 9 |
|---|---|---|---|---|---|---|---|---|---|
| Center frequency (kHz) | 0.4 | 0.56 | 0.73 | 0.96 | 1.3 | 1.7 | 2.3 | 3.0 | 3.9 |
| Band | 10 | 11 | 12 | 13 | 14 | 15 | 16 | 17 | 18 |
| Center frequency (kHz) | 5.4 | 7.35 | 10.5 | 14.5 | 22.0 | 30.0 | 40.0 | 52.5 | 70.0 |

Maximum frequency deviation allowed is ±7.5%

A voltage-to-frequency converter is used in the first FM stage to convert each analogue voltage signal into a varying frequency around the center frequency of the subcarrier assigned for that channel. The 18 channels are then mixed into a single signal spanning the frequency range 0.4 to 70 kHz. For transmission, the length of the antenna has to be one-quarter or one-half of the wavelength. At 10 kHz, which is a typical subcarrier frequency in an 18-channel system, the wavelength is 30 km. Hence, an antenna for transmission at this frequency is totally impractical. In consequence, a second FM stage is used to translate the 0.4- to 70-kHz signal into the radio frequency range as modulations on a typical carrier frequency of 217.5 MHz.[†] At this frequency, the wavelength is 1.38 m, so a transmission antenna length of 0.69 or 0.345 m would be suitable. The signal is received by an antenna of identical length some distance away. A frequency converter is then used to change the signal back into the 0.4- to 70-kHz subcarrier frequency spectrum, following which a series of band-pass filters are applied to extract the 18 separate frequency bands containing measurement data. Finally, a demodulator is applied to each channel to return each signal into varying voltage form.

The inaccuracy of radiotelemetry is typically ±1%. Thus, measurement uncertainty in transmitting a temperature measurement signal with a range of 0–100°C over one channel would be ±1%, that is, ±1°C. However, if unused transmission channels are available, the signal could be divided into two ranges (0–50 and 50–100°C) and transmitted over two channels, reducing the measurement uncertainty to ±0.5°C. By using 10 channels for one variable, a maximum measurement uncertainty of ±0.1°C could be achieved.

In theory, radiotelemetry is very reliable because, although the radio frequency waveband is relatively crowded, specific frequencies within it are allocated to specific usages under national agreements normally backed by legislation. Interference is avoided by licensing each frequency to only one user in a particular area, and limiting the transmission range through limits on the power level of transmitted signals, such that there is no interference to other licensed users of the same frequency in other areas. Unfortunately, interference can still occur in practice, due both to adverse atmospheric conditions extending the transmission range beyond that expected into adjoining areas and to unauthorized transmissions by other parties at the wavelengths licensed to registered users. There is a legal solution to this latter problem, although some time may elapse before the offending transmission is stopped successfully.

[†] Particular frequencies are allocated for industrial telemetry. These are subject to national agreements and vary in different countries.

## 10.7 Digital Transmission Protocols

Digital transmission has very significant advantages compared with analogue transmission because the possibility of signal corruption during transmission is reduced greatly. Many different protocols exist for digital signal transmission. However, the protocol normally used for the transmission of data from a measurement sensor or circuit is asynchronous serial transmission, with other forms of transmission being reserved for use in instrumentation and computer networks. Asynchronous transmission involves converting an analogue voltage signal into a binary equivalent, using an analogue-to-digital converter. This is then transmitted as a sequence of voltage pulses of equal width that represent binary "1" and "0" digits. Commonly, a voltage level of either +5 or +6 V is used to represent binary "1," and zero volt represents binary "0." Thus, the transmitted signal takes the form of a sequence of 6-V pulses separated by zero volt pulses. This is often known by the name of *pulse code modulation*. Such transmission in digital format provides very high immunity to noise because noise is typically much smaller than the amplitude of a pulse representing binary 1. At the receiving end of a transmitted signal, any pulse level between 0 and 3 V can be interpreted as a binary "0" and anything greater than 3 V can be interpreted as a binary "1." A further advantage of digital transmission is that other information, such as the status of industrial equipment, can be conveyed as well as parameter values. However, consideration must be given to the potential problems of aliasing and quantization, and the sampling frequency must therefore be chosen carefully.

Many different mediums can be used to transmit digital signals. Electrical cable, in the form of a twisted pair or coaxial cable, is used commonly as the transmission path. However, in some industrial environments, the noise levels are so high that even digital data become corrupted when transmitted as electrical pulses. In such cases, alternative transmission mechanisms have to be used.

One alternative is to modulate the pulses onto a high-frequency carrier, with positive and zero pulses being represented as two distinct frequencies either side of a center carrier frequency. Once in such a frequency-modulated format, a normal mains electricity supply cable operating at mains frequency is often used to carry the data signal. The large frequency difference between the signal carrier and the mains frequency prevents any corruption of data transmitted, and simple filtering and demodulation are able to extract the measurement signal after transmission. The public switched telephone network can also be used to transmit frequency-modulated data at speeds up to 1200 bits/s, using acoustic couplers as shown in Figure 10.10. The transmitting coupler converts each binary "1" into a tone at 1.4 kHz and each binary "0." into a tone at 2.1 kHz, while the receiving coupler converts the tones back into binary digits.

Another solution is to apply the signal to a digital-to-current converter unit and then use current loop transmission, with 4 mA representing binary "0" and 20 mA representing binary "1." This permits baud rates up to 9600 bit/s at transmission distances up to 3 km. Fiber-optic links and radiotelemetry are also widely used to transmit digital data.

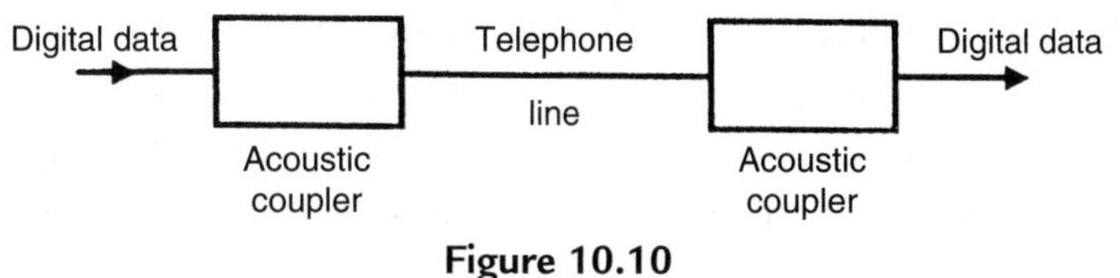


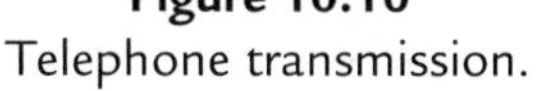
**Figure 10.10**
Telephone transmission.

## 10.8 Summary

This chapter on techniques for transmitting measurement signals over what can, in some circumstances, be quite large distances completes the discussion on the different components that exist in a measurement system. However, as for some other measurement system components such as variable conversion elements, a mechanism for signal transmission is not needed in all measurement systems.

Our discussion began by observing that signals can be transmitted from the sensor at the point of measurement to the rest of the measurement system in a number of alternative ways. As well as alternatives of electric, pneumatic, optical, and radio transmission, both analogue and digital forms exist as alternative transmission formats to carry the transmitted measurement signal.

We started off by noting that electrical transmission is the simplest way to convey measurement signals over some distance, but noted that this has associated problems of attenuation of the measured signal and also a tendency for the measured signal to be corrupted by noise. We therefore went on to look at solutions to these problems. The first solution covered was to amplify the measurement signal to compensate for attenuation during transmission and to shield transmission wires to minimize noise corruption. We went on to look at the alternative electrical transmission method known as current loop transmission, whereby the measurement signal is transmitted as a varying current rather than a varying voltage in order to better protect the measurement signal from induced noise. Finally, we looked at a solution that involved transmitting the measurement signal on a carrier wave using either amplitude modulation or frequency modulation.

Our next subject was pneumatic transmission. We noted that this had the disadvantage of only transmitting measurement signals at relatively slow speeds and, in consequence, was now much less used than it has been in the past. However, we observed that pneumatic transmission is still used in three specific circumstances. First, because it is an intrinsically safe method of transmission, it is still used in some applications where intrinsic safety is required. Second, it provides an alternative to current loop transmission when a high level of noise immunity is required. Finally, it is convenient to use pneumatic transmission in pneumatic control systems where the sensors, actuators, or both are pneumatic.

Our discussion then moved on to fiber-optic transmission, where we noted that fiber optics provided both intrinsically safe operation and gave the transmitted signal immunity to noise corruption by neighboring electromagnetic fields. Attenuation of the transmitted signal along a fiber-optic cable is also much less than for varying voltage transmission along an equivalent length of electric cable.

As well as optical transmission of measurement data along a fiber-optic cable, we noted that it was also possible to transmit data optically across air space rather than along a cable. We observed that this type of transmission existed in three forms, known as point-to-point telemetry, directed telemetry, and diffuse telemetry. Unfortunately, all of these alternatives suffer from the common problem of unreliability in data transmission because the transmission path is susceptible to random interruption when data are transmitted across an open, unprotected air path. In consequence, use of a fiber-optic cable is usually preferred for transmission of data, even though this is much more expensive than transmitting data across an open air path.

We then went on to look at radio transmission. We noted that this is normally used over transmission distances up to 400 miles, but special techniques can allow communication through space over millions of miles. In addition, radiotelemetry is also used commonly over quite short distances to transmit signals where physical electrical or fiber-optic links are difficult to install or maintain, particularly when the measurement signal is mobile. Although obstacles between the energy transmitter and the receiver can cause some attenuation of transmitted measurement data, this problem is far less than that which occurs when attempts are made to transmit data optically over an open air path. While radiotelemetry is generally reliable, two problems can occur, which are both related to transmission frequency. Normally, licensing arrangements give each radio transmission system a unique transmission frequency within a given geographical area. Unfortunately, adverse atmospheric conditions can extend the range of transmission systems into adjoining areas and cause contamination of transmitted signals. A similar problem can occur when there are unauthorized transmissions by other parties at the wavelengths licensed to registered users.

To conclude the chapter, we looked finally at digital transmission protocols. We noted that digital transmission has very significant advantages compared with analogue transmission because the possibility of signal corruption during transmission is reduced greatly. While many different protocols exist for digital signal transmission, we noted that the one normally used for the transmission of data from a measurement sensor or circuit is asynchronous serial transmission. Having looked at how this works, we finished off by looking at the main two alternative means of transmitting digital data, along a "twisted pair" or coaxial electrical cable, and as modulated pulses on a high-frequency carrier.

## 10.9 Problems

10.1. Discuss some reasons why it is necessary in many measurement situations to transmit signals from the point of measurement to some other point.

10.2. Discuss the main features of electrical, pneumatic, fiber-optic, and radio signal transmission. Give examples of the sorts of situations where you would use each of these transmission methods.

10.3. Discuss the different forms of electrical signal transmissions. What are the merits and demerits of each alternative form?

10.4. What is a current loop interface? Discuss some measurement situations where this would be the preferred form of signal transmission.

10.5. Pneumatic transmission is now rarely used in transmission systems. Why has this form of transmission fallen out of favor? What sorts of conditions might cause a system designer to still specify a pneumatic system in a new measurement system?

10.6. Discuss the main features of fiber-optic transmission.

10.7. What are the principal advantages of fiber-optic transmission? Given its significant advantages, why is fiber-optic transmission not used more widely?

10.8. Discuss the three different ways in which fiber-optic cables are used for signal transmission.

10.9. A fiber-optic transmitter converts an electrical measurement signal into light and then injects this into the fiber-optic cable. Discuss the important design features that maximize the proportion of light produced in the transmitter that passes into the cable.

10.10. Discuss the main features of single mode and multimode fiber-optic transmission.

CHAPTER 11

# Intelligent Devices



## 11.1 Introduction

We now find reference to devices with names such as intelligent instruments, smart sensors, and smart transmitters whenever we open a technical magazine or browse through an instrument manufacturer's catalogue. This reflects the fact that intelligent devices have now achieved widespread use in measurement applications. The term *intelligent* is used to denote any measurement device that uses computational power to enhance its measurement performance.

We are probably aware that digital computers have been used in conjunction with measurement systems for many years in the typical control system scenario where a computer uses data on process variables supplied by a measurement system to compute a control signal that is then applied to an actuator in order to modify some aspect of the controlled process. In this case, the computer was not actually part of the measurement system but merely works with it by taking data from the system.

As the cost of computers fell and their power increased, it became common practice to use the computer assigned to a process control function to enhance the quality of measurements by performing various signal processing operations digitally that were carried out previously by analogue electronic circuits. However, in these early applications of digital signal processing, the computer remained as a distinctly separate component within the measurement system.

We have now moved on one stage further to the point where the computer that performs digital signal processing to enhance measurement quality is incorporated into the measurement device. Such devices that incorporate digital signal processing are given the generic name *intelligent devices*. Individual intelligent devices attract various names such as *intelligent instrument*, *intelligent sensor*, *smart sensor*, and *smart transmitter*. There are no hard distinctions between the function of any of these, and which term is used to refer to an intelligent device is largely due to the preference adopted by different manufacturers for one name or another. Similar variation exists in the name used to describe the computational power within the intelligent device, with terms such as *microcomputer* and *microprocessor* being common.

The subject of this chapter is therefore intelligent devices. However, to start off with, we look more generally at some basic principles of digital computation, as this will enable us to better understand how intelligent devices function and what potential difficulties exist in their application and operation.

## 11.2 Principles of Digital Computation

### *11.2.1 Elements of a Computer*

The primary function of a digital computer is the manipulation of data. The three elements essential to the fulfillment of this task are the central processing unit, memory, and input–output interface, as shown in Figure 11.1. These elements are collectively known as computer hardware, and each element exists physically as one or more integrated circuit chips mounted on a printed circuit board. Where the central processing unit (CPU) consists of a single microprocessor, it is usual to regard the system as a microcomputer. The distinction among the terms

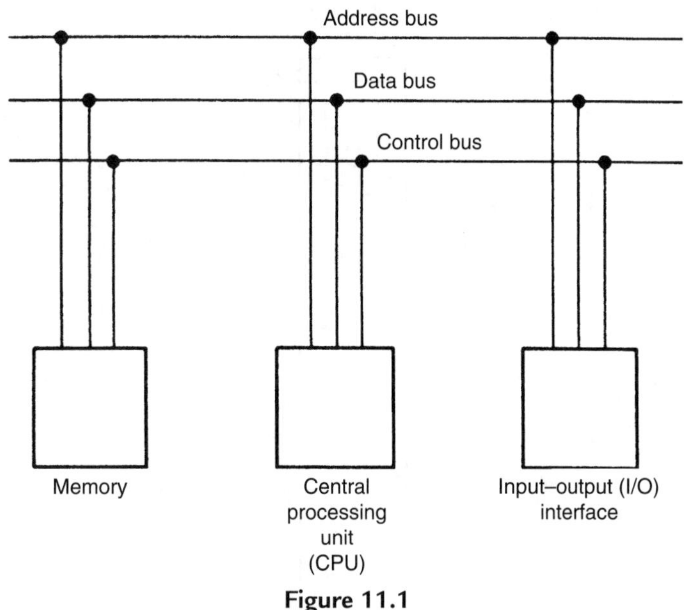


**Figure 11.1**
Elements of a microcomputer.

*microcomputer, minicomputer,* and *mainframe computer* is a very arbitrary division made according to relative computer power. However, this classification has become somewhat meaningless, with present-day *microcomputers* being more powerful than mainframe computers of only a few years ago.

The *central processing unit* part of a computer can be regarded as the brain of the system. A relatively small CPU is commonly called a *microprocessor*. The CPU determines what computational operations are carried out and the sequence in which the operations are executed. During such operation, the CPU makes use of one or more special storage locations within itself known as *registers*. Another part of the CPU is the *arithmetic and logic unit*, which is where all arithmetic operations are evaluated. The CPU operates according to a sequential list of required operations defined by a computer program, known as computer software. This program is held in the second of the three system components known as computer memory.

*Computer memory* also serves several other functions besides this role of holding the computer program. One of these is to provide temporary storage locations that the CPU uses to store variables during execution of the computer program. A further common use of memory is to store data tables used for scaling and variable conversion purposes during program execution.

Memory can be visualized as a consecutive sequence of boxes in which various items are stored, as shown in Figure 11.2 for a typical memory size of 65,536 storage units. If this storage mechanism is to be useful, then it is essential that a means be provided for giving a unique label to each storage box. This is achieved by labeling the first box as 0, the next one as 1, and so on

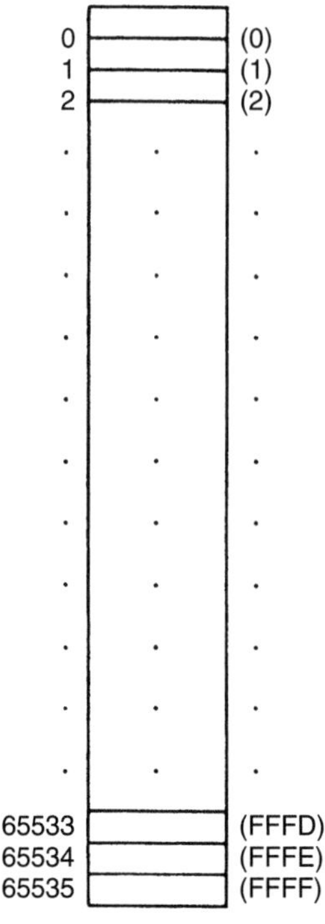


**Figure 11.2**
Schematic representation of computer memory (numbers in parentheses are memory addresses in hexadecimal notation).

for the rest of the storage locations. These numbers are known as *memory addresses.* While these can be labeled by decimal numbers, it is more usual to use hexadecimal notation.

Two main types of computer memory exist and there are important differences between these. The two kinds are *random access memory* (RAM) and *read only memory* (ROM). The CPU can both read from and write to the former, but it can only read from the latter. The importance of ROM becomes apparent if the behavior of each kind of memory when the power supply is turned off is considered. At power-off time, RAM loses its contents but ROM maintains them, and this is the value of ROM. Intelligent devices normally use ROM for storage of the program and data tables and just have a small amount of RAM used by the CPU for temporary variable storage during program execution.

The third essential element of a computer system is the *input–output* (I/O) *interface*, which allows the computer to communicate with the outside world by reading in data values and outputting results after the appropriate computation has been executed. In the case of a microcomputer performing a signal processing function within an intelligent device, this means reading in the values obtained from one or more sensors and outputting a processed value for presentation at the instrument output. All such external peripherals are identified by a unique number, as for memory addresses.

Communication among these three computer elements is provided by three electronic highways known as the *data bus*, *address bus*, and *control bus*. At each data transfer operation executed by the CPU, two items of information must be conveyed along the electronic highway—the item of data being transferred and the address where it is being sent. While both of these items of information could be conveyed along a single bus, it is more usual to use two busses—the data bus and the address bus. The timing of data transfer operations is important, particularly when transfers take place to peripherals such as disc drives and keyboards where the CPU often has to wait until the peripheral is free before it can initialize a data transfer. This timing information is carried by a third highway known as the control bus.

The current trend made possible by advances in very large-scale integration technology is to incorporate all three functions of the central processor unit, memory, and I/O within a single chip (known as a computer on a chip or *microcomputer*). The term *microprocessor* is often used to describe such an integrated unit, but this is strictly incorrect as the device contains more than just processing power.

### 11.2.2 Computer Operation

As has already been mentioned, the fundamental role of a computer is the manipulation of data. Numbers are used both in quantifying items of data and also in the form of codes that define the computational operations that are to be executed. All numbers used for these two purposes must be stored within the computer memory and also transported along the communication busses.

#### *Programming and program execution*

In most modes of usage, including use as part of intelligent devices, computers are involved in manipulating data. This requires data values to be input, processed, and output according to a sequence of operations defined by the computer program. However, in practice, programming the microprocessor within an intelligent device is not normally the province of the instrument user; indeed, there is rarely any provision for the user to create or modify operating programs even if he/she wished to do so. There are several reasons for this. First, the signal processing needed within an intelligent device is usually well defined, and therefore it is more efficient for a manufacturer to produce this rather than to have each individual user produce near identical programs separately. Second, better program integrity and instrument operation is achieved if a standard program produced by the instrument manufacturer is used. Finally, use of a standard program allows it to be burnt into ROM, thereby protecting it from any failure of the instrument power supply. This also facilitates software maintenance and updates by the mechanism of the manufacturer providing a new ROM that simply plugs into the slot occupied previously by the old ROM.

However, even though it is not normally a task undertaken by the user, some appreciation of microprocessor programming for an intelligent device is useful background knowledge. To illustrate the techniques involved in programming, consider a very simple program that reads in a value from a sensor, adds a prestored value to it to compensate for a bias in the sensor measurement, and outputs a corrected reading to a display device.

Let us assume that the addresses of the sensor and output display device are 00C0 and 00C1, respectively, and that the required scaling value has already been stored in memory address 0100. The instructions that follow are formed from the instruction set for a Z80* microprocessor and make use of CPU registers A and B.

```
IN A,C0
IN B,100
ADD A,B
OUT C1,A
```

This list of four instructions constitutes the computer program necessary to execute the required task. The CPU normally executes the instructions one at a time, starting at the top of the list and working downward (although jump and branch instructions change this order). The first instruction (IN A,C0) reads in a value from the sensor at address C0 and places the value in CPU register A (often called the accumulator). The mechanics of the execution of this instruction consist of the CPU putting the required address C0 on the address bus and then putting a command on the control bus that causes the contents of the target address (C0) to be copied onto the data bus and subsequently transferred into the A register. The next instruction (IN B,100) reads in a value from address 100 (the prestored biasing value) and stores it in register B. The following instruction (ADD A,B) adds together the contents of registers A and B and stores the result in register A. Register A now contains the measurement read from the sensor but corrected for bias. The final instruction (OUT C1,A) transfers the contents of register A to the output device on address C1.

### 11.2.3 Computer Input–Output Interface

The input–output interface connects the computer to the outside world and is therefore an essential part of the computer system. When the CPU puts the address of a peripheral onto the address bus, the input–output interface decodes the address and identifies the unique computer peripheral with which a data transfer operation is to be executed. The interface also has to

* The Z80 is now an obsolete 8-bit processor but its simplicity is well suited to illustrating programming techniques. Similar, but necessarily more complex, programming instructions are used with current 16- and 32-bit processors.

interpret the command on the control bus so that the timing of data transfer is correct. One further very important function of the input–output interface is to provide a physical electronic highway for the flow of data between the computer data bus and the external peripheral. In many computer applications, including their use within intelligent devices, the external peripheral requires signals to be in analogue form. Therefore the input–output interface must provide for conversion between these analogue signals and the digital signals required by a digital computer. This is satisfied by analogue-to-digital and digital-to-analogue conversion elements within the input–output interface.

The rest of this section presents some elementary concepts of interfacing in simple terms.

*Address decoding*

A typical address bus in a microcomputer is 16 bits wide,[†] allowing 65,536 separate addresses to be accessed in the range 0000–FFFF (in hexadecimal representation). Special commands on some computers are reserved for accessing the bottom end 256 of these addresses in the range 0000–00FF, and, if these commands are used, only 8 bits are needed to specify the required address. For the purpose of explaining address-decoding techniques, the scheme shown in Figure 11.3 shows that the lower 8 bits of the 16-bit address line are decoded to identify the unique address referenced by one of these special commands. Decoding of all 16 address lines follows a similar procedure but requires a substantially greater number of integrated circuit chips.

Address decoding is performed by a suitable combination of logic gates. Figure 11.3 shows a very simple hardware scheme for decoding 8 address lines. This consists of 256 eight-input NAND gates, which each uniquely decode one of 256 addresses. A NAND gate is a logic element that only gives a logic level 1 output when all inputs are zero and gives a logic level 0 output for any other combination of inputs. Inputs to the NAND gates are connected onto the lower 8 lines of the address bus, and the computer peripherals are connected to the output of the particular gates that decode their unique addresses. There are two pins for each input to the NAND gates that respectively invert and do not invert the input signal. By connecting the 8 address lines appropriately to these two alternative pins at each input, the gate is made to decode a unique address. Consider, for instance, the pin connections shown in Figure 11.4. This NAND gate decodes address C5 (hexadecimal), which is 11000101 in binary. Because of the way in which the input pins to the chip are connected, the NAND gate will see all zeros at its input when 11000101 is on the lower 8 bits of the address bus and therefore will have an output of 1. Any other binary number on the address bus will cause this NAND gate to have a zero output.

[†] Nowadays, 32-bit and even 64-bit address fields are also available.

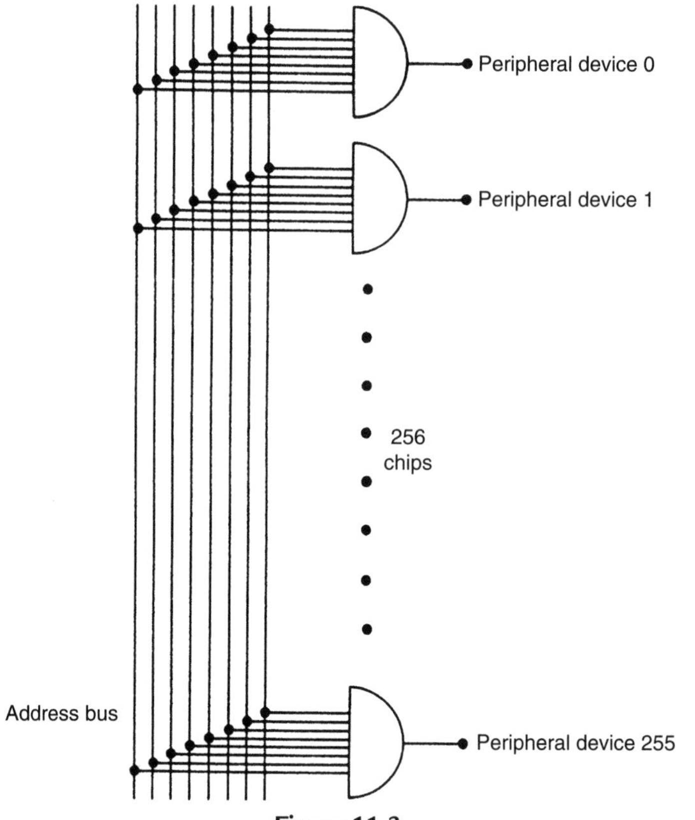


**Figure 11.3**
Simple hardware scheme for decoding eight address lines.

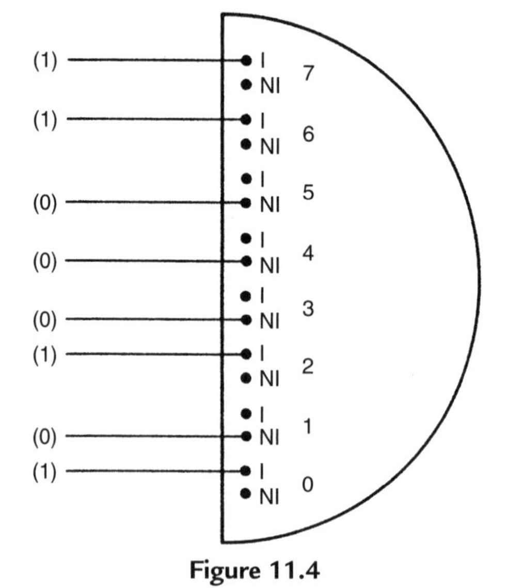


**Figure 11.4**
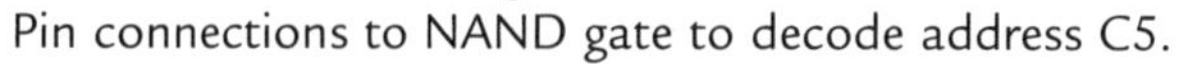
Pin connections to NAND gate to decode address C5.

*Data transfer control*

The transfer of data between the computer and peripherals is managed by control and status signals carried on the control bus that determine the exact sequencing and timing of I/O operations. Such management is necessary because of the different operating speeds of the computer and its peripherals and because of the multitasking operation of many computers. This means that, at any particular instant when a data transfer operation is requested, either the computer or the peripheral may not be ready to take part in the transfer. Typical control and status lines, and their meanings when set at a logic level of 1, are shown here.

- BUSY Peripheral device busy
- READY Peripheral device ready for data transfer
- ENABLE CPU ready for data transfer
- ERROR Malfunction on peripheral device

Similar control signals are set up by both the computer and peripherals, but different conventions are often used to define the status of each device. Differing conventions occur particularly when the computer and peripherals come from different manufacturers and might mean, for instance, that the computer interprets a logic level of 1 as defining a device to be busy but the peripheral device uses logic level 0 to define "device busy" on the appropriate control line. Therefore, translation of control lines between the computer and peripherals is required, which is achieved by a further series of logic gates within the I/O interface.

### 11.2.4 Practical Considerations in Adding Computers to Measurement Systems

The foregoing discussion has presented some of the necessary elements in an input–output interface in a relatively simplistic manner that is just sufficient to give the reader the flavor of what is involved in an interface. Much fine detail has been omitted, and the amount of work involved in the practical design of a real interface should not be underestimated. One significant omission so far is discussion of the scaling that is generally required within the analogue–digital interface of a computer. The raw analogue input and output signals are generally either too large or too small for compatibility with the operating voltage levels of a digital computer and have to be scaled upward or downward. This is normally achieved by operational amplifiers and/or potentiometers. The main features of an operational amplifier are its high gain (typically $\times$ 1,000,000) and its large bandwidth (typically 1 MHz or better). However, when one is used at very high frequencies, the bandwidth becomes significant. The quality of an amplifier is often measured by a criterion called the gain–bandwidth product, which is the product of its gain and bandwidth. Other important attributes of the operational amplifier, particularly when used in a computer input–output interface or within intelligent devices, are its distortion level, overload recovery capacity, and offset level. Special instrumentation amplifiers that are particularly good in these attributes have been developed for instrumentation applications.

Suitable care must always be taken when introducing a computer into a measurement system to avoid creating sources of measurement noise. This applies particularly where one computer is used to process the output of several transducers and is connected to them by signal wires. In such circumstances, the connections and connecting wires can create noise through electrochemical potentials, thermoelectric potentials, offset voltages introduced by common mode impedances, and a.c. noise at power, audio, and radio frequencies. Recognition of all these possible noise sources allows them to be eliminated in most cases by employing good practice when designing and constructing the measurement system.

## 11.3 Intelligent Devices

The term *intelligent device* is used to describe a package containing either a complete measurement system or a component within a measurement system, which incorporates a digital processor. Processing of the output of measurement sensors to correct for errors inherent in the measurement process brings about large improvements in measurement accuracy. Such intelligent devices are known by various names, such as *intelligent instrument*, *intelligent sensor*, *smart sensor*, and *smart transmitter*. There is no formal definition for any of these names, and there is considerable overlap between the characteristics of particular devices and the names given to them. The name used for any particular device depends largely on the whims and style of device manufacturers. The discussion that follows tries to lay out the historical development of intelligent devices and summarizes general understanding of the sorts of characteristics possessed by the various forms of intelligent devices. Their applications to measure particular physical variables are then covered in greater detail in the later chapters of this book.

### 11.3.1 Intelligent Instruments

The first intelligent instrument appeared over 30 years ago, although high prices when such devices first became available meant that their use within measurement systems initially grew very slowly. However, since those early days, there has been a dramatic reduction in the price of all intelligent devices, and the cost differential between intelligent and conventional devices is now very small. Indeed, an intelligent device is sometimes now less expensive than its nonintelligent equivalent because of the greater sales volume for the intelligent version. Thus, intelligent devices are now bought routinely instead of nonintelligent versions in many cases.

The processor within an intelligent instrument allows it to apply preprogrammed signal processing and data manipulation algorithms to measurements. This prewritten software is often known by the name of *embedded software*. One of the main functions performed by the first intelligent instruments to become available was compensation for environmental disturbances to measurements that cause systematic errors. Thus, apart from a primary sensor to measure the variable of interest, intelligent instruments usually have one or more secondary sensors to monitor the value of environmental disturbances. These extra measurements allow the output reading to be corrected for the effects of environmentally induced errors, subject to the following preconditions being satisfied.

(a) The physical mechanism by which a measurement sensor is affected by ambient condition changes must be fully understood and all physical quantities that affect the output must be identified.
(b) The effect of each ambient variable on the output characteristic of the primary sensor must be quantified.
(c) Suitable secondary sensors for monitoring the value of all relevant environmental variables must be available that will be operate satisfactorily in the prevailing environmental conditions.

Condition (a) means that the thermal expansion and contraction of all elements within a sensor must be considered in order to evaluate how it will respond to ambient temperature changes. Similarly, the sensor response, if any, to changes in ambient pressure, humidity, gravitational force, or power supply level (active instruments) must be examined.

Quantification of the effect of each ambient variable on the characteristics of the measurement sensor is then necessary, as stated in condition (b). Analytic quantification of ambient condition changes from purely theoretical consideration of the construction of a sensor is usually extremely complex and so is normally avoided. Instead, the effect is quantified empirically in laboratory tests. In such tests, the output characteristic of the sensor is observed as the ambient environmental conditions are changed in a controlled manner.

One early application of intelligent instruments was in volume flow rate measurement, where the flow rate is inferred by measuring the differential pressure across an orifice plate placed in a fluid-carrying pipe (see Chapter 16). The flow rate is proportional to the square root of the difference in pressure across the orifice plate. For a given flow rate, this relationship is affected both by the temperature and by the mean pressure in the pipe, and changes in the ambient value of either of these cause measurement errors. A typical intelligent flowmeter therefore contains three sensors: a primary one measuring pressure difference across the orifice plate and secondary ones measuring absolute pressure and temperature. The instrument is programmed to correct the output of the primary differential-pressure sensor according to values measured by the secondary sensors, using appropriate physical laws that quantify the effect of ambient temperature and pressure changes on the fundamental relationship between flow and differential pressure. Even 30 years ago, such intelligent flow-measuring instruments achieved typical inaccuracy levels of $\pm 0.1\%$, compared with $\pm 0.5\%$ for their nonintelligent equivalents.

Although automatic compensation for environmental disturbances is a very important attribute of intelligent instruments, many versions of such devices perform additional functions, which was so even in the early days of their development. For example, the orifice-plate flowmeter just discussed usually converts the square root relationship between flow and signal output into a linear one, thus making the output much easier to interpret. Other examples of the sorts of functions performed by intelligent instruments are:

- correction for the loading effect of measurement on the measured system
- signal damping with selectable time constants
- switchable ranges (using several primary sensors within the instrument that each measure over a different range)
- switchable output units (e.g., display in Imperial or SI units)
- linearization of the output
- self-diagnosis of faults
- remote adjustment and control of instrument parameters from up to 1500 meters away via four-way, 20-mA signal lines

These features are discussed in greater detail in Sections 11.3.2 and 11.3.3.

Over the intervening years since their first introduction, the size of intelligent instruments has gradually reduced and the functions performed have steadily increased. One particular development has been inclusion of a microprocessor within the sensor itself in devices that are usually known as *smart sensors*. As further size reduction and device integration has taken place, such smart sensors have been incorporated into packages with other sensors and signal processing circuits. While such a package conforms to the definition of an intelligent instrument given previously, most manufacturers now tend to call the package a *smart transmitter* rather than an intelligent instrument, although the latter term has continued in use in some cases.

### 11.3.2 Smart Sensors

The name *smart sensor* is used most commonly to describe any sensor that has local processing power that enables it to react to local conditions without having to refer back to a central controller. Smart sensors are usually at least twice as accurate as nonsmart devices, have reduced maintenance costs, and require less wiring to the site where they are used. In addition, long-term stability is improved, reducing the required calibration frequency.

Functions possessed by smart sensors vary widely, but consist of at least some of the following:

- Remote calibration capability
- Self-diagnosis of faults
- Automatic calculation of measurement accuracy and compensation for random errors
- Adjustment for measurement nonlinearities to produce a linear output
- Compensation for the loading effect of the measuring process on the measured system

### *Calibration capability*

Self-calibration is very simple in some cases. Sensors with an electrical output can use a known reference voltage level to carry out self-calibration. Also, load cell types of sensors, which are used in weighing systems, can adjust the output reading to zero when there is no applied mass. In the case of other sensors, two methods of self-calibration are possible: use of a look-up table and an interpolation technique. Unfortunately, a *look-up table* requires a large memory capacity to store correction points. Also, a large amount of data has to be gathered from the sensor during calibration. In consequence, the interpolation calibration technique is preferable. This uses an interpolation method to calculate the correction required to any particular measurement and only requires a small matrix of calibration points.

### *Self-diagnosis of faults*

Smart sensors perform self-diagnosis by monitoring internal signals for evidence of faults. While it is difficult to achieve a sensor that can carry out self-diagnosis of all possible faults that might arise, it is often possible to make simple checks that detect many of the more common faults. One example of self-diagnosis in a sensor is measuring the sheath capacitance and resistance in insulated thermocouples to detect breakdown of the insulation. Usually, a specific code is generated to indicate each type of possible fault (e.g., a failing of insulation in a device).

One difficulty that often arises in self-diagnosis is in differentiating between normal measurement deviations and sensor faults. Some smart sensors overcome this by storing multiple measured values around a set point and then calculating minimum and maximum expected values for the measured quantity.

Uncertainty techniques can be applied to measure the impact of sensor fault on measurement quality. This makes it possible in certain circumstances to continue to use a sensor after it has developed a fault. A scheme for generating a validity index has been proposed that indicates the validity and quality of a measurement from a sensor (Henry, 1995).

### *Automatic calculation of measurement accuracy and compensation for random errors*

Many smart sensors can calculate measurement accuracy online by computing the mean over a number of measurements and analyzing all factors affecting accuracy. This averaging process also serves to reduce the magnitude of random measurement errors greatly.

### *Adjustment for measurement nonlinearities*

In the case of sensors that have a nonlinear relationship between the measured quantity and the sensor output, digital processing can convert the output to a linear form, providing that the nature of the nonlinearity is known so that an equation describing it can be programmed into the sensor.

### 11.3.3 Smart Transmitters

In concept, a smart transmitter is almost identical to other intelligent devices described earlier. While the name *smart transmitter* is sometimes used interchangeably with the name *smart sensor,* it is perhaps used more commonly to describe an intelligent device that has greater functionality than just the computer-assisted sensing of a variable that a smart sensor conventionally does, particularly in respect of output functions and ability to compensate for environmental disturbances. In some instances, smart transmitters are known alternatively as *intelligent transmitters.* The term *multivariable transmitter* is also sometimes used, particularly for a device such as a smart flow-measuring instrument. This latter device measures absolute pressure, differential pressure, and process temperature and computes both mass flow rate and volume flow rate of the measured fluid.

Many of the smart transmitters presently available still have an analogue output because of the continuing popularity and investment in 4- to 20-mA current transmission systems. While most devices now available have a digital output, many users convert this to analogue form to maintain compatibility with existing instrumentation systems.

#### *Comparison of performance with other forms of transmitters*

The capabilities of smart transmitters can perhaps best be emphasized by comparing them with the attributes of analogue transmitters and also with devices known as *programmable transmitters.* The latter have computational power but do not have a bidirectional communication ability, meaning that they are not truly intelligent. The respective attributes of these devices are:

(a) Analogue transmitters:
- Require one transmitter for every sensor type and every sensor range.
- Require additional transmitters to correct for environmental changes.
- Require frequent calibration.

(b) Programmable transmitters:
- Include a microprocessor but do not have bidirectional communication (hence are not truly intelligent).
- Require field calibration.

(c) Smart transmitters:
- Include a microprocessor and have bidirectional communication.
- Include secondary sensors that can measure, and so compensate, for environmental disturbances.
- Usually incorporate signal conditioning and analogue to digital conversion.
- Often incorporate multiple sensors covering different measurement ranges and allow automatic selection of required range. The range can be altered readily if initially estimated incorrectly.

- Have a self-calibration capability that allows removal of zero drift and sensitivity drift errors.
- Have a self-diagnostic capability that allows them to report problems or requirements for maintenance.
- Can adjust for nonlinearities to produce a linear output.

### *Summary of advantages of smart transmitters*

The main disadvantage that could be cited for using a smart transmitter instead of a nonsmart one is that it is usually a little larger and heavier than its nonsmart equivalent, but this is not a problem in most applications. There is also normally a greater associated purchase cost. However, these potential disadvantages are minor in most circumstances and are outweighed greatly by the advantages that smart transmitters have, which can be summarized as:

- Improved accuracy and repeatability.
- Automatic calculation of measurement accuracy and compensation for random errors.
- Compensation for the loading effect of the measuring process on the measured system.
- Long-term stability is improved and required recalibration frequency is reduced.
- Adjustment for measurement nonlinearities to produce a linear output.
- Reduced maintenance costs.
- Self-diagnosis of faults.
- Large range coverage, allowing interoperability and giving increased flexibility.
- Remote adjustment of output range on command from a portable keyboard or a PC. This saves on technician time compared with carrying out adjustment manually.
- Reduction in number of spare instruments required, as one spare transmitter can be configured to cover any range and so replace any faulty transmitter.
- Possibility of including redundant sensors, which can be used to replace failed sensors and so improve device reliability.
- Allowing remote recalibration or reranging by sending a digital signal to them.
- Ability to store last calibration date and indicate when next calibration is required.
- Single penetration into the measured process rather than the multiple penetration required by discrete devices, making installation easier and less expensive.
- Ability to store data so that plant and instrument performance can be analyzed. For example, data relating to the effects of environmental variations can be stored and used to correct output measurements over a large range.

### *Self-calibration*

The common use of multiple primary sensors and secondary sensors to measure environmental parameters means that the self-calibration procedure for smart transmitters is more complicated than that for simpler smart sensors. While the general approach to self-calibration remains similar to that explained earlier for smart sensors, the calibration procedure has to be repeated for each primary and secondary sensor within the transmitter. Recommended practice is to use the simplest calibration procedures available for each sensor in the transmitter. However, care has to be taken to ensure that any interaction between measured variables is taken account of. This often means that look-up tables in a smart transmitter have to have a particularly large memory requirement in order to take the cross-sensitivity to other parameters (e.g., temperature) into account because a matrix of correction values has to be stored. This means that interpolation calibration is even more preferable to look-up table calibration than it is in the case of calibrating smart sensors.

#### *Self-diagnosis and fault detection*

Fault diagnosis in the sensors within a smart transmitter is often difficult because it is not easy to distinguish between measurement deviation due to a sensor fault and deviation due to a plant fault. The best theoretical approach to this difficulty is to apply mathematical modeling techniques to the sensor and plant in which it is working, with the aim of detecting inconsistencies in data from the sensor. However, there are very few industrial applications of this approach to fault detection in practice because (1) of the cost of implementation and (2) of the difficulty of obtaining plant models that are robust to plant disturbances. Thus, it is usually necessary to resort to having multiple sensors and using a scheme such as two-out-of-three voting. Further advice on self-checking procedures can be found elsewhere (Bignell and White, 1996).

## 11.4 Communication with Intelligent Devices

The inclusion of computer processing power in intelligent instruments and intelligent actuators creates the possibility of building an instrumentation system where several intelligent devices collaborate together, transmit information to one another, and execute process control functions. Such an arrangement is often known as a *distributed control system*. Additional computer processors can also be added to the system as necessary to provide the necessary computational power when the computation of complex control algorithms is required. Such an instrumentation system is far more fault tolerant and reliable than older control schemes where data from several discrete instruments are carried to a centralized computer controller via long instrumentation cables. This improved reliability arises from the fact that the presence of computer processors in every unit injects a degree of redundancy into the system. Therefore, measurement and control action can still continue, albeit in a degraded form, if one unit fails.

In order to affect the necessary communication when two or more intelligent devices are to be connected together as nodes in a distributed system, some form of electronic highway must be provided between them that permits the exchange of information. Apart from data transfer, a certain amount of control information also has to be transferred. The main purpose of this control information is to make sure that the target device is ready to receive information before data transmission starts. This control information also prevents more than one device trying to send information at the same time.

In modern installations, all communication and data transmission between processing nodes in a distributed instrumentation and control system is carried out digitally along some form of electronic highway. The highway can be a parallel interface, a local area network (LAN), a digital fieldbus, or a combined LAN/fieldbus. A parallel interface protocol is used commonly for connecting a small number of devices spread over a small geographical area, typically a single room. In the case of a large number of devices spread over larger geographical distances, typically a single building or site, an electronic highway in the form of either a LAN or a digital fieldbus is used. Instrumentation networks that are geographically larger than a single building or site can also be built, but these generally require transmission systems that include telephone lines as well as local networks at particular sites within the large system.

Manufacturers normally provide all the hardware and software necessary in order to create an instrumentation network using the various intelligent devices in their product range. However, problems usually occur if the designer of an instrumentation network wishes to use components sourced from different manufacturers where quite serious compatibility issues can arise. To overcome this, the Institute of Electronics and Electrical Engineers (IEEE) has developed IEEE 1451. This is a series of smart device interface standards that allow components from different manufacturers to be connected onto the same network.

### 11.4.1 Input–Output Interface

An *input/output interface* is required to connect each intelligent device onto the electronic highway. Sensors with a digital output pose little interfacing problems. However, many intelligent devices still have an analogue output that uses the standard 4- to 20-mA protocol and requires an analogue-to-digital converter in the input–output interface. For these, a protocol known as *HART* (Highway Addressable Remote Transducer) is the one used most widely to provide the necessary connection of such devices onto a digital network. HART is a bus-based networking protocol that has become a de facto standard for intelligent devices with an analogue sensor output. HART-compatible devices are provided by all major instrument manufacturers.

HART was always intended to be an interim network protocol to satisfy communication needs in the transitional period between the use of analogue communication with nonintelligent devices and fully digital communication with intelligent devices according to the digital fieldbus protocol. Because of this need to support both old and new systems, HART supports two modes of use—a hybrid mode and a fully digital mode.

In *hybrid mode*, status/command signals are digital but data transmission takes place in analogue form (usually in the 4- to 20-mA format). One serious limitation of this mode is that it is not possible to transmit multiple measurement signals on a single bus, as the analogue signals would corrupt each other. Hence, when HART is used in hybrid mode, the network must be arranged in a star configuration using a separate line for each field device rather than a common bus.

In *fully digital mode*, data transmission is digital as well as status/command signals. This enables one cable to carry signals for up to 15 intelligent devices. In practice, the fully digital mode of HART is rarely used, as the data transmission speed is very limited compared with fieldbus protocols. Therefore, the main application of the HART protocol has been to provide a communication capability with intelligent devices when existing analogue measurement signal transmission has to be retained because conversion to fully digital operation would be too expensive.

### 11.4.2 Parallel Data Bus

There are a number of different parallel data buses in existence. All of these have the common feature of transmitting data in parallel, that is, several bits are transmitted simultaneously. They also have separate data and control lines, which means that the data lines are used solely for data transmission and all control signals are routed onto dedicated control lines. This optimizes data transmission speed. However, apart from having this common functionality, there is little compatibility between the different parallel data busses available, with significant differences existing in the number of data lines used, the number of control lines used, the interrupt structure used, the data timing system, and the logic levels used for operation. Equipment manufacturers tend to keep to the same parallel interface protocol for all their range of devices, but different manufacturers use different protocols. Thus, while it will normally be easy to connect together a number of intelligent devices that all come from the same manufacturer, interfacing difficulties are likely to be experienced if devices from different manufacturers are connected together. Fortunately, the situation in the field is not as bad as it sounds because the IEEE 488 bus has now gained prominence as the preferred parallel databus for instrumentation networks and has been adopted by a large number of manufacturers. Since it was first introduced in 1975, the published standard for this bus has been revised on several occasions, with the most recent being in 2004 when the IEEE published a standard jointly with the International Electrotechnical Commission (IEC) as standard IEC 60488 (IEC, 2004a and b). The IEEE 488/IEC 60488 bus provides a parallel interface that facilitates the connection of intelligent instruments, actuators, and controllers within a single room. Physically, the bus consists of a shielded, 24-conductor cable. For a standard IEEE 488 bus, the maximum length of bus allowable is 20 m, with no more than 15 instruments distributed along its length. However, this limit on length and number of instruments can be overcome using an active (i.e., with auxiliary power supply) bus extender. The maximum distance between two particular units on the bus should not exceed about 2 m. The maximum data transfer rate permitted by the bus is 1 Mbit/s in theory, although the maximum data rate achieved in practice over a full 20-m length of bus is more likely to be in the range of 250–500 Kbit/s.

### 11.4.3 Local Area Networks

Local area networks transmit data in digital format along serial transmission lines. Synchronous transmission is normally used because this allows relatively high transmission speeds by transmitting blocks of characters at a time. A typical data block consists of 80 characters: this is preceded by a synchronization sequence and is followed by a stop sequence. The synchronization sequence causes the receiver to synchronize its clock with that of the transmitter. The two main standards for synchronous, serial transmission are RS422 and RS485. These are now formally published by the ANSI Telecommunications Industry Association/Electronic Industries Alliance (TIA/EIA) with the codes ANSI/TIA/EIA-422-B and ANSI/TIA/EIA-485. A useful comparison between the performance and characteristics of each of these and the older RS232 standard (asynchronous serial transmission) can be found in Brook and Herklot (1996).

Local area networks have particular value in the monitoring and control of systems that have a number of separate sensors, actuators, and control units dispersed over a large area. Indeed, for such large instrumentation systems, a local area network is the only viable transmission medium in terms of performance and cost. Parallel data buses, which transmit data in analogue form, suffer from signal attenuation and noise pickup over large distances, and the high cost of the long, multicore cables that they need is prohibitive.

However, the development of instrumentation networks is not without problems. Careful design of the network is required to prevent corruption of data when two or more devices on the network try to access it simultaneously and perhaps put information onto the data bus at the same time. This problem is solved by designing a suitable network protocol that ensures that network devices do not access the network simultaneously, thus preventing data corruption.

In a local area network, the electronic highway can take the form of either copper conductors or fiber-optic cable. Copper conductors are the least expensive option and allow transmission speeds up to 10 Mbit/s, using either a simple pair of twisted wires or a coaxial cable. However, fiber-optic cables are preferred in many networks for a number of reasons. The virtues of fiber-optic cables as a data transmission medium have been expounded in Chapter 10. Apart from the high immunity of the signals to noise, a fiber-optic transmission system can transfer data at speeds up to 240 Mbit/s. The reduction in signal attenuation during transmission also means that much longer transmission distances are possible without repeaters being necessary. For instance, allowable distances between repeaters for a fiber-optic network are quoted as 1 km for a half-duplex operation and up to 3.5 km for a full-duplex operation. In addition, the bandwidth of fiber-optic transmission is higher than for electrical transmission. Some cost savings can be achieved using plastic fiber-optic cables, but these cannot generally be used over distances greater than about 30 m because signal attenuation is too high.

There are many different protocols for local area networks but these are all based on one of three network structures known as star networks, bus networks, and ring networks, as shown in Figure 11.5. A local area network operates within a single building or site and can transmit data over distances up to about 500 m without signal attenuation being a problem. For transmission over greater distances, telephone lines are used in the network. Intelligent devices are interfaced to the telephone line used for data transmission via a modem. The *modem* converts the signal into a frequency-modulated analogue form. In this form, it can be transmitted over either the public switched telephone network or over private lines rented from telephone companies. The latter, being dedicated lines, allow higher data transmission rates.

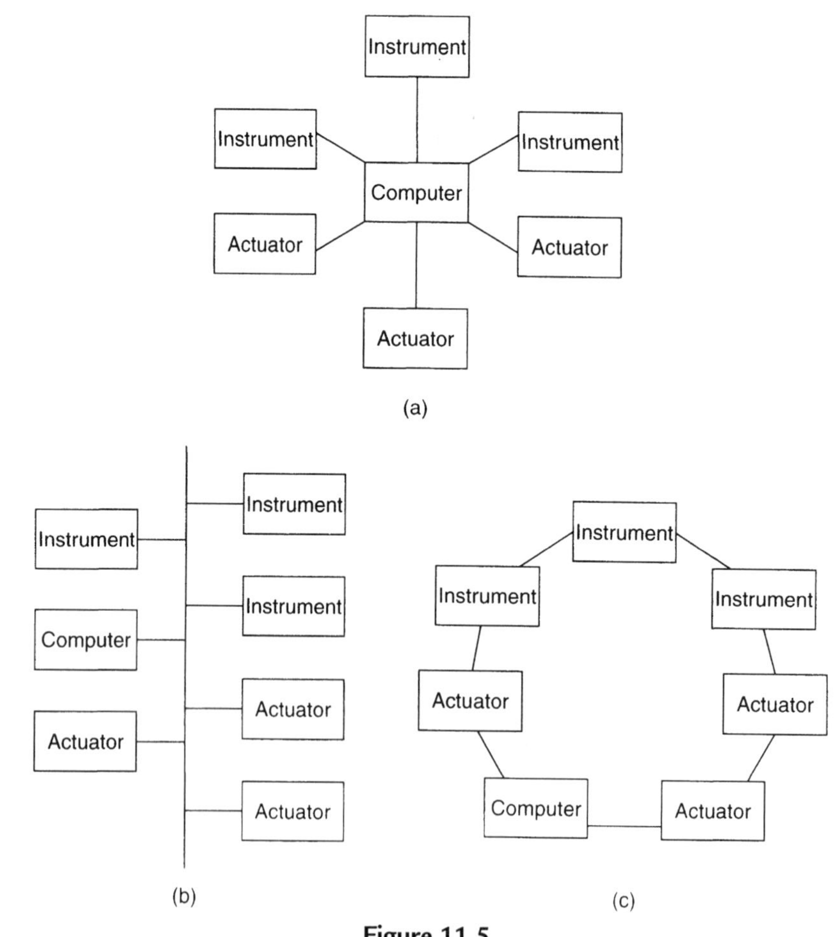


**Figure 11.5**
Network protocols: (a) star, (b) bus, and (c) ring.

### *Star networks*

In a *star network*, each instrument and actuator is connected directly to the supervisory computer by its own signal cable. One apparent advantage of a star network is that data can be transferred if necessary using a simple serial communication protocol such as RS232. This is an industry standard protocol and so compatibility problems do not arise, but it represents very old technology in which data transfer is very slow. Because of this speed problem, parallel communication is usually preferred even for star networks.

While star networks are simple in structure, the central supervisory computer node is a critical point in the system and failure of this means total failure of the whole system. When any device in the network needs to communicate with another device, a request has to be made to the central supervisory computer and all data transferred are routed through this central node. If the central node is inoperational for any reason, then data communication in the network is stopped.

### *Ring and bus networks*

In contrast to star networks, both ring and bus networks have a high degree of resilience in the face of one node breaking down. Hence, they are generally preferred to star networks. If the processor in any node breaks down, data transmission paths in the network are still maintained. Thus, the network can continue to operate, albeit at a degraded performance level, using the remaining computational power in the other processors. Most computer and intelligent instrument/actuator manufacturers provide standard conversion modules that allow their equipment to interface to one of these standard networks.

In a *ring network*, all intelligent devices are connected to a bus that is formed into a continuous ring. The ring protocol sends a special packet (or token) continuously round the ring to control access to the network. A station can only send data when it receives the token. During data transmission, the token is attached to the back of the message sent so that, once the information has been received safely, the token can continue on its journey round the network.

A *bus network* is similar to a ring network but the bus that the devices are connected onto is not continuous. Bus networks are also resilient toward the breakdown of one node in the network. A *contention protocol* is normally used. This allows any station to have immediate access to the network unless another station is using it simultaneously, in which case the protocol manages the situation and prevents data loss/corruption. *Ethernet* is the most common form of a bus network, which has now gained a dominant position in the LAN marketplace.

## 11.4.4 Digital Fieldbuses

*Fieldbus* is a generic word that describes a range of high-speed, bus-based, network protocols that support two-way communication in digital format between a number of intelligent devices in a local area network. All forms of transmission are supported, including twisted pair, coaxial cable, fiber optic, and radio links.

Intelligent devices in an automated system comprise a range of control elements, actuators, information processing devices, storage systems, and operator displays, as well as measurement devices. Hence, any fieldbus protocol must include provision for the needs of all system elements, and the communication requirements of field measurement devices cannot be viewed in isolation from these other elements. The design of a network protocol also has to cater for implementation in both large and small plants. A large plant may contain a number of processors in a distributed control system and have a large number of sensors and actuators. Howeveer, a small plant may be controlled by a single personal computer that provides an operator display on its monitor as well as communicating with plant sensors and actuators.

After fieldbus technology was first introduced in 1988, there was no rapid move to develop an international standard and, in consequence, different manufacturers all developed their own versions. This resulted in more than 50 different fieldbus protocols, with the more prominent ones being Foundation Fieldbus, Profibus, WorldFIP, ControlNet, P-net, and Interbus. Each version supports all devices within the product range of each manufacturer, but there is little compatibility among the different protocols on offer. They differ in many major respects, such as message format, access protocols, and rules for performance prediction. In recognition of the difficulties inherent in attempting to connect devices from different manufacturers that use a variety of incompatible interface standards and network protocols, the International Electrotechnical Commission set up a working part that was charged with defining a standard interface protocol. However, individual manufacturers continued to develop their own versions of fieldbus in parallel with the IEC initiative. The result of this is that although the IEC managed in 1999 to publish a fieldbus standard (IEC 61158, 1999), this had eight different protocol sets defining a standard for each of the eight main fieldbus systems then in operation and involved a document with more than 4000 pages. Since that time, there has been a further increase in the number of standards covered as additional parts of IEC 61158, particularly in respect of high-speed, Ethernet-based fieldbuses.

Despite the failure of IEC 61158 to establish a single fieldbus standard, a consortium of major international instrumentation manufacturers has set up the Fieldbus Foundation in an attempt to move toward one worldwide fieldbus standard. This has resulted in Foundation Fieldbus, which at least provides a common standard for, and interchangeability between, all devices manufactured by members of the consortium. However, competing standards, in particular Profibus, remain.

The basic architecture of Foundation Fieldbus has two levels, an upper and a lower. The lower level provides for communication between field devices and field input–output devices, whereas the upper level enables field input–output devices to communicate with controllers. These two levels have quite different characteristics. The lower level generally requires few connections, only needs a relatively slow data transfer rate, and must support intrinsically safe working. However, the upper level requires numerous connections and fast data transfer, but does not have to satisfy intrinsic safety requirements. Three standard bus speeds are currently specified for the Foundation Fieldbus lower level of 31.25 kbit/s, 1 Mbit/s, and 2.5 Mbit/s. Maximum cable lengths allowed are 1900 m at 31.25 kbit/s, 750 m at 1 Mbit/s, and 500 m at 2.5 Mbit/s. For the upper Foundation Fieldbus layer, a high-speed Ethernet protocol provides a data transfer rate up to 100 Mb/s.

## 11.5 Summary

The primary purpose of this chapter has been to introduce the subject of intelligent devices. However, because computational power is the component that separates intelligent devices from their nonintelligent counterparts, we started the chapter off by reviewing the main principles of digital computation. This led us to study the main elements in a computer, how computers operate particularly in respect of program execution, and how computers interface to outside components. We ended this introduction to digital computation with a review of the practical issues that have to be considered when incorporating computers into measurement devices.

Moving on to the subject of intelligent devices, we found that several different terms are used to describe these. Prominent among these terms are names such as intelligent instrument, intelligent sensor, smart sensor, and smart transmitter. We learned that there are no industry-wide definitions of what any of these names mean, except that all are distinguished by incorporating some form of computational power. As a consequence, the same kind of device, even with very similar attributes, may be known by two or more names. Therefore, the name used to describe a particular intelligent device is subject to the whims and style of its manufacturer.

Having explained this arbitrary nature in the way that intelligent devices are named, we went on to describe some commonly held views about the sorts of functions performed by devices known as smart sensors and the functions typically performed by devices known as smart transmitters. The conclusion drawn from this comparison of functions was that devices called smart transmitters tended to have a greater functionality than those called smart sensors.

Our investigation into the features of intelligent devices led us to conclude that these have significant advantages compared with nonintelligent devices. Perhaps the biggest single benefit is improved measurement accuracy. This is achieved by use of a computer processor within each device that performs actions such as compensation for random errors, adjustment for measurement nonlinearities, and compensation for the loading effect of the measuring process on the measured system. Processing power also enables devices to perform functions such as remote self-calibration and self-diagnosis of faults. Smart transmitters typically have additional features such as incorporation of multiple primary sensors covering different measurement ranges and allowing automatic selection of required range, inclusion of secondary sensors that can measure and compensate for environmental disturbances, plus incorporation of signal conditioning and analogue-to-digital conversion functions. Sometimes smart transmitters also have redundant sensors, which can be used to replace failed sensors and so improve device reliability.

We then went on to look at the issues surrounding communication between intelligent devices and other elements in a measurement/process control system. We noted that all communication and data transmission between processing nodes in a distributed instrumentation and control system required the use of some form of electronic highway, which can be a parallel interface, a local area network, a digital fieldbus, or a combined LAN/fieldbus. We then concluded the chapter with a look in broad detail at the various features in these alternative forms of electronic highways, but observed that there was little point in studying the fine details of any particular form of highway because there were continuing developments in the format of highways, particularly in the protocols used in local area networks and digital fieldbusses. This means that the inclusion of any detailed study in the book would become out of date very quickly.

## 11.6 Problems

11.1. Explain the principal components in the microprocessor contained within an intelligent instrument.

11.2. What are the two main types of computer memory? Which type is used predominantly in intelligent instruments and why?

11.3. What are the mechanisms for programming and program execution within an intelligent instrument?

11.4. Discuss operation of the input–output interface within the processor of an intelligent instrument, mentioning particularly the mechanisms of address decoding and data transfer control.

11.5. What are the practical considerations involved in implementing a computer processor within a measurement system?

11.6. How does an intelligent instrument correct for environmentally induced errors in measurements? What preconditions must be satisfied to allow an intelligent instrument to correct for such errors? How are these preconditions satisfied?

11.7. Explain how adding intelligence to an instrument improves the accuracy of volume flow rate measurements.

11.8. What additional functions does an intelligent instrument typically perfrom apart from the correction of environmentally induced errors in measurements?

11.9. Describe the typical function of devices known as *smart sensors*.

11.10. Describe the mechanisms for communication between an intelligent sensor and other components in a measurement system.

## References

Bignell, J., & White, N. (1996). *Intelligent Sensor Systems.* Institute of Physics Publishing.

Brook, N., & Herklot, T. (1996). *Choosing and implementing a serial interface.* Electronic Engineering.

Henry, M. (1995). *Self-validation improves Coriolis flowmeter.* Control Engineering, May, pp 81–86.

IEC 60488-1. (2004a). *Higher performance protocol for the standard digital interface for programmable instrumentation. Part 1: General.* International Electrotechnical Commission.

IEC 60488-2. (2004b). *Standard digital interface for programmable instrumentation. Part 2: Codes, formats, protocols and common commands.* International Electrotechnical Commission.

IEEE 1451-0. (2007). *Standard for a Smart Transducer Interface for Sensors and Actuators* (related parts of this standard referring to particular aspects of the interface are published separately as IEEE 1451-1, 1999, IEEE 1451-2, 1997, IEEE 1451-3, 2003, IEEE 1451-4, 2004, and IEEE 1451-5, 2007).

IEC 61158 (1999). *Digital data communications for measurement and control - Fieldbus for use in industrial control systems.* International Electrotechnical Commission.

CHAPTER 12

# Measurement Reliability and Safety Systems



## 12.1 Introduction

Previous chapters of this book discussed the design of measurement systems and said a lot about how the performance of measurement systems in respect of parameters such as accuracy can be improved. However, this earlier discussion has mainly been about the attributes of measurement systems when they are new. We have considered the effects of the passage of time only in respect of noting that the characteristics of measurement systems degrade over time and have to be restored back to their starting point by the process of recalibration. What we have not considered so far is the possibility of faults developing in measurement systems. At best, these faults impair the performance of the system, and at worst, they cause the system to stop working entirely. In safety-critical applications, measurement system faults can also have a serious adverse effect on the larger system that the measurement system is part of.

It is therefore appropriate for us to devote this chapter to a study of measurement system reliability issues and their effect on safety. We start off by looking at how reliability is formally defined and will say something about its theoretical principles. This will then lead us to consider ways in which reliability can be quantified. We will look particularly at two laws that quantify the reliability of system components that are in series and in parallel with one another. This will enable us to examine how these laws can be applied to improve the reliability of measurement systems. We will also look at the general precautions that can be taken to reduce the failure rate of instruments, including choosing instruments that can withstand the operating conditions expected, protecting them adequately against damage during use, calibrating them at the prescribed intervals to ensure that measurement inaccuracy remains within acceptable bounds, and duplicating critical measurement system components.

Because software is an important contributor to measurement system reliability, particularly with the current widespread use of intelligent devices, we will extend our treatise on reliability to consider the reliability of the software within a measurement system. We will see that the factors affecting reliability in software are fundamentally different than those affecting the reliability of hardware components. This is because software does not change with time. Therefore, we realize that the reliability of software has to be quantified in terms of the probability of failure of the software because of some undetected error in the software that has existed since it was written. This kind of failure usually occurs when some particular combination of input data is applied to the software and, in consequence, may not occur until the software has been in use for some considerable period of time. Having established a satisfactory way of quantifying software reliability, we go on to consider what can be done to improve reliability.

Our final consideration in this chapter will be system safety. We will see that measurement systems can have an impact on system safety in two main ways. First, failure of the measurement system may cause a dangerous situation to arise in a process because incorrect data are fed into the process control system. Second, the process itself may develop a dangerous fault that the measurement system fails to detect, thus preventing the operation of emergency responses such as the sounding of alarms or the opening of pressure-relief valves. In order to respond to the potential safety problems associated with the malfunction of measurement systems, we will look at the main ways available of designing safety systems.

## 12.2 Reliability

The reliability of measurement systems can be quantified as the mean time between faults occurring in the system. In this context, a fault means the occurrence of an unexpected condition in the system that causes the measurement output to either be incorrect or not to exist at all. The following sections summarize the principles of reliability theory relevant to measurement systems. A fuller account of reliability theory, particularly its application in manufacturing systems, can be found elsewhere (Morris, 1997).

### *12.2.1 Principles of Reliability*

The reliability of a measurement system is defined as the ability of the system to perform its required function within specified working conditions for a stated period of time. Unfortunately, factors such as manufacturing tolerances in an instrument and varying operating conditions conspire to make the faultless operating life of a system impossible to predict. Such factors are subject to random variation and chance, and therefore reliability cannot be defined in absolute terms. The nearest one can get to an absolute quantification of reliability are quasi-absolute terms such as mean time between failures (MTBF), which expresses the average time that the measurement system works without failure. Otherwise, reliability has to be expressed as a statistical parameter that defines the probability that no faults will develop over a specified interval of time.

In quantifying reliability for a measurement system, an immediate difficulty that arises is defining what counts as a fault. Total loss of a measurement output is an obvious fault, but a fault that causes a finite but incorrect measurement is more difficult to identify. The usual approach is to identify such faults by applying statistical process control techniques (Morris, 1997).

#### *Reliability quantification in quasi-absolute terms*

While reliability is essentially probabilistic in nature, it can be quantified in quasi-absolute terms by mean-time-between-failure and mean-time-to-failure (MTTF) parameters. It must be emphasized that these two quantities are usually average values calculated over a number of identical instruments, and therefore the actual values for any particular instrument may vary substantially from the average value.

The *mean time between failures* is a parameter that expresses the average time between faults occurring in an instrument, calculated over a given period of time. For example, suppose that the history of an instrument is logged over a 360-day period and the time interval in days between faults occurring was as follows:

11 23 27 16 19 32 6 24 13 21 26 15 14 33 29 12 17 22

The mean interval is 20 days, which is therefore the mean time between failures. An alternative way of calculating MTBF is simply to count the number of faults occurring over a given period. In the example just given, there were 18 faults recorded over a period of 360 days and so the MTBF can be calculated as $MTBF = 360/18 = 20\,days$.

Unfortunately, in the case of instruments that have a high reliability, such in-service calculation of reliability in terms of the number of faults occurring over a given period of time becomes grossly inaccurate because faults occur too infrequently. In this case, MTBF predictions provided by the instrument manufacturer can be used, as manufacturers have the opportunity to monitor the performance of a number of identical instruments installed in different companies. If there are a total of $F$ faults recorded for $N$ identical instruments in time $T$, the MTBF can be calculated as $MTBF = TN/F$. One drawback of this approach is that it does not take conditions of use, such as the operating environment, into account.

The mean time to failure is an alternative way of quantifying reliability that is normally used for devices such as thermocouples that are discarded when they fail. MTTF expresses the average time before failure occurs, calculated over a number of identical devices. Suppose that a batch of 20 thermocouples is put through an accelerated-use test in the same environment and the time before failure (in months) of each device is as follows:

7 9 13 6 10 11 8 9 14 8 8 12 9 15 11 9 10 12 8 11

The mean of these 20 numbers is 10. Therefore, the simulated MTTF is 10 months.

The final reliability-associated term of importance in measurement systems is the *mean time to repair* (MTTR). This expresses the average time needed for repair of an instrument. MTTR can also be interpreted as the *mean time to replace*, as replacement of a faulty instrument by a spare one is usually preferable in manufacturing systems to losing production while an instrument is repaired. As an example, suppose that the time in hours taken to repair an instrument over a history of 18 breakdowns is recorded, with the following times:

4 1 3 2 1 9 2 1 7 2 3 4 1 3 2 4 4 1

The mean of these values is 3 and the MTTR is therefore 3 hours.

The MTBF and MTTR parameters are often expressed in terms of a combined quantity known as the availability figure. This measures the proportion of total time that an instrument is working, that is, the proportion of total time that it is in an unfailed state. The availability is defined as the ratio:

$$Availability = \frac{MTBF}{MTBF + MTTR}.$$

In measurement systems, the aim must always be to maximize the MTBF figure and minimize the MTTR figure, thereby maximizing the availability. As far as MTBF and MTTF figures are concerned, good design and high-quality control standards during manufacture are the appropriate means of maximizing these figures. Design procedures that mean that faults are easy to repair are also an important factor in reducing the MTTR figure.

### *Failure patterns*

The pattern of failure in an instrument may increase, stay the same, or decrease over its life. In the case of *electronic components*, the failure rate typically changes with time in the manner shown in Figure 12.1a. This form of characteristic is frequently known as a *bathtub curve*. Early in their life, electronic components can have quite a high rate of fault incidence up to time $T_1$ (see Figure 12.1a). After this initial working period, the fault rate decreases to a low level and remains at this low level until time $T_2$ when aging effects cause the fault rate to start to increase again. Instrument manufacturers often "burn in" electronic components for a length of time corresponding to time $T_1$. This means that the components have reached the high-reliability phase of their life before they are supplied to customers.

*Mechanical components* usually have different failure characteristics, as shown in Figure 12.1b. Material fatigue is a typical reason for the failure rate to increase over the life of a mechanical component. In the early part of their life, when all components are relatively new, many

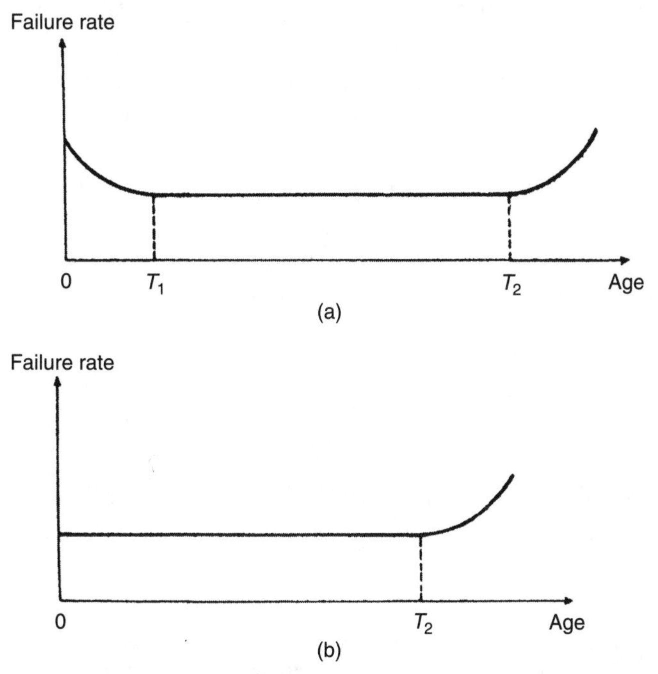

**Figure 12.1**
Typical variation of reliability with component age: (a) electronic components (*bathtub* curve) and (b) mechanical components.

instruments exhibit a low incidence of faults. Then, at a later stage, when fatigue and other aging processes start to have a significant effect, the rate of faults increases and continues to increase thereafter.

*Complex systems* containing many different components often exhibit a constant pattern of failure over their lifetime. The various components within such systems each have their own failure pattern where the failure rate is increasing or decreasing with time. The greater the number of such components within a system, the greater the tendency for the failure patterns in the individual components to cancel out and for the rate of fault incidence to assume a constant value.

### Reliability quantification in probabilistic terms

In probabilistic terms, the reliability, $R(T)$, of instrument $X$ is defined as the probability that the instrument will not fail within a certain period of time, $T$. The unreliability or likelihood of failure, $F(T)$, is a corresponding term that expresses the probability that the instrument will fail within the specified time interval. $R(T)$ and $F(T)$ are related by the expression:

$$F(T) = 1 - R(T). \tag{12.1}$$

To calculate $R(T)$, accelerated lifetime testing* is carried out for a number ($N$) of identical instruments. Providing all instruments have similar conditions of use, times of failure, $t_1, t_2, \cdots, t_n$, will be distributed about the mean time to failure, $t_m$. If the probability density of the time to failure is represented by $f(t)$, the probability that a particular instrument will fail in a time interval, $\delta t$, is given by $f(t)\delta t$, and the probability that the instrument will fail before time $T$ is given by

$$F(T) = \int_0^T f(t)dt.$$

The probability that the instrument will fail in a time interval $\Delta T$ following $T$, assuming that it has survived to time $T$, is given by

$$\frac{F(T + \Delta T) - F(T)}{R(T)},$$

where $R(T)$ is the probability that the instrument will survive to time $T$. Dividing this expression by $\Delta T$ gives the average failure rate in the interval from $T$ to $T+\Delta T$ as

---

* Accelerated lifetime testing means subjecting an instrument to a much greater frequency of use than would normally be expected. If an instrument is normally used 10 times per day, then 100 days of normal use can be simulated by using it 1000 times in a single day.

$$\frac{F(T+\Delta T)-F(T)}{\Delta T R(T)}.$$

In the limit as $\Delta T \rightarrow 0$, the instantaneous failure rate at time $T$ is given by

$$\theta_f = \frac{d[F(T)]}{dt}\frac{1}{R(T)} = \frac{F'(T)}{R(T)}. \tag{12.2}$$

If it is assumed that the instrument is in the constant-failure-rate phase of its life, denoted by the interval between times $t_1$ and $t_2$ in Figure 12.1, then the instantaneous failure rate at $T$ is also the mean failure rate that can be expressed as the reciprocal of the MTBF, that is, mean failure rate $= \theta_f = 1/t_m$.

Differentiating Equation (12.1) with respect to time gives $F'(T) = -R'(T)$. Hence, substituting for $F'(T)$in Equation (12.2) gives

$$\theta_f = -\frac{R'(T)}{R(T)}.$$

This can be solved (Johnson et al., 2010) to give the following expression:

$$R(T) = \exp(-\theta_f T). \tag{12.3}$$

Examination of Equation (12.3) shows that, at time $t = 0$, the unreliability is zero. Also, as $t$ tends to $\propto$, the unreliability tends to a value of 1. This agrees with intuitive expectations that the value of unreliability should lie between values of 0 and 1. Another point of interest in Equation (12.3) is to consider the unreliability when T = MTBF, that is, when $T = t_m$. Then $F(T) = 1 - \exp(-1) = 0.63$, that is, the probability of a product failing after it has been operating for a length of time equal to the MTBF is 63%.

Further analysis of Equation (12.3) shows that, for $T/t_m \leq 0.1$,

$$F(T) \approx T/t_m. \tag{12.4}$$

This is a useful formula for calculating (approximately) the reliability of a critical product that is only used for a time that is a small proportion of its MTBF.

## Example 12.1

If the mean time to failure of an instrument is 50,000 hours, calculate the probability that it will not fail during the first 10,000 hours of operation.

■

## Solution

From Equation (12.3),

$$R(T) = \exp\left(-\theta_f T\right) = \exp(-10{,}000/50{,}000) = 0.8187.$$

## Example 12.2

If the mean time to failure of an instrument is 80,000 hours, calculate the probability that it will not fail during the first 8000 hours of operation.

## Solution

In this case, $T/t_m = 80,000/8000 = 0.1$ and so Equation (12.4) can be applied, giving $R(T) = 1 - F(T) \approx 1 - T/t_m \approx 0.9$. To illustrate the small level of inaccuracy involved in using the approximate expression of Equation (12.4), if we calculate the probability according to Equation (12.3), we get $R(T) = \exp(-0.1) = 0.905$. Thus, there is a small but finite error in applying Equation (12.4) instead of Equation (12.3).

### *12.2.2 Laws of Reliability in Complex Systems*

Measurement systems usually comprise a number of components connected together in series, and hence it is necessary to know how the reliabilities of individual components are aggregated into a reliability figure for the whole system. In some cases, identical measurement components are put in parallel to improve reliability because the measurement system then only fails if all of the parallel components fail. These two cases are covered by particular laws of reliability.

*Reliability of components in series*

A measurement system consisting of several components in series fails when any one of the separate components develops a fault. The reliability of such a system can be quantified as the probability that none of the components will fail within a given interval of time. For a system of $n$ series components, reliability, $R_S$, is the product of separate reliabilities of the individual components according to the joint probability rule (Morris, 1997):

$$R_S = R_1 R_2 \cdots R_n. \tag{12.5}$$

## Example 12.3

A measurement system consists of a sensor, a variable conversion element, and a signal processing circuit, for which the reliability figures are 0.9, 0.95, and 0.99, respectively. Calculate the reliability of the whole measurement system.

## Solution

Applying Equation (12.5), $R_S = 0.9 \times 0.95 \times 0.99 = 0.85$.

*Reliability of components in parallel*

One way of improving the reliability of a measurement system is to connect two or more instruments in parallel. This means that the system only fails if every parallel instrument fails. For such systems, system reliability, $R_S$, is given by

$$R_S = 1 - F_S, \tag{12.6}$$

where $F_S$ is the unreliability of the system. The equation for calculating $F_S$ is similar to Equation (12.5). Thus, for $n$ instruments in parallel, the unreliability is given by

$$F_S = F_1 F_2 \cdots F_n. \tag{12.7}$$

If all the instruments in parallel are identical, then Equation (12.7) can be written in the simpler form:

$$F_S = (F_X)^n, \tag{12.8}$$

where $F_X$ is the unreliability of each instrument.

## Example 12.4

In a particular safety-critical measurement system, three identical instruments are connected in parallel. If the reliability of each instrument is 0.95, calculate the reliability of the measurement system.

## Solution

From Equation (12.1), the unreliability of each instrument $F_X$ is given by $F_X = 1 - R_X = 1 - 0.95 = 0.05$. Applying Equation (12.8), $F_S = (F_X)^3 = (0.05)^3 = 0.000125$. Thus, from Equation (12.6), $R_S = 1 - F_S = 1 - 0.000125 = 0.999875$.

### 12.2.3 Improving Measurement System Reliability

When designing a measurement system, the aim is always to reduce the probability of the system failing to as low a level as possible. An essential requirement in achieving this is to ensure that the system is replaced at or before time $t_2$ in its life shown in Figure 12.1 when the statistical frequency of failures starts to increase. Therefore, the initial aim should be to set the lifetime $T$ equal to $t_2$ and minimize the probability $F(T)$ of the system failing within this specified lifetime. Once all measures to reduce $F(T)$ have been applied, the acceptability of the reliability $R(T)$ has to be assessed against the requirements of the measurement system. Inevitably, cost enters into this, as efforts to increase $R(T)$ usually increase the cost of buying and maintaining the system. Lower reliability is acceptable in some measurement systems where the cost of failure is low, such as in manufacturing systems where the cost of lost production or loss due to making out-of-specification products is not serious. However, in other applications, such as where failure of the measurement system incurs high costs or causes safety problems, high reliability is essential. Some special applications where human access is very difficult or impossible, such as measurements in unmanned spacecraft, satellites, and nuclear power plants, demand especially high reliability because repair of faulty measurement systems is impossible.

The various means of increasing $R(T)$ are considered here. However, once all efforts to increase $R(T)$ have been exhausted, the only solution available if the reliability specified for a working period $T$ is still not high enough is to reduce the period $T$ over which the reliability is calculated by replacing the measurement system earlier than time $t_2$.

#### *Choice of instrument*

The types of components and instruments used within measuring systems has a large effect on the system reliability. Of particular importance in choosing instruments is regarding the type of operating environment in which they will be used. In parallel with this, appropriate protection must be given (e.g., enclosing thermocouples in sheaths) if it is anticipated that the environment may cause premature failure of an instrument. Some instruments are more affected than others, and thus more likely to fail, in certain environments. The necessary knowledge to make informed choices about the suitability of instruments for particular environments, and the correct protection to give them, requires many years of experience, although instrument manufacturers can give useful advice in most cases.

#### *Instrument protection*

Adequate protection of instruments and sensors from the effects of the operating environment is necessary. For example, thermocouples and resistance thermometers should be protected by a sheath in adverse operating conditions.

*Regular calibration*

The most common reason for faults occurring in a measurement system, whereby the error in the measurement goes outside acceptable limits, is drift in the performance of the instrument away from its specified characteristics. Such faults can usually be avoided by ensuring that the instrument is recalibrated at the recommended intervals of time. Types of intelligent instrument and a sensor that perform self-calibration have clear advantages in this respect.

*Redundancy*

Redundancy means the use of two or more measuring instruments or measurement system components in parallel such that any one can provide the required measurement. Example 12.4 showed the use of three identical instruments in parallel to make a particular measurement instead of a single instrument. This increased the reliability from 95 to 99.99%. Redundancy can also be applied in larger measurement systems where particular components within it seriously degrade the overall reliability of the system. Consider the five-component measurement system shown in Figure 12.2a in which the reliabilities of the individual system components are $R_1 = R_3 = R_5 = 0.99$ and $R_2 = R_4 = 0.95$.

Using Equation (12.5), the system reliability is given by $R_S = 0.99 \times 0.95 \times 0.99 \times 0.95 \times 0.99 = 0.876$. Now, consider what happens if redundant instruments are put in parallel with the second and fourth system components, as shown in Figure 12.2b. The reliabilities of these sections of the measurement system are now modified to new values $R_2'$ and $R_4'$, which can be calculated using Equations (12.1), (12.6), and (12.8) as follows: $F_2 = 1 - R_2 = 0.05$. Hence, $F_2' = (0.05)^2 = 0.0025$ and $R_2' = 1 - F_2' = 0.9975$. $R_4' = R_2'$ since $R_4 = R_2$.

Using Equation (12.5) again, the system reliability is now $R_S = 0.99 \times 0.9975 \times 0.99 \times 0.9975 \times 0.99 = 0.965$.

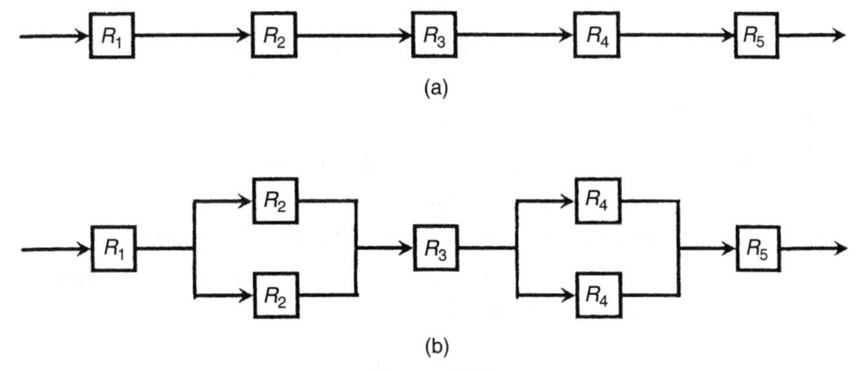


**Figure 12.2**
Improving measurement system reliability: (a) original system and (b) duplicating components that have poor reliability.

Thus, the redundant instruments have improved the system reliability by a large amount. However, this improvement in reliability is only achieved at the cost of buying and maintaining the redundant components that have been added to the measurement system. If this practice of using redundant instruments to improve reliability is followed, provision must be provided for replacing failed components by the standby units. The most efficient way of doing this is to use an automatic switching system, but manual methods of replacement can also work reasonably well in many circumstances.

The principle of increasing reliability by placing components in parallel is often extended to other aspects of measurement systems, such as the connectors in electrical circuits, as bad connections are a frequent cause of malfunction. For example, two separate pairs of plugs and sockets are used frequently to make the same connection. The second pair is redundant, that is, the system can usually function at 100% efficiency without it, but it becomes useful if the first pair of connectors fails.

### 12.2.4 Software Reliability

As computer processors, and the software within them, are found increasingly in most measurement systems, the issue of the reliability of such components has become very important. Computer hardware behaves very much like electronic components in general, and the rules for calculating reliability given earlier can be applied. However, factors affecting reliability in software are fundamentally different. Application of the general engineering definition of reliability to software is not appropriate because characteristics of the error mechanisms in software and in engineering hardware are fundamentally different. Hardware systems that work correctly when first introduced can develop faults at any time in the future, and so the MTBF is a sensible measure of reliability. However, software does not change with time: if it starts off being error free, then it will remain so. Therefore, what we need to know, in advance of its use, is whether faults are going to be found in the software after it has been put into use. Thus, for software, a MTBF reliability figure is of little value. Instead, we must somehow express the probability that errors will not occur in it.

#### *Quantifying software reliability*

A fundamental problem in predicting that errors will not occur in software is that, however exhaustive the testing, it is impossible to say with certainty that all errors have been found and eliminated. Errors can be quantified by three parameters, $D$, $U$, and $T$, where $D$ is the number of errors detected by testing the software, $U$ is the number of undetected errors, and $T$ is the total number of errors (both detected and undetected). Hence,

$$U = T - D. \qquad (12.9)$$

Good program testing can detect most errors and so make $D$ approach $T$ so that $U$ tends toward zero. However, as the value of $T$ can never be predicted with certainty, it is very difficult to predict that software is error free, whatever degree of diligence is applied during testing procedures.

Whatever approach is taken to quantifying reliability, software testing is an essential prerequisite to the quantification methods available. While it is never possible to detect all the errors that might exist, the aim must always be to find and correct as many errors as possible by applying a rigorous testing procedure. Software testing is a particularly important aspect of the wider field of software engineering. However, as it is a subject of considerable complexity, detailed procedures available are outside the scope of this book. A large number of books now cover good software engineering in general and software testing procedures in particular, and the reader requiring further information is referred to referenced texts such as Fenton and Pfleeger (1998), Fenton and Bieman (2010), Naik and Tripathy (2008), Pfleeger and Atlee (2009), and Shooman (2002).

One approach to quantifying software reliability (Fenton and Pfleeger, 1998) is to monitor the rate of error discovery during testing and then extrapolate this into an estimate of the mean time between failures for the software once it has been put into use. Testing can then be extended until the predicted MTBF is greater than the projected time horizon of usage of the software. This approach is rather unsatisfactory because it accepts that errors in the software exist and only predicts that errors will not emerge very frequently.

Confidence in the measurement system is much greater if we can say "there is a high probability that there are zero errors in the software" rather than "there are a finite number of errors in the software but they are unlikely to emerge within the expected lifetime of its usage." One way of achieving this is to estimate the value of $T$ (total number of errors) from initial testing and then carry out further software testing until the predicted value of $T$ is zero in a procedure known as *error seeding* (Mills, 1972). In this method, the programmer responsible for producing the software deliberately puts a number of errors, $E$, into the program, such that the total number of errors in the program increases from $T$ to $T'$, where $T'=T+E$. Testing is then carried out by a different programmer who will identify a number of errors given by $D'$, where $D'=D+E'$ and $E'$ is the number of deliberately inserted errors detected by this second programmer. Normally, the real errors detected ($D$) will be less than $T$ and the seeded errors detected ($E'$) will be less than $E$. However, on the assumption that the ratio of seeded errors detected to the total number of seeded errors will be the same as the ratio of real errors detected to the total number of real errors, the following expression can be written:

$$\frac{D}{T}=\frac{E'}{E}. \tag{12.10}$$

As $E'$ is measured, $E$ is known and $D$ can be calculated from the number of errors $D'$ detected by the second programmer according to $D=D'-E'$, the value of $T$ can then be calculated as

$$T = DE/E'. \tag{12.11}$$

One flaw in Equation (12.11) is the assumption that seeded errors are representative of all real (unseeded) errors in the software, both in proportion and in character. This assumption is never entirely valid in practice because, if errors are unknown, then their characteristics are also unknown. Thus, while this approach may be able to give an approximate indication of the value of $T$, it can never predict its actual value with certainty.

An alternative to error seeding is the *double-testing* approach, where two independent programmers test the same program (Fenton and Pfleeger, 1998). Suppose that the number of errors detected by each programmer is $D_1$ and $D_2$, respectively. Normally, errors detected by the two programmers will be in part common and in part different. Let $C$ be the number of common errors that both programmers find. The error-detection success of each programmer can be quantified as

$$S_1 = D_1/T\ ;\quad S_2 = D_2/T. \tag{12.12}$$

It is reasonable to assume that the proportion of errors $D_1$ that programmer 1 finds out of the total number of errors $T$ is the same proportion as the number of errors $C$ that he/she finds out of the number $D_2$ found by programmer 2, that is:

$$\frac{D_1}{T} = \frac{C}{D_2} = S_1,$$

and hence

$$D_2 = \frac{C}{S_1}. \tag{12.13}$$

From Equation (12.12), $T = D_2/S_2$, and substituting in the value of $D_2$ obtained from Equation (12.13), the following expression for T is obtained:

$$T = C/S_1S_2. \tag{12.14}$$

From Equation (12.13), $S_1 = C/D_2$, and from Equation (12.12), $S_2 = D_2S_1/D_1 = C/D_1$. Thus, substituting for $S_1$ and $S_2$ in Equation (12.14):

$$T = D_1D_2/C. \tag{12.15}$$

Program testing should continue until the number of errors that have been found is equal to the predicted total number of errors $T$. In the case of Example 12.6, this means continuing testing until 30 errors have been found. However, the problem with doing this is that $T$ is only an estimated quantity and there may actually be only 28 or 29 errors in the program. Thus, to continue testing until 30 errors have been found would mean testing forever! Hence, once 28 or 29 errors have been found and continued testing for a significant time after this has detected no more errors, the testing procedure should be terminated, even though the program could still contain 1 or 2 errors. The approximate nature of the calculated value of $T$ also means that its true value could be 31 or 32, and therefore the software may still contain errors if testing is stopped once 30 errors have been found. Thus, the fact that $T$ is only an estimated value means the statement that a program is error free once the number of errors detected is equal to $T$ can only be expressed in probabilistic terms.

To quantify this probability, further testing of the program is necessary. The starting point for this further testing is the stage when the total number of errors $T$ that are predicted have been found (or when the number found is slightly less than $T$ but further testing does not seem to be finding any more errors). The next step is to seed the program with $W$ new errors and then test it until all $W$ seeded errors have been found. Provided that no new errors have been found during this further testing phase, the probability that the program is error free can then be expressed as

$$P = W/(W+1). \tag{12.16}$$

However, if any new error is found during this further testing phase, the error must be corrected and then the seeding and testing procedure must be repeated. Assuming that no new errors are detected, a value of $W = 10$ gives $P = 0.91$ (probability 91% that the program is error free). To get to 99% error-free probability, $W$ has to be 99.

## Example 12.5

The author of a digital signal-processing algorithm that forms a software component within a measurement system adds 12 deliberate faults to the program. The program is then tested by a second programmer, who finds 34 errors. Of these detected errors, the program author recognizes 10 of them as being seeded errors. Estimate the original number of errors present in the software (i.e., excluding seeded errors).

## Solution

The total number of errors detected ($D'$) is 34, and the program author confirms that the number of seeded error amounts for these ($E'$) is 10 and that the total number of seeded errors ($E$) was 12. Because $D' = D + E'$ (see earlier), $D = D' - E' = 24$. Hence, from Equation (12.11), $T = DE/E' = 24 \times 12/10 = 28.8$.

## Example 12.6

A piece of software is tested independently by two programmers, and the number of errors found is 24 and 26, respectively. Of the errors found by programmer 1, 21 are the same as errors found by programmer 2.

## ■ Solution

$D_1 = 24$, $D_2 = 26$, and $C = 21$. Hence, applying Equation (12.15), $T = D_1 D_2 / C = 24 \times 26/21 = 29.7$.

■

*Improving software reliability*

The a priori requirement in achieving high reliability in software is to ensure that it is produced according to sound software engineering principles. Formal standards for achieving high quality in software are set out in BS/ISO/IEC 90003 (2004). Libraries and bookshops, especially academic ones, offer a number of texts on good software design procedures. These differ significantly in their style of approach, but all have the common aim of encouraging the production of error-free software that conforms to the design specification. It is not within the scope of this book to enter into arguments about which software design approach is best, as the choice between different software design techniques largely depends on personal preferences. However, it is essential that software contributing to a measurement system is produced according to good software engineering principles.

The second stage of reliability enhancement is application of a rigorous testing procedure as described in the last section. Because this is a very time-consuming and hence expensive business, testing should only continue until the calculated level of reliability is the minimum needed for requirements of the measurement system. However, if a very high level of reliability is demanded, such rigorous testing becomes extremely expensive and an alternative approach, known as N-version programming, is often used. *N-version programming* requires $N$ different programmers to produce $N$ different versions of the same software according to a common specification. Then, assuming that there are no errors in the specification itself, any difference in the output of one program compared with others indicates an error in that program. Commonly, $N = 3$ is used, that is, three different versions of the program are produced, but $N = 5$ is used for measurement systems that are very critical. In this latter case, a "voting" system is used, which means that up to two out of the five versions can be faulty without incorrect outputs being generated.

Unfortunately, while this approach reduces the chance of software errors in measurement systems, it is not foolproof because the degree of independence between programs cannot be guaranteed. Different programmers, who may be trained in the same place and use the same design techniques, may generate different programs that have the same errors. Thus, this method has the best chance of success if the programmers are trained independently and use different design techniques.

Languages such as ADA also improve the safety of software because they contain special features designed to detect the kind of programming errors that are commonly made. Such languages have been developed specifically with safety-critical applications in mind.

## 12.3 Safety Systems

Measurement system reliability is usually inexorably linked with safety issues, as measuring instruments to detect the onset of dangerous situations that may potentially compromise safety are a necessary part of all safety systems implemented. Statutory safety legislation now exists in all countries around the world. While the exact content of legislation varies from country to country, a common theme is to set out responsibilities for all personnel whose actions may affect the safety of themselves or others. Penalties are prescribed for contravention of the legislation, which can include fines, custodial sentences, or both. Legislation normally sets out duties for both employers and employees.

Duties of employers include:

- To ensure that the process plant is operated and maintained in a safe way so that the health
  and safety of all employees are protected.
- To provide such training and supervision as is necessary to ensure the health and safety of all employees.
- To provide a monitoring and shutdown system (safety system) for any process plant or other equipment that may cause danger if certain conditions arise.
- To ensure the health and safety, as far as is reasonably practical, of all persons who are not employees but who may reasonably be expected to be at risk from operations carried out by a company.

Duties of employees include:

- To take reasonable care for their own safety.
- To take reasonable care for the safety of others.
- To avoid misusing or damaging any equipment or system designed to protect people's safety.

The primary concern of measurement and instrumentation technologists with regard to safety legislation is to (1) ensure that all measurement systems are installed and operated in a safe way and (2) ensure that instruments and alarms installed as part of safety protection systems operate reliably and effectively.

#### *Intrinsic safety*

Intrinsic safety describes the ability of measuring instruments and other systems to operate in explosive or flammable environments without any risk of sparks or arcs causing an explosion or fire. The detailed design of systems to make them intrinsically safe is outside the scope of this book. However, the general principles are either to design electrical systems in a way that avoids any possibility of parts that may spark coming into contact with the operating environment or to avoid using electrical components altogether. The latter point means that pneumatic sensors and actuators continue to find favor in some applications, despite the advantages of electrical devices in most other respects.

*Installation practice*

Good installation practice is necessary to prevent any possibility of people getting electrical shocks from measurement systems. Instruments that have a mains power supply must be subject to normal rules about the condition of supply cables, clamping of wires, and earthing of all metal parts. However, most measurement systems operate at low voltages and so pose no direct threat unless parts of the system come into contact with mains conductors. This should be prevented by applying codes of practice that require all cabling for measurement systems be kept physically separate from that used for carrying mains voltages to equipment. Normally, this prohibits the use of the same trunking to house both signal wires and mains cables, although some special forms of trunking are available that have two separate channels separated by a metal barrier, thus allowing them to be used for both mains cables and signal wires. This subject is covered in depth in the many texts on electrical installation practice.

### 12.3.1 Introduction to Safety Systems

The purpose of safety systems is to monitor parameter values in manufacturing plant and other systems and to make an effective response when plant parameters vary away from normal operating values and cause a potentially dangerous situation to develop. The response can be either to generate an alarm for the plant operator to take action or to take more direct action to shut down the plant automatically. The design and operation of safety systems are now subject to guidelines set by international standard IEC 61508.

*IEC 61508*

IEC 61508 (2005) sets out a code of practice designed to ensure that safety systems work effectively and reliably. Although concerned primarily with electrical, electronic, and programmable electronic safety systems, the principles embodied by the standard can be applied as well to systems with other technologies, such as mechanical, pneumatic, and hydraulic devices.

The IEC 61508 standard is subdivided into three sets of requirements:

- Proper management of design, implementation, and maintenance of safety systems.
- Competence and training of personnel involved in designing, implementing, or maintaining safety systems.
- Technical requirements for the safety system itself.

A full analysis of these various requirements can be found elsewhere (Dean, 1999).

A key feature of IEC 61508 is the *safety integrity level* (SIL), which is expressed as the degree of confidence that a safety system will operate correctly and ensure that there is an adequate response to any malfunctions in the manufacturing plant that may cause a hazard and put human beings at risk. The SIL value is set according to what the tolerable risk is in terms of the rate of failure for a process. The procedure for defining the required SIL value is known as *risk analysis*. What is "tolerable" depends on what the consequences of a dangerous failure are in terms of injury to one or more people or death to one or more people. The acceptable level of tolerance for particular industries and processes is set according to guidelines defined by safety regulatory authorities, expert advice, and legal requirements. The following table gives the SIL value corresponding to various levels of tolerable risk for a continuous operating plant.

| SIL | Probability of dangerous failure per hour | Probability of dangerous failure per year |
|---|---|---|
| 4 | $10^{-9}$ to $10^{-8}$ | $10^{-5}$ to $10^{-4}$ |
| 3 | $10^{-8}$ to $10^{-7}$ | $10^{-4}$ to $10^{-3}$ |
| 2 | $10^{-7}$ to $10^{-6}$ | $10^{-3}$ to $10^{-2}$ |
| 1 | $10^{-6}$ to $10^{-5}$ | $10^{-2}$ to $10^{-1}$ |

The safety system is required to have sufficient reliability to match the rate of dangerous failures in a plant to the SIL value set. This reliability level is known as the *safety integrity* of the system. *Plant reliability* is calculated by identical principles to those set out in Section 12.2 for measurement systems and is based on a count of the number of faults that occur over a certain interval of time. However, it must be emphasized that the frequency of potentially dangerous failures is usually less than the rate of occurrence of faults in general. Thus, the reliability value for a plant cannot be used directly as a prediction of the rate of occurrence of dangerous failures. Hence, the total failures over a period of time must be analyzed and divided between faults that are potentially dangerous and those that are not.

Once risk analysis has been carried out to determine the appropriate SIL value, the required performance of the safety protection system can be calculated. For example, if the maximum allowable probability of dangerous failures per hour is specified as $10^{-8}$ and the actual probability of dangerous failures in a plant is calculated as $10^{-3}$ per hour, then the safety system must have a minimum reliability of $10^{-8}/10^{-3}$, that is, $10^{-5}$ failures for a 1-hour period. A fuller account of calculating safety system requirements is given elsewhere (Simpson and Smith, 1999).

### 12.3.2 Design of a Safety System

A typical safety system consists of a sensor, a trip amplifier, and either an actuator or an alarm generator, as shown in Figure 12.3. For example, in a safety system designed to protect against abnormally high pressures in a process, the sensor would be some form of pressure transducer, and the trip amplifier would be a device that amplifies the measured pressure

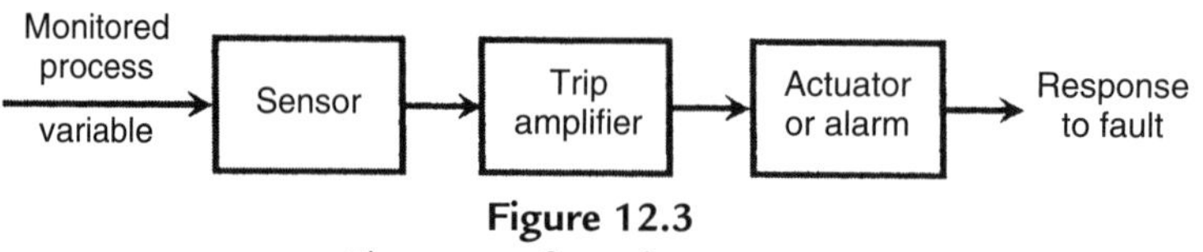


**Figure 12.3**
Elements of a safety system.

signal and generates an output that activates either an actuator or an alarm if the measured pressure signal exceeded a preset threshold value. A typical actuator in this case would be a relief valve.

Software is increasingly embedded within safety systems to provide intelligent interpretation of sensor outputs, such as identifying trends in measurements. Safety systems that incorporate software and a computer processor are known commonly as *microprocessor-based protection systems*. In any system containing software, the reliability of the software is crucial to the overall reliability of the safety system, and the reliability-quantification techniques described in Section 12.3 assume great importance.

To achieve the very high levels of reliability normally specified for safety systems, it is usual to guard against system failure by either triplicating the safety system and implementing two-out-of-three voting or providing a switchable, standby safety system. These techniques are considered next.

### *Two-out-of-three voting system*

This system involves triplicating the safety system, as shown in Figure 12.4. Shutdown action is taken, or an alarm is generated, if two out of the three systems indicate the requirement for action. This allows the safety system to operate reliably if any one of the triplicated systems fails and is often known as a two-out-of-three voting system. The reliability $R_S$ is given by

$$\begin{aligned} R_S &= \text{probability of all three systems operating correctly} \\ &\quad + \text{probability of any two systems operating correctly} \\ &= R_1R_2R_3 + (R_1R_2F_3 + R_1F_2R_3 + F_1R_2R_3), \end{aligned} \tag{12.17}$$

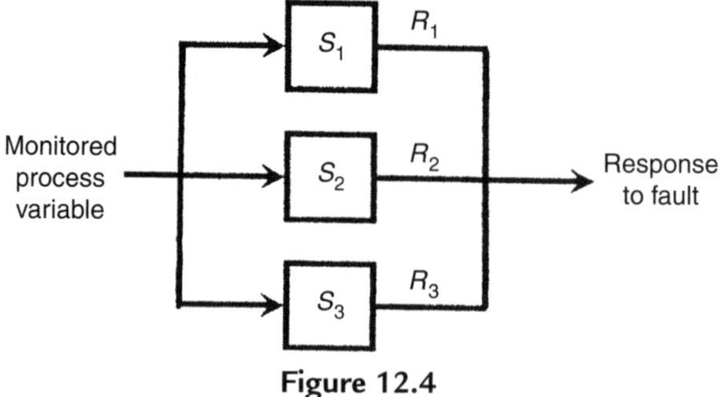


**Figure 12.4**
Two-out-of-three voting system.

where $R_1$, $R_2$, and $R_3$ and $F_1$, $F_2$, and $F_3$ are the reliabilities and unreliabilities of the three systems, respectively. If all of the systems are identical (such that $R_1 = R_2 = R_3 = R$, etc.):

$$R_S = R^3 + 3R^2F = R^3 + 3R^2(1 - R). \quad (12.18)$$

## Example 12.7

In a particular protection system, three safety systems are connected in parallel and a two-out-of-three voting strategy is applied. If the reliability of each of the three systems is 0.95, calculate the overall reliability of the whole protection system.

## Solution

Applying Equation (12.18), $R_S = 0.95^3 + [3 \times 0.95^2 \times (1 - 0.95)] = 0.993$.

*Standby system*

A standby system avoids the cost of providing and running three separate safety systems in parallel. Use of a standby system means that only two safety systems have to be provided. The first system is in continuous use but the second system is normally not operating and is only switched into operation if the first system develops a fault. The flaws in this approach are the necessity for faults in the primary system to be detected reliably and the requirement that the switch must always work correctly. The probability of failure, $F_S$, of a standby system of the form shown in Figure 12.5, assuming no switch failures during normal operation, can be expressed as

$$\begin{aligned} F_S &= \text{probability of systems } S_1 \text{ and } S_2 \text{ both failing, given successful switching} \\ &\quad + \text{probability of } S_1 \text{ and the switching system both failing at the same time} \\ &= F_1F_2R_DR_W + F_1(1 - R_DR_W). \end{aligned}$$

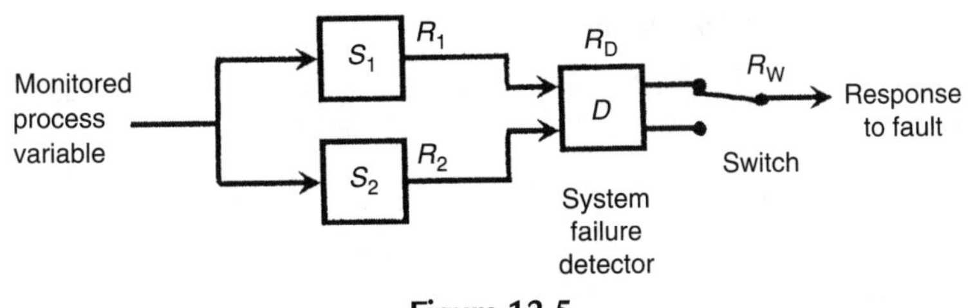


**Figure 12.5**
Standby system.

System reliability is given by

$$R_S = 1 - F_S = 1 - F_1(1 + F_2R_DR_W - R_DR_W), \tag{12.19}$$

where $R_D$ is the reliability of the fault detector and $R_W$ is the reliability of the switching system.

The derivation of Equation (12.19) assumes that there are no switch failures during normal operation of the system, that is, there are no switch failures during the time that the controlled process is operating satisfactorily and there is no need to switch over to the standby system. However, because the switch is subject to a continuous flow of current, its reliability cannot be assumed to be 100%. If the reliability of the switch in normal operation is represented by $R_N$, the expression in Equation (12.19) must be multiplied by $R_N$ and the reliability of the system becomes:

$$R_S = R_N[1 - F_1(1 + F_2R_DR_W - R_DR_W)]. \tag{12.20}$$

The problem of detecting faults in the primary safety system reliably can be solved by operating both safety systems in parallel. This enables faults in the safety system to be distinguished from faults in the monitored process. If only one of the two safety systems indicates a failure, this can be taken to indicate a failure of one of the safety systems rather than a failure of the monitored process. However, if both safety systems indicate a fault, this almost certainly means that the monitored process has developed a potentially dangerous fault. This scheme is known as *one-out-of-two voting*, but it is obviously inferior in reliability to the two-out-of-three scheme described earlier.

## Example 12.8

In a particular protection system, a switchable standby safety system is used to increase reliability. If the reliabilities of the main system, the standby system, and the switching system are 0.95, 0.96,[†] and 0.95, respectively, and the reliability of the switch in normal operation is 0.98, calculate the reliability of the protection system.

## Solution

Applying Equation (12.20), the parameter values are $F_1 = 0.05$, $F_2 = 0.04$, and $R_DR_W = 0.95$ and $R_N = 0.98$. Hence,

$$R_S = 0.98[1 - 0.05(1 + \{0.04 \times 0.95\} - 0.95)] = 0.976.$$

[†] Because the standby system is not subject to normal use, its reliability tends to be higher than the primary system even if the two systems consist of nominally identical components.

### *Actuators and alarms*

The final element in a safety system is either an automatic actuator or an alarm that requires a human response. The reliability of the actuator can be calculated in the same way as all other elements in the system and incorporated into the calculation of the overall system reliability as expressed in Equations (12.17) through (12.20). However, the reliability of alarms cannot be quantified in the same manner. Therefore, safety system reliability calculations have to exclude the alarm element. In consequence, the system designer needs to take steps to maximize the probability that the human operator will take the necessary response to alarms that indicate a dangerous plant condition.

Some useful guidelines for measurement technologists involved in designing alarm systems are provided by Bransby (1999). A very important criterion in system design is that alarms about dangerous conditions in a plant must be much more prominent than alarms about conditions that are not dangerous. Care should also be taken to ensure that the operator of a plant is not bombarded by too many alarms, as this leads the operator to get into the habit of ignoring alarms. Ignoring an alarm indicating that a fault is starting to occur may cause dangerous conditions in the plant to develop. Thus, alarms should be uncommon rather than routine so that they attract the attention of the plant operator. This ensures, as far as possible, that the operator will take the proper action in response to an alarm about a potentially dangerous situation.

## 12.4 Summary

The topic studied in this chapter has been the subject of measurement system reliability and the effect that this can have on other systems associated with the measurement system. We started off by studying the principles of reliability and looking at how the reliability of a measurement system can be quantified. We went on to look at the principal laws of reliability and considered how these could be applied to improve the reliability of a system. We also looked at other precautions that could be taken when designing and operating measurement systems to avoid either failure or impaired performance. These precautions included choosing suitable instruments according to the expected operating conditions, protecting instruments appropriately when using them in adverse environments, recalibrating them at the recommended frequency, and building an element into instruments used in critical parts of a measurement system.

We then went on to consider the subject of software reliability. We noted that this was now very important in view of the widespread use of intelligent instruments containing software. This study showed us that substantial differences exist among the mechanisms contributing to the reliability of software and hardware. The reliability of instrument hardware is related to factors such as mechanical wear, degradation because of the effect of the operating environment, and mechanical failure. These are faults that develop over a period of time. However, the mechanisms of software failure are fundamentally different.

Software does not change with time, and if it is error free when first written, it will remain error free forever. What happens when software is used is that errors that it has had all its life suddenly cause a problem when particular conditions occur, usually when particular combinations of input values are applied. Thus, the usual rules of reliability quantification applied to hardware components in measurement systems are not appropriate for associated software. Instead, we have seen that special procedures have to be applied to quantify software reliability in terms of the probability that it will not fail. Having established appropriate ways of quantifying software reliability, we went on to look at how reliability can be improved.

Our final topic of study in the chapter was that of system safety issues. Having first looked at the definition and quantification of safety levels and the associated IEC 61508 code of practice for safety systems, we went on to look at how safety systems could be designed to address safety issues. We then ended the chapter by looking in some detail at particular designs of safety systems in terms of *two-out-of-three* voting systems, using standby systems and appropriate use of safety actuators and alarms.

## 12.5 Problems

12.1. How is the reliability of a measurement system defined? What is the difference between quantifying reliability in quasi-absolute terms and quantifying it in probabilistic terms?

12.2. Explain the rules for calculating the overall reliability of system components that are connected (a) in series with each other and (b) in parallel with each another.

12.3. Discuss ways in which the reliability of measurement systems can be improved.

12.4. How do mechanisms affecting the reliability of software differ from those affecting the reliability of mechanical and electrical system components?

12.5. How can software reliability be quantified?

12.6. Discuss some ways in which the reliability of software components within measurement systems can be improved.

12.7. What are the principal duties of employers and employees with regard to safety? How do these impact the design and operation of measurement systems?

12.8. Explain the following terms that are met in the design of safety systems: (a) two-out-of-three voting system and (b) standby system.

12.9. The performance of a measuring instrument is monitored over a 1-year (365-day) period, and intervals between faults being recorded are as follows:

27 6 18 41 54 29 46 14 49 38 17 26

Calculate the mean time between failures.

12.10. The days on which an instrument failed were recorded over a 12-month period as follows (such that day 1 = January 1, day 32 = February 1, etc.):

Day number of faults: 18 72 111 173 184 227 286 309 356

Calculate the mean time between failures.

12.11. A manufacturer monitors the performance of a new type of instrument that is installed at 20 different locations. If a total of 9 faults are recorded over a 100-day period, calculate the mean time between failure that should be specified for any one instrument.

12.12. The time before failure of each a platinum resistance thermometer used in a particular location is recorded. The times before failure in days for 10 successive thermometers are recorded as:

405 376 433 425 388 402 445 412 397 366

Calculate the mean time to failure.

12.13. Repair times in hours of an instrument over a history of 10 breakdowns are recorded as follows:

10.5 5.75 8.25 30.0 12.5 15.0 6.5 3.25 14.5 9.25

Calculate the mean time to repair.

12.14. If the mean time between failure for an instrument is 247 days and the mean time to repair is 3 days, calculate its availability.

12.15. If the mean time to failure of an instrument is 100,000 hours, calculate the probability that it will not fail in the first 50,000 hours.

12.16. If the mean time to failure of an instrument is 100,000 hours, calculate the probability that it will not fail in the first 5000 hours.

12.17. Measurement components connected in series have the following reliabilities: 0.98 0.93 0.95 0.99. Calculate the reliability of the whole measurement system.

12.18. In a particular measurement system, two instruments with an individual reliability of 0.95 are connected together in parallel. Calculate the reliability of the measurement system if it can continue to function as long as both of the instruments do not fail at the same time.

12.19. Calculate the reliability of the measurement system shown in Figure 12.2b if the reliabilities of the individual components are $R_1 = R_3 = R_5 = 0.98$ ; $R_2 = R_4 = 0.90$.

12.20. In order to estimate the number of errors in a new piece of software by applying the error-seeding approach, a programmer puts 10 deliberate (seeded) faults into the program. A second programmer then tests the program and finds 27 errors, of which 8 are confirmed by the first programmer to be seeded errors. Estimate the original number of faults in the program (i.e., excluding seeded errors).

12.21. The double-testing approach is applied to test a new computer program and the two programmers who do the testing find 31 and 34 errors, respectively. If 27 of the errors found by programmer 1 are the same as errors in the list produced by programmer 2, estimate the actual number of errors in the program.

12.22. A program is tested, and the total number of errors is estimated as 16. The program is then seeded with 20 deliberate errors and further testing is then carried out until all 20 seeded errors have been found.

(a) If no new (previously undetected) errors are found during this further testing to find all seeded errors, calculate the probability that the program is error free after this further testing.

(b) How many seeded errors would have to be put into the program and then detected to achieve a 98% probability that the program is error free?

12.23. Three safety systems are connected in parallel in a protection system and a two-out-of-three voting strategy is applied. If the reliability of each of the three systems is 0.90, calculate the overall reliability of the whole protection system.

12.24. A switchable standby safety system is used to increase reliability in a protection system. If the reliability of the main system is 0.90, that of the standby system is 0.91, that of the fault detector/switching system is 0.90 and the reliability of the switch in normal operation is 0.96, calculate the reliability of the protection system.

## References and Further Reading


Bransby, M. (1999). The human contribution to safety: Designing alarm systems. *Measurement and control, 32*, 209–213.

BS/ISO/IEC 90003. (2004). *Software engineering: Guideline for application of ISO 9001 to computer software*. British Standards Institute/International Standards Organisation/International Electrotechnical Commission.

Dean, S. (1999). IEC61508—Understanding functional safety assessment. *Measurement and control, 32*, 201–204.

Fenton, N. E., & Bieman, J. (2010). *Software metrics (Software Engineering)*. Chapman and Hall.

Fenton, N. E., & Pfleeger, S. L. (1998). *Software metrics: A rigorous approach*. PWS Publishing.

IEC 61508. (2005). *Functional safety of electrical, electronic and programmable-electronic safety related systems*. Geneva: International Electrotechnical Commission.

Johnson, R., Miller, I. R., & Freund, J. E. (2010). *Miller and Freund's probability and statistics for engineers*. Pearson Education.

Mills, H. D. (1972). *On the statistical validation of computer programs*. Maryland: IBM Federal Systems Division.

Morris, A. S. (1997). *Measurement and calibration requirements for quality assurance to ISO 9000*. John Wiley.

Naik, S., & Tripathy, P. (2008). *Software testing and quality assurance, theory and practice*. John Wiley.

Pfleeger, S. L., & Atlee, J. M. (2009). *Software engineering: Theory and practice*. Prentice Hall.

Shooman, M. L. (2002). *Reliability of computer systems and networks: Fault tolerance, analysis and design*. Wiley-Blackwell.

Simpson, K., & Smith, D. J. (1999). Assessing safety related systems and architectures. *Measurement and control, 32*, 205–208.

CHAPTER 13

# Sensor Technologies

## 13.1 Introduction

We are now moving into the second part of the book, where we look in detail at the range of sensors available for measuring various physical quantities. As we study these sensors, we quickly come to realize that a wide range of different physical principles are involved in their operation. It also becomes apparent that the physical principles on which they operate is often an important factor in choosing a sensor for a given application, as a sensor using a particular principle may perform much better than one using a different principle in given operating conditions. It is therefore prudent to devote this chapter to a study of the various physical principles exploited in measurement sensors before going on to the separate chapters devoted to the measurement of various physical quantities. The physical principles to be examined are capacitance change, resistance change, magnetic phenomena (inductance, reluctance, and eddy currents), Hall effect, properties of piezoelectric materials, resistance change in stretched/strained wires (strain gauges), properties of piezoresistive materials, light transmission (both along an air path and along a fiber-optic cable), properties of ultrasound, transmission of radiation, and properties of micromachined structures (microsensors). It should be noted that the chosen order of presentation of these is arbitrary and does not imply anything about the relative popularity of these various principles. It must also be pointed out that the list of technologies covered in this chapter is not a full list of all the technologies that are used in sensors, but rather a list of technologies common to several different sensors that measure different physical quantities. Many other technologies are used in the measurement of single physical quantities. Temperature measurement is a good example of this, as several of the sensors used are based on technologies not covered in this chapter.

## 13.2 Capacitive Sensors

Capacitive sensors consist of two parallel metal plates in which the dielectric between the plates is either air or some other medium. The capacitance $C$ is given by $C = \epsilon_o \epsilon_r A/d$, where $\epsilon_o$ is the absolute permittivity, $\epsilon_r$ is the relative permittivity of the dielectric medium between the plates, $A$ is the area of the plates, and $d$ is the distance between them. Two forms of capacitive devices exist, which differ according to whether the distance between the plates is fixed or not.

Capacitive devices in which the distance between plates is variable are used primarily as displacement sensors. Motion of the moveable capacitive plate relative to a fixed one changes the capacitance. Such devices can be used directly as a displacement sensor by applying the motion to be measured to the moveable capacitor plate. Capacitive displacement sensors commonly form part of instruments measuring pressure, sound, or acceleration, as explained in later chapters.

In the alternative form of capacitor, the distance between plates is fixed. Variation in capacitance is achieved by changing the dielectric constant of the material between the plates in some way. One application is where the dielectric medium is air and the device is used as a humidity sensor by measuring the moisture content of the air. Another common application is as a liquid level sensor, where the dielectric is part air and part liquid according to the level of the liquid that the device is inserted in. Both of these applications are discussed in greater detail in later chapters. This principle is used in devices to measure moisture content, humidity values, and liquid level, as discussed in later chapters.

## 13.3 Resistive Sensors

Resistive sensors rely on variation of the resistance of a material when the measured variable is applied to it. This principle is applied most commonly in temperature measurement using resistance thermometers or thermistors. It is also used in displacement measurement using strain gauges or piezoresistive sensors. In addition, some moisture meters work on the resistance-variation principle. All of these applications are considered further in later chapters.

## 13.4 Magnetic Sensors

Magnetic sensors utilize the magnetic phenomena of inductance, reluctance, and eddy currents to indicate the value of the measured quantity, which is usually some form of displacement.

*Inductive sensors* translate movement into a change in the mutual inductance between magnetically coupled parts. One example of this is the inductive displacement transducer shown in Figure 13.1. In this, the single winding on the central limb of an "E"-shaped ferromagnetic body is excited with an alternating voltage. The displacement to be measured is applied to a ferromagnetic plate in close proximity to the "E" piece. Movements of the plate alter the flux paths and hence cause a change in the current flowing in the winding. By Ohm's law, the current flowing in the winding is given by $I = V/\omega L$. For fixed values of $w$ and $V$, this equation becomes $I = 1/KL$, where $K$ is a constant. The relationship between $L$ and the

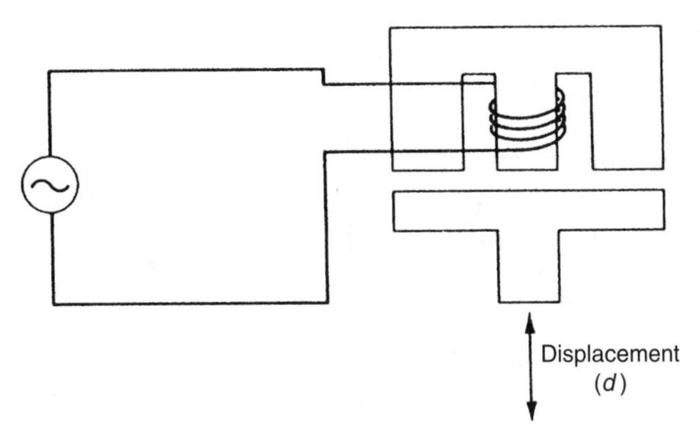


**Figure 13.1**
Inductive displacement sensor.

displacement, *d*, applied to the plate is a nonlinear one, and hence the output-current/displacement characteristic has to be calibrated.

The inductance principle is also used in differential transformers to measure translational and rotational displacements.

In *variable reluctance sensors*, a coil is wound on a permanent magnet rather than on an iron core as in variable inductance sensors. Such devices are used commonly to measure rotational velocities. Figure 13.2 shows a typical instrument in which a ferromagnetic gearwheel is placed next to the sensor. As the tip of each tooth on the gearwheel moves toward and away from the pick-up unit, the changing magnetic flux in the pickup coil causes a voltage to be induced in the coil whose magnitude is proportional to the rate of change of flux. Thus, the output is a sequence of positive and negative pulses whose frequency is proportional to the rotational velocity of the gearwheel.

*Eddy current sensors* consist of a probe containing a coil, as shown in Figure 13.3, that is excited at a high frequency, which is typically 1 MHz. This is used to measure the displacement of the probe relative to a moving metal target. Because of the high frequency of excitation, eddy currents are induced only in the surface of the target, and the current magnitude reduces to

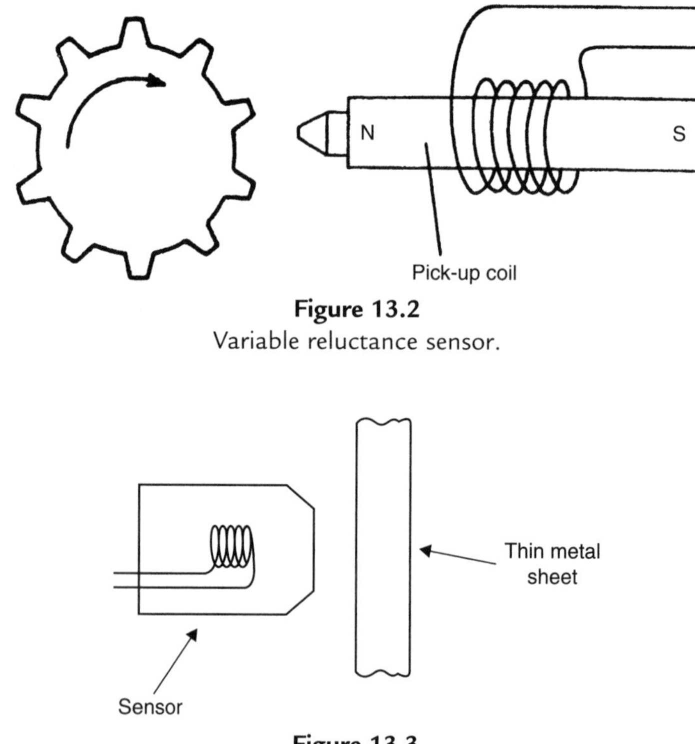


**Figure 13.2**
Variable reluctance sensor.



**Figure 13.3**
Eddy current sensor.

almost zero a short distance inside the target. This allows the sensor to work with very thin targets, such as the steel diaphragm of a pressure sensor. The eddy currents alter the inductance of the probe coil, and this change can be translated into a d.c. voltage output that is proportional to the distance between the probe and the target. Measurement resolution as high as 0.1 μm can be achieved. The sensor can also work with a nonconductive target if a piece of aluminum tape is fastened to it.

## 13.5 Hall-Effect Sensors

Basically, a Hall-effect sensor is a device used to measure the magnitude of a magnetic field. It consists of a conductor carrying a current that is aligned orthogonally with the magnetic field, as shown in Figure 13.4. This produces a transverse voltage difference across the device that is directly proportional to the magnetic field strength. For an excitation current, $I$, and magnetic field strength, $B$, the output voltage is given by $V = KIB$, where $K$ is known as the Hall constant.

The conductor in Hall-effect sensors is usually made from a semiconductor material as opposed to a metal because a larger voltage output is produced for a magnetic field of a given size. In one common use of the device as a proximity sensor, the magnetic field is provided by a permanent magnet built into the device. The magnitude of this field changes when the device comes close to any ferrous metal object or boundary. The Hall effect is also used commonly in computer keyboard push buttons. When a button is depressed, a magnet attached underneath the button moves past a Hall-effect sensor. This generates an induced voltage in the sensor, which is converted by a trigger circuit into a digital output. Such push-button switches can operate at high frequencies without contact bounce.

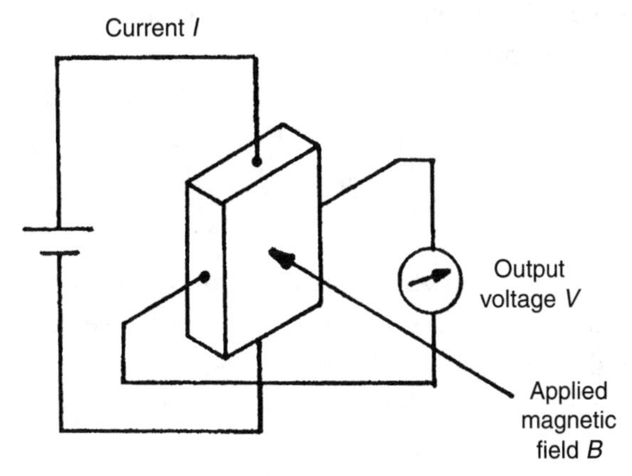


**Figure 13.4**
Principles of Hall-effect sensor.

## 13.6 Piezoelectric Transducers

Piezoelectric transducers produce an output voltage when a force is applied to them. They can also operate in the reverse mode where an applied voltage produces an output force. They are used frequently as ultrasonic transmitters and receivers. They are also used as displacement transducers, particularly as part of devices measuring acceleration, force, and pressure. In ultrasonic receivers, sinusoidal amplitude variations in the ultrasound wave received are translated into sinusoidal changes in the amplitude of the force applied to the piezoelectric transducer. In a similar way, the translational movement in a displacement transducer is caused by mechanical means to apply a force to the piezoelectric transducer. Piezoelectric transducers are made from piezoelectric materials. These have an asymmetrical lattice of molecules that distorts when a mechanical force is applied to it. This distortion causes a reorientation of electric charges within the material, resulting in a relative displacement of positive and negative charges. The charge displacement induces surface charges on the material of opposite polarity between the two sides. By implanting electrodes into the surface of the material, these surface charges can be measured as an output voltage. For a rectangular block of material, the induced voltage is given by

$$V = \frac{kFd}{A}, \tag{13.1}$$

where $F$ is the applied force in g, $A$ is the area of the material in mm, $d$ is the thickness of the material, and $k$ is the piezoelectric constant. The polarity of the induced voltage depends on whether the material is compressed or stretched.

The input impedance of the instrument used to measure the induced voltage must be chosen carefully. Connection of the measuring instrument provides a path for the induced charge to leak away. Hence, the input impedance of the instrument must be very high, particularly where static or slowly varying displacements are being measured.

Materials exhibiting piezoelectric behavior include natural ones such as quartz, synthetic ones such as lithium sulphate, and ferroelectric ceramics such as barium titanate. The piezoelectric constant varies widely between different materials. Typical values of $k$ are 2.3 for quartz and 140 for barium titanate. Applying Equation (13.1) for a force of 1 $g$ applied to a crystal of area 100 $mm^2$ and a thickness of 1 mm gives an output of 23 μV for quartz and 1.4 mV for barium titanate.

Certain polymeric films such as polyvinylidine also exhibit piezoelectric properties. These have a higher voltage output than most crystals and are very useful in many applications where displacement needs to be translated into voltage. However, they have very limited mechanical strength and are unsuitable for applications where resonance might be generated in the material.

The piezoelectric principle is invertible, and therefore distortion in a piezoelectric material can be caused by applying a voltage to it. This is used commonly in ultrasonic transmitters, where application of a sinusoidal voltage at a frequency in the ultrasound range causes sinusoidal variations in the thickness of the material and results in a sound wave being emitted at the chosen frequency. This is considered further in the section on ultrasonic transducers.

## 13.7 Strain Gauges

Strain gauges are devices that experience a change in resistance when they are stretched or strained. They are able to detect very small displacements, usually in the range of 0–50 μm, and are typically used as part of other transducers, for example, diaphragm pressure sensors that convert pressure changes into small displacements of the diaphragm. Measurement inaccuracies as low as ±0.15% of full-scale reading are achievable, and the quoted life expectancy is usually three million reversals. Strain gauges are manufactured to various nominal values of resistance, of which 120, 350, and 1000 Ω are very common. The typical maximum change of resistance in a 120-Ω device would be 5 Ω at maximum deflection.

The traditional type of strain gauge consists of a length of metal resistance wire formed into a zigzag pattern and mounted onto a flexible backing sheet, as shown in Figure 13.5a. The wire is nominally of circular cross section. As strain is applied to the gauge, the shape of the cross section of the resistance wire distorts, changing the cross-sectional area. As the resistance of the wire per unit length is inversely proportional to the cross-sectional area, there is a consequential change in resistance. The input–output relationship of a strain gauge is expressed by the *gauge factor*, which is defined as the change in resistance ($R$) for a given value of strain ($S$), that is, *gauge factor* $= \delta R/\delta S$.

In recent years, wire-type gauges have largely been replaced either by metal-foil types, as shown in Figure 13.5b, or by semiconductor types. Metal-foil types are very similar to metal-wire types except that the active element consists of a piece of metal foil cut into a zigzag pattern. Cutting a foil into the required shape is much easier than forming a piece of resistance wire into the required shape, which makes the devices less expensive to manufacture. A popular

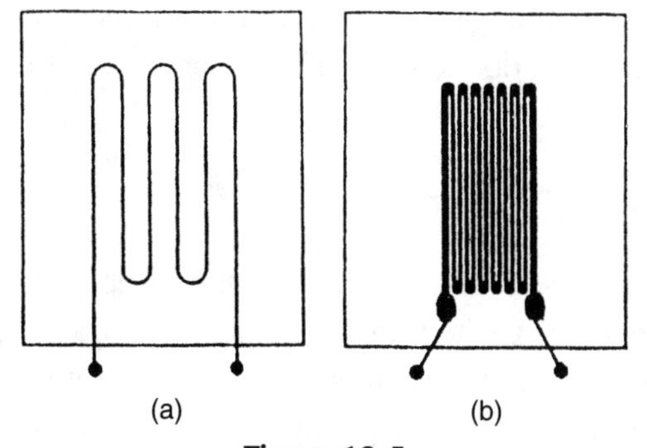


**Figure 13.5**
Strain gauges: (a) wire type and (b) foil type.

material in metal strain gauge manufacture is a copper–nickel–manganese alloy, which is known by the trade name of "Advance." Semiconductor types have piezoresistive elements, which are considered in greater detail in the next section. Compared with metal gauges, semiconductor types have a much superior gauge factor (up to 100 times better) but are more expensive. Also, while metal gauges have an almost zero temperature coefficient, semiconductor types have a relatively high temperature coefficient.

In use, strain gauges are bonded to the object whose displacement is to be measured. The process of bonding presents a certain amount of difficulty, particularly for semiconductor types. The resistance of the gauge is usually measured by a d.c. bridge circuit, and the displacement is inferred from the bridge output measured. The maximum current that can be allowed to flow in a strain gauge is in the region of 5 to 50 mA depending on the type. Thus, the maximum voltage that can be applied is limited and, consequently, as the resistance change in a strain gauge is typically small, the bridge output voltage is also small and amplification has to be carried out. This adds to the cost of using strain gauges.

## 13.8 Piezoresistive Sensors

A piezoresistive sensor is made from semiconductor material in which a p-type region has been diffused into an n-type base. The resistance of this varies greatly when the sensor is compressed or stretched. This is used frequently as a strain gauge, where it produces a significantly higher gauge factor than that given by metal wire or foil gauges. Also, measurement uncertainty can be reduced down to $\pm 0.1\%$. It is also used in semiconductor-diaphragm pressure sensors and in semiconductor accelerometers.

It should also be mentioned that the term *piezoresistive sensor* is sometimes used to describe all types of strain gauges, including metal types. However, this is incorrect as only about 10% of the output from a metal strain gauge is generated by piezoresistive effects, with the remainder arising out of the dimensional cross-sectional change in the wire or foil. Proper piezoelectric strain gauges, which are alternatively known as *semiconductor strain gauges*, produce most (about 90%) of their output through piezoresistive effects, and only a small proportion of the output is due to dimensional changes in the sensor.

## 13.9 Optical Sensors

Optical sensors are based on the transmission of light between a light source and a light detector, as shown in Figure 13.6. The transmitted light can travel along either an air path or a fiber-optic cable. Either form of transmission gives immunity to electromagnetically induced noise and also provides greater safety than electrical sensors when used in hazardous environments.

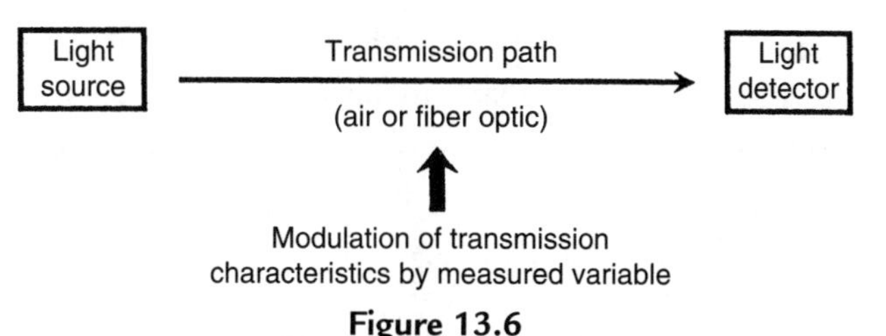


**Figure 13.6**
Operating principles of optical sensors.

### 13.9.1 Optical Sensors (Air Path)

Air path optical sensors are used commonly to measure proximity, translational motion, rotational motion, and gas concentration. These uses are discussed in more detail in later chapters. A number of different types of light sources and light detectors are used.

*Light sources*

Light sources suitable for transmission across an air path include tungsten-filament lamps, laser diodes, and light-emitting diodes (LEDs). However, as the light from tungsten lamps is usually in the visible part of the light frequency spectrum, it is prone to interference from the sun and other sources. Hence, infrared LEDs or infrared laser diodes are usually preferred. These emit light in a narrow frequency band in the infrared region and are not affected by sunlight.

*Light detectors*

The main forms of light detectors used with optical systems are photoconductors (photoresistors), photovoltaic devices (photocells), phototransistors, and photodiodes.

*Photoconductive devices* are sometimes known by the alternative name of *photoresistors*. They convert changes in incident light into changes in resistance, with resistance reducing according to the intensity of light to which they are exposed. They are made from various materials, such as cadmium sulfide, lead sulfide, and indium antimonide.

*Photovoltaic devices* are often called *photocells*. They also are known commonly as *solar cells* when a number of them are used in an array as a means of generating energy from sunlight. They are made from various types of semiconductor material. Their basic mode of operation is to generate an output voltage whose magnitude is a function of the magnitude of the incident light that they are exposed to.

*Photodiodes* are devices where the output current is a function of the amount of incident light. Again, they are made from various types of semiconductor material.

A *phototransistor* is effectively a standard bipolar transistor with a transparent case that allows light to reach its base-collector junction. It has an output in the form of an electrical current and could be regarded as a photodiode with an internal gain. This gain makes it more sensitive to light than a photodiode, particularly in the infrared region, but has a slower response time. It is an ideal partner for infrared LED and laser diode light sources.

### 13.9.2 Optical Sensors (Fiber Optic)

Instead of using air as the transmission medium, optical sensors can use fiber-optic cable to transmit light between a source and a detector. Fiber-optic cables can be made from plastic fibers, glass fibers, or a combination of the two, although it is now rare to find cables made only from glass fibers as these are very fragile. Cables made entirely from plastic fibers have particular advantages for sensor applications because they are inexpensive and have a relatively large diameter of 0.5–1.0 mm, making connection to the transmitter and receiver easy. However, plastic cables cannot be used in certain hostile environments where they would be severely damaged. The cost of fiber-optic cable itself is insignificant for sensing applications, as the total cost of the sensor is dominated by the cost of the transmitter and receiver.

Fiber-optic sensors characteristically enjoy long life. For example, the life expectancy of reflective fiber-optic switches is quoted at 10 million operations. Their accuracy is also good, with $\pm 1\%$ of full-scale reading being quoted as a typical inaccuracy level for a fiber-optic pressure sensor. Further advantages are their simplicity, low cost, small size, high reliability, and capability of working in many kinds of hostile environments. The only significant difficulty in designing a fiber-optic sensor is in ensuring that the proportion of light entering the cable is maximized. This is the same difficulty described earlier when discussing the use of fiber-optic cables for signal transmission.

Two major classes of fiber-optic sensors exist—intrinsic and extrinsic. In *intrinsic sensors*, the fiber-optic cable itself is the sensor, whereas in *extrinsic sensors*, the fiber-optic cable is only used to guide light to/from a conventional sensor.

#### *Intrinsic sensors*

In intrinsic sensors, the physical quantity being measured causes some measurable change in the characteristics of the light transmitted by the cable. The modulated light parameters are one or more of the following:

- intensity
- phase
- polarization
- wavelength
- transit time

Sensors that modulate light intensity tend to use mainly multimode fibers, but only monomode cables are used to modulate other light parameters. A particularly useful feature of intrinsic fiber-optic sensors is that they can, if required, provide distributed sensing over distances of up to 1 meter.

Light intensity is the simplest parameter to manipulate in intrinsic sensors because only a simple source and detector are required. The various forms of switches shown in Figure 13.7 are perhaps the simplest forms of these, as the light path is simply blocked and unblocked as the switch changes state. Modulation of the intensity of transmitted light also takes place in various simple forms of proximity, displacement, pressure, pH, and smoke sensors. Some of these are sketched in Figure 13.8. In proximity and displacement sensors (the latter are sometimes given the special name *fotonic sensors*), the amount of reflected light varies with the distance between the fiber ends and a boundary. In pressure sensors, the refractive index of the fiber, and hence the intensity of light transmitted, varies according to the mechanical deformation of the fibers caused by pressure. In the pH probe, the amount of light reflected

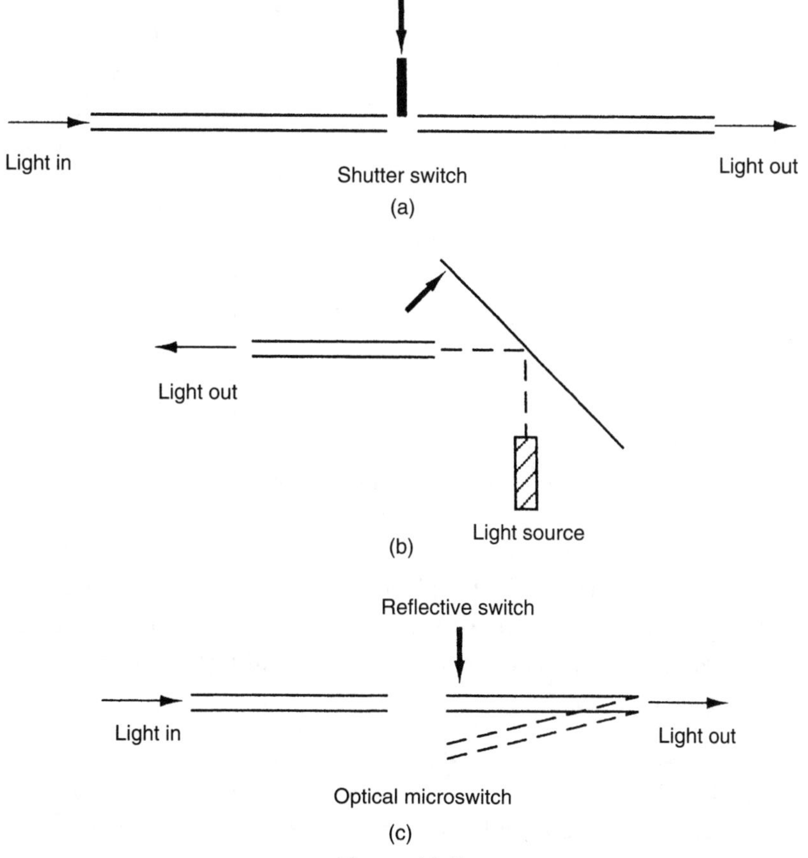


**Figure 13.7**
Intrinsic fiber-optic sensors.

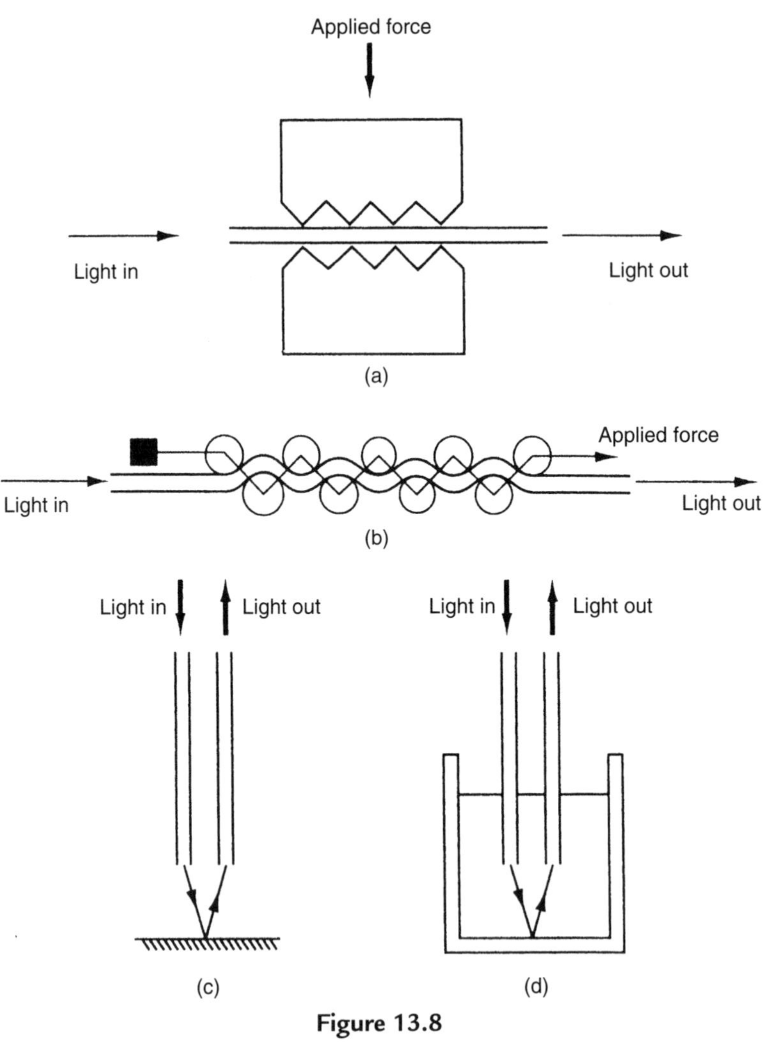


**Figure 13.8**
Intensity-modulating sensors.

back into the fibers depends on the pH-dependent color of the chemical indicator in the solution around the probe tip. Finally, in a form of smoke detector, two fiber-optic cables placed either side of a space detect any reduction in the intensity of light transmission between them caused by the presence of smoke.

A simple form of accelerometer can be made by placing a mass subject to acceleration on a multimode fiber. The force exerted by the mass on the fiber causes a change in the intensity of light transmitted, hence allowing the acceleration to be determined. The typical inaccuracy quoted for this device is $\pm 0.02$ g in the measurement range of $\pm 5$ g and $\pm 2\%$ in the measurement range up to 100 g.

A similar principle is used in probes that measure the internal diameter of tubes. The probe consists of eight strain-gauged cantilever beams that track changes in diameter, giving a measurement resolution of 20 μm.

A slightly more complicated method of affecting light intensity modulation is the variable shutter sensor shown in Figure 13.9. This consists of two fixed fibers with two collimating lenses and a variable shutter between them. Movement of the shutter changes the intensity of light transmitted between the fibers. This is used to measure the displacement of various devices such as Bourdon tubes, diaphragms, and bimetallic thermometers.

Yet another type of intrinsic sensor uses cable where the core and cladding have similar refractive indices but different temperature coefficients. This is used as a temperature sensor. Temperature rises cause the refractive indices to become even closer together and losses from the core to increase, thus reducing the quantity of light transmitted.

Refractive index variation is also used in a form of intrinsic sensor used for cryogenic leak detection. The fiber used for this has a cladding whose refractive index becomes greater than that of the core when it is cooled to cryogenic temperatures. The fiber-optic cable is laid in the location where cryogenic leaks might occur. If any leaks do occur, light traveling in the core is transferred to the cladding, where it is attenuated. Cryogenic leakage is thus indicated by monitoring the light transmission characteristics of the fiber.

A further use of refractive index variation is found in devices that detect oil in water. These use a special form of cable where the cladding used is sensitive to oil. Any oil present diffuses into the cladding and changes the refractive index, thus increasing light losses from the core. Unclad fibers are used in a similar way. In these, any oil present settles on the core and allows light to escape.

A *cross-talk sensor* measures several different variables by modulating the intensity of light transmitted. It consists of two parallel fibers that are close together and where one or more short lengths of adjacent cladding are removed from the fibers. When immersed in a transparent liquid, there are three different effects that each cause a variation in the intensity of light transmitted. Thus, the sensor can perform three separate functions. First, it can measure

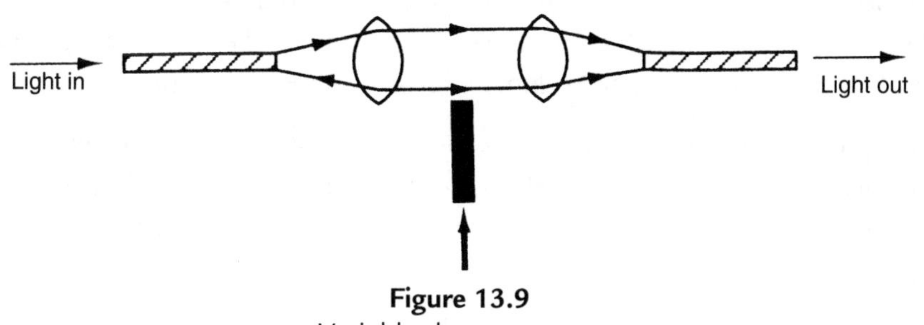


**Figure 13.9**
Variable shutter sensor.

temperature according to the temperature-induced variation in the refractive index of the liquid. Second, it can act as a level detector, as the transmission characteristics between the fibers change according to the depth of the liquid. Third, it can measure the refractive index of the liquid itself when used under controlled temperature conditions.

The refractive index of a liquid can be measured in an alternative way using an arrangement where light travels across the liquid between two cable ends that are fairly close together. The angle of the cone of light emitted from the source cable, and hence the amount of light transmitted into the detector, is dependent on the refractive index of the liquid.

The use of materials where the fluorescence varies according to the value of the measurand can also be used as part of intensity-modulating intrinsic sensors. Fluorescence-modulating sensors can give very high sensitivity and are potentially very attractive in biomedical applications where requirements exist to measure very small quantities, such as low oxygen and carbon monoxide concentrations, and low blood pressure levels. Similarly, low concentrations of hormones, steroids, and so on may be measured.

Further examples of intrinsic fiber-optic sensors that modulate light intensity are described later in Chapter 17 (level measurement) and Chapter 19 (measuring small displacements).

As mentioned previously, phase, polarization, wavelength, and transit time can be modulated as well as intensity in intrinsic sensors. Monomode cables are used almost exclusively in these types of intrinsic sensors.

Phase modulation normally requires a coherent (laser) light source. It can provide very high sensitivity in displacement measurement, but cross sensitivity to temperature and strain degrades its performance. Additional problems are maintaining the frequency stability of the light source and manufacturing difficulties in coupling the light source to the fiber. Various versions of this class of instrument exist to measure temperature, pressure, strain, magnetic fields, and electric fields. Field-generated quantities such as electric current and voltage can also be measured. In each case, the measurand causes a phase change between a measuring and a reference light beam that is detected by an interferometer.

The principle of phase modulation has also been used in the fiber-optic accelerometer (where a mass subject to acceleration rests on a fiber) and in fiber strain gauges (where two fibers are fixed on the upper and lower surfaces of a bar under strain). The fiber-optic gyroscope described in Chapter 20 is a further example of a phase-modulating device.

Devices using polarization modulation require special forms of fibers that maintain polarization. Polarization changes can be affected by electrical fields, magnetic fields, temperature changes, and mechanical strain. Each of these parameters can therefore be measured by polarization modulation.

Various devices that modulate the wavelength of light are used for special purposes. However, the only common wavelength-modulating fiber-optic device is the form of laser Doppler flowmeter that uses fiber-optic cables, as described in Chapter 16.

Fiber-optic devices using modulation of the transit time of light are uncommon because of the speed of light. Measurement of the transit time for light to travel from a source, be reflected off an object, and travel back to a detector is only viable for extremely large distances. However, a few special arrangements have evolved that use transit time modulation. These include instruments such as the optical resonator, which can measure both mechanical strain and temperature. Temperature-dependent wavelength variation also occurs in semiconductor crystal beads (e.g., aluminium gallium arsenide). This is bonded to the end of a fiber-optic cable and excited from an LED at the other end of the cable. Light from the LED is reflected back along the cable by the bead at a different wavelength. Measurement of the wavelength change allows temperatures in the range up to 200°C to be measured accurately. A particular advantage of this sensor is its small size, typically 0.5 mm diameter at the sensing tip. Finally, to complete the catalogue of transit time devices, the frequency modulation in a piezoelectric quartz crystal used for gas sensing can also be regarded as a form of time domain modulation.

#### *Extrinsic sensors*

Extrinsic fiber-optic sensors use a fiber-optic cable, normally a multimode one, to transmit modulated light from a conventional sensor such as a resistance thermometer. A major feature of extrinsic sensors, which makes them so useful in such a large number of applications, is their ability to reach places that are otherwise inaccessible. One example of this is the insertion of fiber-optic cables into the jet engines of aircraft to measure temperature by transmitting radiation into a radiation pyrometer located remotely from the engine. Fiber-optic cable can be used in the same way to measure the internal temperature of electrical transformers, where the extreme electromagnetic fields present make other measurement techniques impossible.

An important advantage of extrinsic fiber-optic sensors is the excellent protection against noise corruption that they give to measurement signals. Unfortunately, the output of many sensors is not in a form that can be transmitted by a fiber-optic cable, and conversion into a suitable form must therefore take place prior to transmission. For example, a platinum resistance thermometer (PRT) translates temperature changes into resistance changes. The PRT therefore needs electronic circuitry to convert the resistance changes into voltage signals and then into a modulated light format, which in turn means that the device needs a power supply. This complicates the measurement process and means that low-voltage power cables must be routed with the fiber-optic cable to the transducer. One particular adverse effect of this is that the advantage of intrinsic safety is lost. One solution to this problem is to use a power source in the form of electronically generated pulses driven by a lithium battery. Alternatively, power can be generated by transmitting light down the fiber-optic cable to a photocell. Both of these solutions provide intrinsically safe operation.

Piezoelectric sensors lend themselves particularly to use in extrinsic sensors because the modulated frequency of a quartz crystal can be transmitted readily into a fiber-optic cable by fitting electrodes to the crystal that are connected to a low-power LED. Resonance of the crystal can be created either by electrical means or by optical means using the photothermal effect. The photothermal effect describes the principle where, if light is pulsed at the required oscillation frequency and directed at a quartz crystal, the localized heating and thermal stress caused in the crystal results in it oscillating at the pulse frequency. Piezoelectric extrinsic sensors can be used as part of various pressure, force, and displacement sensors. At the other end of the cable, a phase locked loop is typically used to measure the transmitted frequency.

Fiber-optic cables are also now commonly included in digital encoders, where the use of fibers to transmit light to and from the discs allows the light source and detectors to be located remotely. This allows the devices to be smaller, which is a great advantage in many applications where space is at a premium.

*Distributed sensors*

A number of discrete sensors can be distributed along a fiber-optic cable to measure different physical variables along its length. Alternatively, sensors of the same type, which are located at various points along a cable, provide distributed sensing of a single measured variable.

## 13.10 Ultrasonic Transducers

Ultrasonic devices are used in many fields of measurement, particularly for measuring fluid flow rates, liquid levels, and translational displacements. Details of such applications can be found in later chapters.

Ultrasound is a band of frequencies in the range above 20 kHz, that is, above the sonic range that humans can usually hear. Measurement devices that use ultrasound consist of one device that transmits an ultrasound wave and another device that receives the wave. Changes in the measured variable are determined either by measuring the change in time taken for the ultrasound wave to travel between the transmitter and receiver or, alternatively, by measuring the change in phase or frequency of the transmitted wave.

The most common form of ultrasonic element is a piezoelectric crystal contained in a casing, as illustrated in Figure 13.10. Such elements can operate interchangeably as either a transmitter or a receiver. These are available with operating frequencies that vary between 20 kHz and 15 MHz. The principles of operation, by which an alternating voltage generates an ultrasonic wave and vice versa, have already been covered in the section on piezoelectric transducers.

**Figure 13.10**
Ultrasonic sensor.

For completeness, mention should also be made of capacitive ultrasonic elements. These consist of a thin, dielectric membrane between two conducting layers. The membrane is stretched across a backplate and a bias voltage is applied. When a varying voltage is applied to the element, it behaves as an ultrasonic transmitter and an ultrasound wave is produced. The system also works in the reverse direction as an ultrasonic receiver. Elements with resonant frequencies in the range between 30 kHz and 3 MHz can be obtained.

### *13.10.1 Transmission Speed*

The transmission speed of ultrasound varies according to the medium through which it travels. Transmission speeds for some common media are given in Table 13.1.

When transmitted through air, the speed of ultrasound is affected by environmental factors such as temperature, humidity, and air turbulence. Of these, temperature has the largest effect. The velocity of sound through air varies with temperature according to

$$V = 331.6 + 0.6T \text{ m/s}, \tag{13.2}$$

where $T$ is the temperature in °C. Thus, even for a relatively small temperature change of 20° from 0 to 20°C, the velocity changes from 331.6 to 343.6 m/s.

Humidity changes have a much smaller effect on speed. If the relative humidity increases by 20%, the corresponding increase in the transmission velocity of ultrasound is 0.07% (corresponding to an increase from 331.6 to 331.8 m/s at 0°C).

Changes in air pressure itself have a negligible effect on the velocity of ultrasound. Similarly, air turbulence normally has no effect. However, if turbulence involves currents of air at

**Table 13.1 Transmission Speed of Ultrasound through Different Media**

| Medium | Velocity (m/s) |
|---|---|
| Air | 331.6 |
| Water | 1440 |
| Wood (pine) | 3320 |
| Iron | 5130 |
| Rock (granite) | 6000 |

different temperatures, then random changes in ultrasound velocity occur according to Equation (13.2).

### 13.10.2 Directionality of Ultrasound Waves

An ultrasound element emits a spherical wave of energy, although the peak energy is always in a particular direction. The magnitude of energy emission in any direction is a function of the angle made with respect to the direction that is normal to the face of the ultrasonic element. Peak emission occurs along a line that is normal to the transmitting face of the ultrasonic element, which is loosely referred to as the *direction of travel.* At any angle other than the "normal" one, the magnitude of transmitted energy is less than the peak value. Figure 13.11 shows characteristics of the emission for a range of ultrasonic elements. This is shown in terms of the attenuation of the transmission magnitude (measured in dB) as the angle with respect to "normal" direction increases. For many purposes, it is useful to treat the transmission as a conical volume of energy, with edges of the cone defined as the transmission angle where the amplitude of the energy in the transmission is $-6$ dB compared with the peak value (i.e., where the amplitude of the energy is half that in the normal direction). Using this definition, a 40-kHz ultrasonic element has a transmission cone of $\pm 50^\circ$ and a 400-kHz element has a transmission cone of $\pm 3^\circ$.

It should be noted that air currents can deflect ultrasonic waves such that the peak emission is no longer normal to the face of the ultrasonic element. It has been shown experimentally that an air current moving with a velocity of 10 km/h deflects an ultrasound wave by 8 mm over a distance of 1 m.

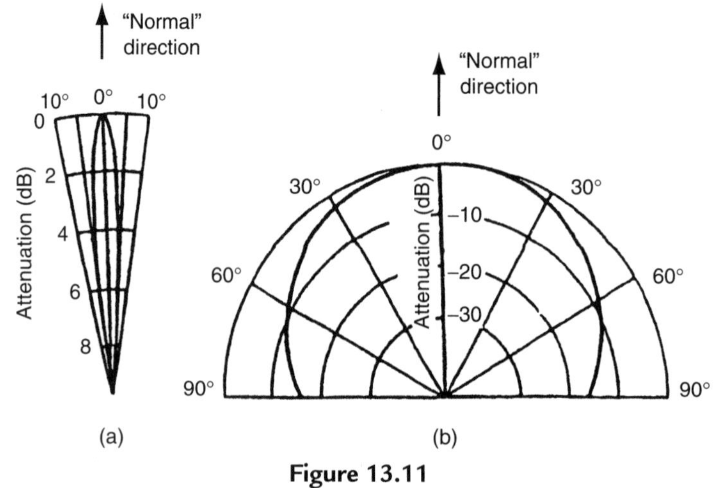


**Figure 13.11**
Ultrasonic emission characteristics.

### *13.10.3 Relationship Between Wavelength, Frequency, and Directionality of Ultrasound Waves*

The frequency and wavelength of ultrasound waves are related according to

$$\lambda = v/f, \tag{13.3}$$

where $\lambda$ is the wavelength, $v$ is the velocity, and $f$ is the frequency of the ultrasound waves.

This shows that the relationship between $\lambda$ and $f$ depends on the velocity of the ultrasound and hence varies according to the nature and temperature of the medium through which it travels. Table 13.2 compares nominal frequencies, wavelengths, and transmission cones (−6 dB limits) for three different types of ultrasonic elements.

It is clear from Table 13.2 that the directionality (cone angle of transmission) reduces as the nominal frequency of the ultrasound transmitter increases. However, the cone angle also depends on factors other than nominal frequency, particularly on the shape of the transmitting horn in the element, and different models of ultrasonic element with the same nominal frequency can have substantially different cone angles.

### *13.10.4 Attenuation of Ultrasound Waves*

Ultrasound waves suffer attenuation in the amplitude of the transmitted energy according to the distance traveled. The amount of attenuation also depends on the nominal frequency of the ultrasound and the adsorption characteristics of the medium through which it travels. The amount of adsorption depends not only on the type of transmission medium but also on the level of humidity and dust in the medium.

The amplitude $X_d$ of the ultrasound wave at a distance $d$ from the emission point can be expressed as

$$\frac{X_d}{X_0} = \frac{\sqrt{e^{-\alpha d}}}{fd}, \tag{13.4}$$

where $X_0$ is the magnitude of the energy at the point of emission, $f$ is the nominal frequency of the ultrasound, and $\alpha$ is the attenuation constant that depends on the ultrasound frequency,

**Table 13.2 Comparison of Frequency, Wavelength, and Cone Angle for Various Ultrasonic Transmitters**

| | | | |
|---|---|---|---|
| Nominal Frequency (kHz) | 23 | 40 | 400 |
| Wavelength (in air at 0°C) | 14.4 | 8.3 | 0.83 |
| Cone angle of transmission (−6 dB limits) | ±80° | ±50° | ±3° |

the medium that the ultrasound travels through, and any pollutants in the medium, such as dust or water particles.

### 13.10.5 Ultrasound as a Range Sensor

The basic principles of an ultrasonic range sensor are to measure the time between transmission of a burst of ultrasonic energy from an ultrasonic transmitter and receipt of that energy by an ultrasonic receiver. Then, the distance $d$ can be calculated from

$$d = vt, \tag{13.5}$$

where $v$ is the ultrasound velocity and $t$ is the measured energy transit time. An obvious difficulty in applying this equation is the variability of $v$ with temperature according to Equation (13.2). One solution to this problem is to include an extra ultrasonic transmitter/receiver pair in the measurement system in which the two elements are positioned a known distance apart. Measurement of the transmission time of energy between this fixed pair provides the necessary measurement of velocity and hence compensation for any environmental temperature changes.

The degree of directionality in the ultrasonic elements used for range measurement is unimportant as long as the receiver and transmitter are positioned carefully so as to face each other exactly (i.e., such that the "normal" lines to their faces are coincident). Thus, directionality imposes no restriction on the type of element suitable for range measurement. However, element choice is restricted by the attenuation characteristics of different types of elements, and relatively low-frequency elements have to be used for the measurement of large ranges.

*Measurement resolution and accuracy*

The best measurement resolution that can be obtained with an ultrasonic ranging system is equal to the wavelength of the transmitted wave. As wavelength is inversely proportional to frequency, high-frequency ultrasonic elements would seem to be preferable. For example, while the wavelength and hence resolution for a 40-kHz element is 8.6 mm at room temperature (20°C), it is only 0.86 mm for a 400-kHz element. However, the choice of element also depends on the required range of measurement. The range of higher frequency elements is much reduced compared with low-frequency ones due to the greater attenuation of the ultrasound wave as it travels away from the transmitter. Hence, the choice of element frequency has to be a compromise between measurement resolution and range.

The best measurement accuracy obtainable is equal to the measurement resolution value, but this is only achieved if the electronic counter used to measure the transmission time starts and stops at exactly the same point in the ultrasound cycle (usually the point in the cycle corresponding to peak amplitude is used). However, the sensitivity of the ultrasonic receiver also affects measurement accuracy. The amplitude of the ultrasound wave generated in the

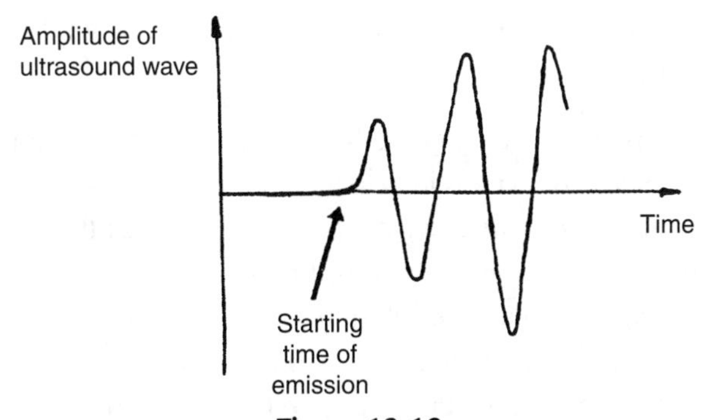


**Figure 13.12**
Ramp up of ultrasonic wave after emission.

transmitter ramps up to full amplitude in the manner shown in Figure 13.12. The receiver has to be sensitive enough to detect the peak of the first cycle, which can usually be arranged. However, if the range of measurement is large, attenuation of the ultrasound wave may cause the amplitude of the first cycle to become less than the threshold level that the receiver is set to detect. In this case, only the second cycle will be detected and there will be an additional measurement error equal to one wavelength. For large transmission distances, even the second cycle may be undetected, meaning that the receiver only "sees" the third cycle.

### 13.10.6 Effect of Noise in Ultrasonic Measurement Systems

Signal levels at the output of ultrasonic measurement systems are usually of low amplitude and are therefore prone to contamination by electromagnetic noise. Because of this, it is necessary to use special precautions such as making ground (earth) lines thick, using shielded cables for transmission of the signal from the ultrasonic receiver, and locating the signal amplifier as close to the receiver as possible.

Another potentially serious form of noise is background ultrasound produced by manufacturing operations in the typical industrial environment in which many ultrasonic range measurement systems operate. Analysis of industrial environments has shown that ultrasound at frequencies up to 100 kHz is generated by many operations, and some operations generate ultrasound at higher frequencies up to 200 kHz. There is not usually any problem if ultrasonic measurement systems operate at frequencies above 200 kHz, but these often have insufficient range for the needs of the measurement situation. In these circumstances, any objects likely to generate energy at ultrasonic frequencies should be covered in sound-absorbing material such that interference with ultrasonic measurement systems is minimized. The placement of sound-absorbing material around the path that the measurement ultrasound wave travels along contributes further toward reducing the effect of background noise. A natural solution to the problem is also partially provided by the fact that the same processes of distance traveled and adsorption that attenuate the amplitude of ultrasound waves traveling between the transmitter and the receiver in the measurement system also attenuate ultrasound noise generated by manufacturing operations.

Because ultrasonic energy is emitted at angles other than the direction that is normal to the face of the transmitting element, a problem arises in respect of energy that is reflected off some object in the environment around the measurement system and back into the ultrasonic receiver. This has a longer path than the direct one between the transmitter and the receiver and can cause erroneous measurements in some circumstances. One solution to this is to arrange for the transmission-time counter to stop as soon as the receiver first detects the ultrasound wave. This will usually be the wave that has traveled along the direct path, and so no measurement error is caused as long as the rate at which ultrasound pulses are emitted is such that the next burst is not emitted until all reflections from the previous pulse have died down. However, in circumstances where the direct path becomes obstructed by some obstacle, the counter will only be stopped when the reflected signal is detected by the receiver, giving a potentially large measurement error.

### 13.10.7 Exploiting Doppler Shift in Ultrasound Transmission

The Doppler effect is evident in all types of wave motion and describes the apparent change in frequency of the wave when there is relative motion between the transmitter and the receiver. If a continuous ultrasound wave with velocity $v$ and frequency $f$ takes $t$ seconds to travel from source $S$ to receiver $R$, then $R$ will receive $ft$ cycles of sound during time $t$ (see Figure 13.13).

Suppose now that $R$ moves toward $S$ at velocity $r$ (with $S$ stationary). $R$ will receive $rt/\lambda$ extra cycles of sound during time $t$, increasing the total number of sound cycles received to $(ft + rt/\lambda)$. With $(ft + rt/\lambda)$ cycles received in $t$ seconds, the apparent frequency $f'$ is given by

$$f' = \frac{ft + rt/\lambda}{t} = f + r/\lambda = f + \frac{rf}{v} = \frac{f(r+v)}{v}$$

[using the relation $\frac{1}{\lambda} = \frac{f}{v}$ from Equation (13.3)].

Frequency difference $\Delta f$ can be expressed as

$$\Delta f = f' - f = \frac{f(v+r)}{v} - f = \frac{fr}{v},$$

from which the velocity of receiver ($r$) can be expressed as $r = v\Delta f/f$.

Similarly, it can be shown that if $R$ moves away from $S$ with velocity $r$, $f'$ is given by

$$f' = \frac{f(v-r)}{v}$$

and

$$\Delta f = -\frac{fr}{v}.$$

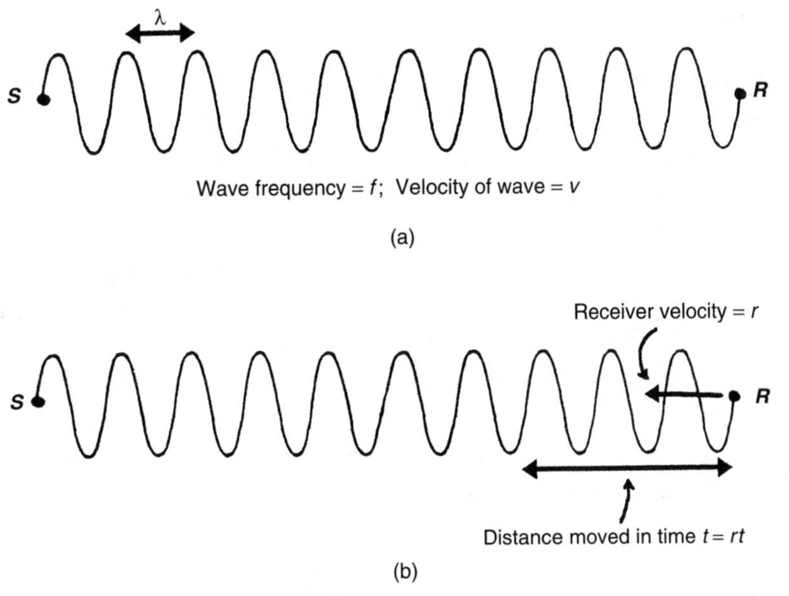


**Figure 13.13**
Illustration of Doppler effect.

If the ultrasound source moves toward the stationary receiver at velocity $s$, it will move a distance $st$ in time $t$, and the $ft$ cycles that are emitted during time $t$ will be compressed into a distance $(vt - st)$.

Hence, the apparent wavelength $\lambda'$ will be given by

$$\lambda' = \frac{vt - st}{ft} = \frac{v - s}{f}.$$

Using Equation (13.3), this can be expressed alternatively as

$$f' = \frac{v}{\lambda'} = \frac{vf}{v - s}.$$

Similarly, with $S$ moving away from $R$, it can be shown that

$$f' = \frac{vf}{v + s}.$$

Thus, the velocity of an ultrasound receiver moving with respect to an ultrasound source can be calculated from the measured ratio between real and apparent frequencies of the wave. This is used in devices such as the Doppler shift flowmeter.

## 13.11 Nuclear Sensors

Nuclear sensors are uncommon measurement devices, partly because of the strict safety regulations that govern their use and partly because they are usually expensive. Some very low-level radiation sources are now available that largely overcome the safety problems, but measurements are then prone to contamination by background radiation. The principle of operation of nuclear sensors is very similar to optical sensors in that radiation is transmitted between a source and a detector through some medium in which the magnitude of transmission is attenuated according to the value of the measured variable. Caesium-137 is used commonly as a $\gamma$-ray source, and a sodium iodide device is used commonly as a $\gamma$-ray detector. The latter gives a voltage output that is proportional to the radiation incident upon it. One current use of nuclear sensors is in a noninvasive technique for measuring the level of liquid in storage tanks (see Chapter 17). They are also used in mass flow rate measurement (see Chapter 16) and in medical-scanning applications (Webster, 1998).

## 13.12 Microsensors

Microsensors are millimeter-sized two- and three-dimensional micromachined structures that have smaller size, improved performance, better reliability, and lower production costs than many alternative forms of sensors. Currently, devices used to measure temperature, pressure, force, acceleration, humidity, magnetic fields, radiation, and chemical parameters are either in production or at advanced stages of research.

Microsensors are usually constructed from a silicon semiconductor material, but are sometimes fabricated from other materials, such as metals, plastics, polymers, glasses, and ceramics deposited on a silicon base. Silicon is an ideal material for sensor construction because of its excellent mechanical properties. Its tensile strength and Young's modulus are comparable to that of steel, while its density is less than that of aluminium. Sensors made from a single crystal of silicon remain elastic almost to the breaking point, and mechanical hysteresis is very small. In addition, silicon has a very low coefficient of thermal expansion and can be exposed to extremes of temperature and most gases, solvents, and acids without deterioration.

Microengineering techniques are an essential enabling technology for microsensors, which are designed so that their electromechanical properties change in response to a change in the measured parameter. Many of the techniques used for integrated circuit (IC) manufacture are also used in sensor fabrication, with common techniques being crystal growing and polishing, thin film deposition, ion implantation, wet and dry chemical and laser etching, and photolithography. However, apart from standard IC production techniques, some special techniques are also needed in addition to produce the three-dimensional structures that are unique to some types of microsensors. The various manufacturing techniques are used to form sensors directly in silicon crystals and films. Typical structures have forms such as thin diaphragms, cantilever beams, and bridges.

While the small size of a microsensor is of particular benefit in many applications, it also leads to some problems that require special attention. For example, microsensors typically have very low capacitance. This makes the output signals very prone to noise contamination. Hence, it is usually necessary to integrate microelectronic circuits that perform signal processing in the device, which therefore becomes a *smart microsensor*. Another problem is that microsensors generally produce output signals of very low magnitude. This requires the use of special types of analogue-to-digital converters that can cope with such low-amplitude input signals. One suitable technique is sigma-delta conversion. This is based on charge balancing techniques and gives better than 16-bit accuracy in less than 20 ms (Riedijk and Huijsing, 1997). Special designs can reduce conversion time down to less than 0.1 ms if necessary.

Microsensors are used most commonly for measuring pressure, acceleration, force, and chemical parameters. They are used in particularly large numbers in the automotive industry, where unit prices can be very low. Microsensors are also widely used in medical applications, particularly for blood pressure measurement.

Mechanical microsensors transform measured variables such as force, pressure, and acceleration into a displacement. The displacement is usually measured by capacitive or piezoresistive techniques, although some devices use other technologies, such as resonant frequency variation, resistance change, inductance change, piezoelectric effect, and changes in magnetic or optical coupling. The design of a cantilever silicon microaccelerometer is shown in Figure 13.14. The proof mass within this is about 100 μm across, and the typical deflection measured is of the order of 1 μm ($10^{-3}$ mm).

An alternative capacitive microaccelerometer provides a calibrated, compensated, and amplified output. It has a capacitive silicon microsensor to measure displacement of the proof mass. This is integrated with a signal processing chip and is protected by a plastic enclosure. The capacitive element has a three-dimensional structure, which gives higher measurement sensitivity than surface-machined elements.

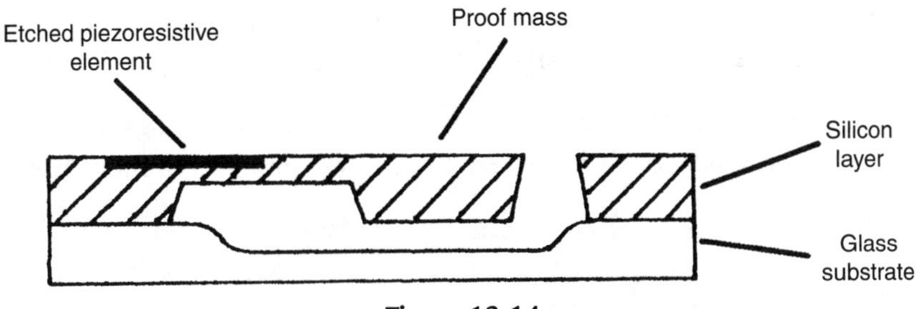


**Figure 13.14**
Silicon microaccelerometer.

Microsensors to measure many other physical variables are either in production or at advanced stages of research. Microsensors measuring magnetic field are based on a number of alternative technologies, such as the Hall effect, magnetoresistors, magnetodiodes, and magnetotransistors. Radiation microsensors are made from silicon p-n diodes or avalanche photodiodes and can detect radiation over wavelengths from the visible spectrum to infrared. Microsensors in the form of a microthermistor, a p-n thermodiode, or a thermotransistor are used as digital thermometers. Microsensors have also enabled measurement techniques that were previously laboratory-based ones to be extended into field instruments. Examples are spectroscopic instruments and devices to measure viscosity.

## 13.13 Summary

This chapter revealed 11 different physical principles used in measurement sensors. As noted in the introduction to the chapter, the chosen order of presentation of these principles has been arbitrary and is not intended to imply anything about the relative popularity of these various principles.

The first principle covered was capacitance change, which we found was based on two capacitor plates with either a variable or a fixed distance between them. We learned that sensors with a variable distance between plates are used primarily for displacement measurement, either as displacement sensors in their own right or to measure the displacement within certain types of pressure, sound, and acceleration sensors. The alternative type of capacitive sensor where the distance between plates is fixed is used typically to measure moisture content, humidity values, and liquid level.

Moving on to the resistance change principle, we found that this is used in a wide range of devices for temperature measurement (resistance thermometers and thermistors) and displacement measurement (strain gauges and piezoresistive sensors). We also noted that some moisture meters work on the resistance variation principle.

We then looked at sensors that use the magnetic phenomena of inductance, reluctance, and eddy currents. We saw that the principle of inductance change was used mainly to measure translational and rotational displacements, reluctance change was used commonly to measure rotational velocities, and the eddy current effect was used typically to measure the displacement between a probe and a very thin metal target, such as the steel diaphragm of a pressure sensor.

Next we looked at the Hall effect. This measures the magnitude of a magnetic field and is used commonly in a proximity sensor. It is also employed in computer keyboard push buttons.

We then moved on to piezoelectric transducers. These generate a voltage when a force is applied to them. Alternatively, if a voltage is applied to them, an output force is produced. A common application is in ultrasonic transmitters and receivers. They are also used as displacement transducers, particularly as part of devices measuring acceleration, force, and pressure.

Our next subject of study was strain gauges. These devices exploit the physical principle of a change in resistance when the metal wire that they are made from is stretched or strained. They detect very small displacements and are used typically within devices such as diaphragm pressure sensors to measure the small displacement of the diaphragm when a pressure is applied to it. We looked into some of the science involved in strain gauge design, particularly in respect to the alternative materials used for the active element.

Moving on, we then looked at piezoresistive sensors. We saw that these could be regarded as a semiconductor strain gauge, as they consist of a semiconductor material whose resistance varies when it is compressed or stretched. They are used commonly to measure the displacement in diaphragm pressure sensors where the resistance change for a given amount of diaphragm displacement is much greater than is obtained in metal strain gauges, thus leading to better measurement accuracy. They are also used as accelerometers. Before concluding this discussion, we also observed that the term *piezoresistive sensor* is sometimes (but incorrectly) used to describe metal strain gauges as well as semiconductor ones.

In our discussion of optical sensors that followed, we observed first of all that these could involve both transmission of light through air and transmission along a fiber-optic cable. Air path optical sensors exploit the transmission of light from a source to a detector across an open air path and are used commonly to measure proximity, translational motion, rotational motion, and gas concentration.

Sensors that involve the transmission of light along a fiber-optic cable are commonly called fiber-optic sensors. Their principle of operation is to translate the measured physical quantity into a change in the intensity, phase, polarization, wavelength, or transmission time of the light carried along the cable. We went on to see that two kinds of fiber-optic sensors can be distinguished, known as intrinsic sensors and extrinsic sensors. In intrinsic sensors, the fiber-optic cable itself is the sensor, whereas in extrinsic sensors, the fiber-optic cable is merely used to transmit light to/from a conventional sensor. Our look at intrinsic sensors revealed that different forms of these are used to measure a very wide range of physical variables, including proximity, displacement, pressure, pH, smoke intensity, acceleration, temperature, cryogenic leakage, oil content in water, liquid level, refractive index of a liquid, parameters in biomedical applications (oxygen concentration, carbon monoxide concentrations, blood pressure level, hormone concentration, steroid concentration), mechanical strain, magnetic field strength, electric field strength, electrical current, electrical voltage, angular position and acceleration in gyroscopes, liquid flow rate, and gas presence. In comparison, the range of physical variables measured by extrinsic sensors is much less, being limited mainly to the measurement of temperature, pressure, force, and displacement (both linear and angular).

This then led to a discussion of ultrasonic sensors. These are used commonly to measure range, translational displacements, fluid flow rate, and liquid level. We learned that ultrasonic sensors work in one of two ways, either by measuring the change in time taken for an ultrasound wave to travel between a transmitter and a receiver or by measuring the change in phase or frequency of the transmitted wave. While both of these principles are simple in concept, we went on to see that the design and use of ultrasonic sensors suffer from a number of potential problems. First, the transmission speed can be affected by environmental factors, with temperature changes being a particular problem and humidity changes to a lesser extent. The nominal operating frequency of ultrasonic elements also has to be chosen carefully according to the intended application, as this affects the effective amount of spread of the transmitted energy either side of the direction normal to the face of the transmitting element. Attenuation of the transmitted wave can cause problems. This is particularly so when ultrasonic elements are used as range sensors. This follows from the start-up nature of a transmitted ultrasonic wave, which exhibits increasing amplitude over the first two or three cycles of the emitted energy wave. Attenuation of the wave as it travels over a distance may mean that the detector fails to detect the first or even second cycle of the transmitted wave, causing an error that is equal to one or two times the ultrasound wavelength. Noise can also cause significant problems in the operation of ultrasonic sensors, as they are contaminated easily by electromagnetic noise and are particularly affected by noise generated by manufacturing operations at a similar frequency to that of the ultrasonic-measuring system. Because there is some emission of ultrasonic energy at angles other than the normal direction to the face of the ultrasonic element, stray reflections of transmissions in these nonnormal directions by structures in the environment around the ultrasonic system may interfere with measurements.

The next type of sensor discussed was nuclear sensors. We learned that these did not enjoy widespread use, with main applications being in the noninvasive measurement of liquid level, mass flow rate measurement, and in some medical scanning applications. This limited number of applications is partly due to the health dangers posed to users by the radiation source that they use and partly due to their relatively high cost. Danger to users can largely be overcome by using low-level radiation sources but this makes measurements prone to contamination by background radiation.

Finally, we looked at microsensors, which we learned were millimeter-sized, two- and three-dimensional micromachined structures usually made from silicon semiconductor materials but sometimes made from other materials. These types of sensors have smaller size, improved performance, better reliability, and lower production costs than many alternative forms of sensors and are used to measure temperature, pressure, force, acceleration, humidity, magnetic fields, radiation, chemical parameters, and some parameters in medical applications such as blood pressure. Despite being superior in many ways to larger sensors, they are affected by several problems. One such problem is their very low capacitance, which makes their output signals very prone to noise contamination. To counteract this, it is normally necessary to integrate microelectronic circuits in the device to perform signal processing. Another problem is the very low magnitude of the output signal, which requires the use of special types of analogue-to-digital converters.

## 13.14 Problems

13.1. Describe the general working principles of capacitive sensors and discuss some applications of them.
13.2. Discuss some applications of resistive sensors.
13.3. What types of magnetic sensors exist and what are they mainly used for? Describe the mode of operation of each.
13.4. What are Hall-effect sensors? How do they work and what are they used for?
13.5. How does a piezoelectric transducer work and what materials are typically used in their construction? Discuss some common applications of this type of device.
13.6. What is a strain gauge and how does it work? What are the problems in making and using a traditional metal-wire strain gauge and how have these problems been overcome in new types of strain gauges?
13.7. Discuss some applications of strain gauges.
13.8. What are piezoresistive sensors and what are they typically used for?
13.9. What is the principal advantage of an optical sensor? Discuss the mode of operation of the two main types of optical sensors.
13.10. What are air path optical sensors? Discuss their mode of operation, including details of light sources and light detectors used. What are their particular advantages and disadvantages over fiber-optic sensors?
13.11. How do fiber-optic sensors work? Discuss their use in intrinsic and extrinsic sensors.
13.12. Explain the basic principles of operation of ultrasonic sensors and discuss what they are typically used for.
13.13. What factors govern the transmission speed and directionality of ultrasonic waves? How do these factors affect the application of ultrasonic sensors?
13.14. Discuss the use of ultrasonic sensors in range-measuring systems, mentioning the effect of attenuation of the wave as it travels. How can measurement resolution and accuracy be optimized?
13.15. Discuss the effects of extraneous noise in ultrasonic measurement systems. How can these effects be reduced?
13.16. Discuss the phenomenon of Doppler shift in ultrasound transmission and explain how this can be used in sensors.
13.17. Why are nuclear sensors not in common use?
13.18. What are microsensors? How are they made and what applications are they used in?

## References

Riedijk, F. R., & Huijsing, J. H. (1997). Sensor interface environment based on a serial bus interface. *Measurement and Control*, *30*, 297–299.
Webster, J. G. (1998). *Medical instrumentation*. John Wiley.

CHAPTER 14

# Temperature Measurement



Measurement and Instrumentation: Theory and Application

## 14.1 Introduction

We are probably well aware that temperature measurement is very important in all spheres of life. In engineering applications, it is particularly important in process industries, where it is the most commonly measured process variable. It is therefore appropriate for us to devote this first chapter on measurement of individual physical variables to the subject of temperature measurement.

Unfortunately, temperature measurement poses some interesting theoretical difficulties because of its rather abstract nature. These difficulties become especially apparent when we come to consider the calibration of temperature-measuring instruments, particularly when we look for a primary reference standard at the top of the calibration chain. Foremost among these difficulties is the fact that any given temperature cannot be related to a fundamental standard of temperature in the same way that the measurement of other quantities can be related to the primary standards of mass, length, and time. If two bodies of lengths, $l_1$ and $l_2$, are connected together end to end, the result is a body of length $l_1 + l_2$. A similar relationship exists between separate masses and separate times. However, if two bodies at the same temperature are connected together, the joined body has the same temperature as each of the original bodies.

This is a root cause of the fundamental difficulties that exist in establishing an absolute standard for temperature in the form of a relationship between it and other measurable quantities for which a primary standard unit exists. In the absence of such a relationship, it is necessary to establish fixed, reproducible reference points for temperature in the form of freezing and triple points of substances where the transition among solid, liquid, and gaseous states is sharply defined. The *International Practical Temperature Scale* (IPTS)* uses this philosophy and defines a number of *fixed points* for reference temperatures. Three examples are:

- The triple point of hydrogen: −259.35°C
- The freezing point of zinc: 419.53°C
- The freezing point of gold: 1064.18°C

A full list of fixed points defined in the IPTS can be found in Section 14.14.

If we start writing down the physical principles affected by temperature, we will get a relatively long list. Many of these physical principles form the basis for temperature-measuring instruments. It is therefore reasonable for us to study temperature measurement by dividing

* The IPTS is subject to periodic review and improvement as research produces more precise fixed reference points. The latest version was published in 1990.

the instruments used to measure temperature into separate classes according to the physical principle on which they operate. This gives us 10 classes of instrument based on the following principles:

- Thermoelectric effect
- Resistance change
- Sensitivity of semiconductor device
- Radiative heat emission
- Thermography
- Thermal expansion
- Resonant frequency change
- Sensitivity of fiber-optic devices
- Color change
- Change of state of material

We consider each of these in the following sections.

## 14.2 Thermoelectric Effect Sensors (Thermocouples)

Thermoelectric effect sensors rely on the physical principle that, when any two different metals are connected together, an e.m.f., which is a function of the temperature, is generated at the junction between the metals. The general form of this relationship is

$$e = a_1T + a_2T^2 + a_3T^3 + \cdots + a_nT^n, \tag{14.1}$$

where $e$ is the e.m.f. generated and $T$ is the absolute temperature.

This is clearly nonlinear, which is inconvenient for measurement applications. Fortunately, for certain pairs of materials, terms involving squared and higher powers of T ($a_2T^2$, $a_3T^3$, etc.) are approximately zero, and the e.m.f.–temperature relationship is approximately linear according to

$$e \approx a_1T. \tag{14.2}$$

Wires of such pairs of materials are connected together at one end, and in this form are known as *thermocouples*. Thermocouples are a very important class of device as they provide the most commonly used method of measuring temperatures in industry.

Thermocouples are manufactured from various combinations of the base metals copper and iron; the base metal alloys of alumel (Ni/Mn/Al/Si), chromel (Ni/Cr), constantan (Cu/Ni), nicrosil (Ni/Cr/Si), nisil (Ni/Si/Mn), nickel–molybdenum, and nickel–cobalt; the noble metals platinum and tungsten; and the noble metal alloys of platinum–rhodium, tungsten–rhenium, and gold–iron. Only certain combinations of these are used as thermocouples, and most standard combinations are known by internationally recognized type letters, for example, type K is chromel–alumel. The e.m.f.–temperature characteristics for some of these standard thermocouples are shown in Figure 14.1: these show reasonable linearity over at least part of their temperature-measuring ranges.

A typical thermocouple, made from one chromel wire and one constantan wire, is shown in Figure 14.2a. For analysis purposes, it is useful to represent the thermocouple by its equivalent electrical circuit, shown in Figure 14.2. The e.m.f. generated at the point where the different wires are connected together is represented by a voltage source, $E_1$, and the point is known as the *hot junction.* The temperature of the hot junction is customarily shown as $T_h$ on the diagram. The e.m.f. generated at the hot junction is measured at the open ends of the thermocouple, which is known as the *reference junction.*

In order to make a thermocouple conform to some precisely defined e.m.f.–temperature characteristic, it is necessary that all metals used are refined to a high degree of pureness and all alloys are manufactured to an exact specification. This makes the materials used expensive; consequently, thermocouples are typically only a few centimeters long. It is clearly impractical to connect a voltage-measuring instrument at the open end of the thermocouple to measure its output in such close proximity to the environment whose temperature is being measured, and therefore *extension leads* up to several meters long are normally connected between the thermocouple and the measuring instrument. This modifies the equivalent circuit to that

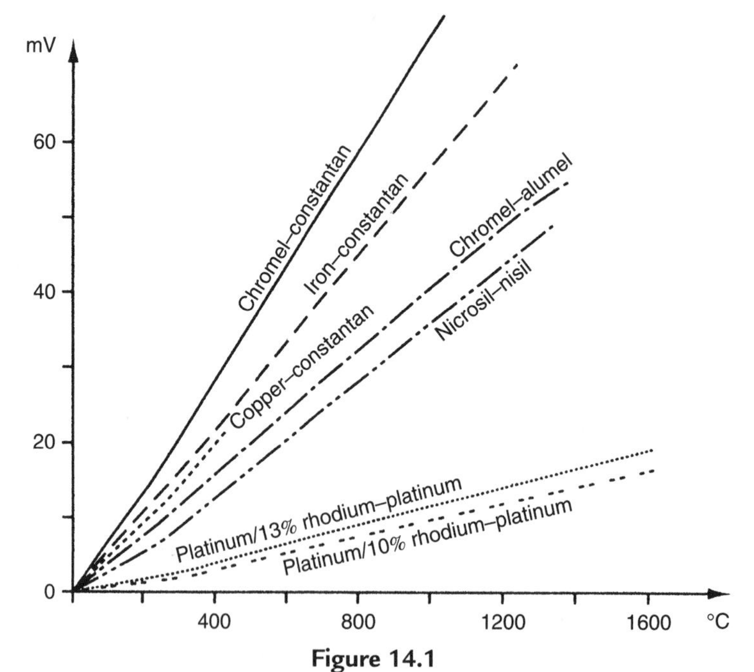


**Figure 14.1**
The e.m.f. temperature characteristics for some standard thermocouple materials.

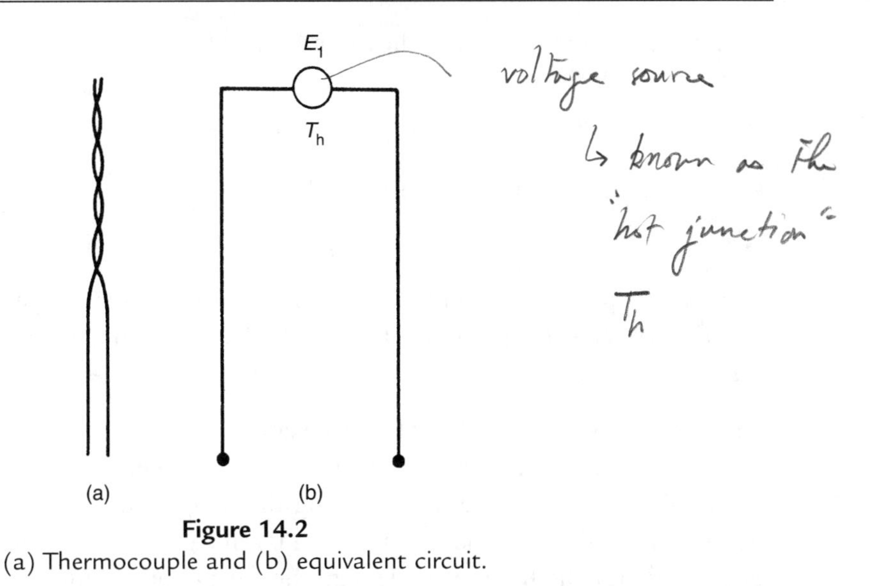


**Figure 14.2**
(a) Thermocouple and (b) equivalent circuit.

shown in Figure 14.3a. There are now three junctions in the system and consequently three voltage sources, $E_1$, $E_2$, and $E_3$, with the point of measurement of the e.m.f. (still called the reference junction) being moved to the open ends of the extension leads.

The measuring system is completed by connecting the extension leads to the voltage-measuring instrument. As the connection leads will normally be of different materials to those of the thermocouple extension leads, this introduces two further e.m.f.-generating junctions,

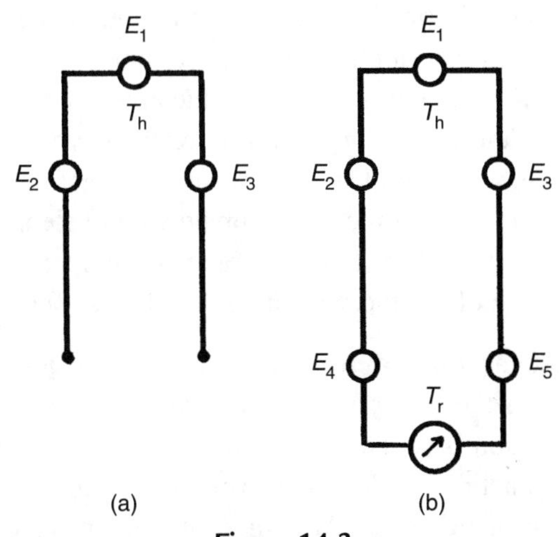


**Figure 14.3**
(a) Equivalent circuit for a thermocouple with extension leads and (b) equivalent circuit for a thermocouple and extension leads connected to a meter.

$E_4$ and $E_5$, into the system, as shown in Figure 14.3b. The net output e.m.f. measured ($E_m$) is then given by

$$E_m = E_1 + E_2 + E_3 + E_4 + E_5 \tag{14.3}$$

and this can be reexpressed in terms of $E_1$ as

$$E_1 = E_m - E_2 - E_3 - E_4 - E_5. \tag{14.4}$$

In order to apply Equation (14.1) to calculate the measured temperature at the hot junction, $E_1$ has to be calculated from Equation (14.4). To do this, it is necessary to calculate the values of $E_2$, $E_3$, $E_4$, and $E_5$.

It is usual to choose materials for the extension lead wires such that the magnitudes of $E_2$ and $E_3$ are approximately zero, irrespective of the junction temperature. This avoids the difficulty that would otherwise arise in measuring the temperature of the junction between the thermocouple wires and the extension leads, and also in determining the e.m.f./temperature relationship for the thermocouple/extension lead combination.

A near-zero junction e.m.f. is achieved most easily by choosing the extension leads to be of the same basic materials as the thermocouple, but where their cost per unit length is reduced greatly by manufacturing them to a lower specification. As an alternative to using lower specification materials of the same basic type as the thermocouple, copper compensating leads are also sometimes used with certain types of base metal thermocouples. In this case, the law of intermediate metals has to be applied to compensate for the e.m.f. at the junction between the thermocouple and compensating leads.

Unfortunately, the use of extension leads of the same basic materials as the thermocouple but manufactured to a lower specification is still prohibitively expensive in the case of noble metal thermocouples. It is necessary in this case to search for base metal extension leads that have a similar thermoelectric behavior to the noble metal thermocouple. In this form, the extension leads are usually known as *compensating leads*. A typical example of this is the use of nickel/copper–copper extension leads connected to a platinum/rhodium–platinum thermocouple. It should be noted that the approximately equivalent thermoelectric behavior of compensating leads is only valid for a limited range of temperatures that is considerably less than the measuring range of the thermocouple that they are connected to.

To analyze the effect of connecting extension leads to the voltage-measuring instrument, a thermoelectric law known as the *law of intermediate metals* can be used. This states that the e.m.f. generated at the junction between two metals or alloys $A$ and $C$ is equal to the sum of the e.m.f. generated at the junction between metals or alloys $A$ and $B$ and the e.m.f. generated at the junction between metals or alloys $B$ and $C$, where all junctions are at the same temperature. This can be expressed more simply as

$$e_{AC} = e_{AB} + e_{BC}. \tag{14.5}$$

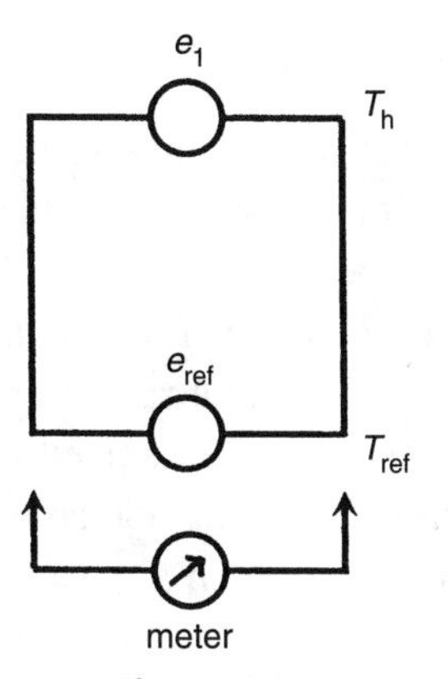


**Figure 14.4**
Effective e.m.f. sources in a thermocouple measurement system.

Suppose we have an iron–constantan thermocouple connected by copper leads to a meter. We can express $E_4$ and $E_5$ in Figure 14.4 as

$$E_4 = e_{iron-copper} \quad ; \quad E_5 = e_{copper-constantan}.$$

The sum of $E_4$ and $E_5$ can be expressed as $E_4 + E_5 = e_{iron-copper} + e_{copper-constantan}$

Applying Equation (14.5): $e_{iron-copper} + e_{copper-constantan} = e_{iron-constantan}$.

Thus, the effect of connecting the thermocouple extension wires to the copper leads to the meter is canceled out, and the actual e.m.f. at the reference junction is equivalent to that arising from an iron–constantan connection at the reference junction temperature, which can be calculated according to Equation (14.1). Hence, the equivalent circuit in Figure 14.3b becomes simplified to that shown in Figure 14.4. The e.m.f. $E_m$ measured by the voltage-measuring instrument is the sum of only two e.m.f.s, consisting of the e.m.f. generated at the hot junction temperature, $E_1$, and the e.m.f. generated at the reference junction temperature, $E_{ref}$. The e.m.f. generated at the hot junction can then be calculated as

$$E_1 = E_m + E_{ref}.$$

$E_{ref}$ can be calculated from Equation (14.1) if the temperature of the reference junction is known. In practice, this is often achieved by immersing the reference junction in an ice bath to maintain it at a reference temperature of 0°C. However, as discussed in the following section on thermocouple tables, it is very important that the ice bath remains exactly at 0°C if this is to be the reference temperature assumed, as otherwise significant measurement errors can arise. For this reason, refrigeration of the reference junction at a temperature of 0°C is often preferred.

### 14.2.1 Thermocouple Tables

Although the preceding discussion has suggested that the unknown temperature, $T$, can be evaluated from the calculated value of the e.m.f., $E_1$, at the hot junction using Equation (14.1), this is very difficult to do in practice because Equation (14.1) is a high-order polynomial expression. An approximate translation between the value of $E_1$ and temperature can be achieved by expressing Equation (14.1) in graphical form as in Figure 14.1. However, this is not usually of sufficient accuracy, and it is normal practice to use tables of e.m.f. and temperature values known as *thermocouple tables*. These include compensation for the effect of the e.m.f. generated at the reference junction ($E_{ref}$), which is assumed to be at 0°C. Thus, the tables are only valid when the reference junction is exactly at this temperature. Compensation for the case where the reference junction temperature is not at zero is considered later in this section.

Tables for a range of standard thermocouples are given in Appendix 3. In these tables, a range of temperatures is given in the left-hand column, and the e.m.f. output for each standard type of thermocouple is given in the columns to the right. In practice, any general e.m.f. output measurement taken at random will not be found exactly in the tables, and interpolation will be necessary between the values shown in the table.

## Example 14.1

If the e.m.f. output measured from a chromel–constantan thermocouple is 14.419 mV with the reference junction at 0°C, the appropriate column in the tables shows that this corresponds to a hot junction temperature of 200°C.

■

## Example 14.2

If the measured output e.m.f. for a chromel–constantan thermocouple (reference junction at 0°C) was 10.65 mV, it is necessary to carry out linear interpolation between the temperature of 160°C corresponding to an e.m.f. of 10.501 mV shown in the tables and the temperature of 170°C corresponding to an e.m.f. of 11.222 mV. This interpolation procedure gives an indicated hot junction temperature of 162°C.

■

### 14.2.2 Nonzero Reference Junction Temperature

If the reference junction is immersed in an ice bath to maintain it at a temperature of 0°C so that thermocouple tables can be applied directly, the ice in the bath must be in a state of just melting. This is the only state in which ice is exactly at 0°C, and otherwise it will be either colder or hotter than this temperature. Thus, maintaining the reference junction at 0°C is not a straightforward matter, particularly if the environmental temperature around the measurement system is relatively hot. In consequence, it is common practice in many practical applications of thermocouples to maintain the reference junction at a nonzero temperature by putting it into a controlled environment maintained by an electrical heating element. In order to still be able to apply thermocouple tables, correction then has to be made for this nonzero reference junction temperature using a second thermoelectric law known as the *law of intermediate temperatures*. This states that

$$E_{(T_h,T_0)} = E_{(T_h,T_r)} + E_{(T_r,T_0)}, \tag{14.6}$$

where $E_{(T_h,T_0)}$ is the e.m.f. with junctions at temperatures $T_h$ and $T_0$, $E_{(T_h,Tr)}$ is the e.m.f. with junctions at temperatures $T_h$ and $T_r$, $E_{(T_r,T_0)}$ is the e.m.f. with junctions at temperatures $T_r$ and $T_0$, $T_h$ is the hot-junction measured temperature, $T_0$ is 0°C, and $T_r$ is the nonzero reference junction temperature that is somewhere between $T_0$ and $T_h$.

## Example 14.3

Suppose that the reference junction of a chromel–constantan thermocouple is maintained at a temperature of 80°C and the output e.m.f. measured is 40.102 mV when the hot junction is immersed in a fluid.

The quantities given are $T_r = 80°C$ and $E_{(T_h,T_r)} = 40.102$ mV
From the tables, $E_{(T_r,T_0)} = 4.983$ mV
Now applying Equation (14.6), $E_{(T_h,T_0)} = 40.102 + 4.983 = 45.085$ mV

Again referring to the tables, this indicates a fluid temperature of 600°C.

In using thermocouples, it is essential that they are connected correctly. Large errors can result if they are connected incorrectly, for example, by interchanging the extension leads or by using incorrect extension leads. Such mistakes are particularly serious because they do not prevent some sort of output being obtained, which may look sensible even though it is incorrect, and so the mistake may go unnoticed for a long period of time. The following examples illustrate the sorts of errors that may arise.

■

## Example 14.4

This example is an exercise in the use of thermocouple tables, but it also serves to illustrate the large errors that can arise if thermocouples are used incorrectly. In a particular industrial situation, a chromel–alumel thermocouple with chromel–alumel extension wires is used to measure the temperature of a fluid. In connecting up this measurement system, the instrumentation engineer responsible has inadvertently interchanged the extension wires from the thermocouple. The ends of the extension wires are held at a reference temperature of 0°C and the output e.m.f. measured is 14.1 mV. If the junction between the thermocouple and extension wires is at a temperature of 40°C, what temperature of fluid is indicated and what is the true fluid temperature?

■

## Solution

The initial step necessary in solving a problem of this type is to draw a diagrammatical representation of the system and to mark on this the e.m.f. sources, temperatures, etc., as shown in Figure 14.5. The first part of the problem is solved very simply by looking up in thermocouple tables what temperature the e.m.f. output of 12.1 mV indicates for a chromel–alumel thermocouple. This is 297.4°C. Then, summing e.m.f.s around the loop:

$$V = 12.1 = E_1 + E_2 + E_3 \text{ or } E_1 = 12.1 - E_2 - E_3$$

$$E_2 = E_3 = Emf_{(alumel-chromel)_{40}} = -Emf_{(chromel-alumel)_{40}}{}^{\dagger} = -1.611 \text{ mV}$$

Hence $E_1 = 12.1 + 1.611 + 1.611 = 15.322$ mV.

Interpolating from the thermocouple tables, this indicates that the true fluid temperature is 374.5°C.

## Example 14.5

This example also illustrates the large errors that can arise if thermocouples are used incorrectly. An iron–constantan thermocouple measuring the temperature of a fluid is connected by mistake with copper–constantan extension leads (such that the two constantan wires are connected together and the copper extension wire is connected to the iron thermocouple wire). If the fluid temperature was actually 200°C and the junction between the thermocouple and extension wires was at 50°C, what e.m.f. would be measured at the open ends of the extension wires if the reference junction is maintained at 0°C? What fluid temperature would be deduced from this (assuming that the connection mistake was not known)?

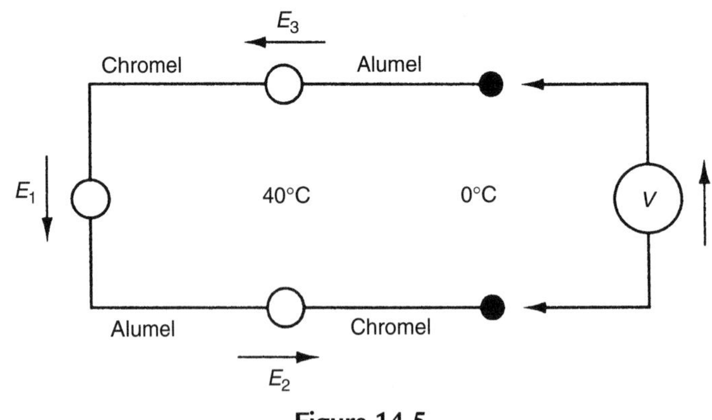


**Figure 14.5**
Diagram for solution of Example 14.5.

† The thermocouple tables quote e.m.f.s using the convention that going from chromel to alumel is positive. Hence, the e.m.f. going from alumel to chromel is minus the e.m.f. going from chromel to alumel.

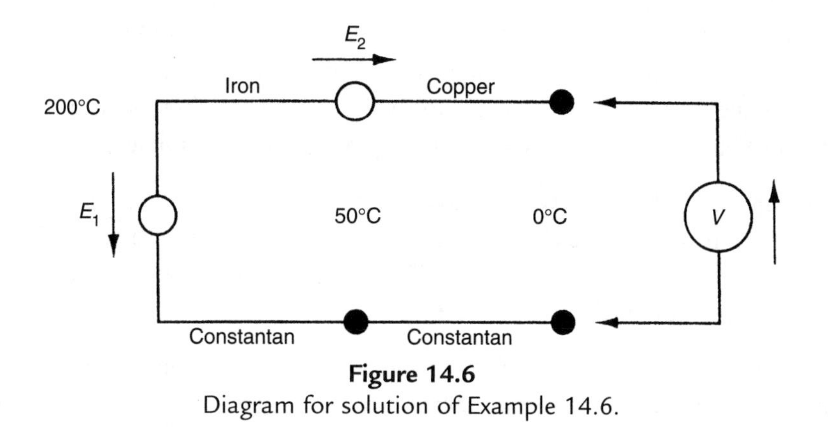


**Figure 14.6**
Diagram for solution of Example 14.6.

## Solution

Again, the initial step necessary is to draw a diagram showing the junctions, temperatures, and e.m.f.s, as shown in Figure 14.6. The various quantities can then be calculated:

$$E_2 = Emf_{(iron-copper)_{50}}$$

By the law of intermediate metals:

$$Emf_{(iron-copper)_{50}} = Emf_{(iron-constantan)_{50}} - Emf_{(copper-constantan)_{50}}$$

$$= 2.585 - 2.035 \text{ [from thermocouple tables]} = 0.55 \text{ mV}$$

$$E_1 = Emf_{(iron-constantan)_{200}} = 10.777 \text{ [from thermocouple tables]}$$

$$V = E_1 - E_2 = 10.777 - 0.55 = 10.227$$

Using tables and interpolating, 10.227 mV indicates a temperature of

$$\left(\frac{10.227 - 10.222}{10.777 - 10.222}\right)10 + 190 = 190.1^\circ\text{C}.$$

### 14.2.3 Thermocouple Types

The five standard base metal thermocouples are chromel–constantan (type E), iron–constantan (type J), chromel–alumel (type K), nicrosil–nisil (type N), and copper–constantan (type T). These are all relatively inexpensive to manufacture but become inaccurate with age and have a short life. In many applications, performance is also affected through contamination by the working environment. To overcome this, the thermocouple can be enclosed in a *protective sheath*, but this has the adverse effect of introducing a significant time constant, making the thermcouple slow to respond to temperature changes. Therefore, as far as possible, thermocouples are used without protection.

*Chromel–constantan thermocouples (type E)* give the highest measurement sensitivity of 68 μV/°C, with an inaccuracy of ±0.5% and a useful measuring range of −200°C up to 900°C. Unfortunately, while they can operate satisfactorily in oxidizing environments when unprotected, their performance and life are seriously affected by reducing atmospheres.

*Iron–constantan thermocouples (type J)* have a sensitivity of 55 μV/°C and are the preferred type for general-purpose measurements in the temperature range of −40 to +750°C, where the typical measurement inaccuracy is ±0.75%. Their performance is little affected by either oxidizing or reducing atmospheres.

*Copper–constantan thermocouples (type T)* have a measurement sensitivity of 43 μV/°C and find their main application in measuring subzero temperatures down to −200°C, with an inaccuracy of ±0.75%. They can also be used in both oxidizing and reducing atmospheres to measure temperatures up to 350°C.

*Chromel–alumel thermocouples (type K)* are widely used, general-purpose devices with a measurement sensitivity of 41 μV/°C. Their output characteristic is particularly linear over the temperature range between 700 and 1200°C and this is therefore their main application, although their full measurement range is −200 to +1300°C. Like chromel–constantan devices, they are suitable for oxidizing atmospheres but not for reducing ones unless protected by a sheath. Their measurement inaccuracy is ±0.75%.

*Nicrosil–nisil thermocouples (type N)* were developed with the specific intention of improving on the lifetime and stability of chromel–alumel thermocouples. They therefore have similar thermoelectric characteristics to the latter but their long-term stability and life are at least three times better. This allows them to be used in temperatures up to 1300°C. Their measurement sensitivity is 39 μV/°C and they have a typical measurement uncertainty of ±0.75%. A detailed comparison between type K and N devices can be found in Brooks (1985).

*Nickel/molybdenum–nickel–cobalt thermocouples (type M)* have one wire made from a nickel–molybdenum alloy with 18% molybdenum and the other wire made from a nickel–cobalt alloy with 0.8% cobalt. They can measure at temperatures up to 1400°C, which is higher than other types of base metal thermocouples. Unfortunately, they are damaged in both oxidizing and reducing atmospheres. This means that they are rarely used except for special applications such as temperature measurement in vacuum furnaces.

*Noble metal thermocouples* are expensive, but they enjoy high stability and long life even when used at high temperatures, although they cannot be used in reducing atmospheres. Unfortunately, their measurement sensitivity is relatively low. Because of this, their use is mainly restricted to measuring high temperatures unless the operating environment is particularly aggressive in low-temperature applications. Various combinations of the metals platinum and tungsten and the metal alloys of platinum–rhodium, tungsten–rhenium, and gold–iron are used.

*Platinum thermocouples (type B)* have one wire made from a platinum–rhodium alloy with 30% rhodium and the other wire made from a platinum–rhodium alloy with 6% rhodium. Their quoted measuring range is +50 to +1800°C, with a measurement sensitivity of 10 μV/°C.

*Platinum thermocouples (type R)* have one wire made from pure platinum and the other wire made from a platinum–rhodium alloy with 13% rhodium. Their quoted measuring range is 0 to +1700°C, with a measurement sensitivity of 10 μV/°C and quoted inaccuracy of ±0.5%.

*Platinum thermocouples (type S)* have one wire made from pure platinum and the other wire made from a platinum–rhodium alloy with 10% rhodium. They have similar characteristics to type R devices, with a quoted measuring range of 0 to +1750°C, measurement sensitivity of 10 μV/°C, and inaccuracy of ±0.5%.

*Tungsten thermocouples (type C)* have one wire made from pure tungsten and the other wire made from a tungsten/rhenium alloy. Their measurement sensitivity of 20 μV/°C is double that of platinum thermocouples, and they can also operate at temperatures up to 2300°C. Unfortunately, they are damaged in both oxidizing and reducing atmospheres. Therefore, their main application is temperature measurement in vacuum furnaces.

*Chromel-gold/iron thermocouples* have one wire made from chromel and the other wire made from a gold/iron alloy which is, in fact, almost pure gold but with a very small iron content (typically 0.15%). These are rare, special-purpose thermocouples with a typical measurement sensitivity of 15 μV/°K designed specifically for cryogenic (very low temperature) applications. The lowest temperature measureable is 1.2°K. Several versions are available, which differ according to the iron content and consequent differences in the measurement range and sensitivity. Because of this variation in iron content, and also because of their rarity, these do not have an international type letter.

### 14.2.4 Thermocouple Protection

Thermocouples are delicate devices that must be treated carefully if their specified operating characteristics are to be maintained. One major source of error is induced strain in the hot junction. This reduces the e.m.f. output, and precautions are normally taken to minimize induced strain by mounting the thermocouple horizontally rather than vertically. It is usual to cover most of the thermocouple wire with thermal insulation, which also provides mechanical protection, although the tip is left exposed if possible to maximize the speed of response to changes in the measured temperature. However, thermocouples are prone to contamination in some operating environments. This means that their e.m.f./temperature characteristic varies from that published in standard tables. Contamination also makes them brittle and shortens their life.

**Table 14.1 Common Sheath Materials for Thermocouples**

| Material | Maximum Operating Temperature (°C)* |
|---|---|
| Mild steel | 900 |
| Nickel-chromium | 900 |
| Fused silica | 1000 |
| Special steel | 1100 |
| Mullite | 1700 |
| Recrystallized alumina | 1850 |
| Beryllia | 2300 |
| Magnesia | 2400 |
| Zicronia | 2400 |
| Thoria | 2600 |

*The maximum operating temperatures quoted assume oxidizing or neutral atmospheres. For operation in reducing atmospheres, the maximum allowable temperature is usually reduced.

Where they are prone to contamination, thermocouples have to be protected by enclosing them entirely in an insulated sheath. Some common sheath materials and their maximum operating temperatures are shown in Table 14.1. While the thermocouple is a device that has a naturally first-order type of step response characteristic, the time constant is usually so small as to be negligible when the thermocouple is used unprotected. However, when enclosed in a sheath, the time constant of the combination of thermocouple and sheath is significant. The size of the thermocouple and hence the diameter required for the sheath have a large effect on the importance of this. The time constant of a thermocouple in a 1-mm-diameter sheath is only 0.15 s and this has little practical effect in most measurement situations, whereas a larger sheath of 6 mm diameter gives a time constant of 3.9 s that cannot be ignored so easily.

### *14.2.5 Thermocouple Manufacture*

Thermocouples are manufactured by connecting together two wires of different materials, where each material is produced so as to conform precisely with some defined composition specification. This ensures that its thermoelectric behavior accurately follows that for which standard thermocouple tables apply. The connection between the two wires is affected by welding, soldering, or, in some cases, just by twisting the wire ends together. Welding is the most common technique used generally, with silver soldering being reserved for copper–constantan devices.

The diameter of wire used to construct thermocouples is usually in the range between 0.4 and 2 mm. Larger diameters are used where ruggedness and long life are required, although these advantages are gained at the expense of increasing the measurement time constant. In the case of noble metal thermocouples, the use of large diameter wire incurs a substantial cost penalty. Some special applications have a requirement for a very fast response time in the measurement of temperature, and in such cases wire diameters as small as 0.1 μm can be used.

### *14.2.6 Thermopile*

The thermopile is the name given to a temperature-measuring device that consists of several thermocouples connected together in series, such that all the reference junctions are at the same cold temperature and all the hot junctions are exposed to the temperature being measured, as shown in Figure 14.7. The effect of connecting $n$ thermocouples together in series is to increase the measurement sensitivity by a factor of $n$. A typical thermopile manufactured by connecting together 25 chromel–constantan thermocouples gives a measurement resolution of 0.001°C.

### *14.2.7 Digital Thermometer*

Thermocouples are also used in digital thermometers, of which both simple and intelligent versions exist (for a description of the latter, see Section 14.12). A simple digital thermometer is a combination of a thermocouple, a battery-powered, dual-slope digital voltmeter to measure the thermocouple output, and an electronic display. This provides a low noise, digital output that can resolve temperature differences as small as 0.1°C. The accuracy achieved is dependent on the accuracy of the thermocouple element, but reduction of measurement inaccuracy to ±0.5% is achievable.

### *14.2.8 Continuous Thermocouple*

The continuous thermocouple is one of a class of devices that detect and respond to heat. Other devices in this class include the *line-type heat detector* and *heat-sensitive cable*. The basic construction of all these devices consists of two or more strands of wire separated by insulation within a long thin cable. While they sense temperature, they do not in fact provide an output measurement of temperature. Their function is to respond to abnormal temperature rises and thus prevent fires, equipment damage, etc.

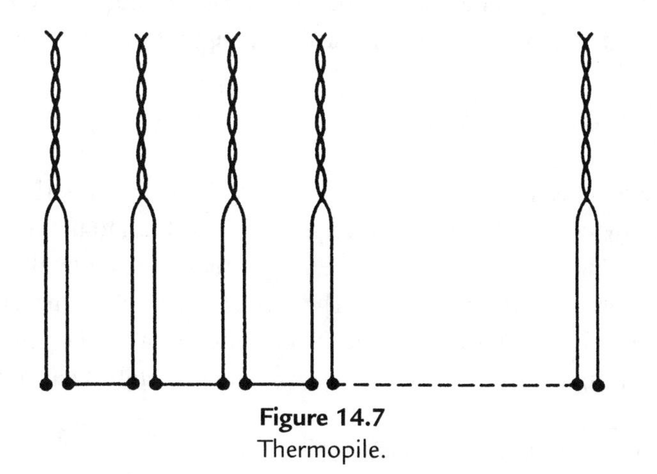

**Figure 14.7**
Thermopile.

The advantages of continuous thermocouples become more apparent if problems with other types of heat detectors are considered. Insulation in the line-type heat detector and heat-sensitive cable consists of plastic or ceramic material with a negative temperature coefficient (i.e., the resistance falls as the temperature rises). An alarm signal can be generated when the measured resistance falls below a certain level. Alternatively, in some versions, the insulation is allowed to break down completely, in which case the device acts as a switch. The major limitation of these devices is that the temperature change has to be relatively large, typically 50–200°C above ambient temperature, before the device responds. Also, it is not generally possible for such devices to give an output that indicates that an alarm condition is developing before it actually happens, and thus allow preventative action. Furthermore, after the device has generated an alarm it usually has to be replaced. This is particularly irksome because there is a large variation in the characteristics of detectors coming from different batches and so replacement of the device requires extensive onsite recalibration of the system.

In contrast, the continuous thermocouple suffers from very few of these problems. It differs from other types of heat detectors in that the two strands of wire inside it are a pair of thermocouple materials[‡] separated by a special, patented mineral insulation and contained within a stainless-steel protective sheath. If any part of the cable is subjected to heat, the resistance of the insulation at that point is reduced and a "hot junction" is created between the two wires of dissimilar metals. An e.m.f. is generated at this hot junction according to normal thermoelectric principles.

The continuous thermocouple can detect temperature rises as small as 1°C above normal. Unlike other types of heat detectors, it can also monitor abnormal rates of temperature rise and provide a warning of alarm conditions developing before they actually happen. Replacement is only necessary if a great degree of insulation breakdown has been caused by a substantial hot spot at some point along the detector's length. Even then, the use of thermocouple materials of standard characteristics in the detector means that recalibration is not needed if it is replaced. Because calibration is not affected either by cable length, a replacement cable may be of a different length to the one it is replacing. One further advantage of continuous thermocouples over earlier forms of heat detectors is that no power supply is needed, thus significantly reducing installation costs.

## 14.3 Varying Resistance Devices

Varying resistance devices rely on the physical principle of the variation of resistance with temperature. The devices are known as either resistance thermometers or thermistors according to whether the material used for their construction is a metal or a semiconductor, and both are common measuring devices. The normal method of measuring resistance is to use a d.c. bridge. The excitation voltage of the bridge has to be chosen very carefully because, although a high value is desirable for achieving high measurement sensitivity,

‡ Normally type E, chromel–constantan, or type K, chromel–alumel.

the self-heating effect of high currents flowing in the temperature transducer creates an error by increasing the temperature of the device and so changing the resistance value.

### *14.3.1 Resistance Thermometers (Resistance Temperature Devices)*

Resistance thermometers, which are alternatively known as *resistance temperature devices*, rely on the principle that the resistance of a metal varies with temperature according to the relationship:

$$R = R_0\left(1 + a_1 T + a_2 T^2 + a_3 T^3 + \cdots + a_n T^n\right). \tag{14.7}$$

This equation is nonlinear and so is inconvenient for measurement purposes. The equation becomes linear if all the terms in $a_2T^2$ and higher powers of $T$ are negligible such that the resistance and temperature are related according to

$$R \approx R_0(1 + a_1 T).$$

This equation is approximately true over a limited temperature range for some metals, notably platinum, copper, and nickel, whose characteristics are summarized in Figure 14.8. Platinum has the most linear resistance/temperature characteristic and also has good chemical inertness. It is therefore far more common than copper or nickel thermocouples. Its resistance–temperature relationship is linear within $\pm 0.4\%$ over the temperature range between $-200$ and $+40$°C. Even at $+1000$°C, the quoted inaccuracy figure is only $\pm 1.2\%$. Platinum thermometers are made in three forms, as a film deposited on a ceramic substrate, as a coil mounted inside a glass or ceramic probe, or as a coil wound on a mandrel, although the last of these are now becoming rare. The nominal resistance at 0°C is typically 100 or 1000 Ω, although 200 and 500 Ω versions also exist. Sensitivity is 0.385 Ω/°C (100 Ω type) or 3.85 Ω/°C (1000 Ω type). A high nominal resistance is advantageous in terms of higher measurement sensitivity, and the resistance of connecting leads has less effect on measurement accuracy. However, cost goes up as the nominal resistance increases.

In addition to having a less linear characteristic, both nickel and copper are inferior to platinum in terms of their greater susceptibility to oxidation and corrosion. This seriously limits their accuracy and longevity. However, because platinum is very expensive compared to nickel and copper, the latter are used in resistance thermometers when cost is important. Another metal, tungsten, is also used in resistance thermometers in some circumstances, particularly for high temperature measurements. The working ranges of each of these four types of resistance thermometers are as shown here:

Platinum: $-270$ to $+1000$°C (although use above 650°C is uncommon)
Copper: $-200$ to $+260$°C
Nickel: $-200$ to $+430$°C
Tungsten: $-270$ to $+1100$°C

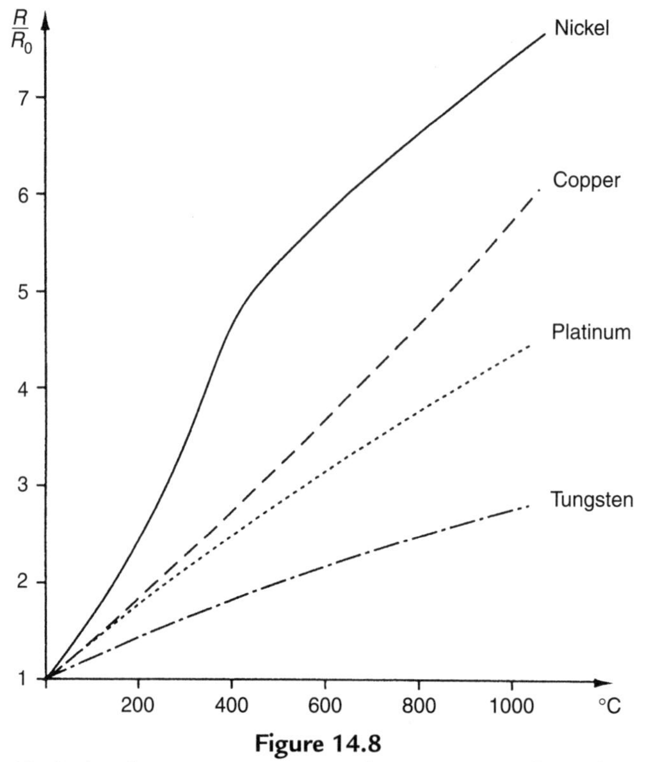


**Figure 14.8**
Typical resistance-temperature characteristics of metals.

In the case of noncorrosive and nonconducting environments, resistance thermometers are used without protection. In all other applications, they are protected inside a sheath. As in the case of thermocouples, such protection reduces the speed of response of the system to rapid changes in temperature. A typical time constant for a sheathed platinum resistance thermometer is 0.4 seconds. Moisture buildup within the sheath can also impair measurement accuracy.

The frequency at which a resistance thermometer should be calibrated depends on the material it is made from and on the operating environment. Practical experimentation is therefore needed to determine the necessary frequency and this must be reviewed if the operating conditions change.

### *14.3.2 Thermistors*

Thermistors are manufactured from beads of semiconductor material prepared from oxides of the iron group of metals such as chromium, cobalt, iron, manganese, and nickel. Normally, thermistors have a negative temperature coefficient, that is, resistance decreases as temperature increases, according to:

$$R = R_0 e^{[\beta (1/T - 1/T_0)]}. \qquad (14.8)$$

This relationship is illustrated in Figure 14.9. However, alternative forms of heavily doped thermistors are now available (at greater cost) that have a positive temperature coefficient. The form of Equation (14.8) is such that it is not possible to make a linear approximation to the curve over even a small temperature range, and hence the thermistor is very definitely a nonlinear sensor. However, the major advantages of thermistors are their relatively low cost and their small size. This size advantage means that the time constant of thermistors operated in sheaths is small, although the size reduction also decreases its heat dissipation capability and so makes the self-heating effect greater. In consequence, thermistors have to be operated at generally lower current levels than resistance thermometers and so the measurement sensitivity is less.

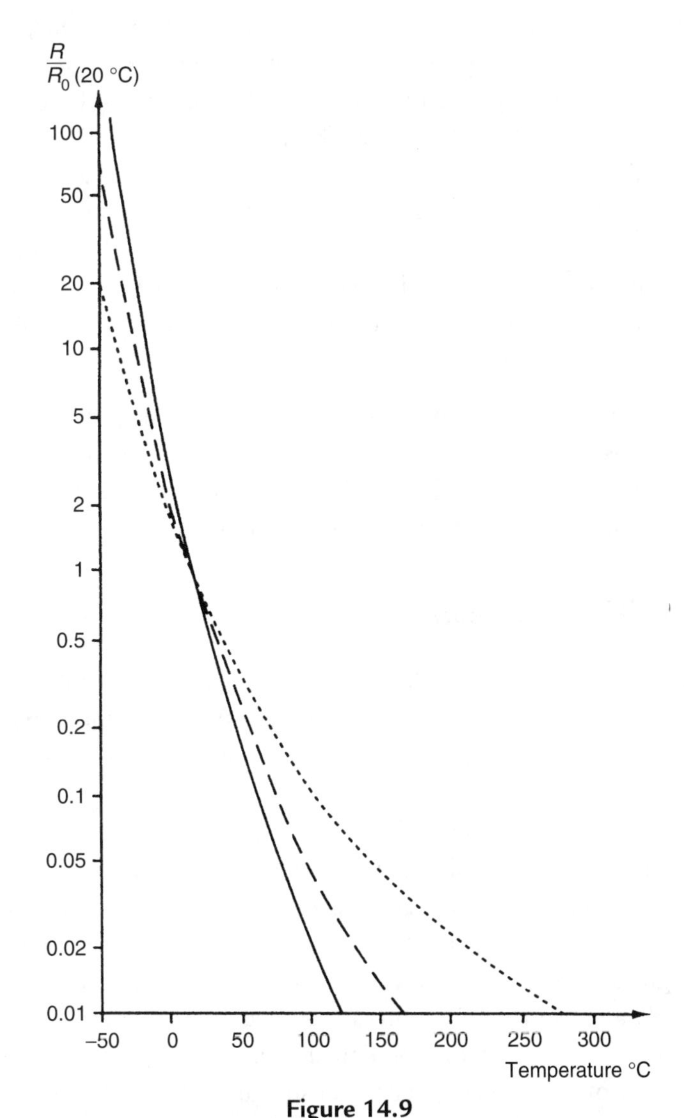


**Figure 14.9**
Typical resistance-temperature characteristics of thermistor materials.

As in the case of resistance thermometers, some practical experimentation is needed to determine the necessary frequency at which a thermistor should be calibrated and this must be reviewed if the operating conditions change.

## 14.4 Semiconductor Devices

Semiconductor devices, consisting of either diodes or integrated circuit transistors, have only been commonly used in industrial applications for a few years, but they were first invented several decades ago. They have the advantage of being relatively inexpensive, but one difficulty that affects their use is the need to provide an external power supply to the sensor.

Integrated circuit transistors produce an output proportional to the absolute temperature. Different types are configured to give an output in the form of either a varying current (typically 1 μA°K) or a varying voltage (typically 10 mV°K). Current forms are normally used with a digital voltmeter that detects the current output in terms of the voltage drop across a 10-KΩ resistor. Although the devices have a very low cost (typically a few dollars) and a better linearity than either thermocouples or resistance thermometers, they only have a limited measurement range from −50 to +150°C. Their inaccuracy is typically ±3%, which limits their range of application. However, they are widely used to monitor pipes and cables, where their low cost means that it is feasible to mount multiple sensors along the length of the pipe/cable to detect hot spots.

In diodes, the forward voltage across the device varies with temperature. Output from a typical diode package is in the microamp range. Diodes have a small size, with good output linearity and typical inaccuracy of only ±0.5%. Silicon diodes cover the temperature range from −50 to +200°C and germanium ones from −270 to +40°C.

## 14.5 Radiation Thermometers

All objects emit electromagnetic radiation as a function of their temperature above absolute zero, and radiation thermometers (also known as radiation pyrometers) measure this radiation in order to calculate the temperature of the object. The total rate of radiation emission per second is given by

$$E = KT^4. \tag{14.9}$$

The power spectral density of this emission varies with temperature in the manner shown in Figure 14.10. The major part of the frequency spectrum lies within the band of wavelengths between 0.3 and 40 μm, which corresponds to visible (0.3−0.72 μm) and infrared (0.72−1000 μm) ranges. As the magnitude of the radiation varies with temperature, measurement of the emission from a body allows the temperature of the body to be calculated. Choice of the best method of measuring the emitted radiation depends on the temperature of the body. At low temperatures, the peak of the power spectral density function (Figure 14.10) lies in the

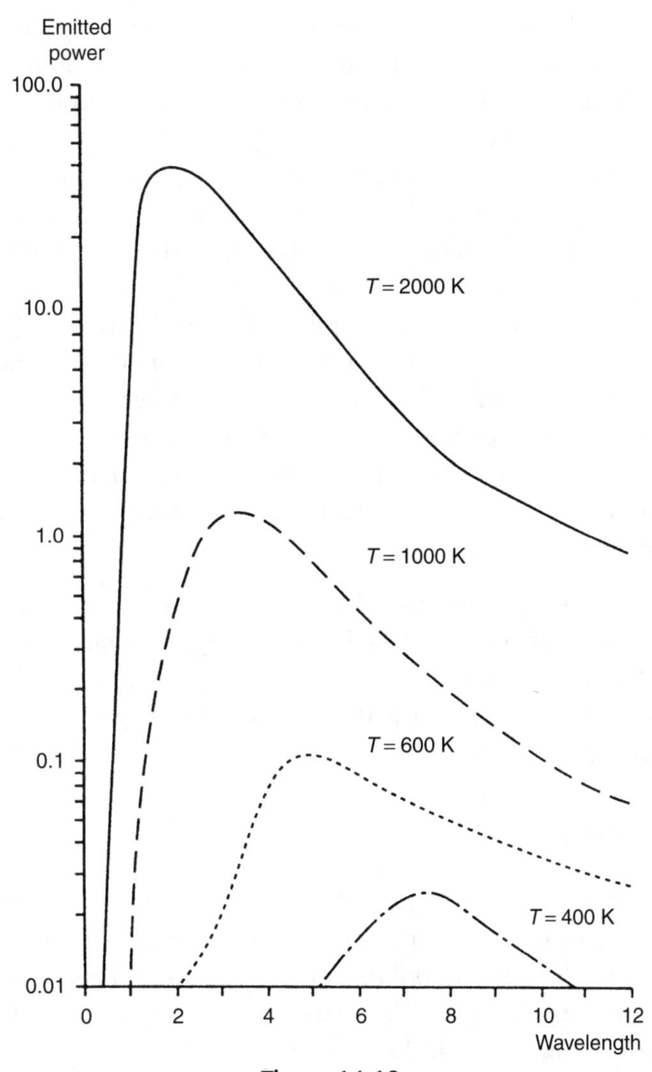


**Figure 14.10**

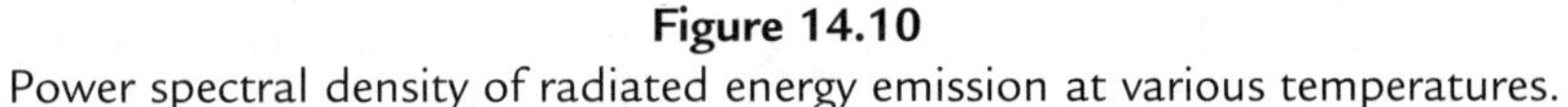
Power spectral density of radiated energy emission at various temperatures.

infrared region, whereas at higher temperatures it moves toward the visible part of the spectrum. This phenomenon is observed as the red glow that a body begins to emit as its temperature is increased beyond 600°C.

Different versions of radiation thermometers are capable of measuring temperatures between −100 and +10,000°C with measurement inaccuracy as low as ±0.05% in the more expensive versions (although this level of accuracy is not obtained when measuring very high temperatures). Portable, battery-powered, hand-held versions are also available, and these are particularly easy to use. The important advantage that radiation thermometers have over other types of temperature-measuring instruments is that there is no contact with the hot body while its temperature is being measured. Thus, the measured system is not disturbed in any way. Furthermore, there is no possibility of contamination, which is particularly important in food, drug, and many other process industries. They are especially suitable for measuring high temperatures beyond the capabilities of contact instruments such as thermocouples, resistance thermometers, and thermistors. They are also capable of measuring moving bodies, for instance, the temperature of steel bars in a rolling mill. Their use is not as straightforward as the discussion so far might have suggested, however, because the radiation from a body varies with the composition and surface condition of the body, as well as with temperature. This dependence on surface condition is quantified by the *emissivity* of the body. The use of radiation thermometers is further complicated by absorption and scattering of the energy between the emitting body and the radiation detector. Energy is scattered by atmospheric dust and water droplets and is absorbed by carbon dioxide, ozone, and water vapor molecules. Therefore, all radiation thermometers have to be calibrated carefully for each particular body whose temperature they are required to monitor.

Various types of radiation thermometers exist, as described next. The optical pyrometer can only be used to measure high temperatures, but various types of radiation pyrometers are available that, between them, cover the whole temperature spectrum. Intelligent versions (see Section 14.12) also now provide full or partial solutions to many of the problems described later for nonintelligent pyrometers.

### 14.5.1 Optical Pyrometer

The optical pyrometer, illustrated in Figure 14.11, is designed to measure temperatures where the peak radiation emission is in the red part of the visible spectrum, that is, where the measured body glows a certain shade of red according to the temperature. This limits the instrument to measuring temperatures above 600°C. The instrument contains a heated tungsten filament within its optical system. The current in the filament is increased until its color is the same as the hot body: under these conditions the filament apparently disappears when viewed against the background of the hot body. Temperature measurement is therefore obtained in terms of the current flowing in the filament. As the brightness of different materials at any particular temperature varies according to the emissivity of the material, the calibration of the optical pyrometer must be adjusted according to the emissivity of the target. Manufacturers provide tables of standard material emissivities to assist with this.

The inherent measurement inaccuracy of an optical pyrometer is ±5°C. However, in addition to this error, there can be a further operator-induced error of ±10°C arising out of the difficulty in judging the moment when the filament "just" disappears. Measurement accuracy can be improved somewhat by employing an optical filter within the instrument that passes a narrow band of frequencies of wavelength around 0.65 μm corresponding to the red part of the visible

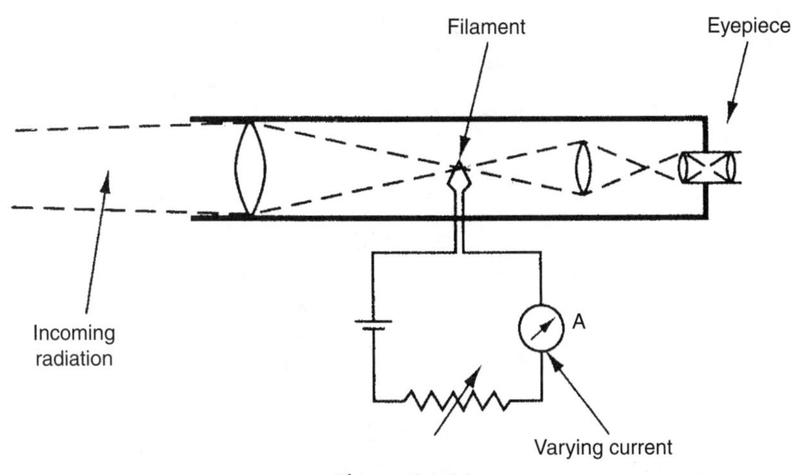


**Figure 14.11**
Optical pyrometer.

spectrum. This also extends the upper temperature measurable from 5000°C in unfiltered instruments up to 10,000°C

The instrument cannot be used in automatic temperature control schemes because the eye of the human operator is an essential part of the measurement system. The reading is also affected by fumes in the sight path. Because of these difficulties and its low accuracy, hand-held radiation pyrometers are rapidly overtaking the optical pyrometer in popularity, although the instrument is still widely used in industry for measuring temperatures in furnaces and similar applications at present.

### *14.5.2 Radiation Pyrometers*

All the alternative forms of radiation pyrometers described here have an optical system similar to that of the optical pyrometer and focus the energy emitted from the measured body. However, they differ by omitting the filament and eyepiece and having instead an energy detector in the same focal plane as the eyepiece was, as shown in Figure 14.12. This principle can be used to measure temperature over a range from −100 to +3600°C The radiation detector is either a thermal detector, which measures the temperature rise in a black body at the focal point of the optical system, or a photon detector.

Thermal detectors respond equally to all wavelengths in the frequency spectrum and consist of thermopiles, resistance thermometers, or thermistors. All of these typically have time constants of several milliseconds because of the time taken for the black body to heat up and the temperature sensor to respond to the temperature change.

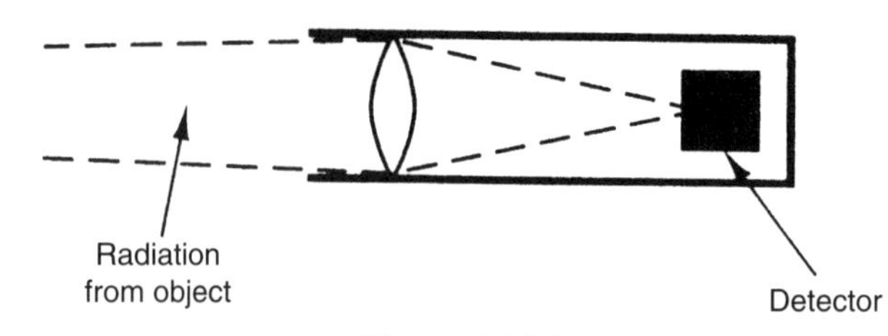


**Figure 14.12**
Structure of a radiation thermometer.

Photon detectors respond selectively to a particular band within the full spectrum and are usually of the photoconductive or photovoltaic type. They respond to temperature changes much faster than thermal detectors because they involve atomic processes, and typical measurement time constants are a few microseconds.

Fiber-optic technology is used frequently in high temperature measurement applications to collect the incoming radiation and transmit it to a detector and processing electronics that are located remotely. This prevents exposure of the processing electronics to potentially damaging, high temperature. Fiber-optic cables are also used to apply radiation pyrometer principles in very difficult applications, such as measuring the temperature inside jet engines by collecting the radiation from inside the engine and transmitting it outside (see Section 14.9).

The size of objects measured by a radiation pyrometer is limited by the optical resolution, which is defined as the ratio of target size to distance. A good ratio is 1:300, which would allow temperature measurement of a 1-mm-sized object at a range of 300 mm. With large distance/target size ratios, accurate aiming and focusing of the pyrometer at the target are essential. It is now common to find "through the lens" viewing provided in pyrometers, using a principle similar to SLR camera technology, as focusing and orientating the instrument for visible light focuses it automatically for infrared light. Alternatively, dual laser beams are sometimes used to ensure that the instrument is aimed correctly toward the target.

Various forms of electrical output are available from the radiation detector: these are functions of the incident energy on the detector and are therefore functions of the temperature of the measured body. While this therefore makes such instruments of use in automatic control systems, their accuracy is often inferior to optical pyrometers. This reduced accuracy arises first because a radiation pyrometer is sensitive to a wider band of frequencies than the optical instrument and the relationship between emitted energy and temperature is less well defined. Second, the magnitude of energy emission at low temperatures gets very small, according to Equation (14.9), increasing the difficulty of accurate measurement.

The forms of radiation pyrometer described here differ mainly in the technique used to measure the emitted radiation. They also differ in the range of energy wavelengths, and hence the temperature range, which each is designed to measure. One further difference is the material used to construct the energy-focusing lens. Outside the visible part of the spectrum, glass becomes almost opaque to infrared wavelengths, and other lens materials such as arsenic trisulfide are used.

### *Broad-band (unchopped) radiation pyrometers*

The broad-band radiation pyrometer finds wide application in industry and has a measurement inaccuracy that varies from ±0.05% of full scale in the best instruments to ±0.5% in the least expensive. However, their accuracy deteriorates significantly over a period of time, and an error of 10°C is common after 1–2 years operation at high temperatures. As its name implies, the instrument measures radiation across the whole frequency spectrum and so uses a thermal detector. This consists of a blackened platinum disc to which a thermopile[§] is bonded. The temperature of the detector increases until the heat gain from the incident radiation is balanced by the heat loss due to convection and radiation. For high-temperature measurement, a two-couple thermopile gives acceptable measurement sensitivity and has a fast time constant of about 0.1 s. At lower measured temperatures, where the level of incident radiation is much less, thermopiles constructed from a greater number of thermocouples must be used to get sufficient measurement sensitivity. This increases the measurement time constant to as much as 2 s. Standard instruments of this type are available to measure temperatures between −20 and +1800°C, although much higher temperatures in theory could be measured by this method.

### *Chopped broad-band radiation pyrometers*

Construction of this form of pyrometer is broadly similar to that shown in Figure 14.12 except that a rotary mechanical device is included that periodically interrupts the radiation reaching the detector. The voltage output from the thermal detector thus becomes an alternating quantity that switches between two levels. This form of a.c. output can be amplified much more readily than the d.c. output coming from an unchopped instrument. This is particularly important when amplification is necessary to achieve an acceptable measurement resolution in situations where the level of incident radiation from the measured body is low. For this reason, this form of instrument is the more common when measuring body temperatures associated with peak emission in the infrared part of the frequency spectrum. For such chopped systems, the time constant of thermopiles is too long. Instead, thermistors are generally used, giving a time constant of 0.01 s. Standard instruments of this type are available to measure temperatures between +20 and +1300°C. This form of pyrometer suffers similar accuracy drift to unchopped forms. Its life is also limited to about 2 years because of motor failures.

[§] Typically manganin–constantan.

### *Narrow-band radiation pyrometers*

Narrow-band radiation pyrometers are highly stable instruments that suffer a drift in accuracy that is typically only 1°C in 10 years. They are also less sensitive to emissivity changes than other forms of radiation pyrometers. They use photodetectors of either the photoconductive or the photovoltaic form whose performance is unaffected by either carbon dioxide or water vapor in the path between the target object and the instrument. A photoconductive detector exhibits a change in resistance as the incident radiation level changes, whereas a photovoltaic cell exhibits an induced voltage across its terminals that is also a function of the incident radiation level. All photodetectors are preferentially sensitive to a particular narrow band of wavelengths in the range of 0.5–1.2 μm and all have a form of output that varies in a highly nonlinear fashion with temperature, and thus a microcomputer inside the instrument is highly desirable. Four commonly used materials for photodetectors are cadmium sulfide, lead sulfide, indium antimonide, and lead tin telluride. Each of these is sensitive to a different band of wavelengths and therefore all find application in measuring the particular temperature ranges corresponding to each of these bands.

Output from the narrow band radiation pyrometer is normally chopped into an a.c. signal in the same manner as used in the chopped broad-band pyrometer. This simplifies amplification of the output signal, which is necessary to achieve an acceptable measurement resolution. The typical time constant of a photon detector is only 5 μs, which allows high chopping frequencies up to 20 kHz. This gives such instruments an additional advantage in being able to measure fast transients in temperature as short as 10 μs.

### *Two-color pyrometer (ratio pyrometer)*

As stated earlier, the emitted radiation–temperature relationship for a body depends on its emissivity. This is very difficult to calculate, and therefore in practice all pyrometers have to be calibrated to the particular body they are measuring. The two-color pyrometer (alternatively known as a ratio pyrometer) is a system that largely overcomes this problem by using the arrangement shown in Figure 14.13. Radiation from the body is split equally into two parts, which are applied to separate narrow-band filters. Outputs from the filters consist of radiation within two narrow bands of wavelengths $\lambda_1$ and $\lambda_2$. Detectors sensitive to these frequencies produce output voltages $V_1$ and $V_2$, respectively. The ratio of these outputs, $(V_1/V_2)$, can be shown (see Dixon, 1987) to be a function of temperature and to be independent of emissivity, provided that the two wavelengths, $\lambda_1$ and $\lambda_2$, are close together.

The theoretical basis of the two-color pyrometer is that output is independent of emissivity because emissivities at the two wavelengths $\lambda_1$ and $\lambda_2$ are equal. This is based on the assumption that $\lambda_1$ and $\lambda_2$ are very close together. In practice, this assumption does not hold and therefore the accuracy of the two-color pyrometer tends to be relatively poor. However, the instrument is still of great use in conditions where the target is obscured by fumes or dust, which is a common

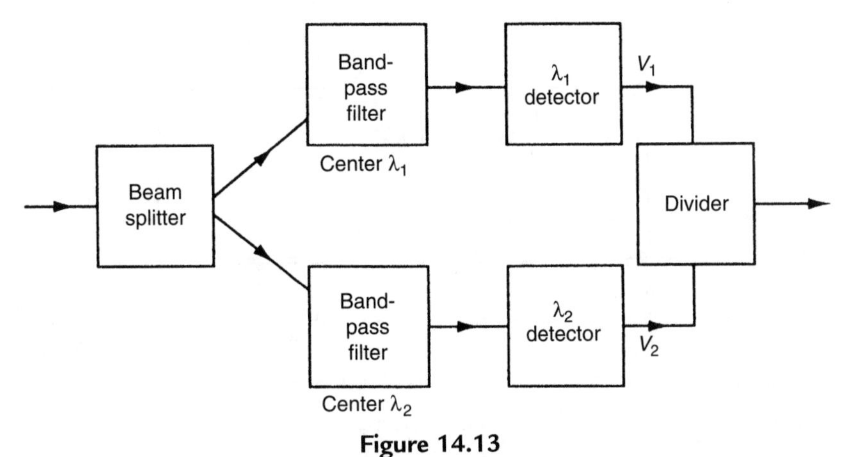


**Figure 14.13**
Two-color pyrometer system.

problem in the cement and mineral processing industries. Two-color pyrometers typically cost 50–100% more than other types of pyrometers.

*Selected waveband pyrometer*

The selected waveband pyrometer is sensitive to one waveband only, for example, 5 μm, and is dedicated to particular, special situations where other forms of pyrometers are inaccurate. One example of such a situation is measuring the temperature of steel billets being heated in a furnace. If an ordinary radiation pyrometer is aimed through the furnace door at a hot billet, it receives radiation from the furnace walls (by reflection off the billet) as well as radiation from the billet itself. If the temperature of the furnace walls is measured by a thermocouple, a correction can be made for the reflected radiation, but variations in transmission losses inside the furnace through fumes and so on make this correction inaccurate. However, if a carefully chosen selected waveband pyrometer is used, this transmission loss can be minimized and the measurement accuracy is thereby improved greatly.

## 14.6 Thermography (Thermal Imaging)

Thermography, or thermal imaging, involves scanning an infrared radiation detector across an object. The information gathered is then processed and an output in the form of the temperature distribution across the object is produced. Temperature measurement over the range from −20°C up to +1500°C is possible. Elements of the system are shown in Figure 14.14.

The radiation detector uses the same principles of operation as a radiation pyrometer in inferring the temperature of the point that the instrument is focused on from a measurement of the incoming infrared radiation. However, instead of providing a measurement of the

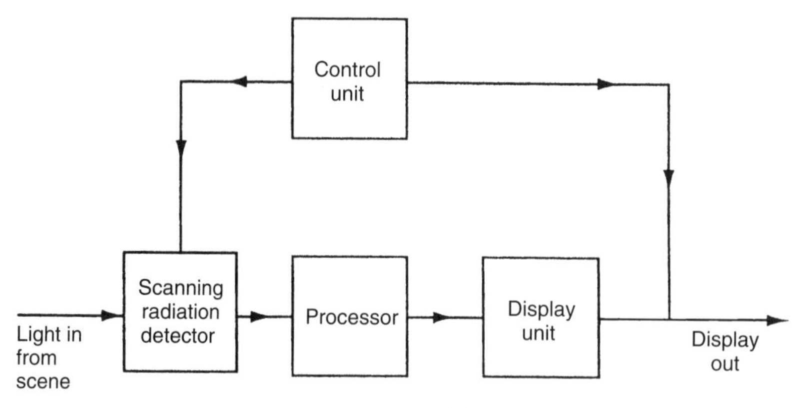


**Figure 14.14**
Thermography (thermal imaging) system.

temperature of a single point at the focal point of the instrument, the detector is scanned across a body or scene, and thus provides information about temperature distributions. Because of the scanning mode of operation of the instrument, radiation detectors with a very fast response are required, and only photoconductive or photovoltaic sensors are suitable. These are sensitive to the portion of the infrared spectrum between wavelengths of 2 and 14 μm.

Simpler versions of thermal imaging instruments consist of hand-held viewers that are pointed at the object of interest. The output from an array of infrared detectors is directed onto a matrix of red light-emitting diodes assembled behind a glass screen, and the output display thus consists of different intensities of red on a black background, with the different intensities corresponding to different temperatures. Measurement resolution is high, with temperature differences as small as 0.1°C being detectable. Such instruments are used in a wide variety of applications, such as monitoring product flows through pipe work, detecting insulation faults, and detecting hot spots in furnace linings, electrical transformers, machines, bearings, etc. The number of applications is extended still further if the instrument is carried in a helicopter, where uses include scanning electrical transmission lines for faults, searching for lost or injured people, and detecting the source and spread pattern of forest fires.

More complex thermal imaging systems comprise a tripod-mounted detector connected to a desktop computer and display system. Multicolor displays are used commonly in such systems, where up to 16 different colors represent different bands of temperature across the measured range. The heat distribution across the measured body or scene is thus displayed graphically as a contoured set of colored bands representing the different temperature levels. Such color thermography systems find many applications, such as inspecting electronic circuit boards and monitoring production processes. There are also medical applications in body scanning.

## 14.7 Thermal Expansion Methods

Thermal expansion methods make use of the fact that the dimensions of all substances, whether solids, liquids, or gases, change with temperature. Instruments operating on this physical principle include the liquid-in-glass thermometer, bimetallic thermometer, and pressure thermometer.

### *14.7.1 Liquid-in-Glass Thermometers*

The liquid-in-glass thermometer is a well-known temperature-measuring instrument used in a wide range of applications. The fluid used is normally either mercury or colored alcohol, which is contained within a bulb and capillary tube, as shown in Figure 14.15a. As the temperature rises, the fluid expands along the capillary tube and the meniscus level is read against a calibrated scale etched on the tube. Industrial versions of the liquid-in-glass thermometer are

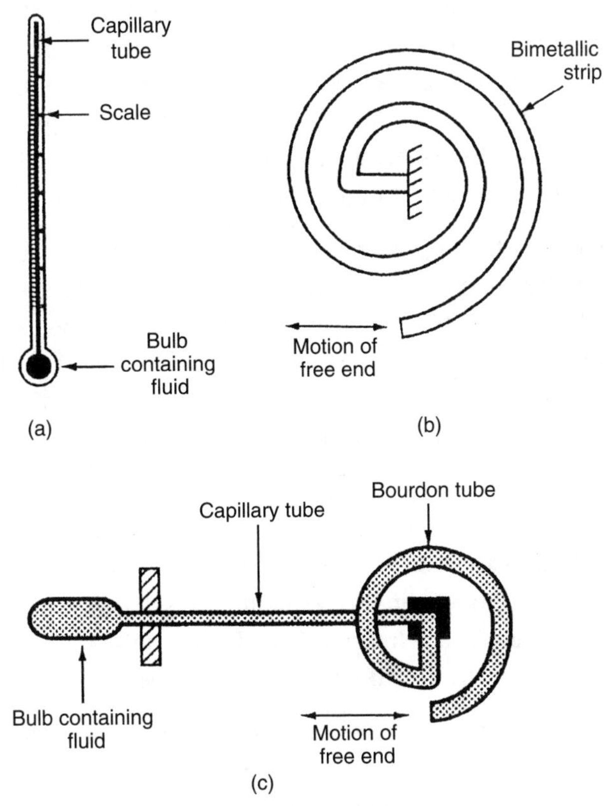

**Figure 14.15**
Thermal expansion devices: (a) liquid-in-glass thermometer, (b) bimetallic thermometer, and (c) pressure thermometer.

normally used to measure temperature in the range between −200 and +1000°C, although instruments are available to special order that can measure temperatures up to 1500°C.

Measurement inaccuracy is typically ±1% of full-scale reading, although an inaccuracy of only ±0.15% can be achieved in the best industrial instruments. The major source of measurement error arises from the difficulty of correctly estimating the position of the curved meniscus of the fluid against the scale. In the longer term, additional errors are introduced due to volumetric changes in the glass. Such changes occur because of creep-like processes in the glass, but occur only over a timescale of years. Annual calibration checks are therefore advisable.

### 14.7.2 Bimetallic Thermometer

The bimetallic principle is probably more commonly known in connection with its use in thermostats. It is based on the fact that if two strips of different metals are bonded together, any temperature change will cause the strip to bend, as this is the only way in which the differing rates of change of length of each metal in the bonded strip can be accommodated. In the bimetallic thermostat, this is used as a switch in control applications. If the magnitude of bending is measured, the bimetallic device becomes a thermometer. For such purposes, the strip is often arranged in a spiral or helical configuration, as shown in Figure 14.15b, as this gives a relatively large displacement of the free end for any given temperature change. The measurement sensitivity is increased further by choosing the pair of materials carefully such that the degree of bending is maximized, with Invar (a nickel-steel alloy) or brass being used commonly.

The system used to measure the displacement of the strip must be designed carefully. Very little resistance must be offered to the end of the strip, as otherwise the spiral or helix will distort and cause a false reading in measurement of the displacement. The device is normally just used as a temperature indicator, where the end of the strip is made to turn a pointer that moves against a calibrated scale. However, some versions produce an electrical output, using either a linear variable differential transformer or a fiber-optic shutter sensor to transduce the output displacement.

Bimetallic thermometers are used to measure temperatures between −75 and +1500°C. The inaccuracy of the best instruments can be as low as ±0.5% but such devices are quite expensive. Many instrument applications do not require this degree of accuracy in temperature measurements, and in such cases much less expensive bimetallic thermometers with substantially inferior accuracy specifications are used.

All such devices are liable to suffer changes in characteristics due to contamination of the metal components exposed to the operating environment. Further changes are to be expected arising from mechanical damage during use, particularly if they are mishandled or dropped. As the magnitude of these effects varies with their application, the required calibration interval must be determined by practical experimentation.

### 14.7.3 Pressure Thermometers

Pressure thermometers have now been superseded by other alternatives in most applications, but they still remain useful in a few applications such as furnace temperature measurement when the level of fumes prevents the use of optical or radiation pyrometers. Examples can also still be found of their use as temperature sensors in pneumatic control systems. The sensing element in a pressure thermometer consists of a stainless-steel bulb containing a liquid or gas. If the fluid were not constrained, temperature rises would cause its volume to increase. However, because it is constrained in a bulb and cannot expand, its pressure rises instead. As such, the pressure thermometer does not strictly belong to the thermal expansion class of instruments but is included because of the relationship between volume and pressure according to Boyle's law: $PV = KT$. The change in pressure of the fluid is measured by a suitable pressure transducer, such as the Bourdon tube (see Chapter 15). This transducer is located remotely from the bulb and is connected to it by a capillary tube as shown in Figure 14.15c.

Pressure thermometers can be used to measure temperatures in the range between $-250$ and $+2000$°C, and their typical inaccuracy is $\pm 0.5\%$ of full-scale reading. However, the instrument response has a particularly long time constant.

The need to protect the pressure-measuring instrument from the environment where the temperature is being measured can require the use of capillary tubes up to 5 m long, and the temperature gradient, and hence pressure gradient, along the tube acts as a modifying input that can introduce a significant measurement error. Errors also occur in the short term due to mechanical damage and in the longer term due to small volumetric changes in the glass components. The rate of increase in these errors is mainly use related and therefore the required calibration interval must be determined by practical experimentation.

## 14.8 Quartz Thermometers

The quartz thermometer makes use of the principle that the resonant frequency of a material such as quartz is a function of temperature, and thus enables temperature changes to be translated into frequency changes. The temperature-sensing element consists of a quartz crystal enclosed within a probe (sheath). The probe usually consists of a stainless-steel cylinder, which makes the device physically larger than devices such as thermocouples and resistance thermometers. The crystal is connected electrically so as to form the resonant element within an electronic oscillator. Measurement of the oscillator frequency therefore allows the measured temperature to be calculated.

The instrument has a very linear output characteristic over the temperature range between $-40$ and $+230$°C, with a typical inaccuracy of $\pm 0.1\%$. Measurement resolution is typically 0.1°C but versions can be obtained with resolutions as small as 0.0003°C. The characteristics of the instrument are generally very stable over long periods of time and therefore only infrequent calibration is necessary. The frequency change form of output means that the device is insensitive to noise. However, it is very expensive and only available from a small number of manufacturers.

## 14.9 Fiber-Optic Temperature Sensors

Fiber-optic cables can be used as either intrinsic or extrinsic temperature sensors, as discussed in Chapter 13, although special attention has to be paid to providing a suitable protective coating when high temperatures are measured. Cost varies from $1000 to $5000, according to type, and the normal temperature range covered is 250 to 3000°C, although special devices can detect down to 100°C and others can detect up to 3600°C. Their main application is measuring temperatures in hard-to-reach locations, although they are also used when very high measurement accuracy is required. Some laboratory versions have an inaccuracy as low as ±0.01%, which is better than a type S thermocouple, although versions used in industry have a more typical inaccuracy of ±1.0%.

While it is often assumed that fiber-optic sensors are intrinsically safe, it has been shown (Johnson, 1994) that flammable gas may be ignited by the optical power levels available from some laser diodes. Thus, the power level used with optical fibers must be chosen carefully, and certification of intrinsic safety is necessary if such sensors are to be used in hazardous environments.

One type of intrinsic sensor uses cable where the core and cladding have similar refractive indices but different temperature coefficients. Temperature rises cause the refractive indices to become even closer together and losses from the core to increase, thus reducing the quantity of light transmitted. Other types of intrinsic temperature sensors include the cross-talk sensor, phase-modulating sensor, and optical resonator, as described in Chapter 13. Research into the use of distributed temperature sensing using fiber-optic cable has also been reported. This can be used to measure things such as the temperature distribution along an electricity supply cable. It works by measuring the reflection characteristics of light transmitted down a fiber-optic cable bonded to the electrical cable. By analyzing back-scattered radiation, a table of temperature versus distance along the cable can be produced, with a measurement inaccuracy of only ±0.5°C.

A common form of extrinsic sensor uses fiber-optic cables to transmit light from a remote targeting lens into a standard radiation pyrometer. This technique can be used with all types of radiation pyrometers, including the two-color version, and a particular advantage is that this method of measurement is intrinsically safe. However, it is not possible to measure very low temperatures because the very small radiation levels that exist at low temperatures are badly attenuated during transmission along the fiber-optic cable. Therefore, the minimum temperature that can be measured is about 50°C, and the light guide for this must not exceed 600 mm in length. At temperatures exceeding 1000°C, lengths of fiber up to 20 m long can be used successfully as a light guide.

One extremely accurate device that uses this technique is known as the Accufibre sensor. This is a form of radiation pyrometer that has a black box cavity at the focal point of the lens system. A fiber-optic cable is used to transmit radiation from the black box cavity to a spectrometric device that computes the temperature. This has a measurement range of 500 to 2000°C, a resolution of $10^{-5}$°C, and an inaccuracy of only ±0.0025% of full scale.

Several other types of devices marketed as extrinsic fiber-optic temperature sensors consist of a conventional temperature sensor (e.g., a resistance thermometer) connected to a fiber-optic cable so that transmission of the signal from the measurement point is free of noise. Such devices must include an electricity supply for the electronic circuit needed to convert the sensor output into light variations in the cable. Thus, low-voltage power cables must be routed with the fiber-optic cable, and the device is therefore not intrinsically safe.

## 14.10 Color Indicators

The color of various substances and objects changes as a function of temperature. One use of this is in the optical pyrometer as discussed earlier. The other main use of color change is in special color indicators that are widely used in industry to determine whether objects placed in furnaces have reached the required temperature. Such color indicators consist of special paints or crayons that are applied to an object before it is placed in a furnace. The color-sensitive component within these is some form of metal salt (usually of chromium, cobalt, or nickel). At a certain temperature, a chemical reaction takes place and a permanent color change occurs in the paint or crayon, although this change does not occur instantaneously but only happens over a period of time.

Hence, the color change mechanism is complicated by the fact that the time of exposure as well as the temperature is important. Such crayons or paints usually have a dual rating that specifies the temperature and length of exposure time required for the color change to occur. If the temperature rises above the rated temperature, then the color change will occur in less than the rated exposure time. This causes little problem if the rate of temperature rise is slow with respect to the specified exposure time required for color change to occur. However, if the rate of rise of temperature is high, the object will be significantly above the rated change temperature of the paint/crayon by the time that the color change happens. In addition to wasting energy by leaving the object in the furnace longer than necessary, this can also cause difficulty if excess temperature can affect the required metallurgical properties of the heated object.

Paints and crayons are available to indicate temperatures between 50 and 1250°C. A typical exposure time rating is 30 minutes, that is, the color change will occur if the paint/crayon is exposed to the rated temperature for this length of time. They have the advantage of low cost, typically a few dollars per application. However, they adhere strongly to the heated object, which can cause difficulty if they have to be cleaned off the object later.

Some liquid crystals also change color at a certain temperature. According to the design of sensors using such liquid crystals, the color change can either occur gradually during a temperature rise of perhaps 50°C or change abruptly at some specified temperature. The latter kinds of sensors are able to resolve temperature changes as small as 0.1°C and, according to type, are used over the temperature range from −20 to +100°C.

## 14.11 Change of State of Materials

Temperature-indicating devices known as Seger cones or pyrometric cones are used commonly in the ceramics industry. They consist of a fused oxide and glass material that is formed into a cone shape. The tip of the cone softens and bends over when a particular temperature is reached. Cones are available that indicate temperatures over the range from 600 to +2000°C.

## 14.12 Intelligent Temperature-Measuring Instruments

Intelligent temperature transmitters have now been introduced into the catalogues of almost all instrument manufacturers, and they bring about the usual benefits associated with intelligent instruments. Such transmitters are separate boxes designed for use with transducers that have either a d.c. voltage output in the millivolt range or an output in the form of a resistance change. They are therefore suitable for use in conjunction with thermocouples, thermopiles, resistance thermometers, thermistors, and broad-band radiation pyrometers. Transmitters normally have nonvolatile memories where all constants used in correcting output values for modifying inputs, etc., are stored, thus enabling the instrument to survive power failures without losing such information. Other facilities in intelligent transmitters include adjustable damping, noise rejection, self-adjustment for zero and sensitivity drifts, and expanded measurement range. These features allow an inaccuracy level of ±0.05% of full scale to be specified.

Mention must be made particularly of intelligent pyrometers, as some versions of these are able to measure the emissivity of the target body and automatically provide an emissivity-corrected output. This particular development provides an alternative to the two-color pyrometer when emissivity measurement and calibration for other types of pyrometers pose difficulty.

Digital thermometers (see Section 14.2.7) also exist in intelligent versions, where inclusion of a microprocessor allows a number of alternative thermocouples and resistance thermometers to be offered as options for the primary sensor.

The cost of intelligent temperature transducers is significantly more than their nonintelligent counterparts, and justification purely on the grounds of their superior accuracy is hard to make. However, their expanded measurement range means immediate savings are made in terms of the reduction in the number of spare instruments needed to cover a number of measurement ranges. Their capability for self-diagnosis and self-adjustment means that they require attention much less frequently, giving additional savings in maintenance costs. Many transmitters are also largely self-calibrating in respect of their signal processing function, although appropriate calibration routines still have to be applied to each sensor that the transmitter is connected to.

## 14.13 Choice between Temperature Transducers

The suitability of different instruments in any particular measurement situation depends substantially on whether the medium to be measured is a solid or a fluid. For measuring the temperature of solids, it is essential that good contact is made between the body and the transducer unless a radiation thermometer is used. This restricts the range of suitable transducers to thermocouples, thermopiles, resistance thermometers, thermistors, semiconductor devices, and color indicators. However, fluid temperatures can be measured by any of the instruments described in this chapter, with the exception of radiation thermometers.

The most commonly used device in industry for temperature measurement is the base metal thermocouple. This is relatively inexpensive, with prices varying widely from a few dollars upward according to the thermocouple type and sheath material used. Typical inaccuracy is ±0.5% of full scale over the temperature range of −250 to +1200°C. Noble metal thermocouples are much more expensive, but are chemically inert and can measure temperatures up to 2300°C with an inaccuracy of ±0.2% of full scale. However, all types of thermocouples have a low-level output voltage, making them prone to noise and therefore unsuitable for measuring small temperature differences.

Resistance thermometers are also in common use within the temperature range of −270 to +650°C, with a measurement inaccuracy of ±0.5%. While they have a smaller temperature range than thermocouples, they are more stable and can measure small temperature differences. The platinum resistance thermometer is generally regarded as offering the best ratio of price to performance for measurement in the temperature range of −200 to +500°C, with prices starting from $20.

Thermistors are another relatively common class of devices. They are small and inexpensive, with a typical cost of around $5. They give a fast output response to temperature changes, with good measurement sensitivity, but their measurement range is quite limited.

Semiconductor devices have a better linearity than thermocouples and resistance thermometers and a similar level of accuracy. Thus, they are a viable alternative to these in many applications.

Integrated circuit transistor sensors are particularly inexpensive (from $10 each), although their accuracy is relatively poor and they have a very limited measurement range (−50 to +150°C). Diode sensors are much more accurate and have a wider temperature range (−270 to +200°C), although they are also more expensive (typical costs are anywhere from $50 to $800).

A major virtue of radiation thermometers is their noncontact, noninvasive mode of measurement. Costs vary from $300 up to $5000 according to type. Although calibration for the emissivity of the measured object often poses difficulties, some instruments now provide automatic calibration. Optical pyrometers are used to monitor temperatures above 600°C in industrial furnaces, etc., but their inaccuracy is typically ±5%. Various forms of radiation pyrometers are used over the temperature range between −20 and +1800°C and can give measurement inaccuracies as low as ±0.05%. One particular merit of narrow-band radiation pyrometers is their ability to measure fast temperature transients of duration as small as 10 μs. No other instrument can measure transients anywhere near as fast as this.

The range of instruments working on the thermal expansion principle are used mainly as temperature-indicating devices rather than as components within automatic control schemes. Temperature ranges and costs are: mercury-in-glass thermometers up to +1000°C (cost from a few dollars), bimetallic thermometers up to +1500°C (cost $50 to $150), and pressure thermometers up to +2000°C (cost $100 to $800). The usual measurement inaccuracy is in the range of ±0.5 to ±1.0%. The bimetallic thermometer is more rugged than liquid-in-glass types but less accurate (however, the greater inherent accuracy of liquid-in-glass types can only be realized if the liquid meniscus level is read carefully).

Fiber-optic devices are more expensive than most other forms of temperature sensors (costing up to $6000) but provide a means of measuring temperature in very inaccessible locations. Inaccuracy varies from ±1% down to ±0.01% in some laboratory versions. Measurement range also varies with type, but up to +3600°C is possible.

The quartz thermometer provides very high resolution (0.0003°C is possible with special versions) but is expensive because of the complex electronics required to analyze the frequency-change form of output. It only operates over the limited temperature range of −40 to +230°C, but gives a low measurement inaccuracy of ±0.1% within this range. It is not used commonly because of its high cost.

Color indicators are used widely to determine when objects in furnaces have reached the required temperature. These indicators work well if the rate of rise of temperature of the object in the furnace is relatively slow but, because temperature indicators only change color over a period of time, the object will be above the required temperature by the time that the indicator responds if the rate of rise of temperature is large. Cost is low; for example, a crayon typically costs $5.

## 14.14 Calibration of Temperature Transducers

The fundamental difficulty in establishing an absolute standard for temperature has already been mentioned in the introduction to this chapter. This difficulty is that there is no practical way in which a convenient relationship can be established that relates the temperature of a body to another measurable quantity expressible in primary standard units. Instead, it is necessary to use a series of reference calibration points for temperatures that are very well defined. These points have been determined by research and international discussion and are published as the *International Practical Temperature Scale*. They provide fixed, reproducible reference points for temperature in the form of freezing points and triple points[‖] of substances where the transition among solid, liquid, and gaseous states is sharply defined. The full set of defined points is:

- Triple point of hydrogen[‖] −259.3467°C
- Triple point of neon[‖] −248.5939°C
- Triple point of oxygen[‖] −218.7916°C
- Triple point of argon[‖] −189.3442°C
- Triple point of mercury[‖] −38.8344°C
- Triple point of water[‖] +0.0100°C
- Melting point of gallium +29.7646°C
- Freezing point of indium +156.5985°C
- Freezing point of tin +231.928°C
- Freezing point of zinc +419.527°C
- Freezing point of aluminum +660.323°C
- Freezing point of silver +961.78°C
- Freezing point of gold +1064.18°C
- Freezing point of copper +1084.62°C

For calibrating intermediate temperatures, interpolation between fixed points is carried out by one of the following reference instruments:

- a helium gas thermometer for temperatures below 24.6°K
- a platinum resistance thermometer for temperatures between 13.8°K and 961.8°C
- a narrow-band radiation thermometer for temperatures above +961.8°C

The triple point method of defining fixed points involves use of a triple point cell. The cell consists of a sealed cylindrical glass tube filled with a highly pure version of the reference substance (e.g., mercury). This must be at least 99.9999 % pure (such that contamination is less

[‖] The triple point of a substance is the temperature and pressure at which the solid, liquid, and gas phases of that substance coexist in thermodynamic equilibrium. For example, in the case of water, the single combination of pressure and temperature at which solid ice, liquid water, and water vapor coexist in a stable equilibrium is a pressure of 611.73 millibars and a temperature of 273.16°K (0.01°C).

than one part in one million). The cell has a well that allows insertion of the thermometer being calibrated. It also has a valve that allows the cell to be evacuated down to the required triple point pressure.

The freezing point method of defining fixed points involves use of an ingot of the reference metal (e.g., tin) that is better than 99.99% pure. This is protected against oxidation inside a graphite crucible with a close-fitting lid. It is heated beyond its melting point and allowed to cool. If its temperature is monitored, an arrest period is observed in its cooling curve at the freezing point of the metal. The melting point method is similar but involves heating the material until it melts (this is only used for materials such as gallium where the melting point is defined more clearly than the freezing point). Electric resistance furnaces are available to carry out these procedures. Up to 1100°C, a measurement uncertainty of less than ±0.5°C is achievable.

The accuracy of temperature calibration procedures is fundamentally dependent on how accurately points on the IPTS can be reproduced. The present limits are:

| | |
|---|---|
| 1°K 0.3% | 800°K 0.001% |
| 10°K 0.1% | 1500°K 0.02% |
| 100°K 0.005% | 4000°K 0.2% |
| 273.15°K 0.0001% | 10,000°K 6.7% |

### *14.14.1 Reference Instruments and Special Calibration Equipment*

The primary reference standard instrument for calibration at the top of the calibration chain is a helium gas thermometer, a platinum resistance thermometer, or a narrow-band radiation thermometer according to the temperature range of the instrument being calibrated, as explained at the end of the last section. However, at lower levels within the calibration chain, almost any instrument from the list of instrument classes given in Section 14.1 may be used for workplace calibration duties in particular circumstances. Where involved in such duties, of course, the instrument used would be one of high accuracy that was reserved solely for calibration duties. The list of instruments suitable for workplace-level calibration therefore includes mercury-in-glass thermometers, base metal thermocouples (type K), noble metal thermocouples (types B, R, and S), platinum resistance thermometers, and radiation pyrometers. However, a subset of this is commonly preferred for most calibration operations. Up to 950°C, the platinum resistance thermometer is often used as a reference standard. Above that temperature, up to about 1750°C, a type S (platinum/rhodium–platinum) thermocouple is usually employed. Type K (chromel–alumel) thermocouples are also used as an alternative reference standard for temperature calibration up to 1000°C.

Although no special types of instruments are needed for temperature calibration, the temperature of the environment within which one instrument is compared with another has to be controlled carefully. This requires purpose-designed equipment, which is available commercially from a number of manufacturers.

For calibration of all temperature transducers other than radiation thermometers above a temperature of 20°C, a furnace consisting of an electrically heated ceramic tube is commonly used. The temperature of such a furnace can typically be controlled within limits of ±2°C over the range from 20 to 1600°C.

Below 20°C, a stirred water bath is used to provide a constant reference temperature, and the same equipment can, in fact, be used for temperatures up to 100°C. Similar stirred liquid baths containing oil or salts (potassium/sodium nitrate mixtures) can be used to provide reference temperatures up to 600°C.

For the calibration of radiation thermometers, a radiation source, which approximates as closely as possible to the behavior of a black body, is required. The actual value of the emissivity of the source must be measured by a surface pyrometer. Some form of optical bench is also required so that instruments being calibrated can be held firmly and aligned accurately.

The simplest form of radiation source is a hot plate heated by an electrical element. The temperature of such devices can be controlled within limits of ±1°C over the range from 0 to 650°C and the typical emissivity of the plate surface is 0.85. Type R noble metal thermocouples embedded in the plate are normally used as the reference instrument.

A black body cavity provides a heat source with a much better emissivity. This can be constructed in various alternative forms according to the temperature range of the radiation thermometers to be calibrated, although a common feature is a blackened conical cavity with a cone angle of about 15°.

For calibrating low-temperature radiation pyrometers (measuring temperatures in the range of 20 to 200°C), the black body cavity is maintained at a constant temperature (±0.5°C) by immersing it in a liquid bath. The typical emissivity of a cavity heated in this way is 0.995. Water is suitable for the bath in the temperature range of 20–90°C, and a silicone fluid is suitable for the range of 80–200°C. Within these temperature ranges, a mercury-in-glass thermometer is used commonly as the standard reference calibration instrument, although a platinum resistance thermometer is used when better accuracy is required.

Another form of black body cavity is one lined with a refractory material and heated by an electrical element. This gives a typical emissivity of 0.998 and is used for calibrating radiation pyrometers at higher temperatures. Within the range of 200–1200°C, temperatures can be controlled within limits of ±0.5°C, and a type R thermocouple is generally used as the reference instrument. At the higher range of 600–1600°C, temperatures can be controlled within limits of ±1°C, and a type B thermocouple (30% rhodium–platinum/6%

rhodium–platinum) is normally used as the reference instrument. As an alternative to thermocouples, radiation thermometers can also be used as a standard within ±0.5°C over the temperature range from 400 to 1250°C.

To provide reference temperatures above 1600°C, a carbon cavity furnace is used. This consists of a graphite tube with a conical radiation cavity at its end. Temperatures up to 2600°C can be maintained with an accuracy of ±5°C. Narrow-band radiation thermometers are used as the reference standard instrument.

Again, the aforementioned equipment merely provides an environment in which radiation thermometers can be calibrated against some other reference standard instrument. To obtain an absolute reference standard of temperature, a fixed-point, black body furnace is used. This has a radiation cavity consisting of a conical-ended cylinder that contains a crucible of 99.999% pure metal. If the temperature of the metal is monitored as it is heated up at a constant rate, an arrest period is observed at the melting point of the metal where the temperature ceases to rise for a short time interval. Thus the melting point, and hence the temperature corresponding to the output reading of the monitoring instrument at that instant, is defined exactly. Measurement uncertainty is of the order of ±0.3°C. The list of metals, and their melting points, was presented earlier at the beginning of Section 14.14.

In the calibration of radiation thermometers, knowledge of the emissivity of the hot plate or black body furnace used as the radiation source is essential. This is measured by special types of surface pyrometer. Such instruments contain a hemispherical, gold-plated surface that is supported on a telescopic arm that allows it to be put into contact with the hot surface. The radiation emitted from a small hole in the hemisphere is independent of the surface emissivity of the measured body and is equal to that which would be emitted by the body if its emissivity value was 100. This radiation is measured by a thermopile with its cold junction at a controlled temperature. A black hemisphere is also provided with the instrument, which can be inserted to cover the gold surface. This allows the instrument to measure the normal radiation emission from the hot body and so allows the surface emissivity to be calculated by comparing the two radiation measurements.

Within this list of special equipment, mention must also be made of standard tungsten strip lamps, which are used for providing constant known temperatures in the calibration of optical pyrometers. The various versions of these provide a range of standard temperatures between 800 and 2300°C to an accuracy of ±2°C.

### 14.14.2 Calculating Frequency of Calibration Checks

The manner in which the appropriate frequency for calibration checks is determined for the various temperature-measuring instruments available was discussed in the instrument review presented in Section 14.1. The simplest instruments from a calibration point of view are liquid-in-glass thermometers. The only parameter able to change within these is the volume of the glass used in their construction. This only changes very slowly with time, and hence only infrequent (e.g., annual) calibration checks are required.

The required frequency for calibration of all other instruments is either (a) dependent on the type of operating environment and the degree of exposure to it or (b) use related. In some cases, both of these factors are relevant.

Resistance thermometers and thermistors are examples of instruments where the drift in characteristics depends on the environment they are operated in and on the degree of protection they have from that environment. Devices such as gas thermometers and quartz thermometers suffer characteristics drift, which is largely a function of how much they are used (or misused!), although in the case of quartz thermometers, any drift is likely to be small and only infrequent calibration checks will be required. Any instruments not mentioned so far suffer characteristics drift due to both environmental and use-related factors. The list of such instruments includes bimetallic thermometers, thermocouples, thermopiles, and radiation thermometers. In the case of thermocouples and thermopiles, it must be remembered that error in the required characteristics is possible even when the instruments are new, as discussed in Section 14.1, and therefore their calibration must be checked before use.

As the factors responsible for characteristics drift vary from application to application, the required frequency of calibration checks can only be determined experimentally. The procedure for doing this is to start by checking the calibration of instruments used in new applications at very short intervals of time and then to progressively lengthen the interval between calibration checks until a significant deterioration in instrument characteristics is observed. The required calibration interval is then defined as that time interval predicted to elapse before the characteristics of the instrument have drifted to the limits allowable in that particular measurement application.

Working and reference standard instruments and ancillary equipment must also be calibrated periodically. An interval of 2 years is usually recommended between such calibration checks, although monthly checks are advised for the black body cavity furnaces used to provide standard reference temperatures in pyrometer calibration. Standard resistance thermometers and thermocouples may also need more frequent calibration checks if the conditions (especially of temperature) and frequency of use demand them.

### *14.14.3 Procedures for Calibration*

The standard way of calibrating temperature transducers is to put them into a temperature-controlled environment together with a standard instrument or to use a radiant heat source of controlled temperature with high emissivity in the case of radiation thermometers. In either case, the controlled temperature must be measured by a standard instrument whose calibration is traceable to reference standards. This is a suitable method for most instruments in the calibration chain but is not necessarily appropriate or even possible for process instruments at the lower end of the chain.

In the case of many process instruments, their location and mode of fixing make it difficult or sometimes impossible to remove them to a laboratory for calibration checks to be carried out. In this event, it is standard practice to calibrate them in their normal operational position, using a reference instrument that is able to withstand whatever hostile environment may be present. If this practice is followed, it is imperative that the working standard instrument is checked regularly to ensure that it has not been contaminated.

Such *in situ calibration* may also be required where process instruments have characteristics that are sensitive to the environment in which they work so that they are calibrated under their usual operating conditions and are therefore accurate in normal use. However, the preferred way of dealing with this situation is to calibrate them in a laboratory with ambient conditions (of pressure, humidity, etc.) set up to mirror those of the normal operating environment. This alternative avoids having to subject reference calibration instruments to harsh chemical environments that are commonly associated with manufacturing processes.

For instruments at the lower end of the calibration chain, that is, those measuring process variables, it is common practice to calibrate them against an instrument that is of the same type but of higher accuracy and reserved only for calibration duties. If a large number of different types of instruments have to be calibrated, however, this practice leads to the need to keep a large number of different calibration instruments. To avoid this, various reference instruments are available that can be used to calibrate all process instruments within a given temperature-measuring range. Examples are the liquid-in-glass thermometer (0 to +200°C), platinum resistance thermometer (−200 to +1000°C), and type S thermocouple (+600 to +1750°C). The optical pyrometer is also often used as a reference instrument at this level for the calibration of other types of radiation thermometers.

For calibrating instruments further up the calibration chain, particular care is needed with regard to both the instruments used and the conditions they are used under. It is difficult and expensive to meet these conditions and hence this function is subcontracted by most companies to specialist laboratories. The reference instruments used are the platinum resistance thermometer in the temperature range of −200 to +1000°C, the platinum–platinum/10% rhodium (type S) thermocouple in the temperature range of +1000 to +1750°C, and a narrow-band radiation thermometer at higher temperatures. An exception is *optical pyrometers*, which are calibrated as explained in the final paragraph of this chapter. A particular note of caution must be made where platinum–rhodium thermocouples are used as a standard. These are very prone to contamination, and if they need to be handled at all, this should be done with very clean hands.

Before ending this chapter, it is appropriate to mention one or two special points that concern the calibration of thermocouples. The mode of construction of thermocouples means that their characteristics can be incorrect even when they are new, due to faults in either the homogeneity of the thermocouple materials or the construction of the device. Therefore, calibration checks should be carried out on all new thermocouples before they are put into use. The procedure for this is to immerse both junctions of the thermocouple in an ice bath and measure its output with a high-accuracy digital voltmeter ($\pm 5\ \mu V$). Any output greater than $5\ \mu V$ would indicate a fault in the thermocouple material and/or its construction. After this check on thermocouples when they are brand new, the subsequent rate of change of thermoelectric characteristics with time is entirely dependent on the operating environment and the degree of exposure to it. Particularly relevant factors in the environment are the type and concentration of trace metal elements and the temperature (the rate of contamination of thermocouple materials with trace elements of metals is a function of temperature). A suitable calibration frequency can therefore only be defined by practical experimentation, and this must be reviewed whenever the operating environment and conditions of use change. A final word of caution when calibrating thermocouples is to ensure that any source of electrical or magnetic fields is excluded, as these will induce erroneous voltages in the sensor.

Special comments are also relevant regarding calibration of a *radiation thermometer*. As well as normal accuracy checks, its long-term stability must also be verified by testing its output over a period that is 1 hour longer than the manufacturer's specified "warm-up" time. This shows up any in components within the instrument that are suffering from temperature-induced characteristics drift. It is also necessary to calibrate radiation thermometers according to the emittance characteristic of the body whose temperature is being measured and according to the level of energy losses in the radiation path between the body and measuring instrument. Such emissivity calibration must be carried out for every separate application that the instrument is used for, using a surface pyrometer.

Finally, it should be noted that the usual calibration procedure for *optical pyrometers* is to sight them on the filament of a tungsten strip lamp in which the current is measured accurately. This method of calibration can be used at temperatures up to 2500°C. Alternatively, they can be calibrated against a standard radiation pyrometer.

## 14.15 Summary

Our review at the start of the chapter revealed that there are 10 different physical principles used commonly as the basis for temperature-measuring devices. During the course of the chapter, we then looked at how each of these principles is exploited in various classes of temperature-measuring devices.

We started off by looking at the thermoelectric effect and its use in thermocouples and thermopiles, and also the derived devices of digital thermometers and continuous thermocouples. Thermocouples are the most commonly used devices for industrial applications of temperature measurement. However, despite their relatively simple operating concept of generating an output voltage as a function of the temperature they are exposed to, proper use of thermocouples requires an understanding of two thermoelectric laws. These laws were presented and their application was explained by several examples. We also saw how the output of a thermocouple has to be interpreted by thermocouple tables. We went on to look at the different types of thermocouples available, which range from a number of inexpensive, base metal types to expensive ones based on noble metals. We looked at the typical characteristics of these and discussed typical applications. Moving on, we noted that thermocouples are quite delicate devices that can suffer from both mechanical damage and chemical damage in certain operating environments, and we discussed ways of avoiding such problems. We also looked briefly at how thermocouples are made.

Our next subject of study concerned resistance thermometers and thermistors, both of these being devices that convert a change in temperature into a change in the resistance of the device. We noted that both of these are also very commonly used measuring devices. We looked at the theoretical principles of each of these and discussed the range of materials used in each class of device. We also looked at the typical device characteristics for each construction material.

Next, we looked at semiconductor devices in the form of diodes and transistors and discussed their characteristics and mode of operation. This discussion revealed that although these devices are less expensive and more linear than both thermocouples and resistance thermometers, their typical measurement range is relatively low. This limits their applicability and means that they are not used as widely as they would be if their measurement range was greater.

Moving on, we looked at the class of devices known as radiation thermometers (alternatively known as radiation pyrometers), which exploit the physical principle that the peak wavelength of radiated energy emission from a body varies with temperature. The instrument is used by pointing it at the body to be measured and analyzing the radiation emitted from the body. This has the advantage of being a noncontact mode of temperature measurement, which is highly attractive in the food and drug industries and any other application where contamination of the measured quantity has to be avoided. We also observed that different versions of radiation thermometers are capable of measuring temperatures between $-100$ and $+10{,}000°C$, with measurement inaccuracy as low as $\pm 0.05\%$ in the more expensive versions. Despite these obvious merits, careful calibration of the instrument to the type of body being measured is essential, as the characteristics of radiation thermometers are critically dependent on the emissivity of the measured body, which varies widely between different materials.

This stage in the chapter marked the end of discussion of the four most commonly used types of temperature-measuring devices. The remaining techniques all have niche applications but none of these are "large volume" uses. The first one covered of these "other techniques" was thermography. Also known as thermal imaging, this involves scanning an infrared radiation detector across either a single object or a scene containing several objects. The information gathered is then processed and an output in the form of the temperature distribution across the object is produced. It thus differs from other forms of temperature sensors in providing information on temperature distribution across an object or scene rather than the temperature at a single discrete point. Temperature measurement over the range from −20°C up to +1500°C is possible.

Our next subject of study concerned the liquid-in-glass thermometer, bimetallic thermometer, and pressure thermometer. These are all usually classed as thermal expansion-based devices, although this is not strictly true in the case of the last one, which is based on the change in pressure of a fluid inside a fixed-volume stainless-steel bulb. The characteristics and typical applications of each of these were discussed.

The quartz thermometer then formed the next subject of study. This uses the principle that the resonant frequency of a material such as quartz changes with temperature. Such devices have very high specifications in terms of linearity, long life, accuracy, and measurement resolution. Unfortunately, their relatively high cost severely limits their application.

Moving on, we then looked at fiber-optic temperature sensors. We saw that their main application is measuring temperatures in hard-to-reach locations, although they are also used when very high measurement accuracy is required.

Next, we discussed color indicators. These consist mainly of special paints or crayons that change color at a certain temperature. They are used primarily to determine when the temperature of objects placed in a furnace reach a given temperature. They are relatively inexpensive, and different paints and crayons are available to indicate temperatures between 50 and 1250°C. In addition, certain liquid crystals that change color at a certain temperature are also used as color indicators. These have better measurement resolution than paints and crayons and, while some versions can indicate low temperatures down to –20°C, the highest temperature that they can indicate is +100°C.

Finally, our discussion of the application of different physical principles in temperature measurement brought us to Seger cones. Also known as pyrometric cones, these have a conical shape where the tip melts and bends over at a particular temperature. They are used commonly in the ceramics industry to detect a given temperature is reached in a furnace.

The chapter then continued with a look at intelligent measuring devices. These are designed for use with various sensors such as thermocouples, thermopiles, resistance thermometers, thermistors, and broad-band radiation pyrometers. Intelligence within the device gives them features such as adjustable damping, noise rejection, self-adjustment for zero and sensitivity drifts, self-fault diagnosis, self-calibration, reduced maintenance requirement, and an expanded measurement range. These features reduce typical measurement inaccuracy down to ±0.05% of full scale.

This completion of the discussion on all types of intelligent and nonintelligent devices allowed us to go on to consider the mechanisms by which a temperature-measuring device is chosen for a particular application. We reviewed the characteristics of each type of device in turn and looked at the sorts of circumstances in which each might be used.

Our final subject of study in the chapter was that of calibrating temperature-measuring devices. We noted first of all that a fundamental difficulty exists in establishing an absolute standard for temperature and that, in the absence of such a standard, fixed reference points for temperature were defined in the form of freezing points and triple points of certain substances. We then went on to look at the calibration instruments and equipment used in workplace calibration. We also established some guidelines about how the frequency of calibration should be set. Finally, we looked in more detail at the appropriate practical procedures for calibrating various types of sensors.

## 14.16 Problems

14.1. Discuss briefly the different physical principles used in temperature-measuring instruments and give examples of instruments that use each of these principles.

14.2. (a) How are thermocouples manufactured? (b) What are the main differences between base metal and noble metal thermocouples? (c) Give six examples of the materials used to make base metal and noble metal thermocouples. (d) Specify the international code letter used to designate thermocouples made from each pair of materials that you give in your answer to part (c).

14.3. Explain what each of the following are in relation to thermocouples: (a) extension leads, (b) compensating leads, (c) law of internediate metals, and (d) law of intermediate temperature.

14.4. What type of base metal thermocouple would you recommend for each of the following applications?
(a) measurement of subzero temperatures
(b) measurement in oxidizing atmospheres
(c) measurement in reducing atmospheres
(d) where high sensitivity measurement is required

14.5. Why do thermocouples need protection from some operating environments and how is this protection given? Discuss any differences between base metal and noble metal thermocouples in the need for protection.

14.6. The temperature of a fluid is measured by immersing an iron–constantan thermocouple in it. The reference junction of the thermocouple is maintained at 0°C in an ice bath and an output e.m.f. of 5.812 mV is measured. What is the indicated fluid temperature?

14.7. The temperature of a fluid is measured by immersing a type K thermocouple in it. The reference junction of the thermocouple is maintained at 0°C in an ice bath and an output e.m.f. of 6.435 mV is measured. What is the indicated fluid temperature?

14.8. The output e.m.f. from a chromel–alumel thermocouple (type K), with its reference junction maintained at 0°C, is 12.207 mV. What is the measured temperature?

14.9. The output e.m.f. from a nicrosil–nisil thermocouple (type N), with its reference junction maintained at 0°C, is 4.21 mV. What is the measured temperature?

14.10. The output e.m.f. from a chromel–constantan thermocouple whose hot junction is immersed in a fluid is measured as 18.25 mV. The reference junction of the thermocouple is maintained at 0°C. What is the temperature of the fluid?

14.11. A copper–constantan thermocouple is connected to copper–constantan extension wires and the reference junction is exposed to a room temperature of 20°C. If the output voltage measured is 6.537 mV, what is the indicated temperature at the hot junction of the thermocouple?

14.12. A platinum/10% rhodium–platinum (type S) thermocouple is used to measure the temperature of a furnace. Output e.m.f., with the reference junction maintained at 50°C, is 5.975 mV. What is the temperature of the furnace?

14.13. In a particular industrial situation, a nicrosil–nisil thermocouple with nicrosil–nisil extension wires is used to measure the temperature of a fluid. In connecting up this measurement system, the instrumentation engineer responsible has inadvertently interchanged the extension wires from the thermocouple. The ends of the extension wires are held at a reference temperature of 0°C and the output e.m.f. measured is 21.0 mV. If the junction between the thermocouple and extension wires is at a temperature of 50°C, what temperature of fluid is indicated and what is the true fluid temperature?

14.14. A copper–constantan thermocouple measuring the temperature of a hot fluid is connected by mistake with chromel–constantan extension wires (such that the two constantan wires are connected together and the chromel extension wire is connected to the copper thermocouple wire). If the actual fluid temperature was 150°C, the junction between the thermocouple and extension wires was at 80°C, and the reference junction was at 0°C, calculate the e.m.f. measured at the open ends of the extension wires. What fluid temperature would be deduced from this measured e.m.f. (assuming that the error of using the wrong extension wires was not known)? (Hint: Apply the law of intermediate metals for the thermocouple-extension lead junction.)

14.15. This question is similar to the last one but involves a chromel–constantan thermocouple rather than a copper–constantan one. In this case, an engineer installed a chromel–constantan thermocouple but used copper–constantan extension leads (such that the two constantan wires were connected together and the copper extension wire was connected to the chromel thermocouple wire). If the thermocouple was measuring a hot fluid whose real temperature is 150°C, the junction between the thermocouple and the extension leads was at 80°C, and the reference junction was at 0°C:

(a) Calculate the e.m.f. (voltage) measured at the open ends of the extension wires.

(b) What fluid temperature would be deduced from this measured e.m.f., assuming that the error in using the incorrect leads was not known?

14.16. While installing a chromel–constantan thermocouple to measure the temperature of a fluid, it is connected by mistake with copper–constantan extension leads (such that the two constantan wires are connected together and the copper extension wire is connected to the chromel thermocouple wire). If the fluid temperature was actually 250°C and the junction between the thermocouple and extension wires was at 80°C, what e.m.f. would be measured at the open ends of the extension wires if the reference junction is maintained at 0°C? What fluid temperature would be deduced from this (assuming that the connection mistake was not known)?

14.17. In connecting extension leads to a chromel–alumel thermocouple, which is measuring the temperature of a fluid, a technician connects the leads the wrong way round (such that the chromel extension lead is connected to the alumel thermocouple lead and vice versa). The junction between the thermocouple and extension leads is at a temperature of 100°C and the reference junction is maintained at 0°C in an ice bath. The technician measures an output e.m.f. of 12.212 mV at the open ends of the extension leads.

(a) What fluid temperature would be deduced from this measured e.m.f.?

(b) What is the true fluid temperature?

14.18. A chromel–constantan thermocouple measuring the temperature of a fluid is connected by mistake with copper–constantan extension leads (such that the two constantan wires are connected together and the copper extension lead wire is connected to the chromel thermocouple wire). If the fluid temperature was actually 250°C and the junction between the thermocouple and extension leads was at 90°C, what e.m.f. would be measured at the open ends of the extension leads if the reference junction is maintained at 0°C? What fluid temperature would be deduced from this (assuming that the connection error was not known)?

14.19. Extension leads used to measure the output e.m.f. of an iron–constantan thermocouple measuring the temperature of a fluid are connected the wrong way round by mistake (such that the iron extension lead is connected to the constantan thermocouple wire and vice versa). The junction between the thermocouple and extension leads is at a temperature of 120°C and the reference junction is at a room temperature of 21°C. The output e.m.f. measured at the open ends of the extension leads is 27.390 mV.

(a) What fluid temperature would be deduced from this measured e.m.f. assuming that the mistake of connecting the extension leads the wrong way round was not known about?

(b) What is the true fluid temperature?

14.20. The temperature of a hot fluid is measured with a copper–constantan thermocouple but, by mistake, this is connected to chromel–constantan extension wires (such that the two constantan wires are connected together and the chromel extension wire is connected to the copper thermocouple wire). If the actual fluid temperature was 200°C, the junction between the thermocouple and extension wires was at 50°C, and the reference junction was at 0°C, calculate the e.m.f. measured at the open ends of the extension wires. What fluid temperature would be deduced from this measured e.m.f. (assuming that the error of using the wrong extension wires was not known)?

14.21. In a particular industrial situation, a chromel–alumel thermocouple with chromel–alumel extension wires is used to measure the temperature of a fluid. In connecting up this measurement system, the instrumentation engineer responsible has inadvertantly interchanged the extension wires from the thermocouple (such that the chromel thermocouple wire is connected to the alumel extension lead wire, etc.). The open ends of the extension leads are held at a reference temperature of 0°C and are connected to a voltmeter, which measures an e.m.f. of 18.75 mV. If the junction between the thermocouple and extension wires is at a temperature of 38°C:

(a) What temperature of fluid is indicated?

(b) What is the true fluid temperature?

14.22. A copper–constantan thermocouple measuring the temperature of a hot fluid is connected by mistake with iron–constantan extension wires (such that the two constantan wires are connected together and the iron extension wire is connected to the copper thermocouple wire). If the actual fluid temperature was 200°C, the junction between the thermocouple and extension wires was at 160°C, and the reference junction was at 0°C, calculate the e.m.f. measured at the open ends of the extension wires. What fluid temperature would be deduced from this measured e.m.f. (assuming that the error of using the wrong extension wires was not known)?

14.23. In a particular industrial situation, a nicrosil–nisil thermocouple with nicrosil–nisil extension wires is used to measure the temperature of a fluid. In connecting up this measurement system, the instrumentation engineer responsible has inadvertantly interchanged the extension wires from the thermocouple (such that the nicrosil thermocouple wire is connected to the nisil extension lead wire, etc.). The open ends of the extension leads are held at a reference temperature of 0°C and are connected to a voltmeter, which measures an e.m.f. of 17.51 mV. If the junction between the thermocouple and extension wires is at a temperature of 140°C:

(a) What temperature of fluid is indicated?

(b) What is the true fluid temperature?

14.24. Explain what the following are: thermocouple, continuous thermocouple, thermopile, and digital thermometer.

14.25. What is the *International Practical Temperature Scale*? Why is it necessary in temperature sensor calibration and how is it used?

14.26. Resistance thermometers and thermistors are both temperature-measuring devices that convert the measured temperature into a resistance change. What are the main differences between these two types of devices in respect of the materials used in their constructions, their cost, and their operating characteristics?

14.27. Discuss the main types of radiation thermometers available. How do they work and what are their main applications?

14.28. Name three kinds of temperature-measuring devices that work on the principle of thermal expansion. Explain how each works and what its typical characteristics are.

14.29. Explain how fiber-optic cables can be used as temperature sensors.

14.30. Discuss the calibration of temperature sensors, mentioning what reference instruments are typically used.

## References


Brookes, C. (1985). Nicrosil-nisil thermocouples. *Journal of measurement and control*, *18*(7), 245–248.

Dixon, J. (1987). Industrial radiation thermometry. *Journal of measurement and control*, *20*(6), 11–16.

Johnson, J. S. (1994). Optical sensors: The OCSA experience. *Measurement and control*, *27*(7), 180–184.

CHAPTER 15

# Pressure Measurement




Measurement and Instrumentation: Theory and Application

## 15.1 Introduction

We are covering pressure measurement next in this chapter because it is required very commonly in most industrial process control systems and is the next-most measured process parameter after temperature. We shall see that many different types of pressure-sensing and pressure measurement systems are available to satisfy this requirement. However, before considering these in detail, it is important for us to understand that pressure can be quantified in three alternative ways in terms of absolute pressure, gauge pressure, or differential pressure. The formal definitions of these are as follows.

*Absolute pressure:* This is the difference between the pressure of the fluid and the absolute zero of pressure.

*Gauge pressure:* This describes the difference between the pressure of a fluid and atmospheric pressure. Absolute and gauge pressures are therefore related by the expression:

$$\text{Absolute pressure} = \text{Gauge pressure} + \text{Atmospheric pressure}.$$

A typical value of atmospheric pressure is 1.013 bar. However, because atmospheric pressure varies with altitude as well as with weather conditions, it is not a fixed quantity. Therefore, because gauge pressure is related to atmospheric pressure, it also is not a fixed quantity.

*Differential pressure:* This term is used to describe the difference between two absolute pressure values, such as the pressures at two different points within the same fluid (often between the two sides of a flow restrictor in a system measuring volume flow rate).

Pressure is a quantity derived from the fundamental quantities of force and area and is usually measured in terms of the force acting on a known area. The SI unit of pressure is the Pascal, which can alternatively be expressed as Newtons per square meter. The bar, which is equal to 10,000 Pascal, is a related metric unit that is more suitable for measuring the most typically met pressure values. The unit of pounds per square inch is not an SI unit, but is still in widespread use, especially in the United State and Canada. Pressures are also sometimes expressed as inches of mercury or inches of water, particularly when measuring blood pressure or pressures in gas pipelines. These two measurement units derive from the height of the liquid column in manometers, which were a very common method of pressure measurement in the past. The torr is a further unit of measurement used particularly to express low pressures (1 torr = 133.3 Pascal).

To avoid ambiguity in pressure measurements, it is usual to append one or more letters in parentheses after the pressure value to indicate whether it is an absolute, gauge, or differential pressure: (a) or (abs) indicates absolute pressure, (g) indicates gauge pressure, and (d) specifies differential pressure. Thus, 2.57 bar (g) means that the pressure is 2.57 bar measured as gauge pressure. In the case of the pounds per square inch unit of pressure measurement, which is still in widespread use, it is usual to express absolute, gauge, and differential pressure as psia, psig, and psid, respectively.

Absolute pressure measurements are made for such purposes as aircraft altitude measurement (in instruments known as altimeters) and when quantifying atmospheric pressure. Very low pressures are also normally measured as absolute pressure values. Gauge pressure measurements are made by instruments such as those measuring the pressure in vehicle tires and those measuring pressure at various points in industrial processes. Differential pressure is measured for some purposes in industrial processes, especially as part of some fluid flow rate-measuring devices.

In most applications, typical values of pressure measured range from 1.013 bar (the mean atmospheric pressure) up to 7000 bar. This is considered to be the "normal" pressure range, and a large number of pressure sensors are available that can measure pressures in this range. Measurement requirements outside this range are much less common. While some of the pressure sensors developed for the "normal" range can also measure pressures that are either lower or higher than this, it is preferable to use special instruments that have been specially designed to satisfy such low- and high-pressure measurement requirements. In the case of low pressures, such special instruments are commonly known as vacuum gauges.

Our discussion summarizes the main types of pressure sensors in use. This discussion is concerned primarily only with the measurement of static pressure, because the measurement of dynamic pressure is a very specialized area that is not of general interest. In general, dynamic pressure measurement requires special instruments, although modified versions of diaphragm-type sensors can also be used if they contain a suitable displacement sensor (usually either a piezoelectric crystal or a capacitive element).

## 15.2 Diaphragms

The diaphragm, shown schematically in Figure 15.1, is one of three types of elastic-element pressure transducers. Applied pressure causes displacement of the diaphragm and this movement is measured by a displacement transducer. Different versions of diaphragm sensors can measure both absolute pressure (up to 50 bar) and gauge pressure (up to 2000 bar) according to whether the space on one side of the diaphragm is, respectively, evacuated or open to the atmosphere. A diaphragm can also be used to measure differential pressure (up to 2.5 bar) by applying the two pressures to the two sides of the diaphragm. The diaphragm can be plastic, metal alloy, stainless steel, or ceramic. Plastic diaphragms are the least expensive, but metal diaphragms give better accuracy. Stainless steel is normally used in high temperature or corrosive environments. Ceramic diaphragms are resistant even to strong acids and alkalis and are used when the operating environment is particularly harsh. The name *aneroid gauge* is sometimes used to describe this type of gauge when the diaphragm is metallic.

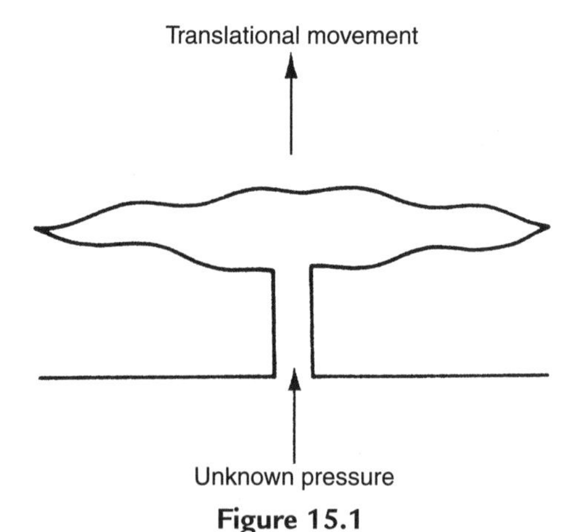


**Figure 15.1**
Schematic representation of a diaphragm pressure sensor.

The typical magnitude of diaphragm displacement is 0.1 mm, which is well suited to a strain-gauge type of displacement-measuring transducer, although other forms of displacement measurements are also used in some kinds of diaphragm-based sensors. If the displacement is measured with strain gauges, it is normal to use four strain gauges arranged in a bridge circuit configuration. The output voltage from the bridge is a function of the resistance change due to the strain in the diaphragm. This arrangement automatically provides compensation for environmental temperature changes. Older pressure transducers of this type used metallic strain gauges bonded to a diaphragm typically made of stainless steel. However, apart from manufacturing difficulties arising from the problem of bonding the gauges, metallic strain gauges have a low gauge factor, which means that the low output from the strain gauge bridge has to be amplified by an expensive d.c. amplifier. The development of semiconductor (piezoresistive) strain gauges provided a solution to the low-output problem, as they have gauge factors up to 100 times greater than metallic gauges. However, the difficulty of bonding gauges to the diaphragm remained and a new problem emerged regarding the highly nonlinear characteristic of the strain–output relationship.

The problem of strain-gauge bonding was solved with the emergence of monolithic piezoresistive pressure transducers. These have a typical measurement uncertainty of $\pm 0.5\%$ and are now the most commonly used type of diaphragm pressure transducer. The monolithic cell consists of a diaphragm made of a silicon sheet into which resistors are diffused during the manufacturing process. Such pressure transducers can be made very small and are often known as *microsensors*. Also, in addition to avoiding the difficulty with bonding, such monolithic silicon-measuring cells have the advantage of being very inexpensive to manufacture in large quantities. Although the inconvenience of a nonlinear characteristic remains, this is normally overcome by processing the output signal with an active linearization circuit or incorporating the cell into a microprocessor-based intelligent measuring transducer. The latter usually provide analogue-to-digital conversion and interrupt facilities within a single chip and give a digital output that is integrated readily into computer control schemes. Such instruments can also offer automatic temperature compensation, built-in diagnostics, and simple calibration procedures. These features allow measurement inaccuracy to be reduced down to a value as low as ±0.1% of full-scale reading.

## 15.3 Capacitive Pressure Sensor

A capacitive pressure sensor is simply a diaphragm-type device in which diaphragm displacement is determined by measuring the capacitance change between the diaphragm and a metal plate that is close to it. Such devices are in common use and are sometimes known as *Baratron gauges*. It is also possible to fabricate capacitive elements in a silicon chip and thus form very small *microsensors*. These have a typical measurement uncertainty of ±0.2%.

## 15.4 Fiber-Optic Pressure Sensors

Fiber-optic sensors, also known as optical pressure sensors, provide an alternative method of measuring displacements in diaphragm and Bourdon tube pressure sensors by optoelectronic means and enable the resulting sensors to have a lower mass and size compared with sensors in which displacement is measured by other methods. The shutter sensor described earlier in Chapter 13 is one form of fiber-optic displacement sensor. Another form is the fotonic sensor shown in Figure 15.2 in which light travels from a light source, down an optical fiber, reflected back from a diaphragm, and then travels back along a second fiber to a photodetector. There is a characteristic relationship between the light reflected and the distance from the fiber ends to the diaphragm, thus making the amount of reflected light dependent on the diaphragm displacement and hence the measured pressure.

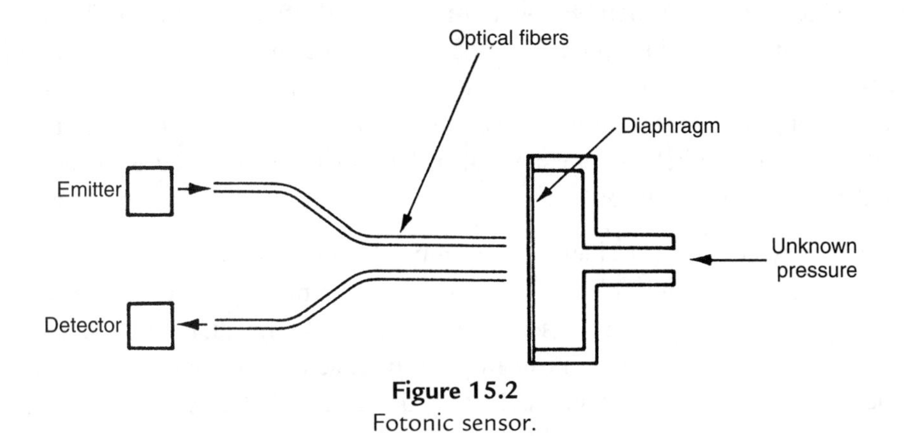


**Figure 15.2**
Fotonic sensor.

Apart from the mass and size advantages of fiber-optic displacement sensors, the output signal is immune to electromagnetic noise. However, measurement accuracy is usually inferior to that provided by alternative displacement sensors, and the choice of such sensors also incurs a cost penalty. Thus, sensors using fiber optics to measure diaphragm or Bourdon tube displacement tend to be limited to applications where their small size, low mass, and immunity to electromagnetic noise are particularly advantageous.

Apart from the limited use with diaphragm and Bourdon tube sensors, fiber-optic cables are also used in several other ways to measure pressure. A form of fiber-optic pressure sensor known as a *microbend sensor* is sketched in Figure 13.8a. In this, the refractive index of the fiber (and hence the intensity of light transmitted) varies according to the mechanical deformation of the fiber caused by pressure. The sensitivity of pressure measurement can be optimized by applying pressure via a roller chain such that bending is applied periodically (see Figure 13.8b). The optimal pitch for the chain varies according to the radius, refractive index, and type of cable involved. Microbend sensors are typically used to measure the small pressure changes generated in Vortex shedding flowmeters. When fiber-optic sensors are used in this flow measurement role, the alternative arrangement shown in Figure 15.3 can be used, where a fiber-optic cable is merely stretched across the pipe. This often simplifies the detection of vortices.

Phase-modulating fiber-optic pressure sensors also exist. The mode of operation of these was discussed in Chapter 13.

## 15.5 Bellows

Bellows, illustrated schematically in Figure 15.4, are another elastic-element type of pressure sensor that operate on very similar principles to the diaphragm pressure sensor. Pressure changes within the bellows, which are typically fabricated as a seamless tube of either metal or metal alloy, produce translational motion of the end of the bellows that can be measured by capacitive, inductive (LVDT), or potentiometric transducers. Different versions can measure either absolute pressure (up to 2.5 bar) or gauge pressure (up to 150 bar). Double-bellows versions also exist that are designed to measure differential pressures of up to 30 bar.

Bellows have a typical measurement uncertainty of only ±0.5%, but have a relatively high manufacturing cost and are prone to failure. Their principle attribute in the past has been their greater measurement sensitivity compared with diaphragm sensors. However, advances in electronics mean that the high-sensitivity requirement can usually be satisfied now by diaphragm-type devices, and usage of bellows is therefore falling.

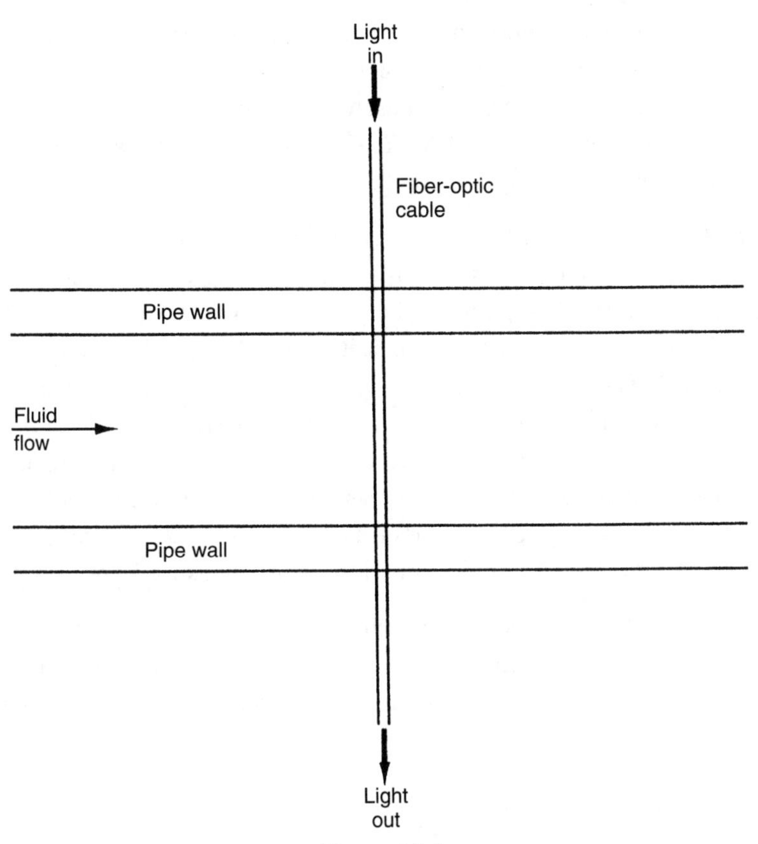


**Figure 15.3**
Use of a fiber-optic pressure sensor in a vortex-shedding flowmeter.

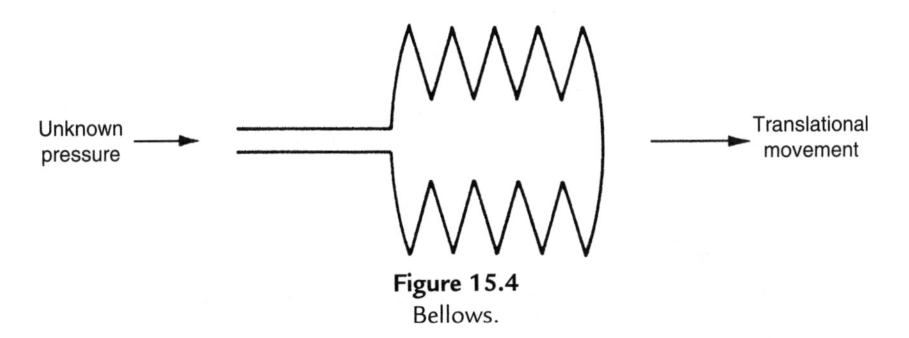


**Figure 15.4**
Bellows.

## 15.6 Bourdon Tube

The Bourdon tube is also an elastic element type of pressure transducer. It is relatively inexpensive and is used commonly for measuring the gauge pressure of both gaseous and liquid fluids. It consists of a specially shaped piece of oval-section, flexible, metal tube that is fixed at one end and free to move at the other end. When pressure is applied at the open, fixed end of the tube, the oval cross section becomes more circular. In consequence, there is displacement of the free end of the tube. This displacement is measured by some form of displacement transducer, which is commonly a potentiometer or LVDT. Capacitive and optical sensors are also sometimes used to measure the displacement.

The three common shapes of Bourdon tubes are shown in Figure 15.5. The maximum possible deflection of the free end of the tube is proportional to the angle subtended by the arc through which the tube is bent. For a C-type tube, the maximum value for this arc is somewhat less than 360°. Where greater measurement sensitivity and resolution are required, spiral and helical tubes are used. These both give much greater deflection at the free end for a given applied pressure. However, this increased measurement performance is only gained at the expense of a substantial increase in manufacturing difficulty and cost compared with C-type tubes and is also associated with a large decrease in the maximum pressure that can be measured. Spiral and helical types are sometimes provided with a rotating pointer that moves against a scale to give a visual indication of the measured pressure.

C-type tubes are available for measuring pressures up to 6000 bar. A typical C-type tube of 25 mm radius has a maximum displacement travel of 4 mm, giving a moderate level of measurement resolution. Measurement inaccuracy is typically quoted at $\pm 1\%$ of full-scale

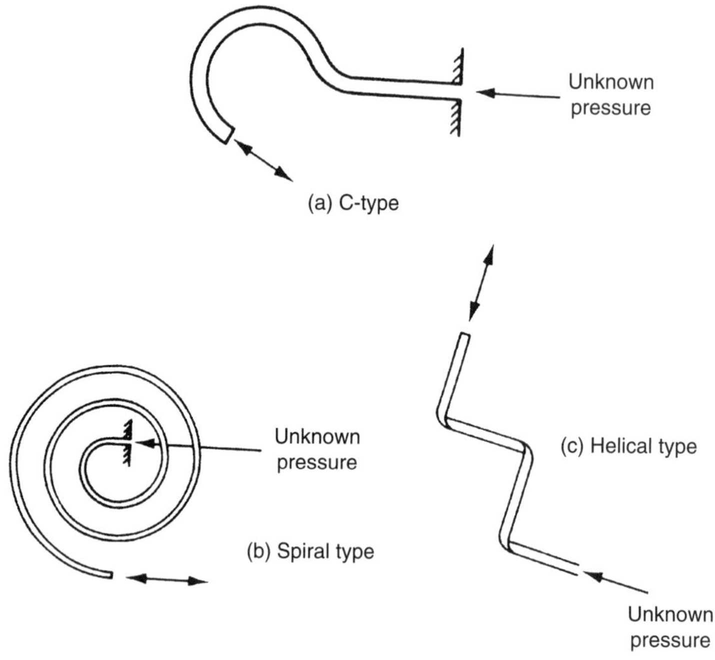


**Figure 15.5**
Three forms of a Bourdon tube.

deflection. Similar accuracy is available from helical and spiral types, but while the measurement resolution is higher, the maximum pressure measurable is only 700 bar.

The existence of one potentially major source of error in Bourdon tube pressure measurement has not been widely documented, and few manufacturers of Bourdon tubes make any attempt to warn users of their products appropriately. The problem is concerned with the relationship between the fluid being measured and the fluid used for calibration. The pointer of Bourdon tubes is normally set at zero during manufacture, using air as the calibration medium. However, if a different fluid, especially a liquid, is subsequently used with a Bourdon tube, the fluid in the tube will cause a nonzero deflection according to its weight compared with air, resulting in a reading error of up to 6%. This can be avoided by calibrating the Bourdon tube with the fluid to be measured instead of with air, assuming of course that the user is aware of the problem. Alternatively, correction can be made according to the calculated weight of the fluid in the tube. Unfortunately, difficulties arise with both of these solutions if air is trapped in the tube, as this will prevent the tube being filled completely by the fluid. Then, the amount of fluid actually in the tube, and its weight, will be unknown.

In conclusion, therefore, Bourdon tubes only have guaranteed accuracy limits when measuring gaseous pressures. Their use for accurate measurement of liquid pressures poses great difficulty unless the gauge can be totally filled with liquid during both calibration and measurement, a condition that is very difficult to fulfill practically.

## 15.7 Manometers

Manometers are passive instruments that give a visual indication of pressure values. Various types exist.

### *15.7.1 U-Tube Manometer*

The U-tube manometer, shown in Figure 15.6a, is the most common form of manometer. Applied pressure causes a displacement of liquid inside the U-shaped glass tube, and output pressure reading $P$ is made by observing the difference, $h$, between the level of liquid in the two halves of the tube $A$ and $B$, according to the equation $P=h\rho g$, where $\rho$ is the specific gravity of the fluid. If an unknown pressure is applied to side $A$, and side $B$ is open to the atmosphere, the output reading is gauge pressure. Alternatively, if side $B$ of the tube is sealed and evacuated, the output reading is absolute pressure. The U-tube manometer also measures the differential pressure, $(p_1-p_2)$, according to the expression $(p_1-p_2)=h\rho g$, if two unknown pressures $p_1$ and $p_2$ are applied, respectively, to sides $A$ and $B$ of the tube.

Output readings from U-tube manometers are subject to error, principally because it is very difficult to judge exactly where the meniscus levels of the liquid are in the two halves of the

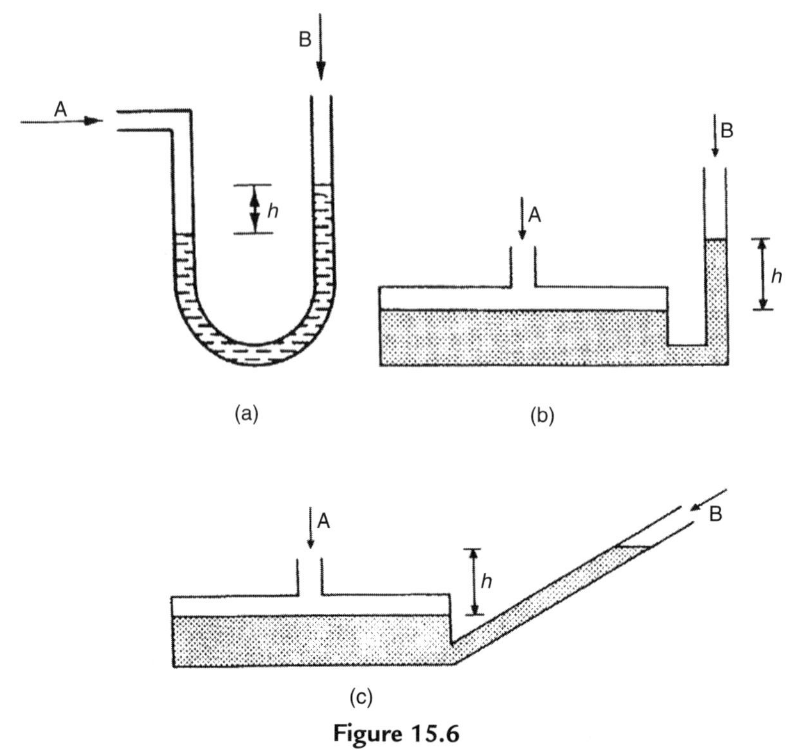


**Figure 15.6**
Three forms of manometer: (a) U tube, (b) well type, and (c) inclined type.

tube. In absolute pressure measurement, an additional error occurs because it is impossible to totally evacuate the closed end of the tube.

U-tube manometers are typically used to measure gauge and differential pressures up to about 2 bar. The type of liquid used in the instrument depends on the pressure and characteristics of the fluid being measured. Water is an inexpensive and convenient choice, but it evaporates easily and is difficult to see. Nevertheless, it is used extensively, with the major obstacles to its use being overcome by using colored water and by regularly topping up the tube to counteract evaporation. However, water is definitely not used when measuring the pressure of fluids that react with or dissolve in water. Water is also unsuitable when high pressure measurements are required. In such circumstances, liquids such as aniline, carbon tetrachloride, bromoform, mercury, or transformer oil are used instead.

### *15.7.2 Well-Type Manometer (Cistern Manometer)*

The well-type or cistern manometer, shown in Figure 15.6b, is similar to a U-tube manometer but one-half of the tube is made very large so that it forms a well. The change in the level of the well as the measured pressure varies is negligible. Therefore, the liquid level in only one tube has to be measured, which makes the instrument much easier to use than the U-tube manometer. If an unknown pressure, $p_1$, is applied to port $A$ and port $B$ is open to the atmosphere, the gauge pressure is given by $p_1 = h\rho$. It might appear that the instrument would give better measurement accuracy than the U-tube manometer because the need to subtract two liquid level measurements in order to arrive at the pressure value is avoided. However, this benefit is swamped by errors that arise due to typical cross-sectional area variations in the glass used to make the tube. Such variations do not affect the accuracy of the U-tube manometer to the same extent.

### *15.7.3 Inclined Manometer (Draft Gauge)*

The inclined manometer or draft gauge shown in Figure 15.6c is a variation on the well-type manometer in which one leg of the tube is inclined to increase measurement sensitivity. However, similar comments to those given earlier apply about accuracy.

## 15.8 Resonant Wire Devices

A typical resonant wire device is shown schematically in Figure 15.7. Wire is stretched across a chamber containing fluid at unknown pressure subjected to a magnetic field. The wire resonates at its natural frequency according to its tension, which varies with pressure. Thus

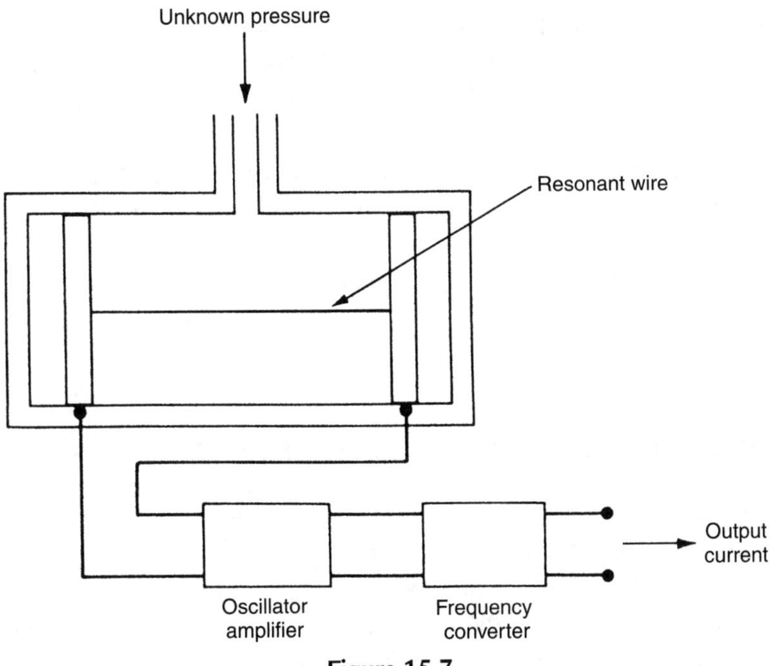


**Figure 15.7**
Resonant wire device.

pressure is calculated by measuring the frequency of vibration of the wire. Such frequency measurement is normally carried out by electronics integrated into the cell. Such devices are highly accurate, with a typical inaccuracy figure being $\pm 0.2\%$ full-scale reading. They are also particularly insensitive to ambient condition changes and can measure pressures between 5 mbar and 2 bar.

## 15.9 Electronic Pressure Gauges

This section is included because many instrument manufacturers' catalogues have a section entitled "electronic pressure gauges." However, in reality, electronic pressure gauges are merely special forms of the pressure gauges described earlier in which electronic techniques are applied to improve performance. All of the following commonly appear in instrument catalogues under the heading "electronic pressure gauges."

*Piezoresistive pressure transducer:* This diaphragm-type sensor uses piezoresistive strain gauges to measure diaphragm displacement.

*Piezoelectric pressure transducer:* This diaphragm-type sensor uses a piezoelectric crystal to measure diaphragm displacement.

*Magnetic pressure transducer:* This class of diaphragm-type device measures diaphragm displacement magnetically using inductive, variable reluctance, or eddy current sensors.

*Capactive pressure transducer:* This diaphragm-type sensor measures variation in capacitance between the diaphragm and a fixed metal plate close to it.

*Fiber-optic pressure sensor:* Known alternatively as an optical pressure sensor, this uses a fiber-optic sensor to measure the displacement of either a diaphragm or a Bourdon tube pressure sensor.

*Potentiometric pressure sensor:* This is a device where the translational motion of a bellows-type pressure sensor is connected to the sliding element of an electrical potentiometer

*Resonant pressure transducer:* This is a form of resonant wire pressure-measuring device in which the pressure-induced frequency change is measured by electronics integrated into the device.

## 15.10 Special Measurement Devices for Low Pressures

The term *vacuum gauge* is applied commonly to describe any pressure sensor designed to measure pressures in the vacuum range (pressures less than atmospheric pressure, i.e., below 1.013 bar). Many special versions of the types of pressure transducers described earlier have been developed for measurement in the vacuum gauge. The typical minimum pressure measureable by these special forms of "normal" pressure-measuring instruments are 10 mbar (Bourdon tubes), 0.1 mbar (manometers and bellows-type instruments), and 0.001 mbar (diaphragms). However, in addition to these special versions of normal instruments, a number of other devices have been specifically developed for measurement of pressures below atmospheric pressure. These special devices include the thermocouple gauge, the Pirani gauge, the thermistor gauge, the McLeod gauge, and the ionization gauge, and they are covered in more detail next. Unfortunately, all of these specialized instruments are quite expensive.

### 15.10.1 Thermocouple Gauge

The thermocouple gauge is one of a group of gauges working on the thermal conductivity principle. At low pressure, the kinematic theory of gases predicts a linear relationship between pressure and thermal conductivity. Thus measurement of thermal conductivity gives an indication of pressure. Figure 15.8 shows a sketch of a thermocouple gauge. Operation of the gauge depends on the thermal conduction of heat between a thin hot metal strip in the center and the cold outer surface of a glass tube (that is normally at room temperature). The metal strip is heated by passing a current through it and its temperature is measured by a thermocouple. The temperature measured depends on the thermal conductivity of the gas in the tube and hence on its pressure. A source of error in this instrument is the fact that heat is also transferred by radiation as well as conduction. This error is of a constant magnitude, independent of pressure. Hence, it can be measured, and thus correction can be made for it. However, it is usually more convenient to design for low radiation loss by choosing a heated element with low emissivity. Thermocouple gauges are typically used to measure pressures in the range $10^{-4}$ mbar up to 1 mbar.

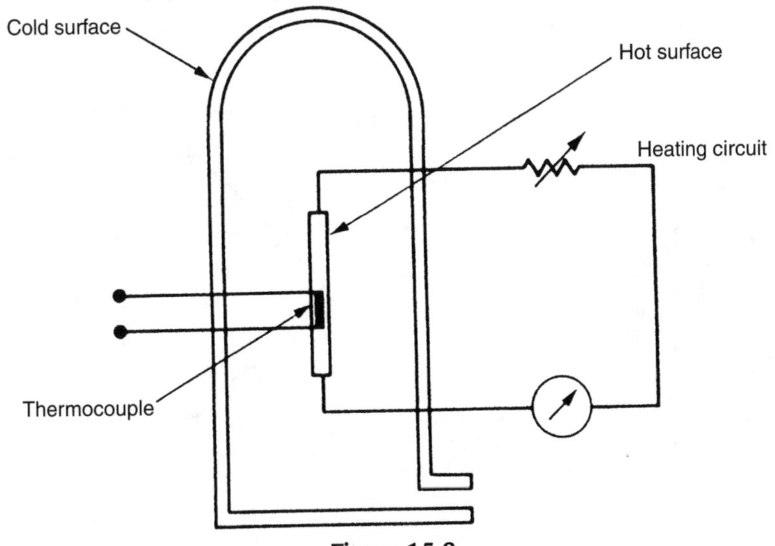


**Figure 15.8**
Thermocouple gauge.

### 15.10.2 Thermistor Gauge

This is identical in its mode of operation to a thermocouple gauge except that a thermistor is used to measure the temperature of the metal strip rather than a thermocouple. It is commonly marketed under the name *electronic vacuum gauge* in a form that includes a digital light-emitting diode display and switchable output ranges.

### 15.10.3 Pirani Gauge

A typical form of Pirani gauge is shown in Figure 15.9a. This is similar to a thermocouple gauge but has a heated element that consists of four coiled tungsten wires connected in parallel. Two identical tubes are normally used, connected in a bridge circuit, as shown in Figure 15.9b, with one containing the gas at unknown pressure and the other evacuated to a very low pressure. Current is passed through the tungsten element, which attains a certain temperature according to the thermal conductivity of the gas. The resistance of the element changes with temperature and causes an imbalance of the measurement bridge. Thus, the Pirani gauge avoids the use of a thermocouple to measure temperature (as in the thermocouple gauge) by effectively using a resistance thermometer as the heated element. Such gauges cover the pressure range $10^{-5}$ to 1 mbar.

### 15.10.4 McLeod Gauge

Figure 15.10a shows the general form of a McLeod gauge in which low-pressure fluid is compressed to a higher pressure that is then read by manometer techniques. In essence, the gauge can be visualized as a U-tube manometer that is sealed at one end and where the bottom

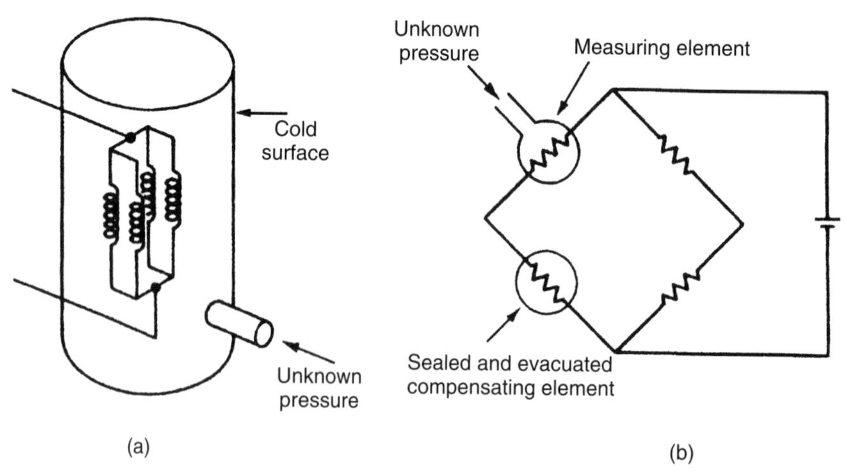


**Figure 15.9**
(a) Pirani gauge and (b) Wheatstone bridge circuit used to measure output.

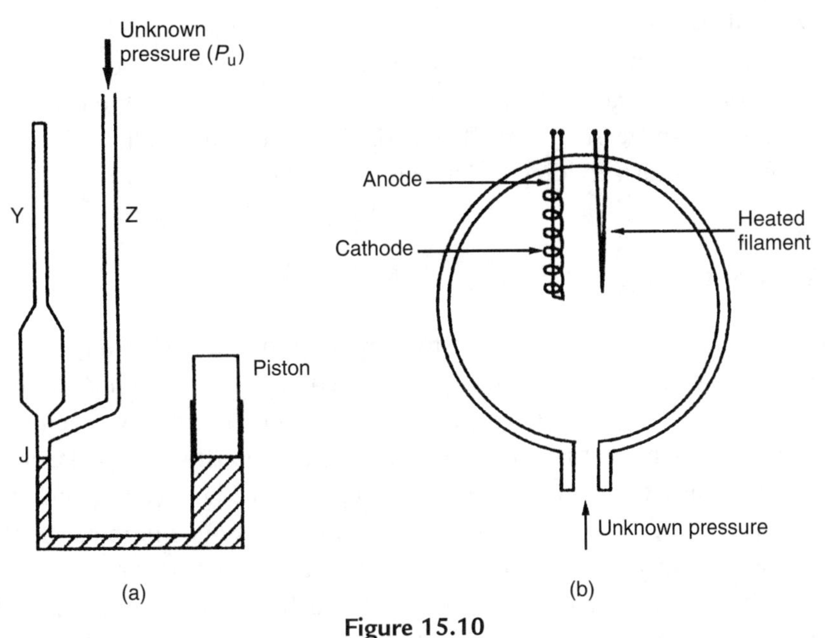


**Figure 15.10**
Other low-pressure gauges: (a) McLeod gauge and (b) ionization gauge.

of the U can be blocked at will. To operate the gauge, the piston is first withdrawn. This causes the level of mercury in the lower part of the gauge to fall below the level of junction J between the two tubes marked Y and Z in the gauge. Fluid at unknown pressure $P_u$ is then introduced via the tube marked Z, from where it also flows into the tube of cross-sectional area A marked Y. Next, the piston is pushed in, moving the mercury level up to block junction J. At the stage where J is just blocked, the fluid in tube Y is at pressure $P_u$ and is contained in a known volume, $V_u$. Further movement of the piston compresses the fluid in tube Y and this process continues until the mercury level in tube Z reaches a zero mark. Measurement of the height ($h$) above the mercury column in tube Y then allows calculation of the compressed volume of the fluid, $V_c$, as $V_c = hA$.

Then, by Boyle's law: $P_u V_u = P_c V_c$.

Also, applying the normal manometer equation, $P_c = P_u + h\rho g$, where ρ is the mass density of mercury, the pressure, $P_u$, can be calculated as

$$P_u = \frac{Ah^2 \rho g}{V_u - Ah} \tag{15.1}$$

Compressed volume $V_c$ is often very much smaller than the original volume, in which case Equation (15.1) approximates to

$$P_u = \frac{Ah^2 \rho g}{V_u} \quad \text{for } Ah << V_u \tag{15.2}$$

Although the smallest inaccuracy achievable with McLeod gauges is ±1%, this is still better than that achievable with most other gauges available for measuring pressures in this range. Therefore, the McLeod gauge is often used as a standard against which other gauges are calibrated. The minimum pressure normally measurable is $10^{-1}$ mbar, although lower pressures can be measured if pressure-dividing techniques are applied.

### 15.10.5 Ionization Gauge

The ionization gauge is a special type of instrument used for measuring very low pressures in the range $10^{-10}$ to 1 mbar. Normally, they are only used in laboratory conditions because their calibration is very sensitive to the composition of the gases in which they operate, and use of a mass spectrometer is often necessary to determine the gas composition around them. They exist in two forms known as a hot cathode and a cold cathode. The hot cathode form is shown schematically in Figure 15.10b. In this, gas of unknown pressure is introduced into a glass vessel containing free electrons discharged from a heated filament, as shown in Figure 15.10b. Gas pressure is determined by measuring the current flowing between an anode and a cathode within the vessel. This current is proportional to the number of ions per unit volume, which in turn is proportional to the gas pressure. Cold cathode ionization gauges operate in a similar fashion except that the stream of electrons is produced by a high voltage electrical discharge.

## 15.11 High-Pressure Measurement (Greater than 7000 bar)

Measurement of pressures above 7000 bar is normally carried out electrically by monitoring the change of resistance of wires of special materials. Materials having resistance pressure characteristics that are suitably linear and sensitive include manganin and gold–chromium alloys. A coil of such wire is enclosed in a sealed, kerosene-filled, flexible bellows, as shown in Figure 15.11. The unknown pressure is applied to one end of the bellows, which transmit pressure to the coil. The magnitude of the applied pressure is then determined by measuring the coil resistance. Devices are often named according to the metal used in them, for example, *manganin wire pressure sensor* and *gold–chromium wire pressure sensor*. Pressures up to 30,000 bar can be measured by devices such as the manganin wire pressure sensor, with a typical inaccuracy of ±0.5%.

## 15.12 Intelligent Pressure Transducers

Adding microprocessor power to pressure transducers brings about substantial improvements in their characteristics. Measurement sensitivity improvement, extended measurement range, compensation for hysteresis and other nonlinearities, and correction for ambient temperature and pressure changes are just some of the facilities offered by intelligent pressure transducers.

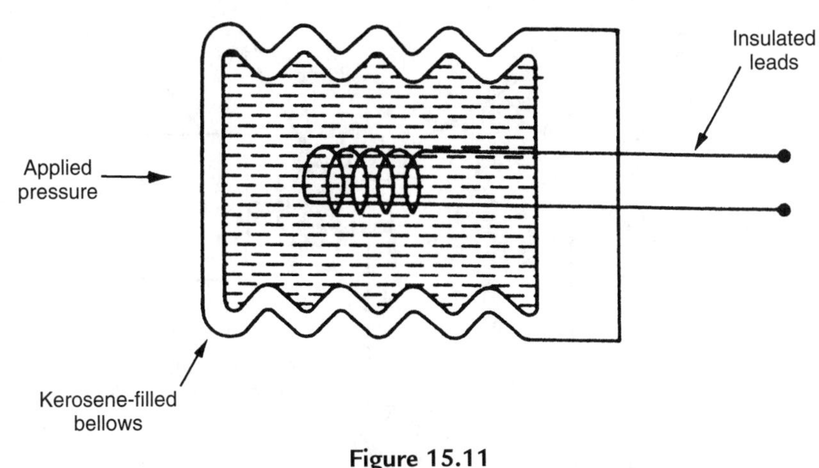


**Figure 15.11**
High-pressure measurement; wire coil in bellows.

For example, inaccuracy values as low as ±0.1% can be achieved with silicon piezoresistive-bridge devices.

Inclusion of microprocessors has also enabled the use of novel techniques of displacement measurement, for example, the optical method of displacement measurement shown in Figure 15.12. In this, the motion is transmitted to a vane that progressively shades one of two monolithic photodiodes exposed to infrared radiation. The second photodiode acts as a reference, enabling the microprocessor to compute a ratio signal that is linearized and is available as either an analogue or a digital measurement of pressure. The typical measurement inaccuracy is ±0.1%. Versions of both diaphragms and Bourdon tubes that use this technique are available.

## 15.13 Differential Pressure-Measuring Devices

Differential pressure-measuring devices have two input ports. One unknown pressure is applied to each port, and instrument output is the difference between the two pressures. An alternative way to measure differential pressure would be to measure each pressure with a separate instrument and then subtract one reading from the other. However, this would produce a far less accurate measurement of the differential pressure because of the well-known problem that the process of subtracting measurements amplifies the inherent inaccuracy in each individual measurement. This is a particular problem when measuring differential pressures of low magnitude.

Differential pressure can be measured by special forms of many of the pressure-measuring devices described earlier. Diaphragm pressure sensors, and their piezoresistive, piezoelectric, magnetic, capacitive, and fiber-optic named variants, are all commonly available in a

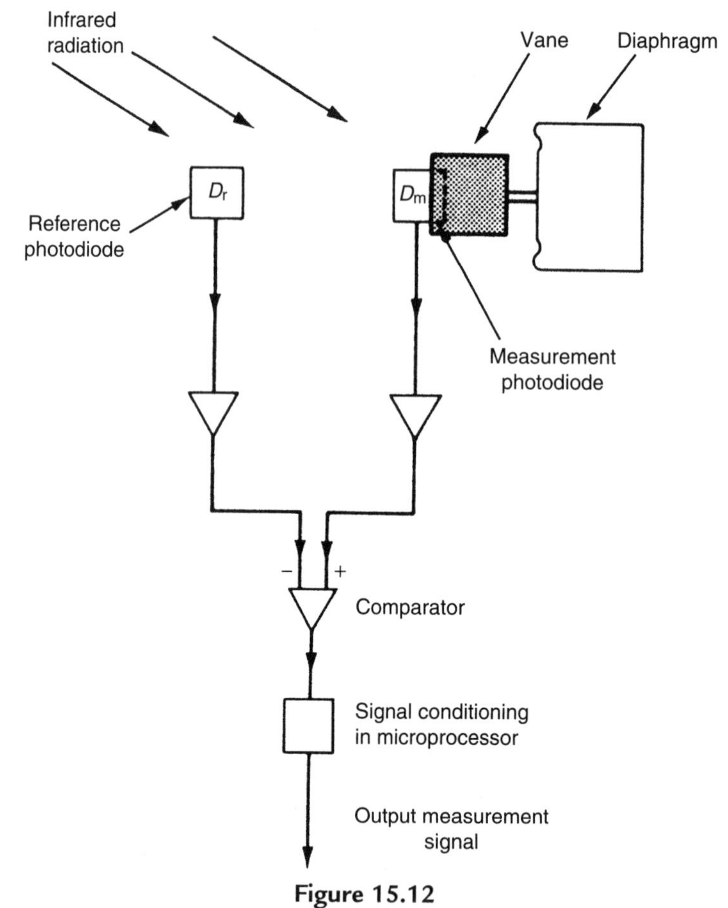


**Figure 15.12**
Example of an intelligent pressure-measuring instrument.

differential-pressure-measuring form in which the two pressures to be subtracted are applied to either side of the diaphragm. Double-bellows pressure transducers (including devices known as potentiometric pressure transducers) are also used, but are much less common than diaphragm-based sensors. A special form of U-tube manometer is also sometimes used when a visual indication of differential pressure values is required. This has the advantage of being a passive instrument that does not require a power supply and it is used commonly in liquid flow-rate indicators.

## 15.14 Selection of Pressure Sensors

Choice between the various types of instruments available for measuring midrange pressures (1.013–7000 bar) is usually strongly influenced by the intended application. Manometers are used commonly when just a visual indication of pressure level is required, and dead-weight gauges, because of their superior accuracy, are used in calibration procedures of other pressure-measuring devices. When an electrical form of output is required, the choice is usually either one out of the several types of diaphragm sensors (strain gauge, piezoresistive, piezoelectric, magnetic, capacitive, or fiber optic) or, less commonly, a Bourdon tube. Bellows-type instruments are also sometimes used for this purpose, but much less frequently. If very high measurement accuracy is required, the resonant wire device is a popular choice.

In the case of pressure measurement in the vacuum range (less than atmospheric pressure, i.e., below 1.013 bar), adaptations of most of the types of pressure transducers described earlier can be used. Special forms of Bourdon tubes measure pressures down to 10 mbar, manometers and bellows-type instruments measure pressures down to 0.1 mbar, and diaphragms can be designed to measure pressures down to 0.001 mbar. However, a number of more specialized instruments have also been developed to measure vacuum pressures, as discussed in Section 15.10. These generally give better measurement accuracy and sensitivity compared with instruments that are designed primarily for measuring midrange pressures. This improved accuracy is particularly evident at low pressures. Therefore, only the special instruments described in Section 15.10 are used to measure pressures below $10^{-4}$ mbar.

At high pressures ($>7000$ bar), the only devices in common use are the manganin wire sensor and similar devices based on alternative alloys to manganin.

For differential pressure measurement, diaphragm-type sensors are the preferred option, with double-bellows sensors being used occasionally. Manometers are also sometimes used to give visual indication of differential pressure values (especially in liquid flow-rate indicators). These are passive instruments that have the advantage of not needing a power supply.

## 15.15 Calibration of Pressure Sensors

Different types of reference instruments are used according to the range of the pressure-measuring instrument being calibrated. In the midrange of pressures from 0.1 mbar to 20 bar, U-tube manometers, dead-weight gauges, and barometers can all be used as reference instruments for calibration purposes. The vibrating cylinder gauge also provides a very accurate reference standard over part of this range. At high pressures above 20 bar, a gold–chrome alloy resistance reference instrument is normally used. For low pressures in the range of $10^{-1}$ to $10^{-3}$ mbar, both the McLeod gauge and various forms of micromanometers are used as a pressure-measuring standard. At even lower pressures below $10^{-3}$ mbar, a pressure-dividing technique is used to establish calibration. This involves setting up a series of orifices of an accurately known pressure ratio and measuring the upstream pressure with a McLeod gauge or micromanometer.

The limits of accuracy with which pressure can be measured by presently known techniques are as follows:

$10^{-7}$ mbar ±4%
$10^{-5}$ mbar ±2%
$10^{-3}$ mbar ±1%
$10^{-1}$ mbar ±0.1%
1 bar ±0.001%
$10^{4}$ bar ±0.1%

### 15.15.1 Reference Calibration Instruments

*Dead-weight gauge (pressure balance)*

The dead-weight gauge, also known by the alternative names of *piston gauge* and *pressure balance*, is shown in schematic form in Figure 15.13. It is a null-reading type of measuring instrument in which weights are added to the piston platform until the piston is adjacent to a fixed reference mark, at which time the downward force of the weights on top of the piston is balanced by the pressure exerted by the fluid beneath the piston. The fluid pressure is therefore calculated in terms of the weight added to the platform and the known area of the piston. The instrument offers the ability to measure pressures to a high degree of accuracy and is widely used as a reference instrument against which other pressure-measuring instruments are calibrated in the midrange of pressures. Unfortunately, its mode of measurement makes it inconvenient to use and is therefore rarely used except for calibration duties.

Special precautions are necessary in the manufacture and use of dead-weight gauges. Friction between the piston and the cylinder must be reduced to a very low level, as otherwise a significant measurement error would result. Friction reduction is accomplished by designing for a small clearance gap between the piston and the cylinder by machining the cylinder to a slightly greater diameter than the piston. The piston and cylinder are also designed so that they

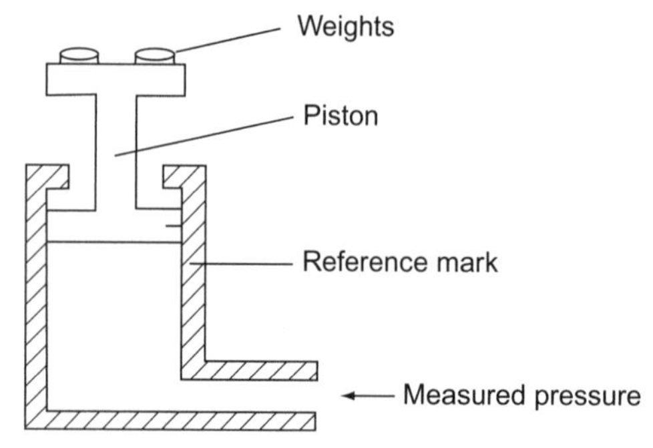


**Figure 15.13**
Schematic representation of a dead-weight gauge.

can be turned relative to one another, which reduces friction still further. Unfortunately, as a result of the small gap between the piston and the cylinder, there is a finite flow of fluid past the seals. This produces a viscous shear force, which partly balances the dead weight on the platform. A theoretical formula exists for calculating the magnitude of this shear force, suggesting that exact correction can be made for it. In practice, however, the piston deforms under pressure and alters the piston/cylinder gap and so shear force calculation and correction can only be approximate.

Despite these difficulties, the instrument gives a typical measurement inaccuracy of only $\pm 0.01\%$. It is normally used for calibrating pressures in the range of 20 mbar up to 20 bar. However, special versions can measure pressures down to 0.1 mbar or up to 7000 bar.

#### *U-tube manometer*

In addition to its use for normal process measurements, the U-tube manometer is also used as a reference instrument for calibrating instruments measuring midrange pressures. Although it is a deflection rather than null type of instrument, it manages to achieve similar degrees of measurement accuracy to the dead-weight gauge because of the error sources noted in the latter. The major source of error in U-tube manometers arises out of the difficulty in estimating the meniscus level of the liquid column accurately. There is also a tendency for the liquid level to creep up the tube by capillary action, which creates an additional source of error.

U tubes for measuring high pressures become unwieldy because of the long lengths of liquid column and tube required. Consequently, U-tube manometers are normally used only for calibrating pressures at the lower end of the midpressure range.

#### *Barometers*

The most commonly used type of barometer for calibration duties is the Fortin barometer. This is a highly accurate instrument that provides measurement inaccuracy levels of between $\pm 0.03$ and $\pm 0.001\%$ of full-scale reading depending on the measurement range. To achieve such levels of accuracy, the instrument has to be used under very carefully controlled conditions of lighting, temperature, and vertical alignment. It must also be manufactured to exacting standards and is therefore very expensive to buy. Corrections have to be made to the output reading according to ambient temperature, local value of gravity, and atmospheric pressure. Because of its expense and difficulties in using it, the barometer is not normally used for calibration other than as a primary reference standard at the top of the calibration chain.

#### *Vibrating cylinder gauge*

The vibrating cylinder gauge, shown in Figure 15.14, acts as a reference standard instrument for calibrating pressure measurements up to 3.5 bar. It consists of a cylinder in which vibrations at the resonant frequency are excited by a current-carrying coil. The pressure-dependent oscillation frequency is monitored by a pickup coil, and this frequency measurement is

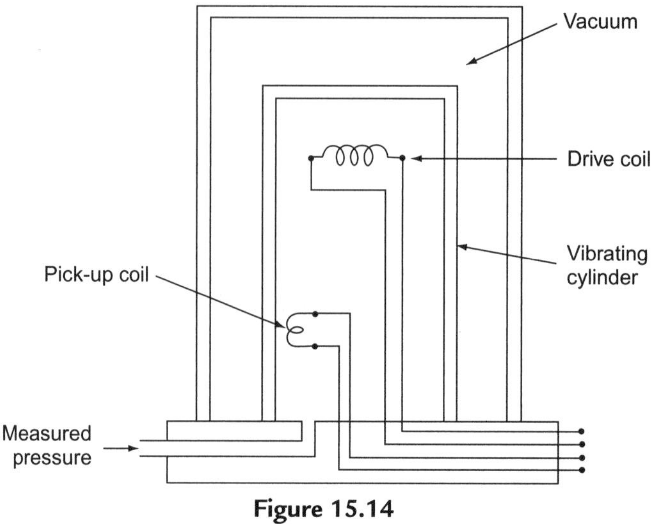


**Figure 15.14**
Vibrating cylinder gauge.

converted to a voltage signal by a microprocessor and signal conditioning circuitry contained within the package. By evacuating the space on the outer side of the cylinder, the instrument is able to measure the absolute pressure of the fluid inside the cylinder. Measurement errors are less than 0.005% over the absolute pressure range up to 3.5 bar.

*Gold-chrome alloy resistance instruments*

For measuring pressures above 7000 bar, an instrument based on measuring the resistance change of a metal coil as the pressure varies is used, and the same type of instrument is also used for calibration purposes. Such instruments commonly use manganin or gold–chrome alloys for the coil. Gold–chrome has a significantly lower temperature coefficient (i.e., its pressure/resistance characteristic is less affected by temperature changes) and is therefore the normal choice for calibration instruments, despite its higher cost. An inaccuracy of only $\pm 0.1\%$ is achievable in such devices.

*McLeod gauge*

The McLeod gauge, which has already been discussed earlier in Section 15.10, can be used for the calibration of instruments designed to measure low pressures between $10^{-4}$ and 0.1 mbar ($10^{-7}$ to $10^{-4}$ bar).

*Ionization gauge*

An ionization gauge is used to calibrate instruments measuring very low pressures in the range $10^{-13}$ to $10^{-3}$ bar. It has the advantage of having a straight-line relationship between output reading and pressure. Unfortunately, its inherent accuracy is relatively poor, and specific points on its output characteristic have to be calibrated against a McLeod gauge.

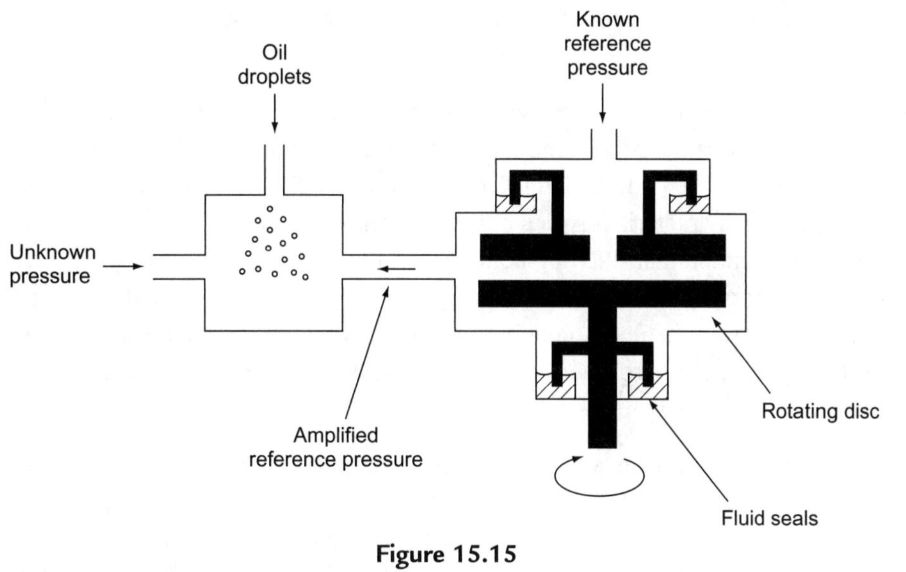


**Figure 15.15**
Centrifugal micromanometer.

*Micromanometers*

Micromanometers are instruments that work on the manometer principle but are specially designed to minimize capillary effects and meniscus reading errors. The centrifugal form of a micromanometer, shown schematically in Figure 15.15, is the most accurate type for use as a calibration standard down to pressures of $10^{-3}$ mbar. In this, a rotating disc serves to amplify a reference pressure, with the speed of rotation being adjusted until the amplified pressure just balances the unknown pressure. This null position is detected by observing when oil droplets sprayed into a glass chamber cease to move. Measurement inaccuracy is $\pm 1\%$.

Other types of micromanometers also exist, which give similar levels of accuracy, but only at somewhat higher pressure levels. These can be used as calibration standards at pressures up to 50 mbar.

### *15.15.2 Calculating Frequency of Calibration Checks*

Some pressure-measuring instruments are very stable and unlikely to suffer drift in characteristics with time. Devices in this class include resonant wire devices, ionization gauges, and high-pressure instruments (those working on the principle of resistance change with pressure). All forms of manometers are similarly stable, although small errors can develop in these through volumetric changes in the glass in the longer term. Therefore, for all these instruments, only annual calibration checks are recommended, unless of course something happens to the instrument that puts its calibration into question.

However, most instruments used to pressure consist of an elastic element and a displacement transducer that measures its movement. Both of these component parts are mechanical in nature. Devices of this type include diaphragms, bellows, and Bourdon tubes. Such instruments can suffer changes in characteristics for a number of reasons. One factor is the characteristics of the operating environment and the degree to which the instrument is exposed to it. Another reason is the amount of mishandling they receive. These parameters are entirely dependent on the particular application the instrument is used in and the frequency with which it is used and exposed to the operating environment. A suitable calibration frequency can therefore only be determined on an experimental basis.

A third class of instrument from the calibration requirements viewpoint is the range of devices working on the thermal conductivity principle. This range includes the thermocouple gauge, Pirani gauge, and thermistor gauge. Such instruments have characteristics that vary with the nature of the gas being measured and must therefore be calibrated each time that they are used.

### 15.15.3 Procedures for Calibration

Pressure calibration requires the output reading of the instrument being calibrated to be compared with the output reading of a reference standard instrument when the same pressure is applied to both. This necessitates designing a suitable leakproof seal to connect the pressure-measuring chambers of the two instruments.

The calibration of pressure transducers used for process measurements often has to be carried out in situ in order to avoid serious production delays. Such devices are often remote from the nearest calibration laboratory and to transport them there for calibration would take an unacceptably long time. Because of this, portable reference instruments have been developed for calibration at this level in the calibration chain. These use a standard air supply connected to an accurate pressure regulator to provide a range of reference pressures. An inaccuracy of $\pm 0.025\%$ is achieved when calibrating midrange pressures in this manner. Calibration at higher levels in the calibration chain must, of course, be carried out in a proper calibration laboratory maintained in the correct manner. However, irrespective of where calibration is carried out, several special precautions are necessary when calibrating certain types of instrument, as described in the following paragraphs.

U-tube manometers must have their vertical alignment set up carefully before use. Particular care must also be taken to ensure that there are no temperature gradients across the two halves of the tube. Such temperature differences would cause local variations in the specific weight of the manometer fluid, resulting in measurement errors. Correction must also be made for the local value of $g$ (acceleration due to gravity). These comments apply similarly to the use of other types of manometers and micromanometers.

The existence of one potentially major source of error in Bourdon tube pressure measurement has not been widely documented, and few manufacturers of Bourdon tubes make any attempt to warn users of their products appropriately. This problem is concerned with the relationship between the fluid being measured and the fluid used for calibration. The pointers of Bourdon tubes are normally set at zero during manufacture, using air as the calibration medium. However, if a different fluid, especially a liquid, is subsequently used with a Bourdon tube, the fluid in the tube will cause a nonzero deflection according to its weight compared with air, resulting in a reading error of up to 6% of full-scale deflection.

This can be avoided by calibrating the Bourdon tube with the fluid to be measured instead of with air. Alternatively, correction can be made according to the calculated weight of the fluid in the tube. Unfortunately, difficulties arise with both of these solutions if air is trapped in the tube, as this will prevent the tube being filled completely by the fluid. Then, the amount of fluid actually in the tube, and its weight, will be unknown. To avoid this problem, at least one manufacturer now provides a bleed facility in the tube, allowing measurement uncertainties of less than 0.1% to be achieved.

When using a McLeod gauge, care must be taken to ensure that the measured gas does not contain any vapor. This would be condensed during the compression process, causing a measurement error. A further recommendation is insertion of a liquid–air cold trap between the gauge and the instrument being calibrated to prevent the passage of mercury vapor into the latter.

## 15.16 Summary

We started the chapter off by looking at the formal definitions of the three ways in which pressure is quantified in terms of absolute, gauge, and differential pressure. We then went on to look at the devices used for measuring pressure in three distinct ranges: normal, midrange pressures between 1.013 bar (the mean atmospheric pressure) and 7000 bar, low or vacuum pressures below 1.013 bar, and high pressures above 7000 bar.

We saw that a large number of devices are available for measurements in the "normal" range. Of these, sensors containing a diaphragm are used most commonly. We looked at the type of material used for the diaphragm in diaphragm-based sensors and also examined the different ways in which diaphragm movement can be measured. These different ways of measuring diaphragm displacement give rise to a number of different names for diaphragm-based sensors, such as capacitive and fiber-optic (optical) sensors.

Moving on, we examined various other devices used to measure midrange pressures. These included bellows sensors, Bourdon tubes, several types of manometers, and resonant wire sensors. We also looked at the range of devices commonly called electronic pressure gauges. Many of these are diaphragm-based sensors that use an electronic means of measuring diaphragm displacement, with names such as piezoresistive pressure sensor, piezoelectric pressure sensor, magnetic pressure sensor, and potentiometric pressure sensor.

We then went on to study the measurement of low pressures. To start with, we observed that special forms of instruments used commonly to measure midrange pressures can measure pressures below atmospheric pressure instruments (Bourdon tubes down to 10 mbar, bellows-type instruments down to 0.1 mbar, manometers down to 0.1 mbar, and diaphragm-based sensors down to 0.001 mbar). As well as these special versions of Bourdon tubes, several other instruments have been specially developed to measure in the low-pressure range. These include thermocouple and thermistor gauges (measure between $10^{-4}$ and 1 mbar), the Pirani gauge (measures between $10^{-5}$ and 1 mbar), the McLeod gauge (measures down to $10^{-1}$ mbar or even lower pressures if it is used in conjunction with pressure-dividing techniques), and the ionization gauge (measures between $10^{-10}$ to 1 mbar).

When we looked at measurement of high pressures, we found that our choice of instrument was much more limited. All currently available instruments for this pressure range involve monitoring the change of resistance in a coil of wire made from special materials. The two most common devices of this type are the manganin wire pressure sensor and gold–chromium wire pressure sensor.

The following three sections were devoted to intelligent pressure-measuring devices, instruments measuring differential pressure, and some guidance about which type of device to use in particular circumstances.

Then our final subject of study in the chapter was the means of calibrating pressure-measuring devices. We looked at various instruments used for calibration, including the dead-weight gauge, special forms of the U-tube manometers, barometers, the vibrating cylinder gauge, gold–chrome alloy resistance instruments, the McLeod gauge, the ionization gauge, and micromanometers. We then considered how the frequency of recalibration should be determined for various kinds of pressure-measuring devices. Finally, we looked in more detail at the appropriate practical procedures and precautions that should be taken for calibrating different types of instruments.

## 15.17 Problems

15.1. Explain the difference among absolute pressure, gauge pressure, and differential pressure. When pressure readings are being written down, what is the mechanism for defining whether the value is a gauge, absolute, or differential pressure?

15.2. Give examples of situations where pressure measurements are normally given as (a) absolute pressure, (b) gauge pressure, and (c) differential pressure.

15.3. Summarize the main classes of devices used for measuring absolute pressure.
15.4. Summarize the main classes of devices used for measuring gauge pressure.
15.5. Summarize the main classes of devices used for measuring differential pressure.
15.6. Explain what a diaphragm pressure sensor is. What are the different materials used in construction of a diaphragm pressure sensor and what are their relative merits?
15.7. Strain gauges are used commonly to measure displacement in a diaphragm pressure sensor. What are the difficulties in bonding a standard strain gauge to the diaphragm and how are these difficulties usually solved?
15.8. What are the advantages in using a monolithic piezoresistive displacement transducer with diaphragm pressure sensors?
15.9. What other types of devices apart from strain gauges are used to measure displacement in a diaphragm strain gauge? Summarize the main features of each of these alternative types of displacement sensors.
15.10. Discuss the mode of operation of fiber-optic pressure sensors. What are their principal advantages over other forms of pressure sensors?
15.11. What are bellows pressure sensors? How do they work? Describe some typical applications.
15.12. How does a Bourdon tube work? What are the three common shapes of Bourdon tubes and what is the typical measurement range of each type?
15.13. Describe the three types of manometers available. What is the typical measurement range of each type?
15.14. What is a resonant wire pressure-measuring device and what is it typically used for?
15.15. What is an electronic pressure gauge? Discuss the different types of electronic gauges that exist.
15.16. Discuss the range of instruments available for measuring very low pressures (pressures below atmospheric pressure).
15.17. How are high pressures (pressures above 7000 bar) normally measured?
15.18. What advantages do intelligent pressure transducers have over their nonintelligent counterparts?
15.19. A differential pressure can be measured by subtracting the readings from two separate pressure transducers. What is the problem with this? Suggest a better way of measuring differential pressures.
15.20. How are pressure transducers calibrated? How is a suitable frequency of calibration determined?
15.21. Which instruments are used as a reference standard in the calibration of pressure sensors?

CHAPTER 16

# Flow Measurement

## 16.1 Introduction

We now move on to look at flow measurement in this chapter. Flow measurement is concerned with quantifying the rate of flow of materials. Such measurement is quite a common requirement in the process industries. The material measured may be in a solid, liquid, or gaseous state. When the material is in a solid state, flow can only be quantified as the mass flow rate, this being the mass of material that flows in one unit of time. When the material is in a liquid or gaseous state, flow can be quantified as either the mass flow rate or the volume flow rate, with the latter being the volume of material that flows in one unit of time. Of the two, a flow measurement in terms of mass flow rate is preferred if very accurate measurement is required. The greater accuracy of mass flow measurement arises from the fact that mass is invariant whereas volume is a variable quantity.

A particular complication in the measurement of flow rate of liquids and gases flowing in pipes is the need to consider whether the flow is laminar or turbulent. Laminar flow is characterized by a motion of the fluid being in a direction parallel to the sides of the pipe, and it occurs in straight lengths of pipe when the fluid is flowing at a low velocity. However, it should be noted that even laminar flow is not uniform across the cross section of the pipe, with the velocity being greatest at the center of the pipe and decreasing to zero immediately next to the wall of the pipe. In contrast, turbulent flow involves a complex pattern of flow that is not in a uniform direction. Turbulent flow occurs in nonstraight sections of pipe and also occurs in straight sections when the fluid velocity exceeds a critical value. Because of the difficulty in measuring turbulent flow, the usual practice is to restrict flow measurement to places where the flow is laminar, or at least approximately laminar. This can be achieved by measuring the flow in the center of a long, straight length of pipe if the flow velocity is below the critical velocity for turbulent flow. In the case of high mean fluid velocity, it is often possible to find somewhere within the flow path where a larger diameter pipe exists and therefore the flow velocity is lower.

## 16.2 Mass Flow Rate

The method used to measure mass flow rate is determined by whether the measured quantity is in a solid, liquid, or gaseous state, as different techniques are appropriate for each. The main techniques available for measuring mass flow rate are summarized here.

### *16.2.1 Conveyor-Based Methods*

Conveyor-based methods are appropriate for measuring the flow of solids in the form of powders or small granular particles. Such powders and particles are produced commonly by crushing or grinding procedures in process industries, and a conveyor is a very suitable means of transporting materials in this form. Transporting materials on a conveyor allows the mass flow rate to be calculated in terms of the mass of material on a given length of conveyor multiplied by the speed of the conveyor. Figure 16.1 shows a typical measurement system. A load cell measures the mass, $M$, of material distributed over a length, $L$, of the conveyor. If the conveyor velocity is $v$, the mass flow rate, $Q$, is given by

$$Q = Mv/L.$$

As an alternative to weighing flowing material, a *nuclear mass-flow sensor* can be used, in which a $\gamma$-ray source is directed at the material being transported along the conveyor. The material absorbs some radiation, and the amount of radiation received by a detector on the other side of the material indicates the amount of material on the conveyor. This technique has obvious safety concerns and is therefore subject to licensing and strict regulation.

### *16.2.2 Coriolis Flowmeter*

As well as sometimes being known by the alternative name of *inertial flowmeter*, the Coriolis flowmeter is often referred to simply as a *mass flowmeter* because of its dominance in the mass flowmeter market. However, this assumption that a mass flowmeter always refers to a Coriolis meter is wrong, as several other types of devices are available to measure mass flow, although it is true to say that they are much less common than Coriolis meters.

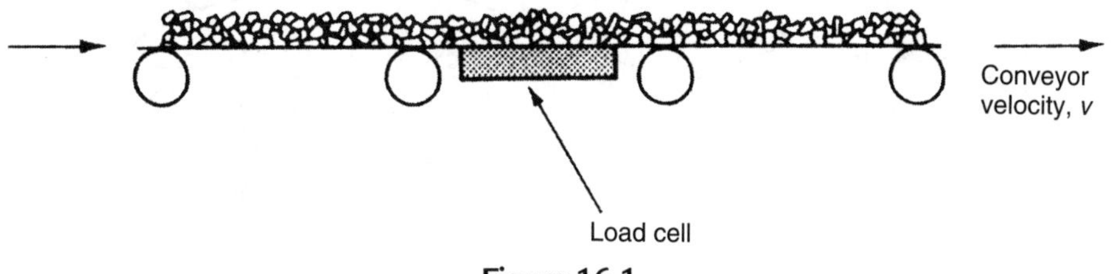


**Figure 16.1**
Conveyor-based mass-flow-rate measurement.

Coriolis meters are used primarily to measure the mass flow rate of liquids, although they have also been used successfully in some gas-flow measurement applications. The flowmeter consists either of a pair of parallel vibrating tubes or as a single vibrating tube that is formed into a configuration that has two parallel sections. The two vibrating tubes (or the two parallel sections of a single tube) deflect according to the mass flow rate of the measured fluid that is flowing inside. Tubes are made of various materials, of which stainless steel is the most common. They are also manufactured in different shapes, such as B shaped, D shaped, U shaped, triangular shaped, helix shaped, and straight. These alternative shapes are sketched in Figure 16.2a, and a U-shaped tube is shown in more detail in Figure 16.2b. The tubes are anchored at two points. An electromechanical drive unit, positioned midway between the two anchors, excites vibrations in each tube at the tube resonant frequency. Vibrations in the two tubes, or the two parallel sections of a single tube, are 180 degrees out of phase.

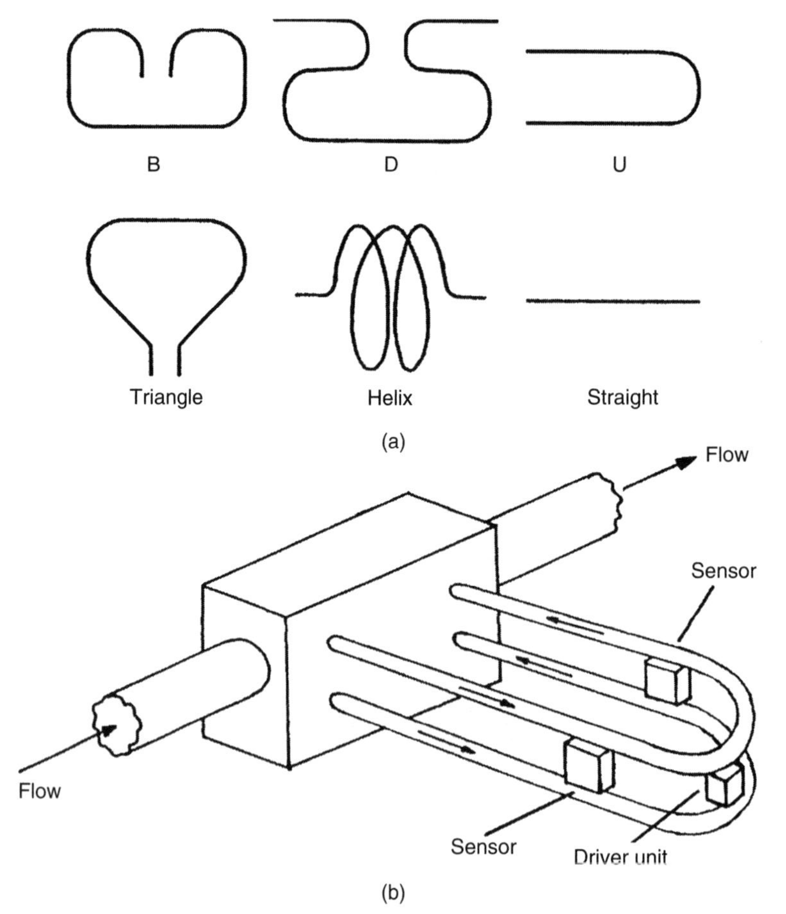


**Figure 16.2**
(a) Coriolis flowmeter shapes; (b) detail of U-shaped Coriolis flowmeter.

The vibratory motion of each tube causes forces on the particles in the flowing fluid. These forces induce motion of the fluid particles in a direction that is orthogonal to the direction of flow, which produces a Coriolis force. This Coriolis force causes a deflection of the tubes that is superimposed on top of the vibratory motion. The net deflection of one tube relative to the other is given by $d = kfR$, where $k$ is a constant, $f$ is the frequency of the tube vibration, and $R$ is the mass flow rate of the fluid inside the tube. This deflection is measured by a suitable sensor.

Coriolis meters give excellent accuracy, with measurement uncertainties of $\pm 0.2\%$ being typical. They also have low maintenance requirements. However, apart from being expensive (typical cost is \$6000), they suffer from a number of operational problems. Failure may occur after a period of use because of mechanical fatigue in the tubes. Tubes are also subject to both corrosion caused by chemical interaction with the measured fluid and abrasion caused by particles within the fluid. Diversion of the flowing fluid around the flowmeter causes it to suffer a significant pressure drop, although this is much less evident in straight tube designs.

### 16.2.3 Thermal Mass Flow Measurement

Thermal mass flowmeters are used primarily to measure the flow rate of gases. The principle of operation is to direct the flowing material past a heated element. The mass flow rate is inferred in one of two ways: (a) by measuring the temperature rise in the flowing material or (b) by measuring the heater power required to achieve a constant set temperature in the flowing material. In both cases, the specific heat and density of the flowing fluid must be known. Typical measurement uncertainty is $\pm 2\%$. Standard instruments require the measured gas to be clean and noncorrosive. However, versions made from special alloys can cope with more aggressive gases. Tiny versions of thermal mass flowmeters have been developed that can measure very small flow rates in the range of nanoliters ($10^{-9}$ liters) or microliters ($10^{-6}$ liters) per minute.

### 16.2.4 Joint Measurement of Volume Flow Rate and Fluid Density

Before the advent of the Coriolis meter, the usual way of measuring the mass flow rate was to compute this from separate, simultaneous measurements of the volume flow rate and the fluid density. In many circumstances, this is still the least expensive option, although measurement accuracy is substantially inferior to that provided by a Coriolis meter.

## 16.3 Volume Flow Rate

Volume flow rate is an appropriate way of quantifying the flow of all materials that are in a gaseous, liquid, or semiliquid slurry form (where solid particles are suspended in a liquid host), although measurement accuracy is inferior to mass flow measurement as noted earlier. Materials in these forms are usually carried in pipes, and various instruments can be used to measure the volume flow rate as described later. As noted in the introduction, these all assume laminar flow. In addition, flowing liquids are sometimes carried in an open channel, in which case the volume flow rate can be measured by an open channel flowmeter.

### 16.3.1 Differential Pressure (Obstruction-Type) Meters

Differential pressure meters involve the insertion of some device into a fluid-carrying pipe that causes an obstruction and creates a pressure difference on either side of the device. Such meters are sometimes known as obstruction-type meters or flow restriction meters. Devices used to obstruct the flow include the *orifice plate*, *Venturi tube*, *flow nozzle*, and *Dall flow tube*, as illustrated in Figure 16.3. When such a restriction is placed in a pipe, the velocity of the fluid through the restriction increases and the pressure decreases. The volume flow rate is then proportional to the square root of the pressure difference across the obstruction. The manner in which this pressure difference is measured is important. Measuring the two pressures with different instruments and calculating the difference between the two measurements is not satisfactory because of the large measurement error that can arise when the pressure difference is small, as explained in Chapter 3. Therefore, the normal procedure is to use a differential pressure transducer, which is commonly a diaphragm-type device.

The *pitot static tube* is another device that measures flow by creating a pressure difference within a fluid-carrying pipe. However, in this case, there is negligible obstruction of flow in the pipe. The pitot tube is a very thin tube that obstructs only a small part of the flowing fluid and thus measures flow at a single point across the cross section of the pipe. This measurement

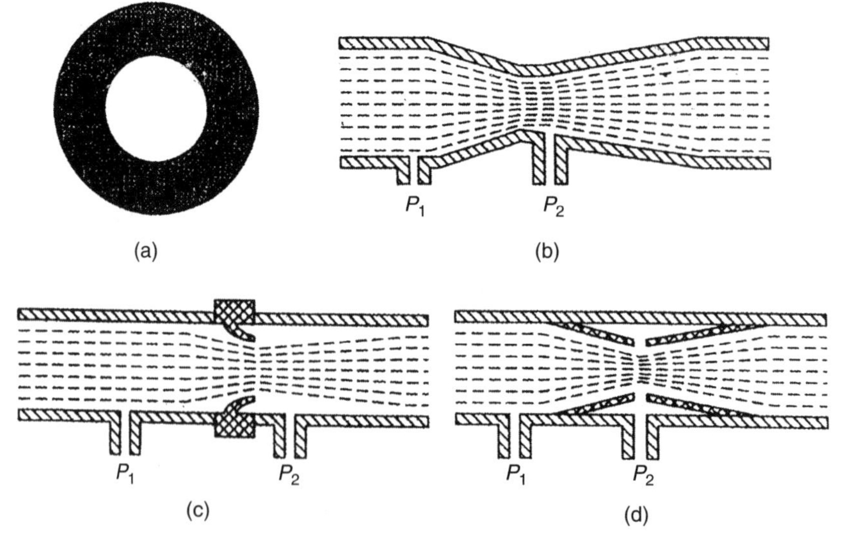


**Figure 16.3**
Obstruction devices: (a) orifice plate, (b) Venturi, (c) flow nozzle, and (d) Dall flow tube.

only equates to average flow velocity in the pipe for the case of uniform flow. The *annubar* is a type of multiport pitot tube that measures the average flow across the cross section of the pipe by forming the mean value of several local flow measurements across the cross section of the pipe.

All applications of this method of flow measurement assume laminar flow by ensuring that the flow conditions upstream of the obstruction device are in steady state; a certain minimum length of straight run of pipe ahead of the flow measurement point is specified to achieve this. The minimum lengths required for various pipe diameters are specified in standards tables. However, a useful rule of thumb widely used in process industries is to specify a length of 10 times the pipe diameter. If physical restrictions make this impossible to achieve, special flow-smoothing vanes can be inserted immediately ahead of the measurement point.

Flow restriction-type instruments are popular because they have no moving parts and are therefore robust, reliable, and easy to maintain. However, one significant disadvantage of this method is that the obstruction causes a permanent loss of pressure in the flowing fluid. The magnitude and hence importance of this loss depend on the type of obstruction element used, but where the pressure loss is large, it is sometimes necessary to recover the lost pressure by an auxiliary pump further down the flow line. This class of device is not normally suitable for measuring the flow of slurries, as the tappings into the pipe to measure the differential pressure are prone to blockage, although the Venturi tube can be used to measure the flow of dilute slurries.

Figure 16.4 illustrates approximately the way in which the flow pattern is interrupted when an orifice plate is inserted into a pipe. Other obstruction devices also have a similar effect to this, although the magnitude of pressure loss is smaller. Of particular interest is the fact that the minimum cross-sectional area of flow occurs not within the obstruction but at a point downstream of there. Knowledge of the pattern of pressure variation along the pipe, as shown in Figure 16.5,

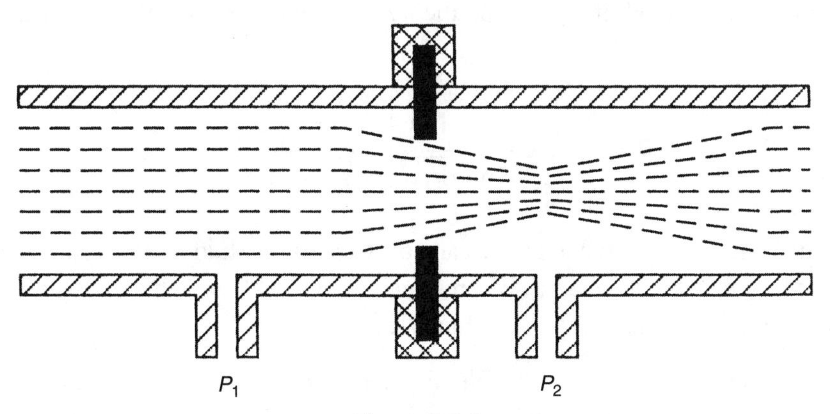


**Figure 16.4**
Profile of flow across orifice plate.

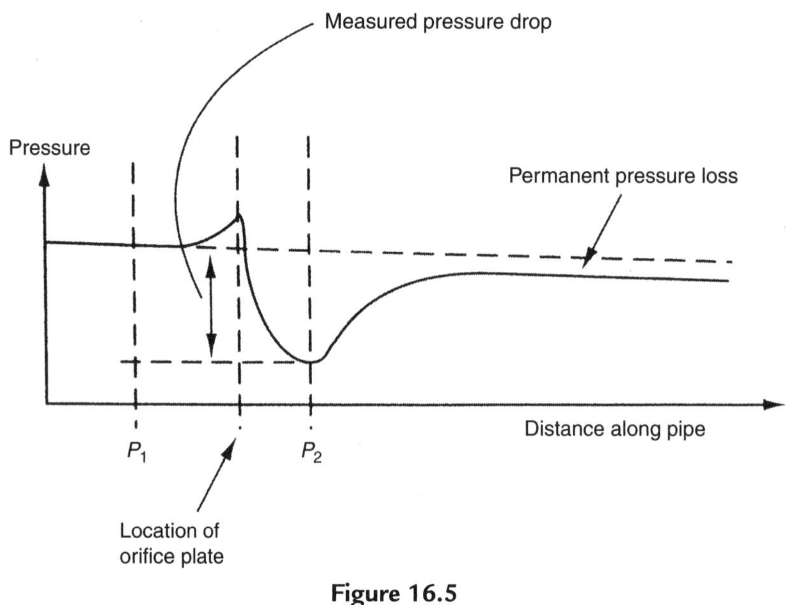


**Figure 16.5**
Pattern of pressure variation either side of orifice plate.

is also of importance in using this technique of volume–flow-rate measurement. This shows that the point of minimum pressure coincides with the point of minimum cross-section flow a little way downstream of the obstruction. Figure 16.5 also shows that there is a small rise in pressure immediately before the obstruction. It is therefore important not only to position the instrument measuring $P_2$ exactly at the point of minimum pressure, but also to measure the pressure $P_1$ at a point upstream of the point where the pressure starts to rise before the obstruction.

In the absence of any heat transfer mechanisms, and assuming frictionless flow of an incompressible fluid through the pipe, the theoretical volume flow rate of the fluid, $Q$, is given by

$$Q = \left[\frac{A_2}{\sqrt{1-(A_2/A_1)^2}}\right]\left[\sqrt{\frac{2(P_1-P_2)}{\rho}}\right], \tag{16.1}$$

where $A_1$ and $P_1$ are the cross-sectional area and pressure of the fluid flow before the obstruction, $A_2$ and $P_2$ are the cross-sectional area and pressure of the fluid flow at the narrowest point of the flow beyond the obstruction, and $\rho$ is the fluid density.

Equation (16.1) is never entirely applicable in practice for two main reasons. First, the flow is always impeded by a friction force, which varies according to the type of fluid and its velocity and is quantified by a constant known as the Reynold's number. Second, the cross-sectional area of the fluid flow ahead of the obstruction device is less than the diameter of the pipe carrying it, and the minimum cross-sectional area of the fluid after the obstruction is less than the diameter of the obstruction. This latter problem means that neither $A_1$ nor $A_2$ can be measured accurately. Fortunately, provided the pipe is smooth and therefore the friction force is small, these two problems can be accounted for adequately by applying a constant called the discharge coefficient. This modifies Equation (16.1) to the following:

$$Q = \left[\frac{C_D A'_2}{\sqrt{1-(A'_2/A'_1)^2}}\right]\left[\sqrt{\frac{2(P_1-P_2)}{\rho}}\right], \qquad (16.2)$$

where $A'_1$ and $A'_2$ are the actual pipe diameters before and at the obstruction and $C_D$ is the discharge coefficient that corrects for the friction force and the difference between the pipe and flow cross-section diameters.

Before Equation (16.2) can be evaluated, the discharge coefficient must be calculated. As this varies between each measurement situation, it would appear at first sight that the discharge coefficient must be determined by practical experimentation in every case. However, provided that certain conditions are met, standard tables can be used to obtain the value of the discharge coefficient appropriate to the pipe diameter and fluid involved.

One particular problem with all flow restriction devices is that the pressure drop, $(P_1 - P_2)$, varies as the square of the flow rate $Q$ according to Equation (16.2). The difficulty of measuring small pressure differences accurately has already been noted earlier. In consequence, the technique is only suitable for measuring flow rates that are between 30 and 100% of the maximum flow rate that a given device can handle. This means that alternative flow measurement techniques have to be used in applications where the flow rate can vary over a large range that can drop to below 30% of the maximum rate.

### *Orifice plate*

The orifice plate is a metal disc with a concentric hole in it, which is inserted into the pipe carrying the flowing fluid. Orifice plates are simple, inexpensive, and available in a wide range of sizes. In consequence, they account for almost 50% of the instruments used in industry for measuring volume flow rate. One limitation of the orifice plate is that its inaccuracy is typically at least $\pm 2\%$ and may approach $\pm 5\%$. Also, the permanent pressure loss caused in the measured fluid flow is between 50 and 90% of the magnitude of the pressure difference, $(P_1 - P_2)$. Other problems with the orifice plate are a gradual change in the discharge coefficient over a period of time as the sharp edges of the hole wear away and a tendency for any particles in the flowing fluid to stick behind the hole, thereby reducing its diameter gradually as the particles build up. The latter problem can be minimized by using an orifice plate with an eccentric hole. If this hole is close to the bottom of the pipe, solids in the flowing fluid tend to be swept through, and buildup of particles behind the plate is minimized. A very similar problem arises if there are any bubbles of vapor or gas in the flowing fluid when liquid flow is involved. These also tend to build up behind an orifice plate and distort the pattern of flow. This difficulty can be avoided by mounting the orifice plate in a vertical run of pipe.

### *Venturis and similar devices*

A number of obstruction devices are available that are specially designed to minimize pressure loss in the measured fluid. These have various names such as Venturi, flow nozzle, and Dall flow tube. They are all much more expensive than an orifice plate but have better performance. The smooth internal shape means that they are not prone to solid particles or bubbles of gas sticking in the obstruction, as is likely to happen in an orifice plate. The smooth shape also means that they suffer much less wear and, consequently, have a longer life than orifice plates. They also require less maintenance and give greater measurement accuracy.

**Venturi:** The Venturi has a precision-engineered tube of a special shape. This offers measurement uncertainty of only $\pm 1\%$. However, the complex machining required to manufacture it means that it is the most expensive of all the obstruction devices discussed. Permanent pressure loss in the measured system is 10–15% of the pressure difference $(P_1 - P_2)$ across it.

**Dall flow tube:** The Dall flow tube consists of two conical reducers inserted into a fluid-carrying pipe. It has a very similar internal shape to the Venturi, except that it lacks a throat. This construction is much easier to manufacture, which gives the Dall flow tube an advantage in cost over the Venturi, although the typical measurement inaccuracy is a little higher ($\pm 1.5\%$). Another advantage of the Dall flow tube is its shorter length, which makes the engineering task of inserting it into the flow line easier. The Dall tube has one further operational advantage in that the permanent pressure loss imposed on the measured system is only about 5% of the measured pressure difference $(P_1 - P_2)$.

**Flow nozzle:** This nozzle is of simpler construction still and is therefore less expensive than either a Venturi or a Dall flow tube, but the pressure loss imposed on the flowing fluid is 30–50% of the measured pressure difference $(P_1 - P_2)$ across the nozzle.

### *Pitot static tube*

The pitot static tube is used mainly for making temporary measurements of flow, although it is also used in some instances for permanent flow monitoring. It measures the local velocity of flow at a particular point within a pipe rather than the average flow velocity as measured by other types of flowmeters. This may be very useful where there is a requirement to measure local flow rates across the cross section of a pipe in the case of nonuniform flow. Multiple pitot tubes are normally used to do this.

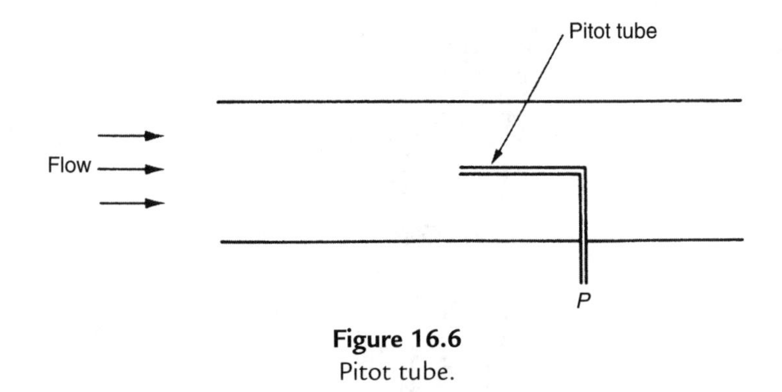


**Figure 16.6**
Pitot tube.

The instrument depends on the principle that a tube placed with its open end in a stream of fluid, as shown in Figure 16.6, will bring to rest that part of the fluid that impinges on it, and the loss of kinetic energy will be converted to a measurable increase in pressure inside the tube. This pressure ($P_1$), as well as the static pressure of the undisturbed free stream of flow ($P_2$), is measured. The flow velocity can then be calculated from the formula:

$$v = C\sqrt{2g(P_1 - P_2)}.$$

The constant $C$, known as the pitot tube coefficient, is a factor that corrects for the fact that not all fluid incident on the end of the tube will be brought to rest: a proportion will slip around it according to the design of the tube. Having calculated $v$, the volume flow rate can then be calculated by multiplying $v$ by the cross-sectional area of the flow pipe, $A$.

Pitot tubes have the advantage that they cause negligible pressure loss in the flow. They are also inexpensive, and the installation procedure consists of the very simple process of pushing them down a small hole drilled in the flow-carrying pipe. Their main failing is that measurement inaccuracy is typically about $\pm 5\%$, although more expensive versions can reduce inaccuracy down to $\pm 1\%$. The *annubar* is a development of the pitot tube that has multiple sensing ports distributed across the cross section of the pipe and thus provides an approximate measurement of the mean flow rate across the pipe.

### *16.3.2 Variable Area Flowmeters (Rotameters)*

In the variable area flowmeter (which is also sometimes known as a rotameter), the differential pressure across a variable aperture is used to adjust the area of the aperture. The aperture area is then a measure of the flow rate. The instrument is reliable, inexpensive, and used extensively throughout industry, accounting for about 20% of all flowmeters sold. Normally, because this type of instrument only gives a visual indication of flow rate, it is of

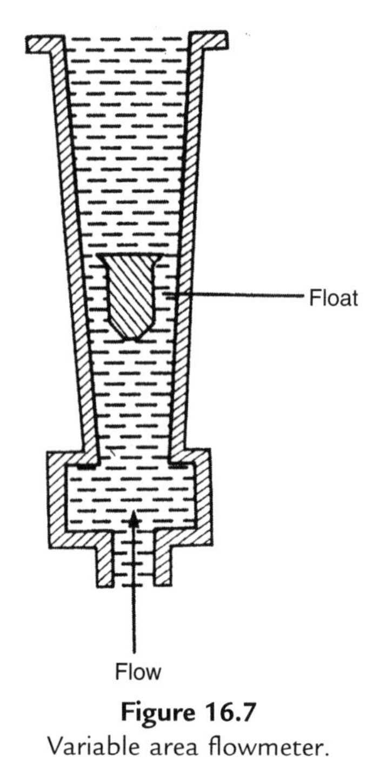


**Figure 16.7**
Variable area flowmeter.

no use in automatic control schemes. However, special versions of variable area flowmeters are now available that incorporate fiber optics. In these, a row of fibers detects the position of the float by sensing the reflection of light from it, and an electrical signal output can be derived from this.

In its simplest form, shown in Figure 16.7, the instrument consists of a tapered glass tube containing a float that takes up a stable position where its submerged weight is balanced by the up thrust due to the differential pressure across it. The position of the float is a measure of the effective annular area of the flow passage and hence of the flow rate. The inaccuracy of the least expensive instruments is typically $\pm 5\%$, but more expensive versions offer measurement inaccuracies as low as $\pm 0.5\%$.

### 16.3.3 Positive Displacement Flowmeters

Positive displacement flowmeters account for nearly 10% of the total number of flowmeters used in industry and are used in large numbers for metering domestic gas and water consumption. The least expensive instruments have a typical inaccuracy of about $\pm 2\%$, but the inaccuracy in more expensive ones can be as low as $\pm 0.5\%$. These higher quality instruments are used extensively within the oil industry, as such applications can justify the high cost of such instruments.

All positive displacement meters operate using mechanical divisions to displace discrete volumes of fluid successively. While this principle of operation is common, many different mechanical arrangements exist for putting the principle into practice. However, all versions of positive displacement meters are low friction, low maintenance, and long life devices, although they do impose a small permanent pressure loss on the flowing fluid. Low friction is especially important when measuring gas flows, and meters with special mechanical arrangements to satisfy this requirement have been developed.

The *rotary piston meter* is a common type of positive displacement meter used particularly for the measurement of domestic water supplies. It consists, as shown in Figure 16.8, of a slotted cylindrical piston moving inside a cylindrical working chamber that has an inlet port and an outlet port. The piston moves round the chamber such that its outer surface maintains contact with the inner surface of the chamber, and, as this happens, the piston slot slides up and down a fixed division plate in the chamber. At the start of each piston motion cycle, liquid is admitted to volume B from the inlet port. The fluid pressure causes the piston to start to rotate around the chamber, and, as this happens, liquid in volume C starts to flow out of the outlet port, and also liquid starts to flow from the inlet port into volume A. As the piston rotates further, volume B becomes shut off from the inlet port, while liquid continues to be admitted into A and pushed out of C. When the piston reaches the end point of its motion cycle, the outlet port is opened to volume B, and the liquid that has been transported round inside the piston is expelled. After this, the piston pivots about the contact point between the top of its slot and the division plate, and volume A effectively becomes volume C ready for

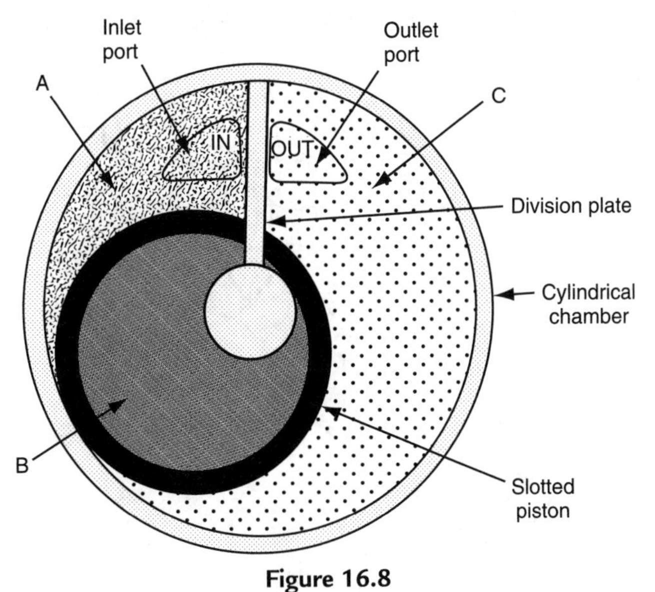


**Figure 16.8**
Rotary piston form of positive displacement flowmeter.

the start of the next motion cycle. A peg on top of the piston causes a reciprocating motion of a lever attached to it. This is made to operate a counter, and the flow rate is therefore determined from the count in unit time multiplied by the quantity (fixed) of liquid transferred between inlet and outlet ports for each motion cycle.

The *nutating disk meter* is another form of positive displacement meter in which the active element is a disc inside a precision-machined chamber. Liquid flowing into the chamber causes the disc to nutate (wobble), and these nutations are translated into a rotary motion by a roller cam. Rotations are counted by a pulse transmitter that provides a measurement of the flow rate. This form of meter is noted for its ruggedness and long life. It has a typical measurement accuracy of $\pm 1.0\%$. It is used commonly for water supply measurement.

The *oval gear meter* is yet another form of positive displacement meter that has two oval-shaped gear wheels. It is used particularly for measuring the flow rate of high viscosity fluids. It can also cope with measuring fluids that have variable viscosity.

### 16.3.4 Turbine Meters

A turbine flowmeter consists of a multibladed wheel mounted in a pipe along an axis parallel to the direction of fluid flow in the pipe, as shown in Figure 16.9. The flow of fluid past the wheel causes it to rotate at a rate proportional to the volume flow rate of the fluid. This rate of rotation has traditionally been measured by constructing the flowmeter such that it behaves as a variable reluctance tachogenerator. This is achieved by fabricating the turbine blades from a ferromagnetic material and placing a permanent magnet and coil inside the meter housing. A voltage pulse is induced in the coil as each blade on the turbine wheel moves past it,

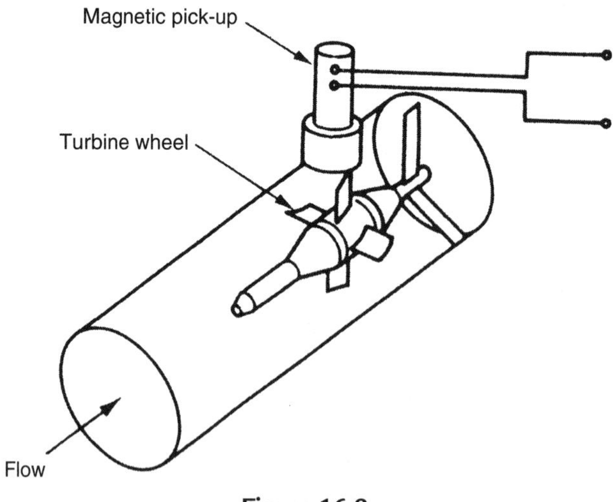


**Figure 16.9**
Turbine flowmeter.

and if these pulses are measured by a pulse counter, the pulse frequency and hence flow rate can be deduced. In recent instruments, fiber optics are also now sometimes used to count the rotations by detecting reflections off the tip of the turbine blades.

Provided that the turbine wheel is mounted in low-friction bearings, measurement inaccuracy can be as low as $\pm 0.2\%$. However, turbine flowmeters are less rugged and reliable than flow restriction-type instruments and are affected badly by any particulate matter in the flowing fluid. Bearing wear is a particular problem, which also imposes a permanent pressure loss on the measured system. Turbine meters are particularly prone to large errors when there is any significant second phase in the fluid measured. For instance, using a turbine meter calibrated on pure liquid to measure a liquid containing 5% air produces a 50% measurement error. As an important application of the turbine meter is in the petrochemical industries, where gas/oil mixtures are common, special procedures are being developed to avoid such large measurement errors.

Readers may find reference in manufacturers' catalogues to a *Woltmann meter*. This is a type of turbine meter that has helical blades and is used particularly for measuring high flow rates. It is also sometimes known as a *helix meter*.

Turbine meters have a similar cost and market share to positive displacement meters and compete for many applications, particularly in the oil industry. Turbine meters are smaller and lighter than the latter and are preferred for low-viscosity, high-flow measurements. However, positive displacement meters are superior in conditions of high viscosity and low flow rate.

### 16.3.5 Electromagnetic Flowmeters

Electromagnetic flowmeters, sometimes known just as *magnetic flowmeters*, are limited to measuring the volume flow rate of electrically conductive fluids. A typical measurement inaccuracy of around $\pm 1\%$ is acceptable in many applications, but the instrument is expensive both in terms of the initial purchase cost and in running costs, mainly due to its electricity consumption. A further reason for its high cost is the need for careful calibration of each instrument individually during manufacture, as there is considerable variation in the properties of the magnetic materials used.

The instrument, shown in Figure 16.10, consists of a stainless-steel cylindrical tube fitted with an insulating liner, which carries the measured fluid. Typical lining materials used are neoprene, polytetrafluoroethylene, and polyurethane. A magnetic field is created in the tube by placing mains-energized field coils either side of it, and the voltage induced in the fluid is measured by two electrodes inserted into opposite sides of the tube. The ends of these electrodes are usually flush with the inner surface of the cylinder. The electrodes are constructed from a material that is unaffected by most types of flowing fluids, such as stainless steel, platinum–iridium alloys,

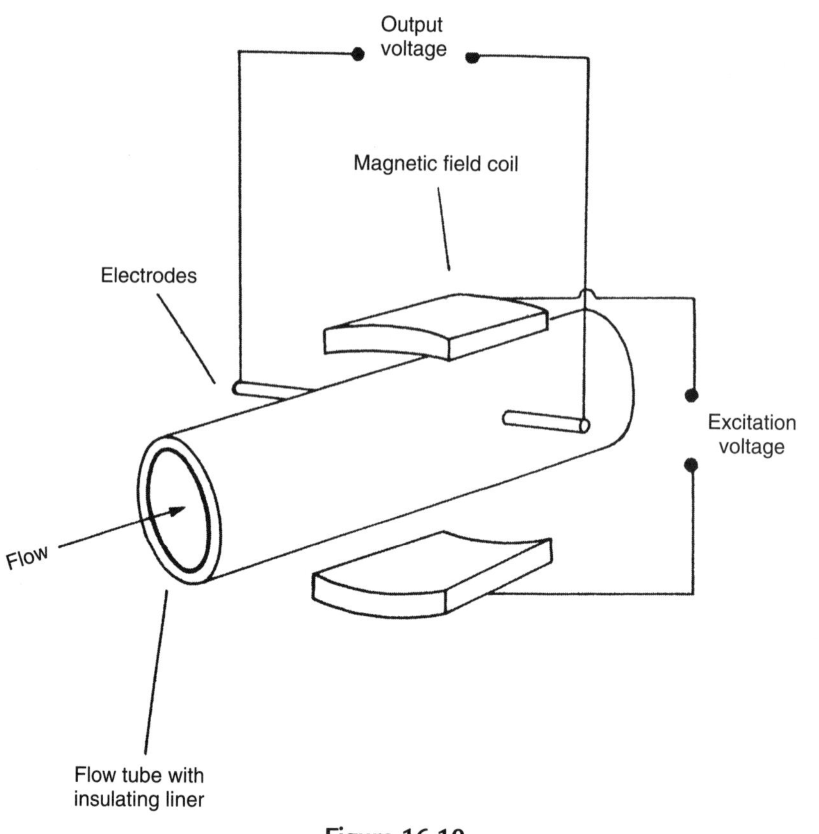


**Figure 16.10**
Electromagnetic flowmeter.

Hastelloy, titanium, and tantalum. In the case of rarer metals in this list, the electrodes account for a significant part of the total instrument cost.

By Faraday's law of electromagnetic induction, the voltage, $E$, induced across a length, $L$, of the flowing fluid moving at velocity, $v$, in a magnetic field of flux density, $B$, is given by

$$E = BLv, \tag{16.3}$$

where $L$ is the distance between the electrodes, which is the diameter of the tube, and $B$ is a known constant. Hence, measurement of voltage $E$ induced across the electrodes allows flow velocity $v$ to be calculated from Equation (16.3). Having thus calculated $v$, it is a simple matter to multiply $v$ by the cross-sectional area of the tube to obtain a value for the volume flow rate. The typical voltage signal measured across the electrodes is 1 mV when the fluid flow rate is 1 m/s.

The internal diameter of electromagnetic flowmeters is normally the same as that of the rest of the flow-carrying pipe work in the system. Therefore, there is no obstruction to fluid flow and consequently no pressure loss is associated with measurement. Like other forms of flowmeters, the electromagnetic type requires a minimum length of straight pipe work immediately prior to the point of flow measurement in order to guarantee the accuracy of measurement, although a length equal to five pipe diameters is usually sufficient.

While the flowing fluid must be electrically conductive, the method is of use in many applications and is particularly useful for measuring the flow of slurries in which the liquid phase is electrically conductive. Corrosive fluids can be handled, providing a suitable lining material is used. At the present time, electromagnetic flowmeters account for about 15% of the new flowmeters sold and this total is slowly growing. One operational problem is that the insulating lining is subject to damage when abrasive fluids are being handled, which can give the instrument a limited life.

New developments in electromagnetic flowmeters are producing instruments that are physically smaller than before. Also, by employing better coil designs, electricity consumption is reduced. This means that battery-powered versions are now available commercially. Also, whereas conventional electromagnetic flowmeters require a minimum fluid conductivity of 10 $\mu$mho/cm$^3$, new versions can cope with fluid conductivities as low as 1 $\mu$mho/cm$^3$.

### *16.3.6 Vortex-Shedding Flowmeters*

The vortex-shedding flowmeter is used as an alternative to traditional differential pressure meters in many applications. The operating principle of the instrument is based on the natural phenomenon of vortex shedding, created by placing an unstreamlined obstacle (known as a bluff body) in a fluid-carrying pipe, as indicated in Figure 16.11. When fluid flows past the obstacle, boundary layers of viscous, slow-moving fluid are formed along

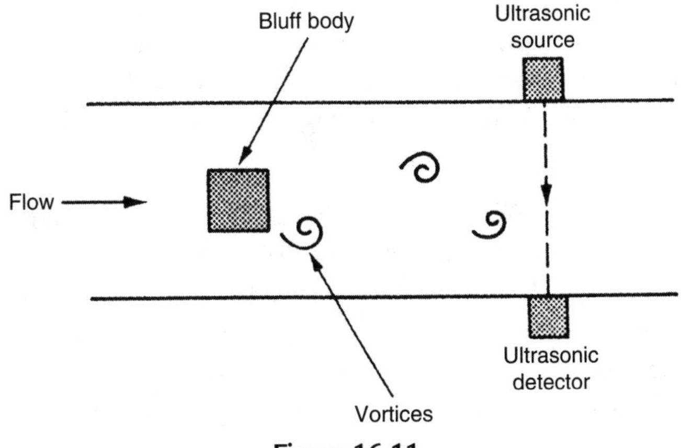


**Figure 16.11**
Vortex-shedding flowmeter.

the outer surface. Because the obstacle is not streamlined, the flow cannot follow the contours of the body on the downstream side, and the separate layers become detached and roll into eddies or vortices in the low-pressure region behind the obstacle. The shedding frequency of these alternately shed vortices is proportional to the fluid velocity past the body. Various thermal, magnetic, ultrasonic, and capacitive vortex detection techniques are employed in different instruments.

Such instruments have no moving parts, operate over a wide flow range, have low power consumption, require little maintenance, and have a similar cost to measurement using an orifice plate. They can measure both liquid and gas flows, and a common inaccuracy value quoted is $\pm 1\%$ of full-scale reading, although this can be seriously downgraded in the presence of flow disturbances upstream of the measurement point and a straight run of pipe before the measurement point of 50 pipe diameters is recommended. Another problem with the instrument is its susceptibility to pipe vibrations, although new designs are becoming available that have a better immunity to such vibrations.

### 16.3.7 Ultrasonic Flowmeters

The ultrasonic technique of volume flow rate measurement is, like the magnetic flowmeter, a noninvasive method. It is not restricted to conductive fluids, however, and is particularly useful for measuring the flow of corrosive fluids and slurries. In addition to its high reliability and low maintenance requirements, a further advantage of an ultrasonic flowmeter over an electromagnetic flowmeter is that the instrument can be clamped externally onto existing pipe work instead of being inserted as an integral part of the flow line. As the procedure of breaking into a pipeline to insert a flowmeter can be as expensive as the cost of the flowmeter itself, the ultrasonic flowmeter has enormous cost advantages. Its clamp-on mode of operation also has significant safety advantages in avoiding the possibility of personnel installing flowmeters coming into contact with hazardous fluids, such as poisonous, radioactive, flammable, or explosive ones. Also, any contamination of the fluid being measured (e.g., food substances and drugs) is avoided. Ultrasonic meters are still less common than differential pressure or electromagnetic flowmeters, although usage continues to expand year by year.

Two different types of ultrasonic flowmeter exist that employ distinct technologies—one based on Doppler shift and the other on transit time. In the past, the existence of these alternative technologies has not always been readily understood and has resulted in ultrasonic technology being rejected entirely when one of these two forms has been found to be unsatisfactory in a particular application. This is unfortunate because the two technologies have distinct characteristics and areas of application, and many situations exist where one form is very suitable and the other is not. To reject both, having only tried out one, is therefore a serious mistake. Ultrasonic flowmeters have become available that combine both Doppler shift and transit time technologies.

Particular care has to be taken to ensure a stable flow profile in ultrasonic flowmeter applications. It is usual to increase the normal specification of the minimum length of straight pipe run prior to the point of measurement, expressed as a number of pipe diameters, from a value of 10 up to 20 or, in some cases, even 50 diameters. Analysis of the reasons for poor performance in many instances of ultrasonic flowmeter application has shown failure to meet this stable flow profile requirement to be a significant factor.

*Doppler shift ultrasonic flowmeter*

The principle of operation of the Doppler shift flowmeter is shown in Figure 16.12. A fundamental requirement of these instruments is the presence of scattering elements within the flowing fluid, which deflect the ultrasonic energy output from the transmitter such that it enters the receiver. These can be provided by solid particles, gas bubbles, or eddies in the flowing fluid. The scattering elements cause a frequency shift between transmitted and reflected ultrasonic energy, and measurement of this shift enables fluid velocity to be inferred.

The instrument consists essentially of an ultrasonic transmitter–receiver pair clamped onto the outside wall of a fluid-carrying vessel. Ultrasonic energy consists of a train of short bursts of sinusoidal waveforms at a frequency between 0.5 and 20 MHz. This frequency range is described as ultrasonic because it is outside the range of human hearing. The flow velocity, $v$, is given by

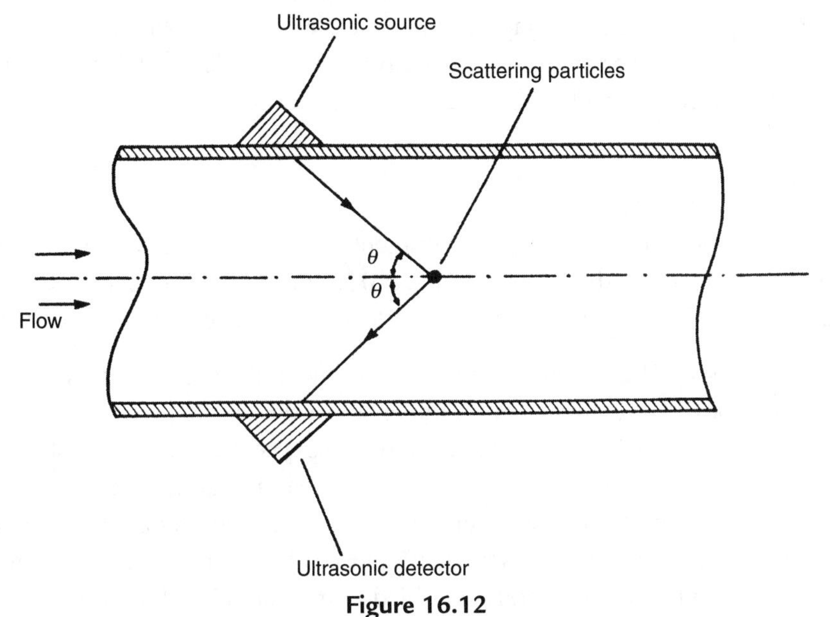


**Figure 16.12**
Doppler shift ultrasonic flowmeter.

$$v = \frac{c(f_t - f_r)}{2f_t \cos(\theta)}, \tag{16.4}$$

where $f_t$ and $f_r$ are the frequencies of the transmitted and received ultrasonic waves, respectively, $c$ is the velocity of sound in the fluid being measured, and $\theta$ is the angle that the incident and reflected energy waves make with the axis of flow in the pipe. Volume flow rate is then calculated readily by multiplying the measured flow velocity by the cross-sectional area of the fluid-carrying pipe.

The electronics involved in Doppler shift flowmeters is relatively simple and therefore inexpensive. Ultrasonic transmitters and receivers are also relatively inexpensive, being based on piezoelectric oscillator technology. Therefore, as all of its components are inexpensive, the Doppler shift flowmeter itself is inexpensive. The measurement accuracy obtained depends on many factors, such as the flow profile; the constancy of pipe wall thickness; the number, size, and spatial distribution of scatterers; and the accuracy with which the speed of sound in the fluid is known. Consequently, accurate measurement can only be achieved by the tedious procedure of carefully calibrating the instrument in each particular flow measurement application. Otherwise, measurement errors can approach $\pm 10\%$ of the reading; for this reason, Doppler shift flowmeters are often used merely as flow indicators rather than for accurate quantification of the volume flow rate.

Versions are now available that are being fitted inside the flow pipe, flush with its inner surface. This overcomes the problem of variable pipe thickness, and an inaccuracy level as small as $\pm 0.5\%$ is claimed for such devices. Other recent developments are the use of multiple path ultrasonic flowmeters that use an array of ultrasonic elements to obtain an average velocity measurement. This reduces error due to nonuniform flow profiles substantially but there is a substantial cost penalty involved in such devices.

#### *Transit time ultrasonic flowmeter*

A transit time ultrasonic flowmeter is an instrument designed for measuring the volume flow rate in clean liquids or gases. It consists of a pair of ultrasonic transducers mounted along an axis aligned at angle $\theta$ with respect to the fluid flow axis, as shown in Figure 16.13.

Each transducer consists of a transmitter–receiver pair, with the transmitter emitting ultrasonic energy that travels across to the receiver on the opposite side of the pipe. These ultrasonic elements are normally piezoelectric oscillators of the same type used in Doppler shift flowmeters. Fluid flowing in the pipe causes a time difference between the transit times of beams traveling upstream and downstream, and measurement of this difference allows the flow velocity to be calculated. The typical magnitude of this time difference is 100 ns in a total transit time of 100 μs, and high-precision electronics are therefore needed to measure the difference. There are three distinct ways of measuring the time shift. These are direct measurement, conversion to a phase

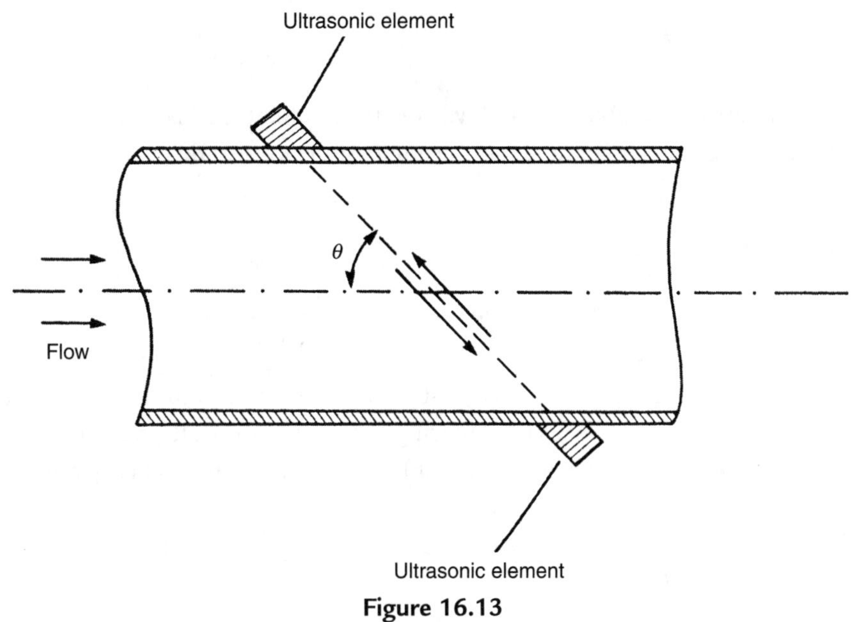


**Figure 16.13**
Transit time ultrasonic flowmeter.

change, and conversion to a frequency change. The third of these options is particularly attractive, as it obviates the need to measure the speed of sound in the measured fluid as required by the first two methods. A scheme applying this third option is shown in Figure 16.14. This also multiplexes the transmitting and receiving functions so that only one ultrasonic element is needed in each transducer. The forward and backward transit times across the pipe, $T_f$ and $T_b$, are given by

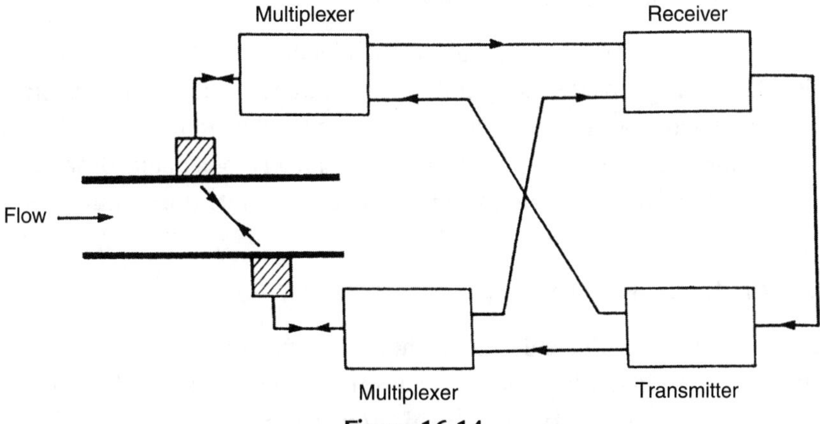


**Figure 16.14**
Transit time measurement system.

$$T_f = \frac{L}{c + v\cos(\theta)} \quad ; \quad T_b = \frac{L}{c - v\cos(\theta)},$$

where $c$ is the velocity of sound in the fluid, $v$ is the flow velocity, $L$ is the distance between the ultrasonic transmitter and receiver, and $\theta$ is the angle of the ultrasonic beam with respect to the fluid flow axis.

The time difference, $\delta T$, is given by

$$\delta T = T_b - T_f = \frac{2vL\cos(\theta)}{c^2 - v^2\cos^2(\theta).}$$

This requires knowledge of $c$ before it can be solved. However, a solution can be found much more simply if the receipt of a pulse is used to trigger transmission of the next ultrasonic energy pulse. Then, the frequencies of the forward and backward pulse trains are given by

$$F_f = \frac{1}{T_f} = \frac{c - v\cos(\theta)}{L} \quad ; \quad F_b = \frac{1}{T_b} = \frac{c + v\cos(\theta)}{L}.$$

If the two frequency signals are now multiplied together, the resulting beat frequency is given by

$$\delta F = F_b - F_f = \frac{2v\cos(\theta)}{L}.$$

$c$ has now been eliminated, and $v$ can be calculated from a measurement of $\delta F$ as

$$v = \frac{L\delta F}{2\cos(\theta)}.$$

This is often known as the *sing-around flowmeter.*

Transit time flowmeters are of more general use than Doppler shift flowmeters, particularly where the pipe diameter involved is large and hence the transit time is consequently sufficiently large to be measured with reasonable accuracy. It is possible then to reduce the inaccuracy value down to $\pm 0.5\%$. However, the instrument costs more than a Doppler shift flowmeter because of the greater complexity of the electronics needed to make accurate transit time measurements.

#### *Combined Doppler shift/transit time flowmeters*

Recently, some manufacturers have developed ultrasonic flowmeters that use a combination of Doppler shift and transit time. The exact mechanism by which these work is rarely, if ever, disclosed, as manufacturers wish to protect details from competitors. However, details of various forms of combined Doppler shift/transit time measurement techniques are filed in patent offices.

### 16.3.8 Other Types of Flowmeters for Measuring Volume Flow Rate

*Gate-type meter*

A gate meter consists of a spring-loaded, hinged flap mounted at right angles to the direction of fluid flow in the fluid-carrying pipe. The flap is connected to a pointer outside the pipe. The fluid flow deflects the flap and pointer, and the flow rate is indicated by a graduated scale behind the pointer. The major difficulty with such devices is in preventing leaks at the hinge point. A variation on this principle is the *air vane meter*, which measures deflection of the flap by a potentiometer inside the pipe. This is used to measure airflow within automotive fuel-injection systems. Another similar device is the *target meter*. This consists of a circular, disc-shaped flap in the pipe. Fluid flow rate is inferred from the force exerted on the disc measured by strain gauges bonded to it. This meter is very useful for measuring the flow of dilute slurries but does not find wide application elsewhere as it has a relatively high cost. Measurement uncertainty in all of these types of meters varies between 1 and 5% according to the cost and design of each instrument.

*Jet meter*

These come in two forms—a single jet meter and a multiple jet meter. In the first, flow is diverted into a single jet, which impinges on the radial vanes of an impeller. The multiple jet form diverts the flow into multiple jets arranged at equal angles around an impeller mounted on a horizontal axis.

A paddle wheel meter is a variation of the single jet meter in which the impeller only projects partially into the flowing fluid.

*Pelton wheel flowmeter*

This uses a similar mechanical arrangement to the old-fashioned water wheels used for power generation at the time of the industrial revolution. Flowing fluid is directed onto the blades of the flowmeter wheel by a jet, and the flow rate is determined from the rate of rotation of the wheel. This type of flowmeter is used to measure the flow rate of a diverse range of materials, including acids, aggressive chemicals, and hot fats at both low and high flow rates. Special versions can measure very small flow rates down to 3 ml/min.

*Laser Doppler flowmeter*

This instrument gives direct measurements of flow velocity for liquids containing suspended particles flowing in a pipe. Light from a laser is focused by an optical system to a point in the flow, with fiber-optic cables being used commonly to transmit the light. The movement of particles causes a Doppler shift of the scattered light and produces a signal in a photodetector that is related to the fluid velocity. A very wide range of flow velocities between 10 μm/s and 105 m/s can be measured by this technique.

Because sufficient particles for satisfactory operation are normally present naturally in most liquid and gaseous fluids, the introduction of artificial particles is rarely needed. The technique is advantageous in measuring flow velocity directly rather than inferring it from a pressure difference. It also causes no interruption in the flow and, as the instrument can be made very small, it can measure velocity in confined areas. One limitation is that it measures local flow velocity in the vicinity of the focal point of the light beam, which can lead to large errors in the estimation of mean volume flow rate if the flow profile is not uniform. However, this limitation can be used constructively in applications of the instrument where the flow profile across the cross section of a pipe is determined by measuring the velocity at a succession of points.

The final comment on this instrument has to be that although it could potentially be used in many applications, it has competition from many other types of instruments that offer similar performance at lower cost. Its main application at the present time is in measuring blood flow in medical applications.

#### *Thermal anemometers*

Thermal anemometry was first used in a *hot-wire anemometer* to measure the volume flow rate of gases flowing in pipes. A hot-wire anemometer consists of a piece of thin (typical diameter 5 μm), electrically heated wire (usually tungsten, platinum of a platinum–iridium alloy) inserted into the gas flow. The flowing gas has a cooling effect on the wire, which reduces its resistance. Measurement of the resistance change (usually by a bridge circuit) allows the volume flow rate of the gas to be calculated. Unfortunately, the device is not robust because of the very small diameter of the wire used in its construction. However, it has a very fast speed of response, which makes it an ideal measurement device in conditions where the flow velocity is changing. It is also insensitive to the direction of gas flow, making it a very useful measuring device in conditions of turbulent flow. Recently, more robust devices have been made by using a thin metal film instead of a wire. In this form, the device is known as a *hot-film anemometer*. Typically, the film is platinum and is deposited on a quartz probe of a typical diameter of 0.05 mm. The increased robustness means that the hot-film anemometer is also used to measure the flow rate of liquids such as water.

#### *Coriolis meter*

While the Coriolis meter is intended primarily to be a mass flow-measuring instrument, it can also be used to measure volume flow rate when high measurement accuracy is required. However, its high cost means that alternative instruments are normally used for measuring volume flow rate.

### *16.3.9 Open Channel Flowmeters*

Open channel flowmeters measure the flow of liquids in open channels and are particularly relevant to measuring the flow of water in rivers as part of environmental management schemes. The normal procedure is to build a weir or flume of constant width across the flow and measure the velocity of flow and the height of liquid immediately before the weir or flume with an ultrasonic or radar level sensor, as shown in Figure 16.15. The volume flow rate can then be calculated from this measured height.

As an alternative to building a weir or flume, electromagnetic flowmeters up to 180 mm wide are available that can be placed across the channel to measure the flow velocity, providing the flowing liquid is conductive. If the channel is wider than 180 mm, two or more electromagnetic meters can be placed side by side. Apart from measuring the flow velocity in this way, the height of the flowing liquid must also be measured, and the width of the channel must also be known in order to calculate the volume flow rate.

As a third alternative, ultrasonic flowmeters are also used to measure flow velocity in conjunction with a device to measure the liquid depth.

## 16.4 Intelligent Flowmeters

All the usual benefits associated with intelligent instruments are applicable to most types of flowmeters. Indeed, all types of mass flowmeters routinely have intelligence as an integral part of the instrument. For volume flow rate measurement, intelligent differential pressure-measuring instruments can be used to good effect in conjunction with obstruction-type flow

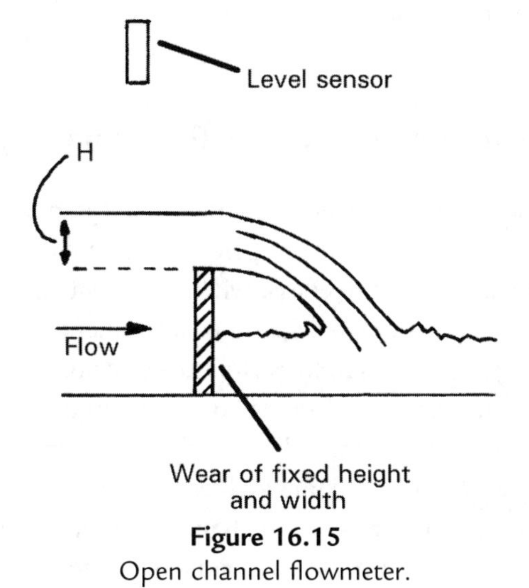


**Figure 16.15**
Open channel flowmeter.

transducers. One immediate benefit of this in the case of the commonest flow restriction device, the orifice plate, is to extend the lowest flow measurable with acceptable accuracy down to 20% of the maximum flow value. In positive displacement meters, intelligence allows compensation for thermal expansion of meter components and temperature-induced viscosity changes. Correction for variations in flow pressure is also provided for. Intelligent electromagnetic flowmeters are also available, and these have a self-diagnosis and self-adjustment capability. The usable instrument range is typically from 3 to 100% of the full-scale reading, and the quoted maximum inaccuracy is $\pm 0.5\%$. It is also normal to include a nonvolatile memory to protect constants used for correcting for modifying inputs and so on against power supply failures. Intelligent turbine meters are able to detect their own bearing wear and also report deviations from initial calibration due to blade damage, etc. Some versions also have a self-adjustment capability.

The ability to carry out digital signal processing has also led to emergence of the *cross-correlation ultrasonic flowmeter*. This is a variant of the transit time form of ultrasonic flowmeter in which a series of ultrasonic signals are injected into the flowing liquid. The ultrasonic receiver stores the echo pattern from each input signal and then cross-correlation techniques are used to produce a map of the profile of the water flow in different layers. Thus, the instrument provides information on the profile of the flow rate across the cross section of the pipe rather than just giving a measurement of the mean flow rate in the pipe.

The trend is now moving toward total flow computers, which can process inputs from almost any type of transducer. Such devices allow user input of parameters such as specific gravity, fluid density, viscosity, pipe diameters, thermal expansion coefficients, and discharge coefficients. Auxiliary inputs from temperature transducers are also catered for. After processing raw flow transducer output with this additional data, flow computers are able to produce measurements of flow to a very high degree of accuracy.

## 16.5 Choice between Flowmeters for Particular Applications

The number of relevant factors to be considered when specifying a flowmeter for a particular application is very large. These include the temperature and pressure of the fluid, its density, viscosity, chemical properties and abrasiveness, whether it contains particles, whether it is a liquid or gas, etc. This narrows the field to a subset of instruments that are physically capable of making the measurement. Next, the required performance factors of accuracy, rangeability, acceptable pressure drop, output signal characteristics, reliability, and service life must be considered. Accuracy requirements vary widely across different applications, with measurement uncertainty of $\pm 5\%$ being acceptable in some and less than $\pm 0.5\%$ being demanded in others. Finally, economic viability must be assessed, which must take into account not only the purchase cost, but also reliability, installation difficulties, maintenance requirements, and service life.

Where only a visual indication of flow rate is needed, the variable area meter is popular. Where a flow measurement in the form of an electrical signal is required, the choice of available instruments is very large. The orifice plate is used extremely commonly for such purposes and accounts for almost 50% of instruments currently in use in industry. Other forms of differential pressure meters and electromagnetic flowmeters are used in significant numbers. Currently, there is a trend away from rotating devices, such as turbine meters and positive displacement meters. At the same time, usage of ultrasonic and vortex meters is expanding.

## 16.6 Calibration of Flowmeters

The first consideration in choosing a suitable way to calibrate flow-measuring instruments is to establish exactly what accuracy level is needed so that the calibration system instituted does not cost more than necessary. In some cases, such as handling valuable fluids or where there are legal requirements as in petrol pumps, high accuracy levels (e.g., error $\leq 0.1\%$) are necessary and the expensive procedures necessary to achieve these levels are justified. However, in other situations, such as in measuring additives to the main stream in a process plant, only low levels of accuracy are needed (e.g., error $\approx 5\%$ is acceptable) and relatively inexpensive calibration procedures are sufficient.

The accuracy of flow measurement is affected greatly by the flow conditions and characteristics of the flowing fluid. Therefore, wherever possible, process flow-measuring instruments are calibrated on-site in their normal measuring position. This ensures that calibration is performed in the actual flow conditions, which are difficult or impossible to reproduce exactly in a laboratory. To ensure the validity of such calibration, it is also normal practice to repeat flow calibration checks until the same reading is obtained in two consecutive tests. However, it has been suggested that even these precautions are inadequate and that statistical procedures are needed.

If on-site calibration is not feasible or is not accurate enough, the only alternative is to send the instrument away for calibration using special equipment provided by instrument manufacturers or other specialist calibration companies. However, this is usually an expensive option. Furthermore, the calibration facility does not replicate the normal operating conditions of the meter tested, and appropriate compensation for differences between calibration conditions and normal use conditions must be applied.

The equipment and procedures used for calibration depend on whether mass, liquid, or gaseous flows are being measured. Therefore, separate sections are devoted to each of these cases. It must also be stressed that all calibration procedures mentioned in the following paragraphs in respect to fluid flow only refer to flows of single phase fluids (i.e., liquids or gases). Where a second or third phase is present, calibration is much more difficult and specialist advice should be sought from the manufacturer of the instrument used for measurement.

### 16.6.1 Calibration Equipment and Procedures for Mass Flow-Measuring Instruments

Where the conveyor method is used for measuring the mass flow of solids in the form of particles or powders, both mass-measuring and velocity-measuring instruments are involved. Suitable calibration techniques for each of these are discussed in later chapters.

In the case of Coriolis and thermal mass flowmeters, the usual method of calibrating these while in situ in their normal measurement position is to provide a diversion valve after the meter. During calibration procedures, the valve is opened for a measured time period to allow some of the fluid to flow into a container that is subsequently weighed. Alternatively, the meter can be removed for calibration using special test rigs normally provided by the instrument manufacturer.

### 16.6.2 Calibration Equipment and Procedures for Instruments Measuring Volume Flow Rate of Liquids

*Calibrated tank*

Probably the simplest piece of equipment available for calibrating instruments measuring liquid flow rates is the calibrated tank. This consists of a cylindrical vessel, as shown in Figure 16.16, with conical ends that facilitate draining and cleaning of the tank. A *sight tube* with a graduated scale is placed alongside the final, upper, cylindrical part of the tank, which allows the volume of liquid in the tank to be measured accurately. Flow rate calibration is performed by measuring the time taken, starting from an empty tank, for a given volume of liquid to flow into the vessel.

Because the calibration procedure starts and ends in zero flow conditions, it is not suitable for calibrating instruments affected by flow acceleration and deceleration characteristics. This therefore excludes instruments such as differential pressure meters (orifice plate, flow nozzle, Venturi, Dall flow tube, pitot tube), turbine flowmeters, and vortex-shedding flowmeters. The technique is further limited to the calibration of low-viscosity liquid flows, although lining the tank with an epoxy coating can allow the system to cope with somewhat higher viscosities. The limiting factor in this case is the drainage characteristics of the tank, which must be such that the residue liquid left after draining has an insufficient volume to affect the accuracy of the next calibration.

*Pipe prover*

The commonest form of pipe prover is the bidirectional type, shown in Figure 16.17, which consists of a U-shaped tube of metal of accurately known cross section. The purpose of the U bend is to give a long flow path within a compact spatial volume. Alternative versions with more than one U bend also exist to cater for situations where an even longer flow path is required. Inside the tube is a hollow, inflatable sphere, which is filled with water until its diameter is about 2% larger than that of the tube. As such, the sphere forms a seal with the sides

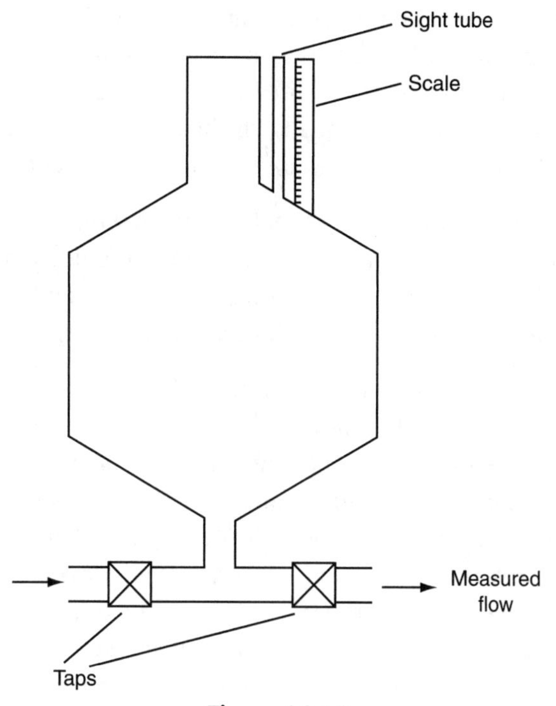


**Figure 16.16**
Calibrated tank.

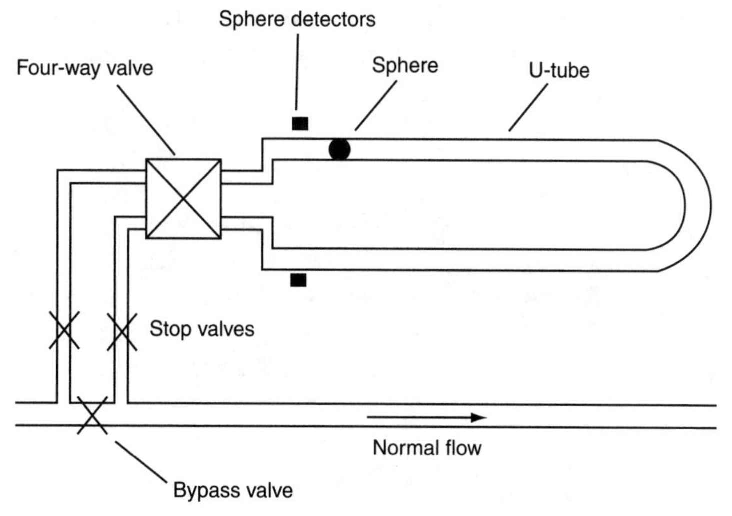


**Figure 16.17**
Bidirectional pipe prover.

of the tube and acts as a piston. The prover is connected into the existing fluid-carrying pipe network via tappings either side of a bypass valve. A four-way valve at the start of the U tube allows fluid to be directed in either direction around it. Calibration is performed by diverting flow into the prover and measuring the time taken for the sphere to travel between two detectors in the tube. The detectors are normally of an electromechanical, plunger type.

Unidirectional versions of the aforementioned also exist in which fluid only flows in one direction around the tube. A special handling valve has to be provided to return the sphere to the starting point after each calibration, but the absence of a four-way flow control valve makes such devices significantly less expensive than bidirectional types.

Pipe provers are particularly suited to the calibration of pressure-measuring instruments that have a pulse type of output, such as turbine meters. In such cases, the detector switches in the tube can be made to gate the instrument's output pulse counter. This enables not only the basic instrument to be calibrated, but also the ancillary electronics within it at the same time. The inaccuracy level of such provers can be as low as $\pm 0.1\%$. This level of accuracy is maintained for high fluid viscosity levels and also at very high flow rates. Even higher accuracy is provided by an alternative form of prover, which consists of a long, straight metal tube containing a metal piston. However, such devices are more expensive than the other types discussed earlier and their large space requirements also often cause great difficulties.

*Compact prover*

The compact prover has an identical operating principle to that of the other pipe provers described earlier but occupies a much smaller spatial volume. It is therefore used extensively in situations where there is insufficient room to use a larger prover. Many different designs of compact prover exist, operating in both unidirectional and bidirectional modes, and one such design is shown in Figure 16.18. Common features of compact provers are an accurately

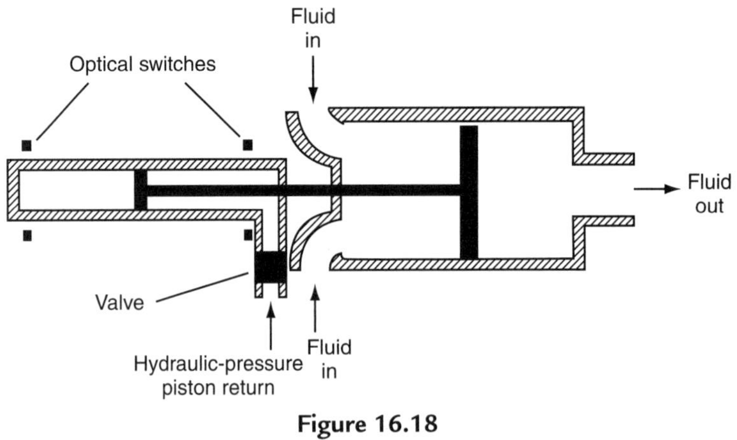


**Figure 16.18**
Compact prover.

machined cylinder containing a metal piston that is driven between two reference marks by the flowing fluid. The instants at which the reference marks are passed are detected by switches, of optical form in the case of the version shown in Figure 16.18. Provision has to be made within these instruments for returning the piston back to the starting point after each calibration and a hydraulic system is used commonly for this. Again, measuring the piston traverse time is made easier if the switches can be made to gate a pulse train, and therefore compact provers are also most suited to instruments having a pulse-type output such as turbine meters. Measurement uncertainty levels down to $\pm 0.1\%$ are possible.

The main technical difficulty in compact provers is measuring the traverse time, which can be as small as 1 second. The pulse count from a turbine meter in this time would typically be only about 100, making the possible measurement error 1%. To overcome this problem, electronic pulse interpolation techniques have been developed that can count fractions of pulses.

*Positive displacement meter*

High-quality versions of the positive displacement flowmeter can be used as a reference standard in flowmeter calibration. The general principles of these were explained in Section 16.3.3. Such devices can give measurement inaccuracy levels down to $\pm 0.2\%$.

*Gravimetric method*

A variation on the principle of measuring the volume of liquid flowing in a given time is to weigh the quantity of fluid flowing in a given time. Apart from its applicability to a wider range of instruments, this technique is not limited to low-viscosity fluids, as any residual fluid in the tank before calibration will be detected by the load cells and therefore compensated for. In the simplest implementation of this system, fluid is allowed to flow for a measured length of time into a tank resting on load cells. As before, the stop–start mode of fluid flow makes this method unsuitable for calibrating differential pressure, turbine, and vortex-shedding flowmeters. It is also unsuitable for measuring high flow rates because of the difficulty in bringing the fluid to rest. These restrictions can be overcome by directing the flowing fluid into the tank via diverter valves. In this alternative, it is important that the timing system be synchronized carefully with operation of the diverter valves.

All versions of gravimetric calibration equipment are less robust than volumetric types and so on-site use is not recommended.

*Orifice plate*

A flow line equipped with a certified orifice plate is sometimes used as a reference standard in flow calibration, especially for high flow rates through large-bore pipes. While measurement uncertainty is of the order of $\pm 1\%$ at best, this is adequate for calibrating many flow-measuring instruments.

*Turbine meter*

Turbine meters are also used as a reference standard for testing flowmeters. Their main application, as for orifice plates, is in calibrating high flow rates through large-bore pipes. Measurement uncertainty down to ±0.2% is attainable.

### 16.6.3 Calibration Equipment and Procedures for Instruments Measuring Volume Flow Rate of Gases

Calibration of gaseous flows poses considerable difficulties compared with calibrating liquid flows. These problems include the lower density of gases, their compressibility, and difficulty in establishing a suitable liquid/air interface as utilized in many liquid flow measurement systems.

In consequence, the main methods of calibrating gaseous flows, as described later, are small in number. Certain other specialized techniques, including the gravimetric method and the pressure–volume–temperature method, are also available. These provide primary reference standards for gaseous flow calibration with measurement uncertainty down to ±0.3%. However, the expense of the equipment involved is such that it is usually only available in National Standards Laboratories.

*Bell prover*

A bell prover consists of a hollow, inverted, metal cylinder suspended over a bath containing light oil, as shown in Figure 16.19. The air volume in the cylinder above the oil is connected, via a tube and a valve, to the flowmeter being calibrated. An air flow through the meter is created by allowing the cylinder to fall downward into the bath, thus displacing the air contained within it. The flow rate, which is measured by timing the rate of fall of the cylinder, can be adjusted by changing the value of counterweights attached via a low-friction pulley system to the cylinder. This is essentially laboratory-only equipment and therefore on-site calibration is not possible.

*Positive displacement meter*

As for liquid flow calibration, positive displacement flowmeters can be used for the calibration of gaseous flows with inaccuracy levels down to ±0.2%.

*Compact prover*

Compact provers of the type used for calibrating liquid flows are unsuitable for application to gaseous flows. However, special designs of compact provers are being developed for gaseous flows, and hence such devices may find application in gaseous flow calibration in the future.

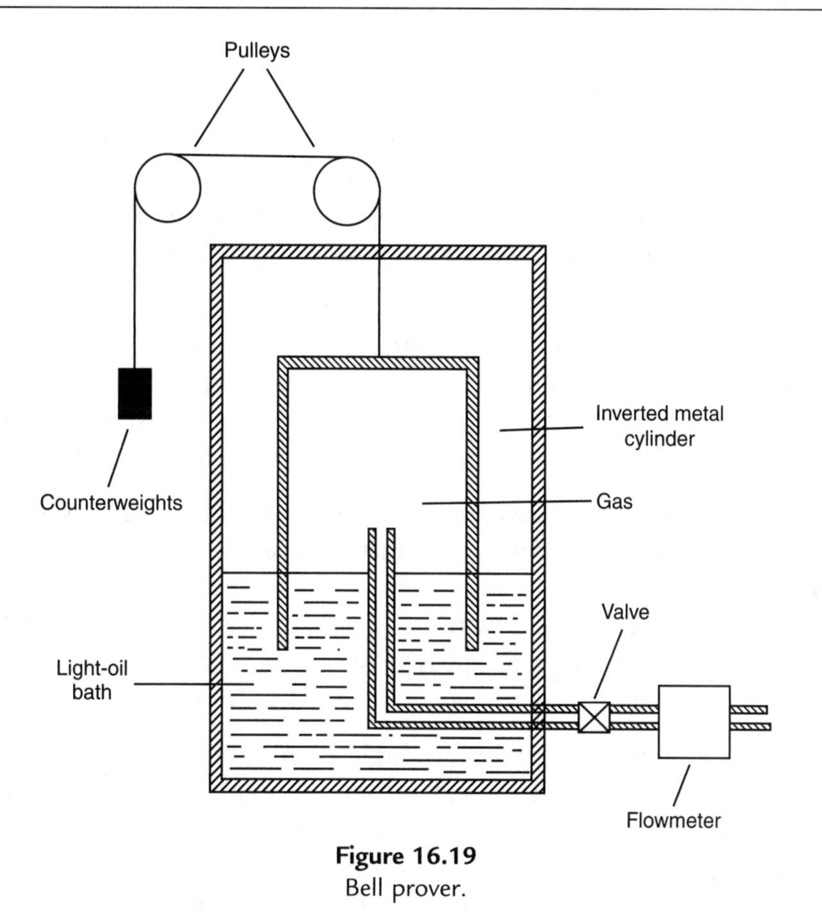


**Figure 16.19**
Bell prover.

### *16.6.4 Reference Standards*

Traceability of flow rate calibration to fundamental standards is provided for by reference to primary standards of the separate quantities that the flow rate is calculated from. Mass measurements are calibrated by comparison with a copy of the international standard kilogram (see Chapter 18), and time is calibrated by reference to a caesium resonator standard. Volume measurements are calibrated against standard reference volumes that are themselves calibrated gravimetrically using a mass measurement system traceable to the standard kilogram.

## 16.7 Summary

We started this chapter off by observing that flow rate could be measured either as mass flow rate or volume flow rate. We also observed that the material being measured could be in solid, liquid, or gaseous form. In the case of solids, we quickly found that this could only be measured in terms of the mass flow rate. However, in the case of liquids and gases, we found that we have the option of measuring either mass flow rate or volume flow rate. Of these two alternatives, we observed that mass flow measurement was the more accurate.

Before proceeding to look at flow measurement in detail, we had a brief look at the differences between laminar flow and turbulent flow. This taught us that the flow rate was difficult to measure in turbulent flow conditions and even in laminar flow at high velocities. Therefore, as far as possible, the measurement was made at a point in the flow where the flow was at least approximately laminar and the flow velocity was as small as possible.

This allowed us to look at flow-measuring instruments in more detail. We started with mass flow and observed that this could be measured in one of three ways—conveyor-based methods, Coriolis flowmeter, and thermal mass flowmeter. We examined the mode of operation of each of these and made some comments about their applicability.

Moving on, we then started to look at volume flow rate measurement and worked progressively through a large number of different instruments that can be used. First, we looked at obstruction devices. These are placed in a fluid-carrying pipe and cause a pressure difference across the obstruction that is a function of the flow rate of the fluid. Various obstruction devices were discussed, from the commonly used inexpensive but less accurate orifice plate to more expensive but more accurate devices such as the Venturi tube, flow nozzle, and Dall flow tube.

After looking at flow obstruction devices, we looked at a number of other instruments for measuring volume flow rate of fluids flowing in pipes, including the variable area flowmeter, positive displacement flowmeter, turbine flowmeter, electromagnetic flowmeter, vortex-shedding flowmeter, and, finally, ultrasonic flowmeters in both transit time and Doppler shift forms. We also looked briefly at several other devices, including gate-type meters, laser Doppler flowmeter, and thermal anemometer. Finally, we also had a brief look at measuring fluid flow in open channels and observed three ways of doing this.

We rounded off our discussion of flow measurement by looking at intelligent devices. We observed that these bring the usual benefits associated with intelligent instruments, including improved measurement accuracy and extended measurement range, with facilities for self-diagnosis and self-adjustment also being common. This led on to some discussion about the most appropriate instrument to use in particular flow measurement situations and applications out of all the instruments covered in the chapter.

We then concluded the chapter by considering the subject of flowmeter calibration. These calibration methods were considered in three parts. First, we looked at the calibration of instruments measuring mass flow. Second, we looked at the calibration of instruments measuring the volume flow rate of liquids. Finally, we looked at the calibration of instruments measuring the volume flow rate of gases.

## 16.8 Problems

16.1. Name and discuss three different kinds of instruments used for measuring the mass flow rate of substances (mass flowing in unit time).

16.2. Instruments used to measure the volume flow rate of fluids (volume flowing in unit time) can be divided into a number of different types. Explain what these different types are and discuss briefly how instruments in each class work, using sketches of instruments as appropriate.

16.3. What is a Coriolis meter? What is it used for and how does it work?

16.4. Name four different kinds of differential pressure meters. Discuss briefly how each one works and explain the main advantages and disadvantages of each type.

16.5. Explain how each of the following works and give typical applications: rotameter and rotary piston meter.

16.6. How does an electromagnetic flowmeter work and what is it typically used for?

16.7. Discuss the mode of operation and applications of each of the following: turbine meter and vortex-shedding flowmeter.

16.8. What are the two main types of ultrasonic flowmeters? Discuss the mode of operation of each.

16.9. How do each of the following work and what are they particularly useful for: gate-type meter, jet meter, Pelton wheel meter, laser Doppler flowmeter, and thermal anemometer.

16.10. What is an open channel flowmeter? Draw a sketch of one and explain how it works.

16.11. What instruments, special equipment, and procedures are used in the calibration of flowmeters used for measuring the flow of liquids?

16.12. What instruments, special equipment, and procedures are used in the calibration of flowmeters used for measuring the flow of gases?

CHAPTER 17

# Level Measurement



## 17.1 Introduction

Level measurement is required in a wide range of applications and can involve the measurement of solids in the form of powders or small particles as well as liquids. While some applications require levels to be measured to a high degree of accuracy, other applications only need an approximate indication of level. A wide variety of instruments are available to meet these differing needs.

Simple devices such as dipsticks or float systems are relatively inexpensive. Although only offering limited measurement accuracy, they are entirely adequate for applications and find widespread use. A number of higher accuracy devices are also available for applications that require a better level of accuracy. The list of devices in common use that offer good measurement accuracy includes pressure-measuring devices, capacitive devices, ultrasonic devices, radar devices, and radiation devices. A number of other devices used less commonly are also available. All of these devices are discussed in more detail in this chapter.


Measurement and Instrumentation: Theory and Application

## 17.2 Dipsticks

Dipsticks offer a simple means of measuring the level of liquids approximately. The *ordinary dipstick* is the least expensive device available. This consists of a metal bar on which a scale is etched, as shown in Figure 17.1a. The bar is fixed at a known position in the liquid-containing vessel. A level measurement is made by removing the instrument from the vessel and reading off how far up the scale the liquid has wetted. As a human operator is required to remove and read the dipstick, this method can only be used in relatively small and shallow vessels. One common use is in checking the remaining amount of beer in an ale cask.

The *optical dipstick*, illustrated in Figure 17.1b, is an alternative form that allows a reading to be obtained without removing the dipstick from the vessel and so is applicable to larger, deeper tanks. Light from a source is reflected from a mirror, passes round the chamfered end of the dipstick, and enters a light detector after reflection by a second mirror. When the chamfered end comes into contact with liquid, its internal reflection properties are altered and light no longer enters the detector. By using a suitable mechanical drive system to move the instrument up and down and measure its position, the liquid level can be monitored.

## 17.3 Float Systems

Float systems are simple and inexpensive and provide an alternative way of measuring the level of liquids approximately that is widely used. The system consists of a float on the surface of the liquid whose position is measured by means of a suitable transducer. They have a typical measurement inaccuracy of $\pm 1\%$. The system using a potentiometer, shown earlier in Figure 2.2, is very common and is well known for its application to monitoring the level in

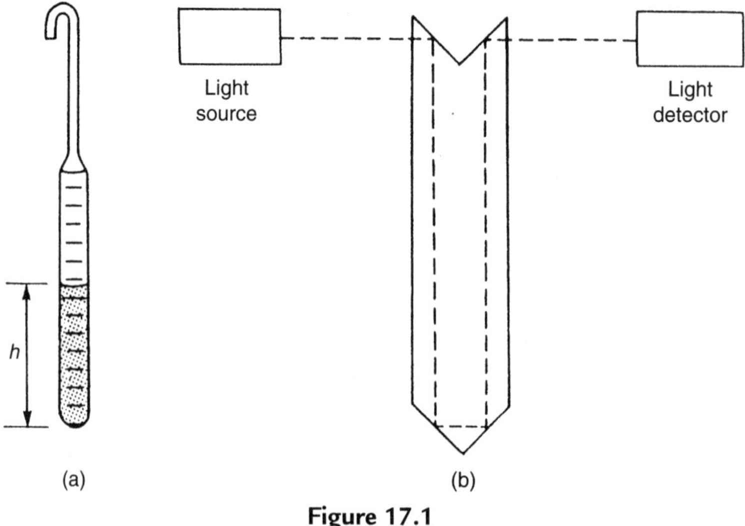


**Figure 17.1**
Dipsticks: (a) simple and (b) optical.

motor vehicle fuel tanks. An alternative system, which is used in greater numbers, is called the *float-and-tape gauge* (or *tank gauge*). This has a tape attached to the float that passes round a pulley situated vertically above the float. The other end of the tape is attached to either a counterweight or a negative-rate counterspring. The amount of rotation of the pulley, measured by either a synchro or a potentiometer, is then proportional to the liquid level. These two essentially mechanical systems of measurement are popular in many applications, but the maintenance requirements of them are always high.

## 17.4 Pressure-Measuring Devices (Hydrostatic Systems)

Pressure-measuring devices measure the liquid level to a better accuracy and use the principle that the hydrostatic pressure due to a liquid is directly proportional to its depth and hence to the level of its surface. Several instruments are available that use this principle and are widely used in many industries, particularly in harsh chemical environments. In the case of open-topped vessels (or covered ones that are vented to the atmosphere), the level can be measured by inserting a pressure sensor at the bottom of the vessel, as shown in Figure 17.2a. The liquid level, $h$, is then related to the measured pressure, $P$, according to $h = P/\rho g$, where $\rho$ is the liquid density and $g$ is

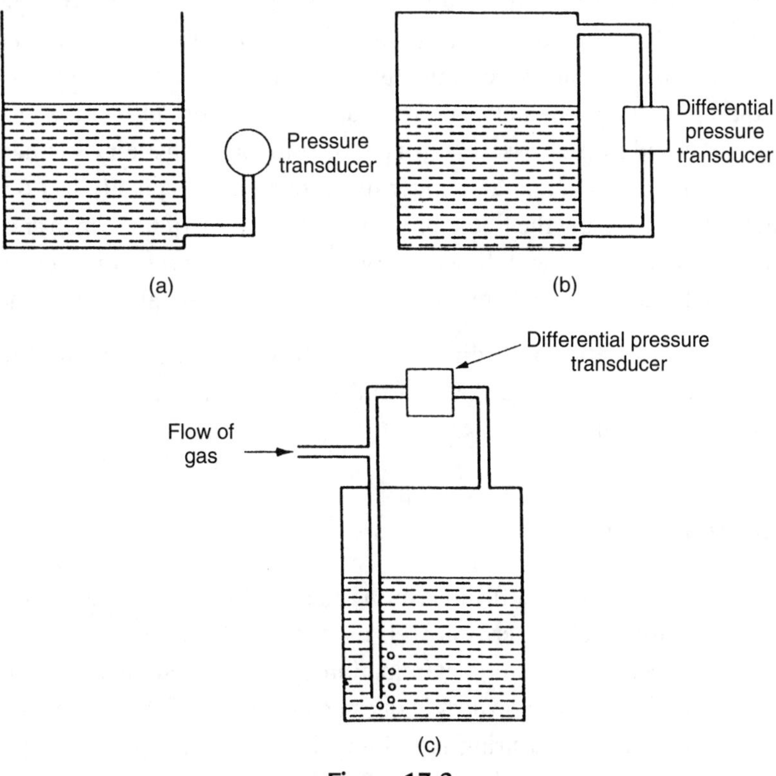


**Figure 17.2**
Hydrostatic systems: (a) open-topped vessel, (b) sealed vessel, and (c) bubbler unit.

acceleration due to gravity. One source of error in this method can be imprecise knowledge of the liquid density. This can be a particular problem in the case of liquid solutions and mixtures (especially hydrocarbons), and in some cases only an estimate of density is available. Even with single liquids, the density is subject to variation with temperature, and therefore temperature measurement may be required if very accurate level measurements are needed.

Where liquid-containing vessels are totally sealed, the liquid level can be calculated by measuring the differential pressure between the top and the bottom of the tank, as shown in Figure 17.2b. The differential pressure transducer used is normally a standard diaphragm type, although silicon-based microsensors are being used in increasing numbers. The liquid level is related to the differential pressure measured, $\delta P$, according to $h = \delta P / \rho g$. The same comments as for the case of the open vessel apply regarding uncertainty in the value of $\rho$. An additional problem that can occur is an accumulation of liquid on the side of the differential pressure transducer measuring the pressure at the top of the vessel. This can arise because of temperature fluctuations, which allow liquid to alternately vaporize from the liquid surface and then condense in the pressure tapping at the top of the vessel. The effect of this on the accuracy of the differential pressure measurement is severe, but the problem is avoided easily by placing a drain pot in the system.

A final pressure-related system of level measurement is the *bubbler unit* shown in Figure 17.2c. This uses a dip pipe that reaches to the bottom of the tank and is purged free of liquid by a steady flow of gas through it. The rate of flow is adjusted until gas bubbles are just seen to emerge from the end of the tube. The pressure in the tube, measured by a pressure transducer, is then equal to the liquid pressure at the bottom of the tank. It is important that the gas used is inert with respect to the liquid in the vessel. Nitrogen, or sometimes just air, is suitable in most cases. Gas consumption is low, and a cylinder of nitrogen may typically last 6 months. This method is suitable for measuring the liquid pressure at the bottom of both open and sealed tanks. It is particularly advantageous in avoiding the large maintenance problem associated with leaks at the bottom of tanks at the site of pressure tappings required by alternative methods.

Measurement uncertainty varies according to the application and condition of the measured material. A typical value would be $\pm 0.5\%$ of full-scale reading, although $\pm 0.1\%$ can be achieved in some circumstances.

## 17.5 Capacitive Devices

Capacitive devices are widely used for measuring the level of both liquids and solids in powdered or granular form. They perform well in many applications, but become inaccurate if the measured substance is prone to contamination by agents that change the dielectric constant. Ingress of moisture into powders is one such example of this. They are also suitable for use in extreme conditions measuring liquid metals (high temperatures), liquid gases (low temperatures), corrosive liquids (acids, etc.), and high-pressure processes. Two versions are

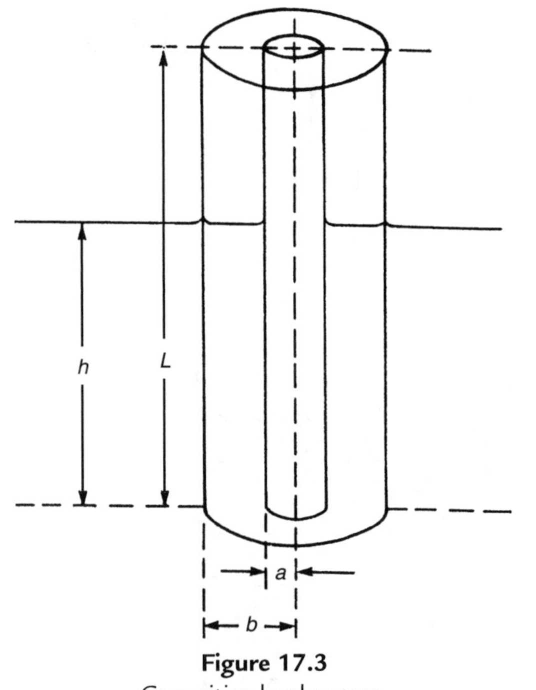


**Figure 17.3**
Capacitive level sensor.

used according to whether the measured substance is conducting or not. For nonconducting substances (less than 0.1 μmho/cm$^3$), two bare-metal capacitor plates in the form of concentric cylinders are immersed in the substance, as shown in Figure 17.3. The substance behaves as a dielectric between the plates according to the depth of the substance. For concentric cylinder plates of radius $a$ and $b$ ($b > a$), and total height $L$, the depth of the substance, $h$, is related to the measured capacitance, $C$, by

$$h = \frac{C \log_e(b/a) - 2\pi\epsilon_o}{2\pi\epsilon_o(\epsilon - 1)} \tag{17.1}$$

where $\epsilon$ is the relative permittivity of the measured substance and $\epsilon_o$ is the permittivity of free space. In the case of conducting substances, exactly the same measurement techniques are applied, but the capacitor plates are encapsulated in an insulating material. The relationship between $C$ and $h$ in Equation (17.1) then has to be modified to allow for the dielectric effect of the insulator. Measurement uncertainty is typically 1–2%.

## 17.6 Ultrasonic Level Gauge

Ultrasonic level measurement is one of a number of noncontact techniques available. It is used primarily to measure the level of materials that are either in a highly viscous liquid form or in solid (powder or granular) form. The principle of the ultrasonic level gauge is that energy

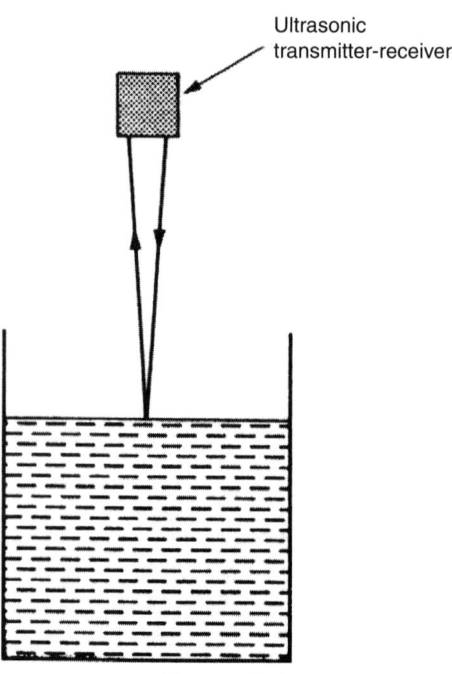


**Figure 17.4**
Ultrasonic level gauge.

from an ultrasonic source above the material is reflected back from the material surface into an ultrasonic energy detector, as illustrated in Figure 17.4. Measurement of the time of flight allows the level of the material surface to be inferred. In alternative versions (only valid for liquids), the ultrasonic source is placed at the bottom of the vessel containing the liquid, and the time of flight between emission, reflection off the liquid surface, and detection back at the bottom of the vessel is measured.

Ultrasonic techniques are especially useful in measuring the position of the interface between two immiscible liquids contained in the same vessel or measuring the sludge or precipitate level at the bottom of a liquid-filled tank. In either case, the method employed is to fix the ultrasonic transmitter–receiver transducer at a known height in the upper liquid, as shown in Figure 17.5. This establishes the level of the liquid/liquid or liquid/sludge level in absolute terms. When using ultrasonic instruments, it is essential that proper compensation is made for the working temperature if this differs from the calibration temperature, as the speed of ultrasound through air varies with temperature (see Chapter 13). Ultrasound speed also has a small sensitivity to humidity, air pressure, and carbon dioxide concentration, but these factors are usually insignificant. Temperature compensation can be achieved in two ways. First, the operating temperature can be measured and an appropriate correction made. Second, and preferably, a comparison method can be used in which the system is calibrated each time it is used by measuring the transit time of ultrasonic energy between two known reference points. This

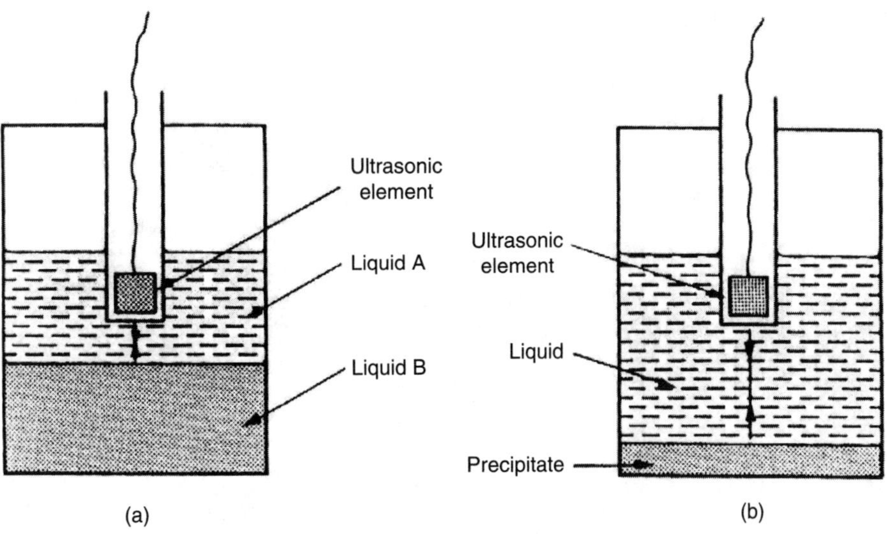


**Figure 17.5**
Measuring interface positions: (a) liquid/liquid interface and (b) liquid/precipitate interface.

second method takes account of humidity, pressure, and carbon dioxide concentration variations as well as providing temperature compensation. With appropriate care, measurement uncertainty can be reduced to about $\pm 1\%$.

## 17.7 Radar (Microwave) Sensors

Level-measuring instruments using microwave radar are an alternative technique for noncontact measurement. Currently, they are still very expensive ($\approx$\$5000), but prices are falling and usage is expanding rapidly. They are able to provide successful level measurement in applications that are otherwise very difficult, such as measurement in closed tanks, where the liquid is turbulent, and in the presence of obstructions and steam condensate. They can also be used for detecting the surface of solids in powder or particulate form. The technique involves directing a constant amplitude, frequency-modulated microwave signal at the liquid surface. A receiver measures the phase difference between the reflected signal and the original signal transmitted directly through air to it, as shown in Figure 17.6. This measured phase difference is linearly proportional to the liquid level. The system is similar in principle to ultrasonic level measurement, but has the important advantage that the transmission time of radar through air is almost totally unaffected by ambient temperature and pressure fluctuations. However, as the microwave frequency is within the band used for radio communications, strict conditions on amplitude levels have to be satisfied, and the appropriate licenses have to be obtained.

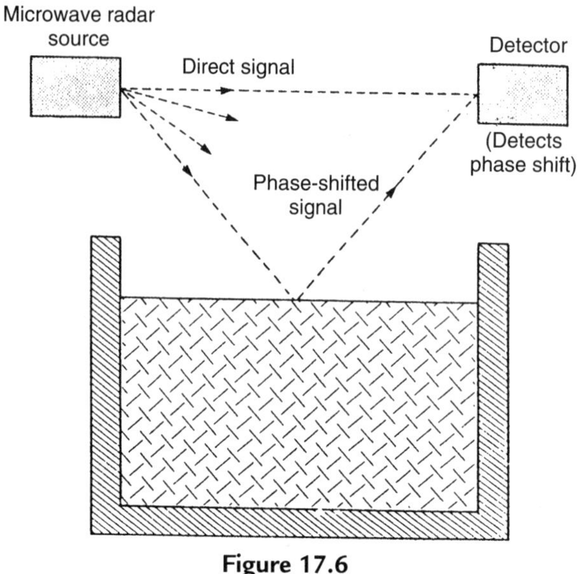


**Figure 17.6**
Radar level detector.

## 17.8 Nucleonic (or Radiometric) Sensors

Nucleonic, sometimes called radiometric, sensors are relatively expensive. They use a radiation source and detector system located outside a tank in the manner shown in Figure 17.7. The noninvasive nature of this technique in using a source and detector system outside the tank is particularly attractive. The absorption of both $\beta$ and $\gamma$ rays varies with the amount of material between the source and the detector, and hence is a function of the level of the material in the tank. Caesium-137 is a commonly used $\gamma$-ray source. The radiation level measured by the detector, $I$, is related to the length of material in the path, $x$, according to

$$I = I_o \exp(-\mu \rho x), \qquad (17.2)$$

where $I_o$ is the intensity of radiation that would be received by the detector in the absence of any material, $\mu$ is the mass absorption coefficient for the measured material, and $\rho$ is the mass density of the measured material.

In the arrangement shown in Figure 17.7, radiation follows a diagonal path across the material, and therefore some trigonometrical manipulation has to be carried out to determine material level $h$ from $x$. In some applications, the radiation source can be located in the center of the bottom of the tank, with the detector vertically above it. Where this is possible, the relationship between radiation detected and material level is obtained by directly substituting $h$ in place of $x$ in Equation (17.2). Apart from use with liquid materials at normal

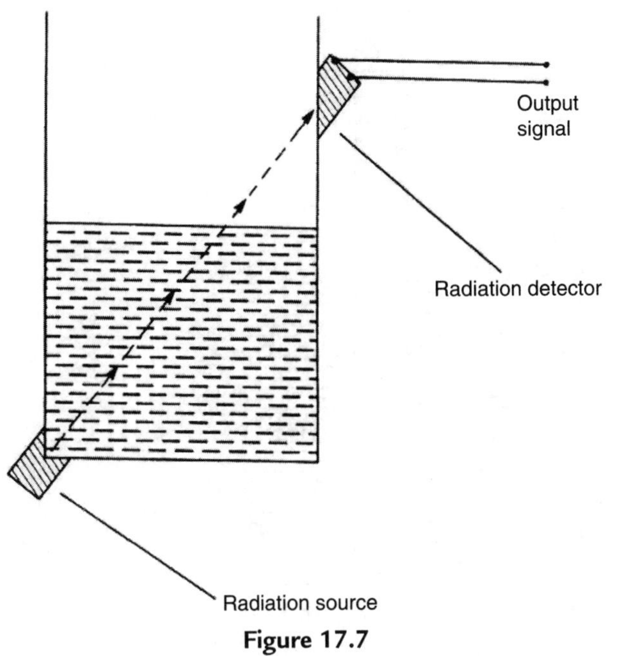


**Figure 17.7**
Using a radiation source to measure the level.

temperatures, this method is used commonly for measuring the level of hot, liquid metals and also solid materials in a powdered granular form.

Unfortunately, because of the obvious dangers associated with using radiation sources, very strict safety regulations have to be satisfied when applying this technique. Very low activity radiation sources are used in some systems to overcome safety problems, but the system is then sensitive to background radiation and special precautions have to be taken regarding the provision of adequate shielding. Because of the many difficulties in using this technique, it is only used in special applications.

## 17.9 Other Techniques

### *17.9.1 Vibrating Level Sensor*

The principle of the vibrating level sensor is illustrated in Figure 17.8. The instrument consists of two piezoelectric oscillators fixed to the inside of a hollow tube that generate flexural vibrations in the tube at its resonant frequency. The resonant frequency of the tube varies according to the depth of its immersion in the liquid. A phase-locked loop circuit is used to track these changes in resonant frequency and adjust the excitation frequency applied to the tube by the piezoelectric oscillators. The liquid level measurement is therefore obtained in terms of the output frequency of the oscillator when the tube is resonating.

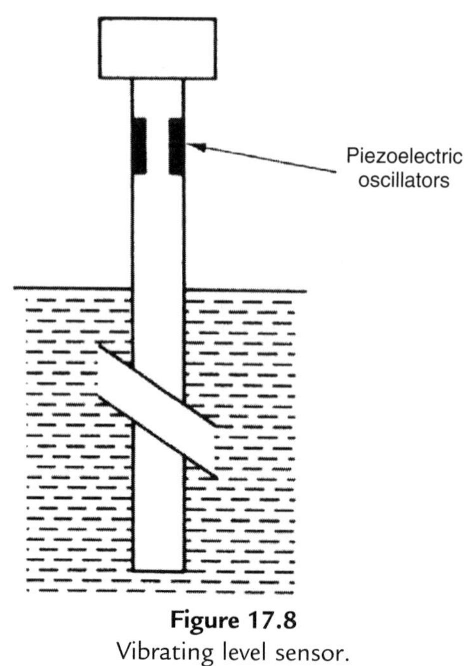


**Figure 17.8**
Vibrating level sensor.

### 17.9.2 Laser Methods

One laser-based method is the *reflective level sensor*. This sensor uses light from a laser source that is reflected off the surface of the measured liquid into a line array of charge-coupled devices, as shown in Figure 17.9. Only one of these will sense light, according to the level of the liquid. An alternative, laser-based technique operates on the same general principles as the radar method described earlier but uses laser-generated pulses of infrared light directed at the liquid surface. This is immune to environmental conditions and can be used with sealed vessels, provided that a glass window is at the top of the vessel.

## 17.10 Intelligent Level-Measuring Instruments

Most types of level gauges are now available in intelligent form. Pressure-measuring devices (Section 17.3) are obvious candidates for inclusion within intelligent level-measuring instruments, and versions claiming $\pm 0.05\%$ inaccuracy are now on the market. Such instruments can also carry out additional functions, such as providing automatic compensation for liquid density variations. Microprocessors are also used to simplify installation and setup procedures.

## 17.11 Choice between Different Level Sensors

The first consideration in choosing a level sensor is whether it is a liquid or a solid that is being measured. The second consideration is the degree of measurement accuracy required.

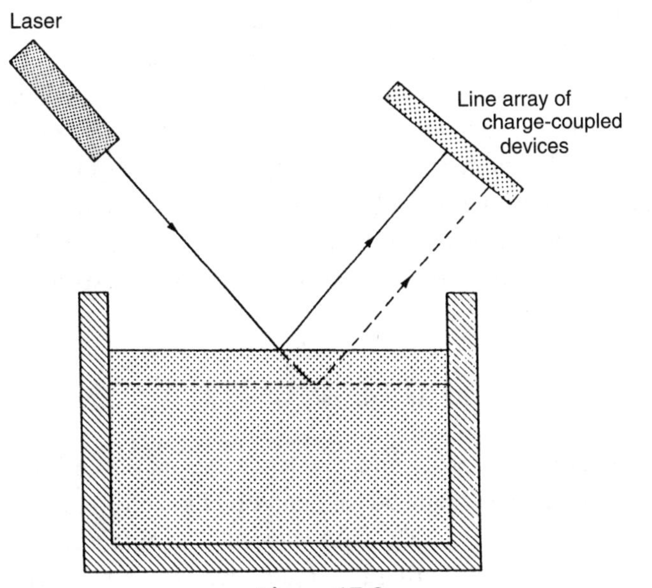


**Figure 17.9**
Reflective level sensor.

If it is liquids being measured and a relatively low level of accuracy is acceptable, dipsticks and float systems would often be used. Of these, dipsticks require a human operator, whereas float systems provide an electrical output that can be recorded or output to an electronic display as required.

Where greater measurement accuracy is required in the measurement of liquid level, a number of different devices can be used. These can be divided into two distinct classes according to whether the instrument does or does not make contact with the material whose level is being measured. The advantage of noncontact devices is that they have a higher reliability than contact devices for a number of reasons. All pressure-measuring devices (hydrostatic systems) fall into the class of a device that does make contact with the measured liquid and are used quite frequently. However, if there is a particular need for high reliability, noncontact devices such as capacitive, ultrasonic, or radiation devices are preferred. Of these, capacitive sensors are used most commonly but are unsuitable for applications where the liquid may become contaminated, as this changes its dielectric constant and hence the capacitance value. Ultrasonic sensors are less affected by contamination of the measured fluid but only work well with highly viscous fluids. Radar (microwave) and radiation sensors have the best immunity to changes in temperature, composition, moisture content, and density of the measured material and so are preferred in many applications. However, both of these are relatively expensive. Further guidance on this can be found in elsewhere.

In the case of measuring the level of solids (which must be in powdered or particle form), the choice of instrument is limited to the options of capacitive, ultrasonic, radar (microwave), and radiation sensors. As for measuring the level of liquids, radar and radiation sensors have the best immunity to changes in temperature, composition, moisture content, and density of the measured material and so are preferred in many applications. However, they both have a high cost. Either capacitive or ultrasonic devices provide a less expensive solution. Capacitive devices generally perform better but become inaccurate if the measured material is contaminated, in which case ultrasonic sensors are preferred out of these two less expensive solutions.

## 17.12 Calibration of Level Sensors

The sophistication of calibration procedures for level sensors depends on the degree of accuracy required. If the accuracy demands are not too high and a tank is relatively shallow, a simple dipstick inserted into a tank will suffice to verify the output reading of any other form of level sensor being used for monitoring the liquid level in the tank. However, this only provides one calibration point. Other calibration points can only be obtained by putting more liquid into the tank or by emptying some liquid from the tank. Such variation of the liquid level may or may not be convenient. However, even if it can be done without too much disturbance to normal use of the tank, the reading from the dipstick is of very limited accuracy because of the ambiguity in determining the exact point of contact between the dipstick and the meniscus of the liquid.

If the dipstick method is not accurate enough or is otherwise unsuitable, an alternative method of calibrating the level is to use a calibration tank that has vertical sides and a flat bottom of known cross-sectional area. Tanks with circular bottoms and rectangular bottoms are both used commonly. With the level sensor in situ, measured quantities of liquid are emptied into the tank. This increases the level of liquid in the tank in steps, and each step creates a separate calibration point. The quantity of liquid added at each stage of the calibration process can be measured either in terms of its volume or in terms of its mass. If the volume of each quantity of liquid added is measured, knowledge of the cross-sectional area of the tank bottom allows the liquid level to be calculated directly. If the mass of each quantity of liquid added is measured, the specific gravity of the liquid has to be known in order to calculate its volume and hence the liquid level. In this case, use of water as the calibration liquid is beneficial because its specific gravity is unity and therefore the calculation of level is simplified.

To measure added water in terms of its volume, calibrated volumetric measures are used. If a 1-liter measure is used, this has a typical inaccuracy of $\pm 0.1\%$. Unfortunately, errors in the measurement of each quantity of water added are cumulative, and therefore the possible error after 10 quantities of water have been added increases 10-fold to $\pm 1.0\%$. If 20 quantites are added to create 20 calibration points, the possible error is $\pm 2.0\%$ and so on.

Better accuracy can be obtained in the calibration process if the added water is measured in terms of its mass. This can be done conveniently by mounting the calibration tank on an electronic load cell. The typical inaccuracy of such a load cell is $\pm 0.05\%$ of its full-scale reading. This means that the inaccuracy of the level measurement when the tank is full is $\pm 0.05\%$ if the load cell is chosen such that it is giving its maximum output mass reading when the tank is full. Because the total mass of water in the tank is measured at each point in the calibration process, measurement errors are not cumulative. However, errors do increase for smaller volumes of water in the tank because measurement uncertainty is expressed as a percentage of the full-scale reading of the load cell. Therefore, when the tank is only 10% full, the possible measurement error is $\pm 0.5\%$. This means that calibration inaccuracy increases for smaller quantities of water in the tank but measurement uncertainty is always less than the case where measured volumes of water are added to the tank even for low levels.

Wherever possible, liquid used in the calibration tank is water, as this avoids the cost involved in using any other liquid and it also makes the calculation of level simpler when the quantities of water added to the tank are measured in terms of their volume. Unfortunately, liquid used in the tank often has to be the same as that which the sensor being calibrated normally measures. For example, the specific gravity of the measured liquid is crucial to the operation of both hydrostatic systems and capacitive level sensors. Another example is level measurement using a radiation source, as the passage of radiation through the liquid between the source and the detector is affected by the nature of the liquid.

## 17.13 Summary

We have seen that level sensors can be used to measure the position of the surface within some type of container of both solid materials in the form of powders and of liquids. We have looked at various types of level sensors, following which we considered how the various forms of level sensors available could be calibrated.

One very important observation made at the start of our discussion was that the accuracy requirements during level measurement vary widely, which has an important effect on the type of sensor used in any given situation and the corresponding calibration requirements. For example, if the surface level of a liquid within a tank used for cooling purposes in an industrial process is being monitored, only a very approximate measurement of level is needed to allow a prediction about how long it will be before the tank needs refilling. However, if the level of liquid of a consumer product within a container is being monitored during the filling process, high accuracy is required in the measurement process.

Where only approximate measurements of liquid level are needed, we saw that dipsticks provide a suitable, low-cost method of measurement, although these require a human operator and so cannot be used as part of an automatic level control system. Float systems are also relatively low-cost instruments and have an electrical form of output that can be used as part of an automatic level control system, although the accuracy is little better than that of dipsticks.

Our discussion then moved on to sensors that provide greater measurement accuracy. First among these were hydrostatic systems. These are widely used in many industries for measuring the liquid level, particularly in harsh chemical environments. Measurement uncertainty is usually about $\pm 0.5\%$ of full-scale reading, although this can be reduced to $\pm 0.1\%$ in the best hydrostatic systems. Because accurate knowledge of the liquid density is important in the operation of hydrostatic systems, serious measurement errors can occur if these systems are used to measure the level of mixtures of liquids, as the density of such mixtures is rarely known to a sufficient degree of accuracy.

Moving on to look at capacitive level sensors, we observed that these were widely used for measuring the level of both liquids and solids in powdered or granular form, with a typical measurement uncertainty of 1–2%. They are particularly useful for measuring the level of difficult materials such as liquid metals (high temperatures), liquid gases (low temperatures), and corrosive liquids (acids, etc.). However, they become inaccurate if the measured substance is prone to contamination by agents that change the dielectric constant.

Next on the list of devices studied was the ultrasonic level sensor. We noted that this is one of a number of noncontact techniques available. It is used primarily to measure the level of materials that are either in a highly viscous liquid form or in solid (powder or granular) form. We also observed that it is particularly useful for measuring the position of the interface between two immiscible liquids contained in the same vessel, and also for measuring the sludge or precipitate level at the bottom of a liquid-filled tank. The lowest measurement uncertainty achievable is $\pm 1\%$, but errors increase if the system is not calibrated properly, particularly in respect of the ambient temperature because of the changes in ultrasound speed that occur when the temperature changes.

The discussion then moved on to radar sensors, another noncontact measurement technique. We saw that this, albeit very expensive, technique provided a method for measuring the level in conditions that are too difficult for most other forms of level sensors. Such conditions include measurement in closed tanks, where the liquid is turbulent, and in the presence of obstructions and steam condensate. Like ultrasonic sensors, they can also measure the level of solids in powder or granular form.

We then looked at nucleonic sensors. These provide yet another means of noncontact level measurement that finds niche applications in measuring the level of hot, molten metals and also in measuring the level of powdered or granular solids. However, apart from the high cost of nucleonic sensors, it is necessary to adhere to very strict safety regulations when using such sensors.

Having then looked briefly at two other less common level sensors, namely the vibrating level sensor and laser-based sensors, we went on to make brief comments about intelligent level sensors. We noted that most of the types of level sensors discussed were now available in an intelligent form that quoted measurement uncertainty values down to $\pm 0.05\%$.

The final subject covered in this chapter was that of level sensor calibration. We noted that devices such as a simple dipstick could be used to calibrate sensors that were only required to provide approximate measurements of level. However, for more accurate calibration, we observed that it was usual to use a calibrated tank in which quantities of liquid were added, measured either by weight or by volume, to create a series of calibration points. We concluded that greater accuracy could be achieved in the calibration points if each quantity of liquid was weighed rather than measured with volumetric measures. We also noted that water was the least expensive liquid to use in the calibration tank but observed that it was necessary to use the same liquid as normally measured for certain sensors.

## 17.14 Problems

17.1. How do dipsticks and float systems work and what are their advantages and disadvantages in liquid level measurement?

17.2. Sketch three different kinds of hydrostatic level measurement systems. Discuss briefly the mode of operation and applications of each.

17.3. Discuss the mode of operation of the following, using a sketch to aid your discussion as appropriate: capacitive level sensor and ultrasonic level sensor.

17.4. What are the merits of microwave and radiometric level sensors? Discuss how each of these devices works.

17.5. What are the main things to consider when choosing a liquid level sensor for a particular application? What types of devices could you use for an application that required (a) low measurement accuracy, (b) high measurement accuracy where contact between the sensor and the measured liquid is acceptable, or (c) high measurement accuracy where there must not be any contact between the sensor and the measured liquid?

17.6. Discuss the range of devices able to measure the level of the surface of solid material in powdered form contained within a hopper.

17.7. What procedures could you use to calibrate a sensor that is only required to provide approximate measurements of liquid level?

17.8. What is the best calibration procedure to use for sensors required to give high accuracy in level measurement?

CHAPTER 18

# Mass, Force, and Torque Measurement

## 18.1 Introduction

Mass, force, and torque are covered together within this chapter because they are closely related quantities. Mass describes the quantity of matter that a body contains. Force is the product of mass times acceleration, according to Newton's second law of motion:

$$Force = Mass \times acceleration.$$

Forces can be applied in either a horizontal or a vertical direction. A force applied in a downward, vertical direction gives rise to the term *weight*, which is defined as the downward force exerted by a mass subject to a gravitational force:

$$Weight = Mass \times acceleration\ due\ to\ gravity.$$

The final quantity covered in this chapter, torque, can be regarded as a rotational force. When applied to a body, torque causes the body to rotate about its axis of rotation. This is analagous to the horizontal motion of a body when a horizontal force is applied to it.

## 18.2 Mass (Weight) Measurement

The *mass* of a body is always quantified in terms of a measurement of the *weight* of the body, this being the downward force exerted by the body when it is subject to gravity. Three methods are used to measure this force.

The first method of measuring the downward force exerted by a mass subject to gravity involves the use of a *load cell*. The load cell measures the downward force $F$, and then the mass $M$ is calculated from the equation:

$$M = F/g,$$

where $g$ is acceleration due to gravity.

Because the value of $g$ varies by small amounts at different points around the earth's surface, the value of $M$ can only be calculated exactly if the value of $g$ is known exactly. Nevertheless, load cells are, in fact, the most common instrument used to measure mass, especially in industrial applications. Several different forms of load cells are available. Most load cells are now electronic, although pneumatic and hydraulic types also exist. These types vary in features and accuracy, but all are easy to use as they are deflection-type instruments that give an output reading without operator intervention.

The second method of measuring mass is to use a spring balance. This also measures the downward force when the measured mass is subject to gravity. Hence, as in the case of load cells, the mass value can only be calculated exactly if the value of $g$ is known exactly. Like a load cell, the spring balance is also a deflection-type instrument and so is easy to use.

The final method of measuring mass is to use some form of mass balance instrument. These provide an absolute measurement, as they compare the gravitational force on the mass being measured with the gravitational force on a standard mass. Because the same gravitational force is applied to both masses, the exact value of $g$ is immaterial. However, being a null-type instrument, any form of balance is tedious to use.

The following paragraphs consider these various forms of mass-measuring instruments in more detail.

### 18.2.1 Electronic Load Cell (Electronic Balance)

The electronic load cell is now the preferred type of load cell in most applications. Within an electronic load cell, the gravitational force on the body being measured is applied to an elastic element. This deflects according to the magnitude of the body mass. Mass measurement is thereby translated into a displacement measurement task.

The elastic elements used are specially shaped and designed, some examples of which are shown in Figure 18.1. The design aims are to obtain a linear output relationship between the applied force and the measured deflection and to make the instrument insensitive to forces that are not applied directly along the sensing axis. Load cells exist in both compression and tension forms. In the compression type, the measured mass is placed on top of a platform resting on the load cell, which therefore compresses the cell. In the alternative tension type, the mass is hung from the load cell, thereby putting the cell into tension.

Various types of displacement transducers are used to measure the deflection of the elastic elements. Of these, the strain gauge is used most commonly, as this gives the best measurement accuracy, with an inaccuracy figure less than $\pm 0.05\%$ of full-scale reading being obtainable. Load cells, including strain gauges, are used to measure masses over a very wide range between 0 and 3000 tonne. The measurement capability of an individual instrument designed to measure masses at the bottom end of this range would typically be 0.1–5 kg, whereas instruments designed for the top of the range would have a typical measurement span of 10–3000 tonne.

Elastic force transducers based on differential transformers (LVDT) to measure defections are used to measure masses up to 25 tonne. Apart from having a lower maximum measuring capability, they are also inferior to strain gauge-based instruments in terms of their $\pm 0.2\%$ inaccuracy value. Their major advantages are their longevity and almost total lack of maintenance requirements.

The final type of displacement transducer used in this class of instrument is the piezoelectric device. Such instruments are used to measure masses in the range of 0 to 1000 tonne. Piezoelectric crystals replace the specially designed elastic member used normally in this class of instrument, allowing the device to be physically small. As discussed previously, such devices can only

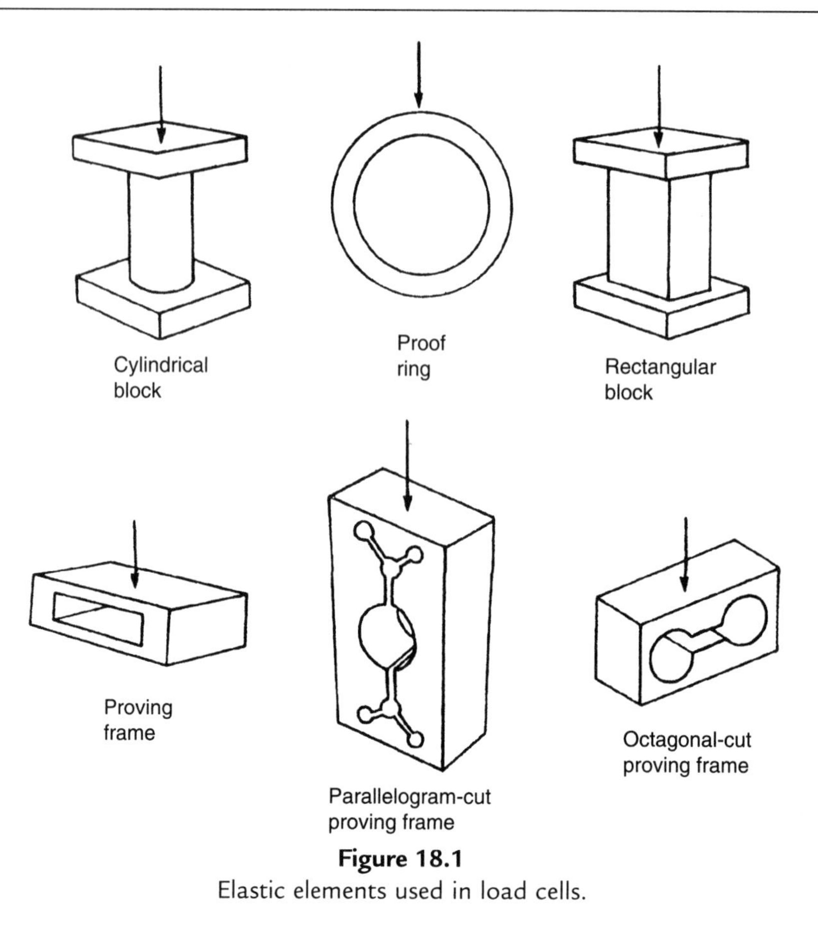


**Figure 18.1**
Elastic elements used in load cells.

measure dynamically changing forces because the output reading results from an induced electrical charge whose magnitude leaks away with time. The fact that the elastic element consists of a piezoelectric crystal means that it is very difficult to design such instruments to be insensitive to forces applied at an angle to the sensing axis. Therefore, special precautions have to be taken in applying these devices. Although such instruments are relatively inexpensive, their lowest inaccuracy is $\pm 1\%$ of full-scale reading and they also have a high temperature coefficient.

Electronic load cells have significant advantages over most other forms of mass-measuring instruments in terms of their relatively low cost, wide measurement range, tolerance of dusty and corrosive environments, remote measurement capability, tolerance of shock loading, and ease of installation. However, one particular problem that can affect their performance is the phenomenon of creep. Creep describes the permanent deformation that an elastic element undergoes after it has been under load for a period of time. This can lead to significant measurement errors in the form of a bias on all readings if the instrument is not recalibrated

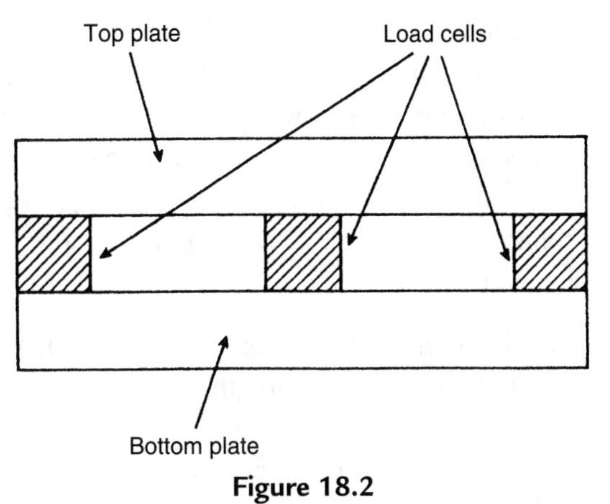


**Figure 18.2**
Load cell-based electronic balance.

from time to time. However, careful design and choice of materials can largely eliminate the problem.

Several compression-type load cells are often used together in a form of instrument known as an *electronic balance*. This is shown schematically in Figure 18.2. Commonly, either three or four load cells are used in the balance, with the output mass measurement being formed from the sum of the outputs of each cell. Where appropriate, the upper platform can be replaced by a tank for weighing liquids, powders, and so on.

### 18.2.2 Pneumatic and Hydraulic Load Cells

Pneumatic and hydraulic load cells translate mass measurement into a pressure measurement task, although they are now less common than the electronic load cell. A pneumatic load cell is shown schematically in Figure 18.3. Application of a mass to the cell causes deflection of a

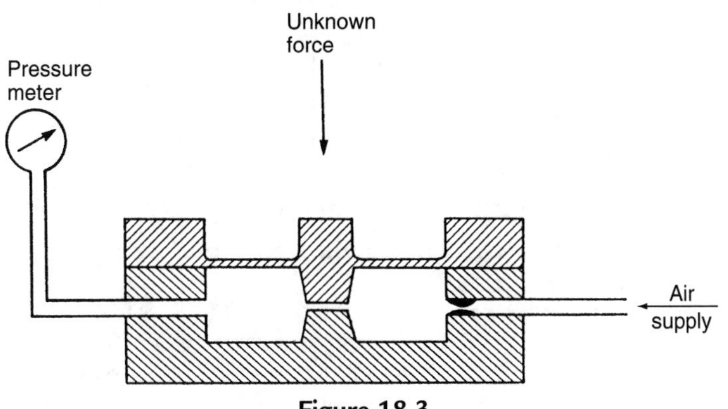


**Figure 18.3**
Pneumatic load cell.

diaphragm acting as a variable restriction in a nozzle-flapper mechanism. The output pressure measured in the cell is approximately proportional to the magnitude of the gravitational force on the applied mass. The instrument requires a flow of air at its input of around 0.25 $m^3/h$ at a pressure of 4 bar. Standard cells are available to measure a wide range of masses. For measuring small masses, instruments are available with a full-scale reading of 25 kg, while instruments with a full-scale reading of 25 tonne are obtainable at the top of the range. Inaccuracy is typically ±0.5% of full scale in pneumatic load cells.

The alternative, hydraulic load cell is shown in Figure 18.4. In this, the gravitational force due to the unknown mass is applied, via a diaphragm, to oil contained within an enclosed chamber. The corresponding increase in oil pressure is measured by a suitable pressure transducer. These instruments are designed for measuring much larger masses than pneumatic cells, with a load capacity of 500 tonne being common. Special units can be obtained to measure masses as large as 50,000 tonne. In addition to their much greater measuring range, hydraulic load cells are much more accurate than pneumatic cells, with an inaccuracy figure of ±0.05% of full scale being typical. However, in order to obtain such a level of accuracy, correction for the local value of *g* (acceleration due to gravity) is necessary. A measurement resolution of 0.02% is attainable.

### 18.2.3 Intelligent Load Cells

Intelligent load cells are formed by adding a microprocessor to a standard cell. This brings no improvement in accuracy because the load cell is already a very accurate device. What it does produce is an intelligent weighing system that can compute total cost from the measured weight, using stored cost per unit weight information, and provide an output in the form of a digital display. Cost per weight values can be prestored for a large number of substances, making such instruments very flexible in their operation.

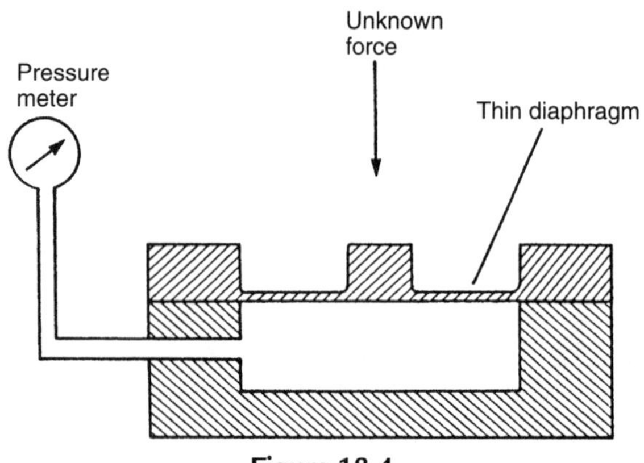


**Figure 18.4**
Hydraulic load cell.

In applications where the mass of an object is measured by several load cells used together (e.g., load cells located at the corners of a platform in an electronic balance), the total mass can be computed more readily if the individual cells have a microprocessor providing digital output. In addition, it is also possible to use significant differences in the relative readings between different load cells as a fault detection mechanism in the system.

### *18.2.4 Mass Balance (Weighing) Instruments*

Mass balance instruments are based on comparing the gravitational force on the measured mass with the gravitational force on another body of known mass. This principle of mass measurement is known commonly as *weighing* and is used in instruments such as the beam balance, weigh beam, pendulum scale, and electromagnetic balance. Various forms of mass balance instruments are available, as discussed next.

#### *Beam balance (equal arm balance)*

In the beam balance, shown in Figure 18.5, standard masses are added to a pan on one side of a pivoted beam until the magnitude of the gravity force on them balances the magnitude of the gravitational force on the unknown mass acting at the other end of the beam. This equilibrium position is indicated by a pointer that moves against a calibrated scale.

Instruments of this type are capable of measuring a wide span of masses. Those at the top end of the range can typically measure masses up to 1000 grams, whereas those at the bottom end of the range can measure masses of less than 0.01 gram. Measurement resolution can be as good as 1 part in $10^7$ of the full-scale reading if the instrument is designed and manufactured very carefully. The lowest measurement inaccuracy value attainable is $\pm 0.002\%$.

One serious disadvantage of this type of instrument is its lack of ruggedness. Continuous use and the inevitable shock loading that will occur from time to time both cause damage to

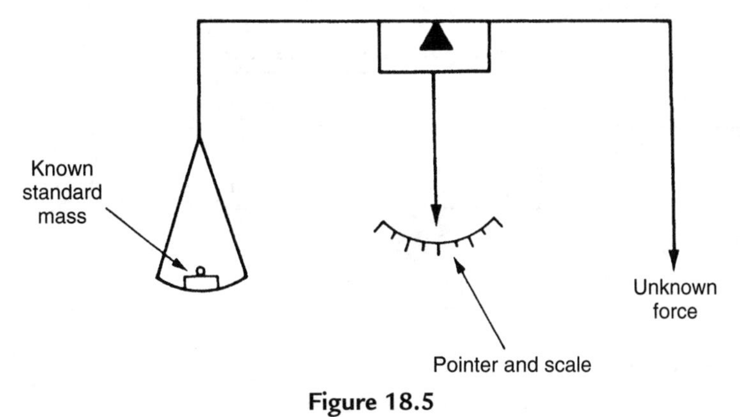


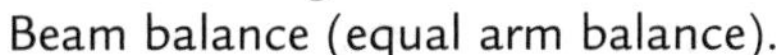
**Figure 18.5**
Beam balance (equal arm balance).

the knife edges, leading to deterioration in measurement accuracy and measurement resolution. A further problem affecting their use in industrial applications is that it takes a relatively long time to make each measurement. For these reasons, the beam balance is normally reserved as a calibration standard and is not used in day-to-day production environments.

*Weigh beam*

The weigh beam, sketched in two alternative forms in Figure 18.6, operates on similar principles to the beam balance but is much more rugged. In the first form, standard masses are added to balance the unknown mass and fine adjustment is provided by a known mass that is moved along a notched, graduated bar until the pointer is brought to the null, balance point. The alternative form has two or more graduated bars (three bars shown in Figure 18.6). Each bar

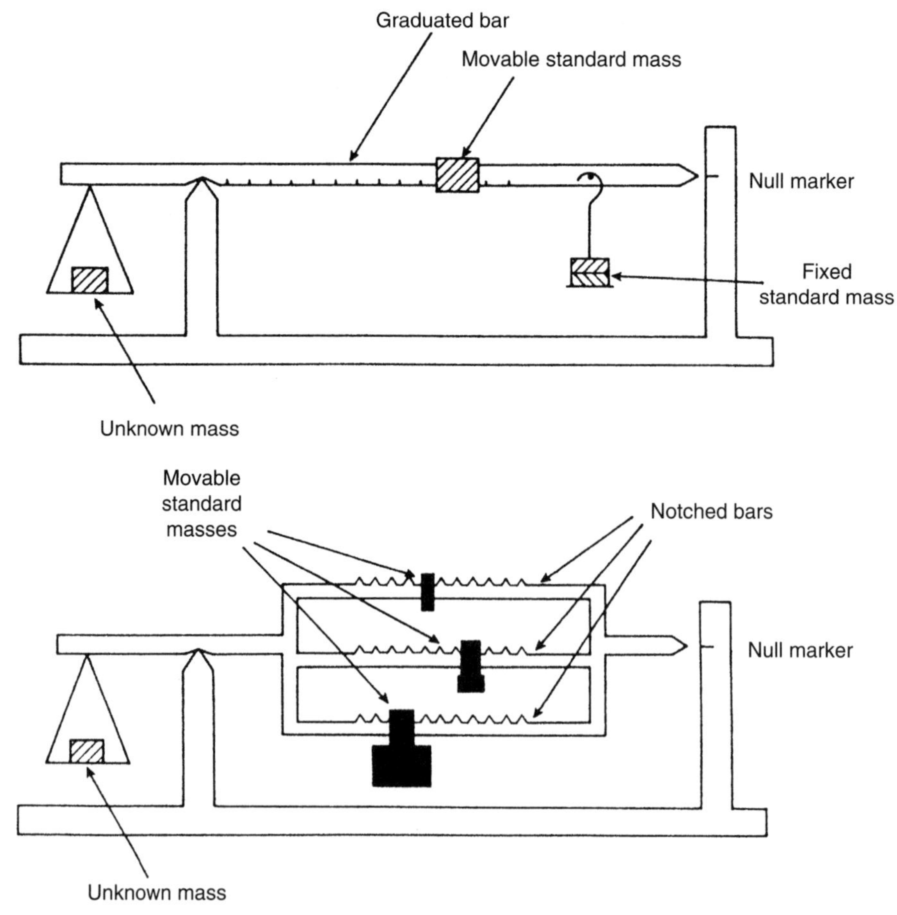


**Figure 18.6**
Two alternative forms of weigh beams.

carries a different standard mass, which is moved to appropriate positions on the notched bar to balance the unknown mass. Versions of these instruments are used to measure masses up to 50 tonne.

### *Pendulum scale*

The pendulum scale is another instrument that works on the mass-balance principle. In one arrangement shown in Figure 18.7, the unknown mass is put on a platform that is attached by steel tapes to a pair of cams. Downward motion of the platform, and hence rotation of the cams, under the influence of the gravitational force on the mass, is opposed by the gravitational force acting on two pendulum-type masses attached to the cams. The amount of rotation of the cams when the equilibrium position is reached is determined by the deflection of a pointer against a scale. The shape of the cams is such that this output deflection is linearly proportional to the applied mass. Other mechanical arrangements also exist that have the same effect of producing an output deflection of a pointer moving against a scale. It is also possible to replace the pointer and scale system by a rotational displacement transducer that gives an electrical output. Various versions of the instrument can measure masses in the range between 1 kg and 500 tonne, with a typical measurement inaccuracy of ±0.1%.

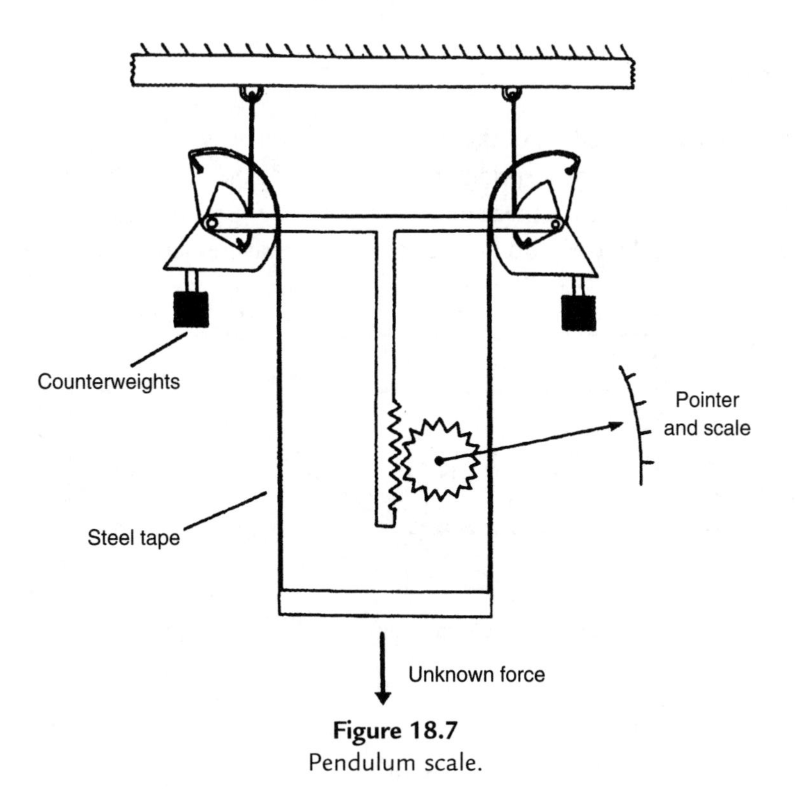


**Figure 18.7**
Pendulum scale.

Recently, the instrument has become much less common because of its inferior performance compared with instruments based on newer technology such as electronic balances. One potential source of difficulty with the instrument is oscillation of the weigh platform when mass is applied. Where necessary, in instruments measuring larger masses, dashpots are incorporated into the cam system to damp out such oscillations. A further possible problem can arise, mainly when measuring large masses, if the mass is not placed centrally on the platform. This can be avoided by designing a second platform to hold the mass, which is hung from the first platform by knife edges. This lessens the criticality of mass placement.

*Electromagnetic balance*

The electromagnetic balance uses the torque developed by a current-carrying coil suspended in a permanent magnetic field to balance the unknown mass against the known gravitational force produced on a standard mass, as shown in Figure 18.8. A light source and detector system is used to determine the null-balance point. The voltage output from the light detector is amplified and applied to the coil, thus creating a servosystem where deflection of the coil in equilibrium is proportional to the applied force. Its advantages over beam balances, weigh beams, and pendulum scales include its smaller size, its insensitivity to environmental changes (modifying inputs), and its electrical form of output. Despite these apparent advantages, it is no longer in common use because of the development of other instruments, particularly electronic balances.

### 18.2.5 Spring Balance

Spring balances provide a method of mass measurement that is both simple and inexpensive. The mass is hung on the end of a spring and deflection of the spring due to the downward gravitational force on the mass is measured against a scale. Because the characteristics of the spring are very susceptible to environmental changes, measurement accuracy is usually

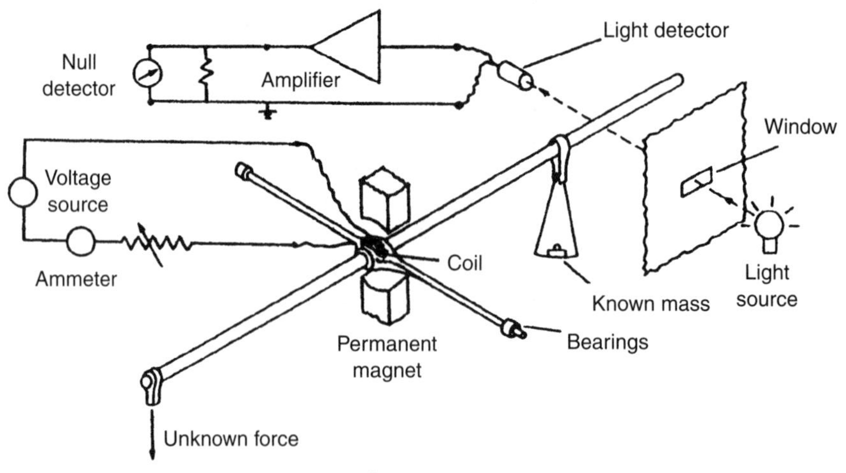


**Figure 18.8**
Electromagnetic balance.

relatively poor. However, if compensation is made for changes in spring characteristics, then a measurement inaccuracy less than ±0.2% is achievable. According to the design of the instrument, masses between 0.5 kg and 10 tonne can be measured.

## 18.3 Force Measurement

This section is concerned with the measurement of horizontal forces that either stretch or compress the body that they are applied to according to the direction of the force with respect to the body. If a force of magnitude, $F$, is applied to a body of mass, $M$, the body will accelerate at a rate, $A$, according to the equation:

$$F = MA.$$

The standard unit of force is the *Newton*, this being the force that will produce an acceleration of 1 meter per second squared in the direction of the force when applied to a mass of 1 kilogram. One way of measuring an unknown force is therefore to measure acceleration when it is applied to a body of known mass. An alternative technique is to measure the variation in the resonant frequency of a vibrating wire as it is tensioned by an applied force. Finally, forms of load cells that deform in the horizontal direction when horizontal forces are applied can also be used as force sensors. These techniques are discussed next.

### *18.3.1 Use of Accelerometers*

The technique of applying a force to a known mass and measuring the acceleration produced can be carried out using any type of accelerometer. Unfortunately, the method is of very limited practical value because, in most cases, forces are not free entities but are part of a system (from which they cannot be decoupled) in which they are acting on some body that is not free to accelerate. However, the technique can be of use in measuring some transient forces and also for calibrating forces produced by thrust motors in space vehicles.

### *18.3.2 Vibrating Wire Sensor*

This instrument, illustrated in Figure 18.9, consists of a wire that is kept vibrating at its resonant frequency by a variable-frequency oscillator. The resonant frequency of a wire under tension is given by

$$f = \frac{0.5}{L}\sqrt{\left(\frac{M}{T}\right)},$$

where $M$ is the mass per unit length of the wire, $L$ is the length of the wire, and $T$ is the tension due to the applied force, $F$. Thus, measurement of the output frequency of the oscillator allows the force applied to the wire to be calculated.

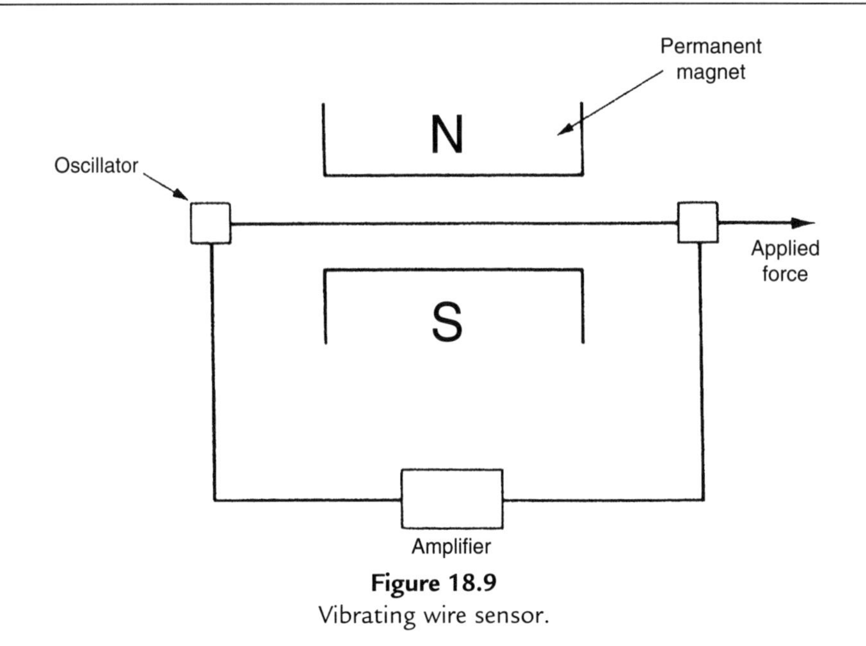


**Figure 18.9**
Vibrating wire sensor.

### 18.3.3 Use of Load Cells

Special forms of electronic load cells designed to deflect in the horizontal direction are used to measure horizontal forces applied to them.

## 18.4 Torque Measurement

Measurement of applied torques is of fundamental importance in all rotating bodies to ensure that the design of the rotating element is adequate to prevent failure under shear stresses. Torque measurement is also a necessary part of measuring the power transmitted by rotating shafts. The four methods of measuring torque consist of (i) measuring the strain produced in a rotating body due to an applied torque, (ii) an optical method, (iii) measuring the reaction force in cradled shaft bearings, and (iv) using equipment known as the *Prony brake*. Of these, the first two should be regarded as "normal" ways of measuring torque at the present time as the latter two are no longer in common use.

### 18.4.1 Measurement of Induced Strain

Measuring the strain induced in a shaft due to an applied torque has been the most common method used for torque measurement in recent years. The method involves bonding four strain gauges onto a shaft as shown in Figure 18.10, where the strain gauges are arranged in a d.c. bridge circuit. The output from the bridge circuit is a function of the strain in the shaft and hence of the torque applied. It is very important that positioning of the strain

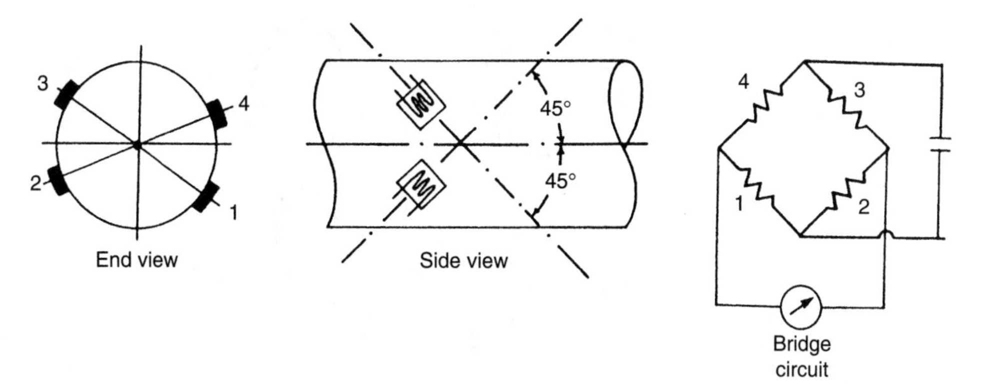


**Figure 18.10**
Position of torque-measuring strain gauges on a shaft.

gauges on the shaft is precise, and the difficulty in achieving this makes the instrument relatively expensive.

This technique is ideal for measuring the stalled torque in a shaft before rotation commences. However, a problem is encountered in the case of rotating shafts because a suitable method then has to be found for making the electrical connections to the strain gauges. One solution to this problem found in many commercial instruments is to use a system of slip rings and brushes for this, although this increases the cost of the instrument still further.

### *18.4.2 Optical Torque Measurement*

Optical techniques for torque measurement have become available recently with the development of laser diodes and fiber-optic light transmission systems. One such system is shown in Figure 18.11. Two black-and-white striped wheels are mounted at either end of the rotating shaft and are in alignment when no torque is applied to the shaft. Light from a laser diode light source is directed by a pair of fiber-optic cables onto the wheels. The rotation of the wheels causes pulses of reflected light, which are transmitted back to a receiver by a second pair of fiber-optic cables. Under zero torque conditions, the two pulse trains of reflected light are in phase with each other. If torque is now applied to the shaft, the reflected light is modulated. Measurement by the receiver of the phase difference between the reflected pulse trains therefore allows the magnitude of torque in the shaft to be calculated. The cost of such instruments is relatively low, and an additional advantage in many applications is their small physical size.

### *18.4.3 Reaction Forces in Shaft Bearings*

Any system involving torque transmission through a shaft contains both a power source and a power absorber where the power is dissipated. The magnitude of the transmitted torque can be measured by cradling either the power source or the power absorber end of the shaft in bearings, and then measuring the reaction force, $F$, and the arm length, $L$, as shown in Figure 18.12.

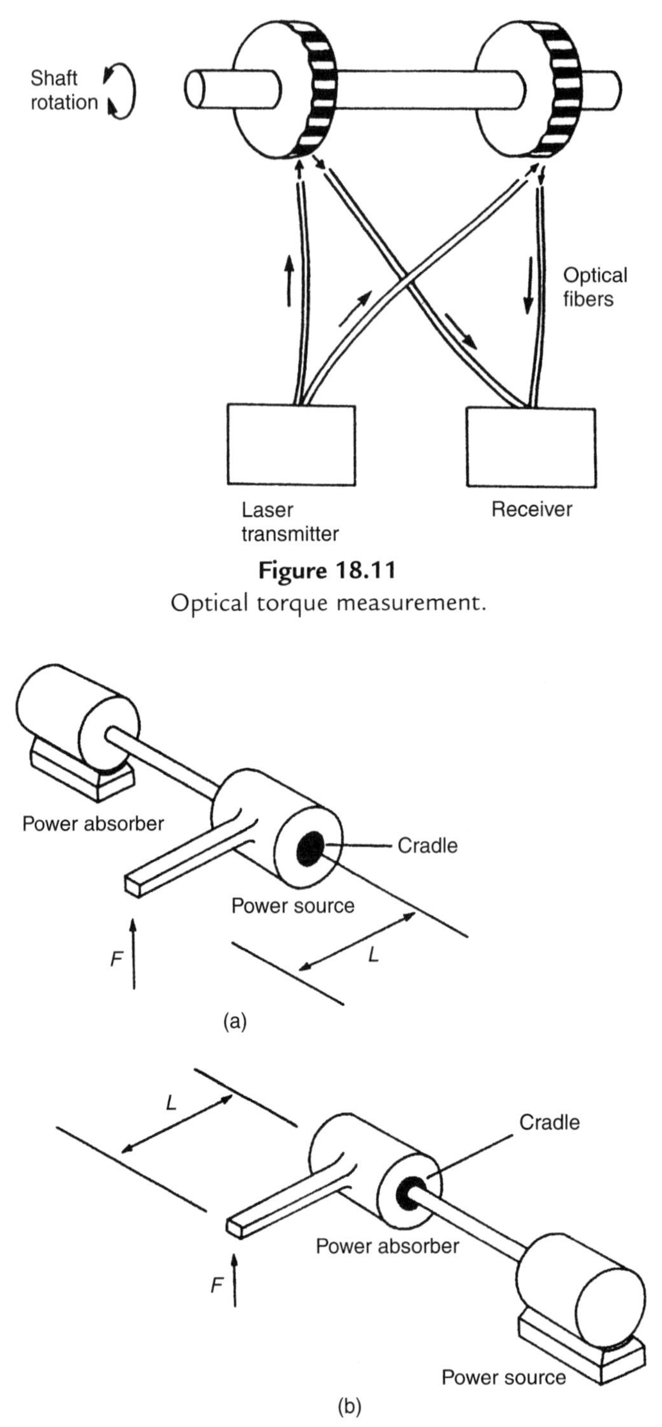


**Figure 18.11**
Optical torque measurement.



**Figure 18.12**
Measuring reaction forces in cradled shaft bearings.

The torque is then calculated as the simple product, $FL$. Pendulum scales are used very commonly for measuring the reaction force. Inherent errors in the method are bearing friction and windage torques. This technique is no longer in common use.

### 18.4.4 Prony Brake

The Prony brake is another torque-measuring system that is now uncommon. It is used to measure the torque in a rotating shaft and consists of a rope wound round the shaft, as illustrated in Figure 18.13. One end of the rope is attached to a spring balance and the other end carries a load in the form of a standard mass, $m$. If the measured force in the spring balance is $F_s$, then the effective force, $F_e$, exerted by the rope on the shaft is given by

$$F_e = mg - F_s.$$

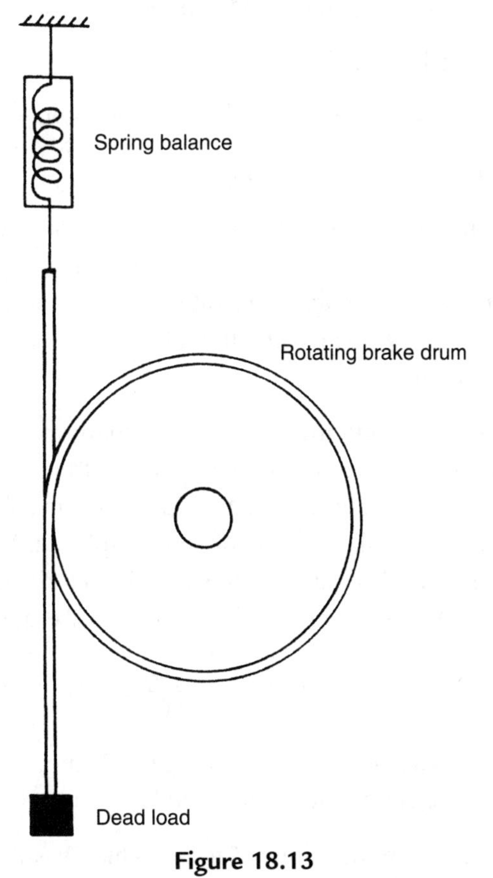


**Figure 18.13**
A Prony brake.

If the radius of the shaft is $R_s$ and that of the rope is $R_r$, then the effective radius, $R_e$, of the rope and drum with respect to the axis of rotation of the shaft is given by

$$R_e = R_s + R_r.$$

The torque in the shaft, $T$, can then be calculated as

$$T = F_e R_e.$$

While this is a well-known method of measuring shaft torque, a lot of heat is generated because of friction between the rope and shaft, and water cooling is usually necessary.

## 18.5 Calibration of Mass, Force, and Torque Measuring Sensors

One particular difficulty that arises in the calibration of mass, force, and torque measuring instruments is variability in the value of $g$ (acceleration due to gravity). Apart from instruments such as the beam balance and pendulum scale, which directly compare two masses, all other instruments have an output reading that depends on the value of $g$.

The value of g is given by Helmert's formula:

$$g = 980.6 - 2.6\cos\phi - 0.000309h,$$

where $\phi$ is the latitude and $h$ is the altitude in meters.

It can be seen from this formula that $g$ varies with both latitude and altitude. At the equator ($\cos\phi = 0°$), $g = 978.0$, whereas at the poles ($\cos\phi = 90°$), g = 983.2. In Britain, a working value of 980.7 is normally used for $g$, and very little error can normally be expected when using this value. Where necessary, the exact value of $g$ can be established by measuring the period and length of a pendulum.

Another difficulty that arises in calibrating mass, force, and torque sensors is the presence of an upward force generated by the air medium in which the instruments are tested and used. According to Archimedes' principle, when a body is immersed in a fluid (air in this case), there is an upward force proportional to the volume of fluid displaced. Even in pure mass-balance instruments, an error is introduced because of this unless both the body of unknown mass and the standard mass have the same density. This error can be quantified as

$$Error = \frac{SG_a}{SG_a} - \frac{SG_a}{SG_m},$$

where $SG_a$ is the specific gravity of air, $SG_u$ is the specific gravity of the substance being measured, and $SG_m$ is the specific gravity of the standard mass.

Fortunately, maximum error due to this upward force (which has the largest magnitude when weighing low-density liquids such as petrol) will not exceed 0.2%. Therefore, in most circumstances, the error due to air buoyancy can be neglected. However, for calibrations at the top of the calibration tree, where the highest levels of accuracy are demanded, either correction must be made for this factor or it must be avoided by carrying out the calibration in vacuum conditions.

### 18.5.1 Mass Calibration

The primary requirement in mass calibration is maintenance of a set of standard masses applied to the mass sensor being calibrated. Provided that this set of standard masses is protected from damage, there is little reason for the value of the masses to change. Despite this, values of the masses must be checked at prescribed intervals, typically annually, in order to maintain the traceability of the calibration to reference standards. The instrument used to provide this calibration check on standard masses is a beam balance, a weigh beam, a pendulum scale, an electromagnetic balance, or a proof ring-based load cell.

*Beam balance*

A beam balance is used for calibrating masses in the range between 10 mg and 1 kg. The measurement resolution and accuracy achieved depend on the quality and sharpness of the knife edge that the pivot is formed from. For high measurement resolution, friction at the pivot must be as close to zero as possible, and hence a very sharp and clean knife edge pivot is demanded. The two halves of the beam on either side of the pivot are normally of equal length and are measured from the knife edge. Any bluntness, dirt, or corrosion in the pivot can cause these two lengths to become unequal, causing consequent measurement errors. Similar comments apply about the knife edges on the beam that the two pans are hung from. It is also important that all knife edges are parallel, as otherwise displacement of the point of application of the force over the line of the knife edge can cause further measurement errors. This last form of error also occurs if the mass is not placed centrally on the pan.

Great care is therefore required in the use of such an instrument, but, provided that it is kept in good condition, particularly with regard to keeping the knife edges sharp and clean, high measurement accuracy is achievable. Such a good condition can be confirmed by applying calibrated masses to each side of the balance. If the instrument is then balanced exactly, all is well.

*Weigh beam*

In order to use it as a calibration standard, a weigh beam has to be manufactured and maintained to a high standard. However, providing these conditions are met, it can be used as a standard for calibrating masses up to 50 tonne.

*Pendulum scale*

Like the weigh beam, the pendulum scale can only be used for calibration if it is manufactured to a high standard and maintained properly, with special attention to the cleanliness and lubrication of moving parts. Provided that these conditions are met, it can be used as a calibration standard for masses between 1 kg and 500 tonne.

*Electromagnetic balance*

Various forms of electromagnetic balance exist as alternatives to the three instruments just described for calibration duties. A particular advantage of the electromagnetic balance is its use of an optical system to magnify motion around the null point, leading to higher measurement accuracy. Consequently, this type of instrument is often preferred for calibration duties, particularly for higher measurement ranges. The actual degree of accuracy achievable depends on the magnitude of the mass being measured. In the range between 100 g and 10 kg, an inaccuracy of $\pm 0.0001\%$ is achievable. Above and below this range, inaccuracy is worse, increasing to $\pm$ 0.002% measuring 5 tonne and $\pm 0.03\%$ measuring 10 mg.

*Proof ring-based load cell*

The proof ring-based load cell is used for calibration in the range between 150 kg and 2000 tonne. When used for calibration, displacement of the proof ring in the instrument is measured by either an LVDT or a micrometer. As the relationship between the applied mass/force and the displacement is not a straight-line one, a force/deflection graph has to be used to interpret the output. The lowest measurement inaccuracy achievable is $\pm 0.1\%$.

### 18.5.2 Force Sensor Calibration

Force sensors are calibrated using special machines that apply a set of known force values to the sensor. The machines involved are very large and expensive. For this reason, force sensor calibration is normally devolved to either specialist calibration companies or manufacturers of the measurement devices being calibrated, who will give advice about the frequency of calibration necessary to maintain the traceability of measurements to national reference standards.

### 18.5.3 Calibration of Torque Measuring Systems

As for the case of force sensor calibration, special machines are required for torque measurement system calibration that can apply accurately known torque values to the system being calibrated. Such machines are very expensive. It is therefore normal to use the services of specialist calibration companies or to use similar services provided by the manufacturer of the torque measurement system. Again, the company to which the calibration task is assigned will give advice on the required frequency of calibration.

## 18.6 Summary

We have covered the measurement of all three quantities—mass, force, and torque—in this chapter as the three quantities are closely related. We also learned that weight was another related quantity as this describes the force exerted on a mass subject to gravity.

Mass is measured in one of three distinct ways, using load cell, using a spring balance, or using one of several instruments working on the mass-balance principle. Of these, load cells and spring balances are deflection-type instruments, whereas the mass balance is a null-type instrument. This means that a balance is somewhat tedious to use compared with other forms of mass-measuring instruments.

In respect of load cells, we looked first at the electronic load cell, as this is now the type of load cell preferred in most applications where masses between 0.1 kg and 3000 tonne in magnitude are measured. We learned that pneumatic and hydraulic load cells represent somewhat older technology that is used much less frequently nowadays. However, special types of hydraulic load cells still find a significant number of applications in measuring large masses, where the maximum capability is 50,000 tonne. We noted that variations in the local value of $g$ (the acceleration due to gravity) have some effect on the accuracy of load cells but observed that the magnitude of this error was usually small. Before leaving the subject of load cells, we also made some mention of intelligent load cells.

Looking next at mass balance instruments, we saw that a particular advantage that they had was their immunity to variations in the value of $g$. We studied the various types of balance available in the form of the beam balance, weigh beam, pendulum scale, and electromagnetic balance.

We then ended the review of mass-measuring instruments by looking at the spring balance. Our conclusion about this was that, while simple and inexpensive, its measurement accuracy is usually relatively poor.

Moving on to force measurement, we noted that transient forces could be measured by an accelerometer. However, static forces were measured either by a vibrating wire sensor or by a special form of load cell.

Looking next at torque measurement, we saw that the main two current methods for measuring torque were to measure the induced strain in a rotating shaft or measure the torque optically. Brief mention of two older techniques was made, in the form of measuring the reaction forces in the bearings supporting a rotating shaft and in using a device called the Prony brake. However, we noted that neither of these is now in common use.

We then concluded the chapter by examining the techniques used for calibrating the measuring devices covered in the chapter. We noted that calibration of mass-measuring sensors involved the use of a set of standard masses. As regarding the calibration of force and torque sensors, we saw that both of these required the use of special machines that generate a set of known force or torque values. Because such machines are very expensive, we noted that it was normal to use the services of either specialist calibration companies or the manufacturers of the measurement devices being calibrated.

## 18.7 Problems

18.1. What is the difference between mass and weight? Discuss briefly the three main methods of measuring the mass of a body.

18.2. Explain, using a sketch as appropriate, how each of the following forms of load cells work: (a) electronic, (b) pneumatic, (c) hydraulic, and (d) intelligent.

18.3. Discuss the main characteristics of the four kinds of load cells mentioned in Problem 18.2. Which form is most common, and why?

18.4. Discuss briefly the working characteristics of each of the following: (a) beam balance, (b) weigh beam, and (c) pendulum scale.

18.5. How does a spring balance work? What are its advantages and disadvantages compared with other forms of mass-measuring instruments?

18.6. What are the available techniques for measuring force acting in a horizontal direction?

18.7. Discuss briefly the four main methods used to measure torque.

18.8. Discuss the general principles of calibrating mass-measuring instruments.

18.9. Which instruments are used as a reference standard in mass calibration? What special precautions have to be taken in manufacturing and using such reference instruments?

CHAPTER 19

# Translational Motion, Vibration, and Shock Measurement




Measurement and Instrumentation: Theory and Application

## 19.1 Introduction

Movement is an integral part of many systems and therefore sensors to measure motion are an important tool for engineers. Motion occurs in many forms. Simple movement causes a *displacement* in the body affected by it, although this can take two alternative forms according to whether it is motion in a straight line (*translational displacement*) or angular motion about an axis (*rotational displacement*). Displacement only describes the fact that a body has moved but does not define the speed at which the motion occurs. Speed is defined by the term *velocity*. As for displacement, velocity occurs in two forms—*translational velocity* describes the speed at which a body changes position when moving in a straight line and *rotational velocity* (sometimes called *angular velocity*) describes the speed at which a body turns about the axis of rotation. Finally, it is clear that changes in velocity occur during the motion of a body. To start with, the body is at rest and the velocity is zero. At the start of motion, there is a change in velocity from zero to some nonzero value. The term *acceleration* is used to describe the rate at which the velocity changes. As for displacement and velocity, acceleration also comes in two forms—*translational acceleration* describes the rate of change of translational velocity and *rotational acceleration* (sometimes called *angular acceleration*) describes the rate of change of rotational velocity.

With motion occurring in so many different forms, a review of the various sensors used to measure these different forms of motion would not fit conveniently within a single chapter. Therefore, this chapter only reviews sensors used for measuring translational motion, with those used for measuring rotational motion being deferred to the next chapter. The following sections therefore look in turn at the measurement of translational displacement, velocity, and acceleration.

The subjects of vibration and shock are also included in final sections of this chapter. Both of these are related to translational acceleration and therefore properly belong within this chapter on translational displacement. Vibrations consist of linear harmonic motion, and measurement of the accelerations involved in this motion is important in many industrial and other environments. Shock is also related to acceleration and characterizes the motion involved when a moving body is suddenly brought to rest, often when a falling body hits the floor. This normally involves large-magnitude deceleration (negative acceleration).

## 19.2 Displacement

Translational displacement transducers are instruments that measure the motion of a body in a straight line between two points. Apart from their use as a primary transducer measuring the motion of a body, translational displacement transducers are also widely used as a secondary component in measurement systems, where some other physical quantity, such as pressure, force, acceleration, or temperature, is translated into a translational motion by the primary measurement transducer. Many different types of translational displacement transducers exist and these, along with their relative merits and characteristics, are discussed in the following sections of this chapter. Factors governing the choice of a suitable type of instrument in any particular measurement situation are considered in the final section at the end of the chapter.

### 19.2.1 Resistive Potentiometer

The resistive potentiometer is perhaps the best-known displacement-measuring device. It consists of a resistance element with a movable contact as shown in Figure 19.1. A voltage, $V_s$, is applied across the two ends *A* and *B* of the resistance element, and an output voltage, $V_O$, is measured between the point of contact *C* of the sliding element and the end of resistance element *A*. A linear relationship exists between the output voltage, $V_O$, and distance *AC*, which can be expressed by

$$\frac{V_O}{V_s} = \frac{AC}{AB}. \tag{19.1}$$

The body whose motion is being measured is connected to the sliding element of the potentiometer so that translational motion of the body causes a motion of equal magnitude of the slider along the resistance element and a corresponding change in the output voltage, $V_O$.

Three different types of potentiometers exist, wire wound, carbon film, and plastic film, so named according to the material used to construct the resistance element. Wire-wound potentiometers consist of a coil of resistance wire wound on a nonconducting former. As the slider moves along the potentiometer track, it makes contact with successive turns of the wire coil. This limits the resolution of the instrument to the distance from one coil to the next. Much better

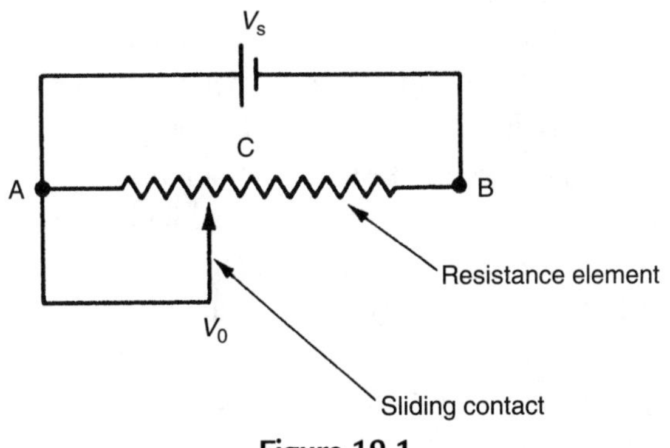


**Figure 19.1**
Resistive potentiometer.

measurement resolution is obtained from potentiometers using either a carbon film or a conducting plastic film for the resistance element. Theoretically, the resolution of these is limited only by the grain size of the particles in the film, suggesting that measurement resolutions up to $10^{-4}$ should be attainable. In practice, resolution is limited by mechanical difficulties in constructing the spring system that maintains the slider in contact with the resistance track, although these types are still considerably better than wire-wound types.

Operational problems of potentiometers all occur at the point of contact between the sliding element and the resistance track. The most common problem is dirt under the slider, which increases the resistance and thereby gives a false output voltage reading or, in the worst case, causes a total loss of output. High-speed motion of the slider can also cause the contact to bounce, giving an intermittent output. Friction between the slider and the track can also be a problem in some measurement systems where the body whose motion is being measured is moved by only a small force of a similar magnitude to these friction forces.

\The life expectancy of potentiometers is normally quoted as a number of reversals, that is, as the number of times the slider can be moved backward and forward along the track. The values quoted for wire-wound, carbon film, and plastic film types are, respectively, 1, 5, and 30 million. In terms of both life expectancy and measurement resolution, therefore, the carbon and plastic film types are clearly superior, although wire-wound types do have one advantage in respect of their lower temperature coefficient. This means that wire-wound types exhibit much less variation in their characteristics in the presence of varying ambient temperature conditions.

A typical inaccuracy value that is quoted for translational motion resistive potentiometers is $\pm 1\%$ of full-scale reading. Manufacturers produce potentiometers to cover a large span of measurement ranges. At the bottom end of this span, instruments with a range of $\pm 2$ mm are available, while instruments with a range of $\pm 1$ m are produced at the top end.

The resistance of the instrument measuring the output voltage at the potentiometer slider can affect the value of the output reading, as discussed in Chapter 3. As the slider moves along the potentiometer track, the ratio of the measured resistance to that of the measuring instrument varies, and thus the linear relationship between the measured displacement and the voltage output is distorted as well. This effect is minimized when the potentiometer resistance is small relative to that of the measuring instrument. This is achieved by (1) using a very high-impedance measuring instrument and (2) keeping the potentiometer resistance as small as possible. Unfortunately, the latter is incompatible with achieving high measurement sensitivity as this requires high potentiometer resistance. A compromise between these two factors is therefore necessary. The alternative strategy of obtaining high measurement sensitivity by keeping the potentiometer resistance low and increasing the excitation voltage is not possible in practice because of the power-rating limitation. This restricts the allowable power loss in the potentiometer to its heat dissipation capacity.

The process of choosing the best potentiometer from a range of instruments that are available, taking into account power rating and measurement linearity considerations, is illustrated in the following example.

## Example 19.1

The output voltage from a translational motion potentiometer of stroke length 0.1 meter is to be measured by an instrument whose resistance is 10 KΩ. The maximum measurement error, which occurs when the slider is positioned two-thirds of the way along the element (i.e., when $AC = 2AB/3$ in Figure 19.1), must not exceed 1% of the full-scale reading. The highest possible measurement sensitivity is also required. A family of potentiometers having a power rating of 1 watt per 0.01 meter and resistances ranging from 100 to 10 KΩ in 100-Ω steps are available. Choose the most suitable potentiometer from this range and calculate the sensitivity of measurement that it gives.

■

## Solution

Referring to the labeling used in Figure 19.1, let the resistance of portion $AC$ of the resistance element $R_i$ and that of the whole length $AB$ of the element be $R_t$. Also, let the resistance of the measuring instrument be $R_m$ and the output voltage measured by it be $V_m$. When the voltage-measuring instrument is connected to the potentiometer, the net resistance across $AC$ is the sum of two resistances in parallel ($R_i$ and $R_m$) given by

$$R_{AC} = \frac{R_i R_m}{R_i + R_m}.$$

Let the excitation voltage applied across the ends $AB$ of the potentiometer be $V$ and the resultant current flowing between $A$ and $B$ be $I$. Then $I$ and $V$ are related by

$$I = \frac{V}{R_{AC} + R_{CB}} = \frac{V}{[R_i R_m / R_i + R_m] + R_t - R_i}.$$

$V_m$ can now be calculated as

$$V_m = IR_{AC} = \frac{VR_i R_m}{\{[R_i R_m/(R_i + R_m)] + R_t - R_i\}\{R_i + R_m\}}.$$

If we express the voltage that exists across $AC$ in the absence of the measuring instrument as $V_0$, then we can express the error due to the loading effect of the measuring instrument as error $= V_0 - V_m$.

From Equation (19.1), $V_o = (R_i V)/R_t$. Thus,

$$Error = V_o - V_m$$
$$= V\left(\frac{R_i}{R_t}\right)\left(\frac{R_i R_m}{\{[R_i R_m/R_i + R_m] + R_t - R_i\}\{R_i + R_m\}}\right)\left(\frac{R_i^2(R_i - R_t)}{R_t[R_i R_t + R_m R_t - R_i^2]}\right). \quad (19.2)$$

Substituting $R_i = 2R_t/3$ into Equation (19.2) to find the maximum error:

$$Maximum\ error = \frac{2R_t}{2R_t + 9R_m}.$$

For a maximum error of 1%,

$$\frac{2R_t}{2R_t + 9R_m} = 0.01. \quad (19.3)$$

Substituting $R_m = 10{,}000\ \Omega$ into Equation (19.3) gives $R_t = 454\ \Omega$. The nearest resistance values in the range of potentiometers available are 400 and 500 Ω. The value of 400 Ω has to be selected, as this is the only one that gives a maximum measurement error of less than 1%.

The thermal rating of the potentiometers is quoted as 1 watt/0.01 m, that is, 10 watts for a total length of 0.1 m. By Ohm's law,
maximum supply voltage =

$$\sqrt{power \times resistance} = \sqrt{10 \times 400} = 63.25\ Volts.$$

Thus, the measurement sensitivity = 63.25/0.1 V/m = 632.5 V/m. ■

### 19.2.2 Linear Variable Differential Transformer (LVDT)

The linear variable differential transformer, which is commonly known by the abbreviation LVDT, consists of a transformer with a single primary winding and two secondary windings connected in the series-opposing manner shown in Figure 19.2. The object whose translational displacement is to be measured is attached physically to the central iron core of the transformer so that all motions of the body are transferred to the core.

For an excitation voltage $V_s$ given by $V_s = V_p \sin(\omega t)$, the e.m.f.s induced in the secondary windings $V_a$ and $V_b$ are given by

$$V_a = K_a \sin(\omega t - \phi) \quad ; \quad V_b = K_b \sin(\omega t - \phi).$$

Parameters $K_a$ and $K_b$ depend on the amount of coupling between the respective secondary and primary windings and hence on the position of the iron core. With the core in the central position, $K_a = K_b$, we have $V_a = V_b = K \sin(\omega t - \phi)$.

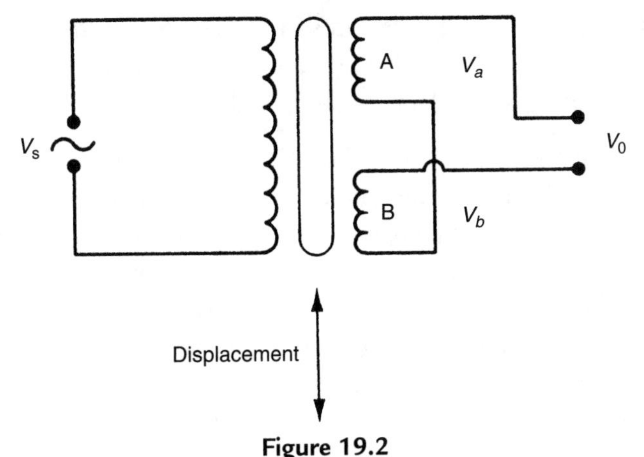


**Figure 19.2**
Linear variable differential transformer.

Because of the series opposition mode of connection of secondary windings, $V_o = V_a - V_b$, and hence with the core in the central position, $V_o = 0$. Suppose now that the core is displaced upward (i.e., toward winding $A$) by distance $x$. If then $K_a = K_1$ and $K_b = K_2$ , we have $V_o = (K_1 - K_2)\sin(\omega t - \phi)$.

If, alternatively, the core were displaced downward from the null position (i.e., toward winding $B$) by distance $x$, the values of $K_a$ and $K_b$ would then be $K_a = K_2$ and $K_b = K_1$, and we would have:

$$V_o = (K_2 - K_1)\sin(\omega t - \phi) = (K_1 - K_2)\sin(\omega t + [\pi - \phi]).$$

Thus for equal magnitude displacements $+x$ and $-x$ of the core away from the central (null) position, the magnitude of the output voltage, $V_o$, is the same in both cases. The only information about the direction of movement of the core is contained in the phase of the output voltage, which differs between the two cases by 180°. If, therefore, measurements of core position on both sides of the null position are required, it is necessary to measure the phase as well as the magnitude of the output voltage. The relationship between the magnitude of the output voltage and the core position is approximately linear over a reasonable range of movement of the core on either side of the null position and is expressed using a constant of proportionality $C$ as $V_o = Cx$.

The only moving part in an LVDT is the central iron core. As the core is only moving in the air gap between the windings, there is no friction or wear during operation. For this reason, the instrument is a very popular one for measuring linear displacements and has a quoted life expectancy of 200 years. The typical inaccuracy is ±0.5% of full-scale reading and measurement resolution is almost infinite. Instruments are available to measure a wide span of measurements from ±100 μm to ±100 mm. The instrument can be made suitable for operation in corrosive environments by enclosing the windings within a nonmetallic barrier, which leaves the magnetic flux paths between the core and windings undisturbed. An epoxy resin is used commonly to encapsulate the coils for this purpose. One further operational advantage of the instrument is its insensitivity to mechanical shock and vibration.

Some problems that affect the accuracy of the LVDT are the presence of harmonics in the excitation voltage and stray capacitances, both of which cause a nonzero output of low magnitude when the core is in the null position. It is also impossible in practice to produce two identical secondary windings, and the small asymmetry that invariably exists between the secondary windings adds to this nonzero null output. The magnitude of this is always less than 1% of the full-scale output and, in many measurement situations, is of little consequence. Where necessary, the magnitude of these effects can be measured by applying known displacements to the instrument. Following this, appropriate compensation can be applied to subsequent measurements.

### 19.2.3 Variable Capacitance Transducers

Like variable inductance, the principle of variable capacitance is used in displacement-measuring transducers in various ways. The three most common forms of variable capacitance transducers are shown in Figure 19.3. In Figure 19.3a, capacitor plates are formed by two concentric, hollow, metal cylinders. The displacement to be measured is applied to the inner cylinder, which alters the capacitance. The second form, Figure 19.3b, consists of two flat, parallel, metal plates, one of which is fixed and one of which is movable. Displacements to be measured are applied to the movable plate, and the capacitance changes as this moves. Both of these first two forms use air as the dielectric medium between the plates. The final form, Figure 19.3c, has two flat, parallel, metal plates with a sheet of solid dielectric material between them. The displacement to be measured causes a capacitance change by moving the dielectric sheet.

Inaccuracies as low as $\pm 0.01\%$ are possible with these instruments, with measurement resolutions of 1 μm. Individual devices can be selected from manufacturers' ranges that measure displacements as small as $10^{-11}$ m or as large as 1 m. The fact that such instruments consist only of two simple conducting plates means that it is possible to fabricate devices that are tolerant to a wide range of environmental hazards, such as extreme temperatures, radiation, and corrosive atmospheres. As there are no contacting moving parts, there is no friction or wear in operation and the life expectancy quoted is 200 years. The major problem with variable capacitance transducers is their high impedance. This makes them very susceptible to noise and means that the length and position of connecting cables need to be chosen very carefully. In addition, very high impedance instruments need to be used to measure the value of the capacitance. Because of these difficulties, use of these devices tends to be limited to those few applications where high accuracy and measurement resolution of the instrument are required.

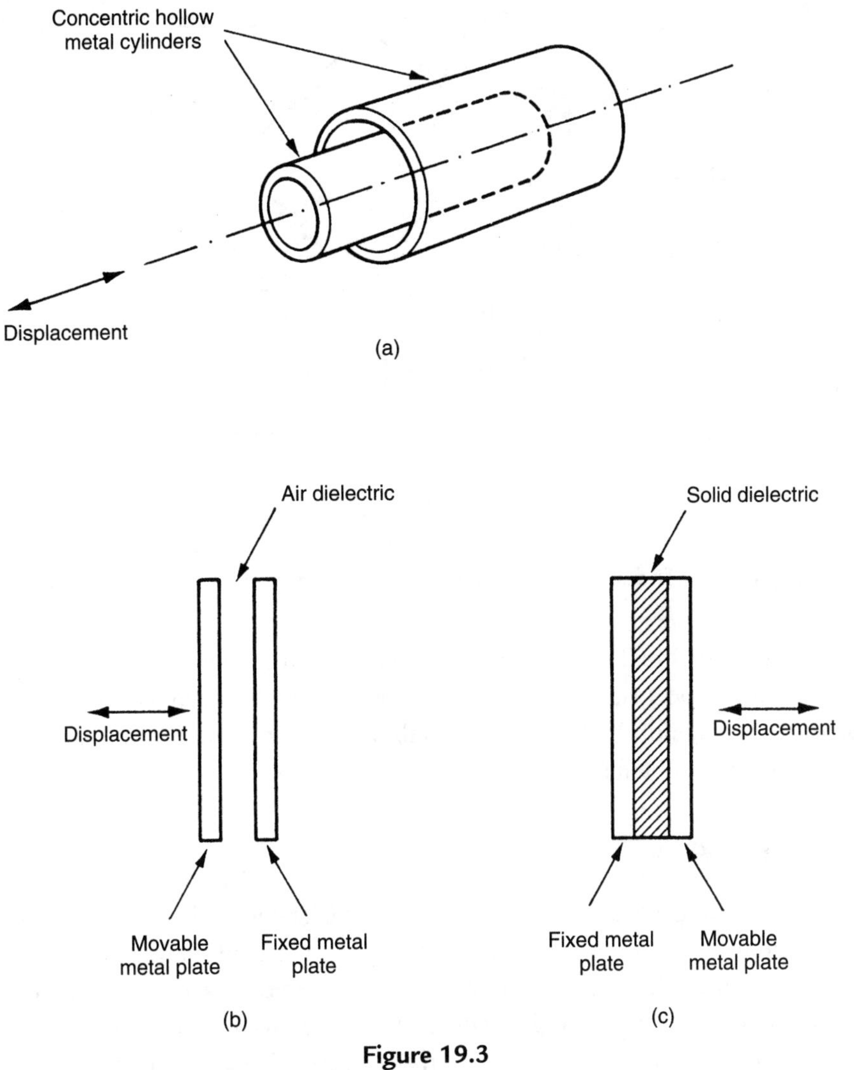


**Figure 19.3**
Variable capacitance transducer.

### *19.2.4 Variable Inductance Transducers*

One simple type of variable inductance transducer was shown earlier in Figure 13.4. This has a typical measurement range of 0–10 mm. An alternative form of variable inductance transducer, shown in Figure 19.4a, has a very similar size and physical appearance to the LVDT, but has a center-tapped single winding. The two halves of the winding are connected, as shown in Figure 19.4b, to form two arms of a bridge circuit that is excited with an alternating voltage. With the core in the central position, the output from the bridge is zero. Displacements of the core either side of the null position cause a net output voltage that is approximately proportional

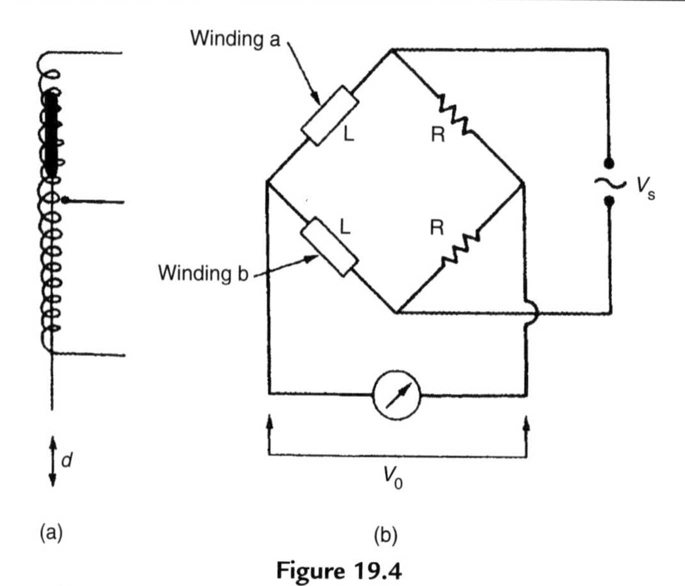


**Figure 19.4**
(a) Variable inductance transducers; (b) connection in a bridge circuit.

to the displacement for small movements of the core. Instruments in this second form are available to cover a wide span of displacement measurements. At the lower end of this span, instruments with a range of 0–2 mm are available, while at the top end, instruments with a range of 0–5 m can be obtained.

### 19.2.5 Strain Gauges

The principles of strain gauges were covered earlier in Chapter 13. Because of their very small range of measurement (typically 0–50 μm), strain gauges are normally only used to measure displacements within devices such as diaphragm-based pressure sensors rather than as a primary sensor in their own right for direct displacement measurement. However, strain gauges can be used to measure larger displacements if the range of displacement measurement is extended by the scheme illustrated in Figure 19.5. In this, the displacement to be measured is applied to a wedge fixed between two beams carrying strain gauges. As the wedge is displaced downward, the beams are forced apart and strained, causing an output reading on the strain gauges. Using this method, displacements up to about 50 mm can be measured.

### 19.2.6 Piezoelectric Transducers

The piezoelectric transducer is effectively a force-measuring device used within many instruments designed to measure either force itself or the force-related quantities of pressure and acceleration. It is included within this discussion of linear displacement transducers because its mode of operation is to generate an e.m.f. proportional to the distance by which

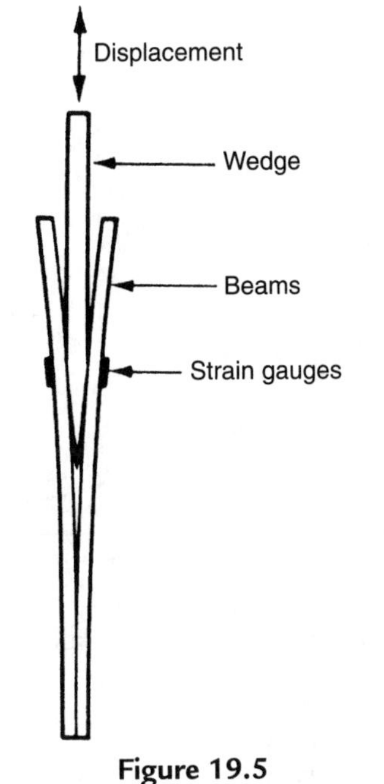


**Figure 19.5**
Strain gauges measuring large displacements.

it is compressed. The device is manufactured from a crystal, which can be either a natural material, such as quartz, or a synthetic material, such as lithium sulfate. The crystal is mechanically stiff (i.e., a large force is required to compress it); consequently, piezoelectric transducers can only be used to measure the displacement of mechanical systems that are stiff enough themselves to be unaffected by the stiffness of the crystal. When the crystal is compressed, a charge is generated on the surface that is measured as the output voltage. Unfortunately, as is normal with any induced charge, the charge leaks away over a period of time. Consequently, the output voltage-time characteristic is as shown in Figure 19.6. Because of this characteristic, piezoelectric transducers are not suitable for measuring static or slowly varying displacements, even though the time constant of the charge-decay process can be lengthened by adding a shunt capacitor across the device.

As a displacement-measuring device, the piezoelectric transducer has a very high sensitivity, about 1000 times better than a strain gauge. Its typical inaccuracy is $\pm 1\%$ of full-scale reading and its life expectancy is three million reversals.

### *19.2.7 Nozzle Flapper*

The nozzle flapper is a displacement transducer that translates displacements into a pressure change. A secondary pressure-measuring device is therefore required within the instrument. The general form of a nozzle flapper is shown schematically in Figure 19.7. Fluid at a known supply

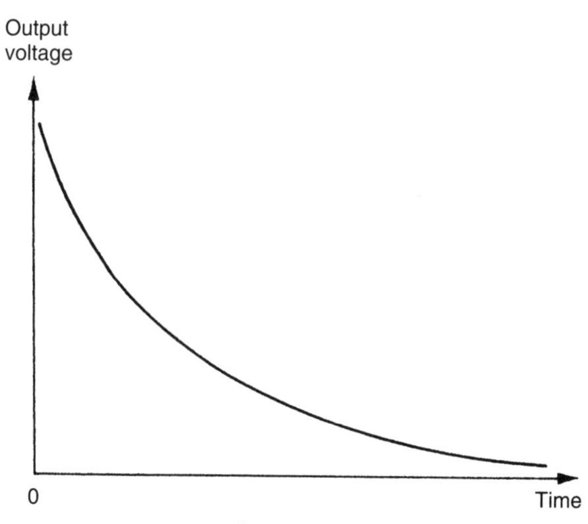


**Figure 19.6**
Voltage-time characteristic of a piezoelectric transducer following step displacement.

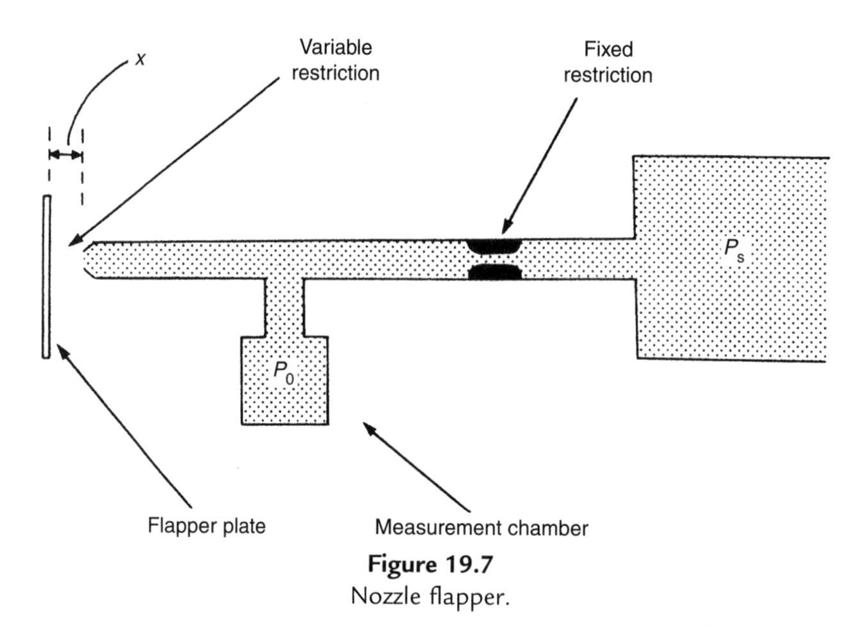


**Figure 19.7**
Nozzle flapper.

pressure, $P_s$, flows through a fixed restriction and then through a variable restriction formed by the gap, $x$, between the end of the main vessel and the flapper plate. The body whose displacement is being measured is connected physically to the flapper plate. The output measurement of the instrument is the pressure, $P_o$, in the chamber shown in Figure 19.7, which is almost proportional to $x$ over a limited range of movement of the flapper plate. The instrument typically has a first-order response characteristic. Air is used very commonly as the working fluid, which gives the instrument a time constant of about 0.1 second. The instrument has extremely high sensitivity but its range of measurement is quite small. A typical measurement range is ±0.05 mm with a measurement resolution of ±0.01 μm. One very common application of nozzle flappers is measuring displacements within a load cell, which are typically very small.

### 19.2.8 Other Methods of Measuring Small/Medium-Sized Displacements

Apart from the methods outlined earlier, several other techniques for measuring small translational displacements exist, as discussed here. Some of these involve special instruments that have a very limited sphere of application, for instance, in measuring machine tool displacements.

*Linear inductosyn*

A linear inductosyn is an extremely accurate instrument widely used for axis measurement and control within machine tools. Typical measurement resolution is 2.5 μm. The instrument consists of two magnetically coupled parts separated by an air gap, typically 0.125 mm wide, as shown in Figure 19.8. One part, the track, is attached to the axis along which displacements are to be

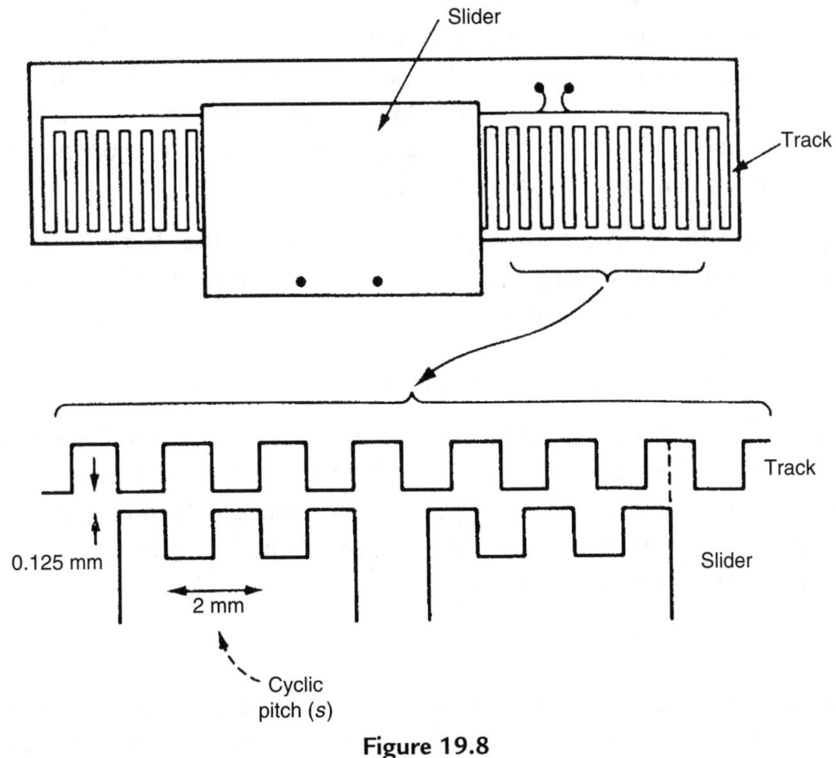


**Figure 19.8**
Linear inductosyn.

measured. This would generally be the bed of a machine tool. The other part, the slider, is attached to the body that is to be measured or positioned. This would usually be a cutting tool.

The track, which may be several meters long, consists of a fine metal wire formed into the pattern of a continuous rectangular waveform and deposited onto a glass base. The typical pitch (cycle length), $s$, of the pattern is 2 mm, which extends over the full length of the track. The slider is usually about 50 mm wide and carries two separate wires formed into continuous rectangular waveforms that are displaced with respect to each other by one-quarter of the cycle pitch, that is, by 90 electrical degrees. The wire waveform on the track is excited by an applied voltage given by $V_s = V\sin(\omega t)$.

This excitation causes induced voltages in the slider windings. When the slider is positioned in the null position such that its first winding is aligned with the winding on the track, the output voltages on the two slider windings are given by $V_1 = 0$ ; $V_2 = V\sin(\omega t)$.

For any other position, slider winding voltages are given by $V_1 = V\sin(\omega t)\sin(2\pi x/s)$ ; $V_2 = V\sin(\omega t)\cos(2\pi x/s)$, where $x$ is displacement of the slider away from the null position.

Consideration of these equations for the slider-winding outputs shows that the pattern of output voltages repeats every cycle pitch. Therefore, the instrument can only discriminate displacements of the slider within one cycle pitch of the windings. This means that the typical measurement range of an inductosyn is only 2 mm. This is of no use in normal applications, and therefore an additional displacement transducer with coarser resolution but larger measurement range has to be used as well. This coarser measurement is made commonly by translating the linear displacements by suitable gearing into rotary motion, which is then measured by a rotational displacement transducer.

One slight problem with the inductosyn is the relatively low level of electromagnetic coupling between the track and slider windings. Compensation for this is made using a high-frequency excitation voltage (5–10 kHz is common).

#### *Translation of linear displacements into rotary motion*

In some applications, it is inconvenient to measure linear displacements directly, either because there is insufficient space to mount a suitable transducer or because it is inconvenient for other reasons. A suitable solution in such cases is to translate the translational motion into rotational motion by suitable gearing. Any of the rotational displacement transducers discussed in the next chapter can then be applied.

#### *Integration of output from velocity transducers and accelerometers*

If velocity transducers or accelerometers already exist in a system, displacement measurements can be obtained by integration of the output from these instruments. However, this only gives information about the relative position with respect to some arbitrary starting point.

It does not yield a measurement of the absolute position of a body in space unless all motions away from a fixed starting point are recorded.

### *Laser interferometer*

The standard interferometer has been used for over 100 years for accurate measurement of displacements. The laser interferometer is a relatively recent development that uses a laser light source instead of the conventional light source used in a standard interferometer. The laser source extends the measurement range of the instrument by a significant amount while maintaining the same measurement resolution found in a standard interferometer. In the particular form of laser interferometer shown in Figure 19.9, a dual-frequency helium–neon (He-Ne) laser is used that gives an output pair of light waves at a nominal frequency of $5 \times 10^{14}$ Hz. The two waves differ in frequency by $2 \times 10^{6}$ Hz and have opposite polarization. This dual-frequency output waveform is split into a measurement beam and a reference beam by the first beam splitter.

The reference beam is sensed by the polarizer and photodetector, *A*, which converts both waves in the light to the same polarization. The two waves interfere constructively and destructively alternately, producing light–dark flicker at a frequency of $2 \times 10^{6}$ Hz. This excites a 2-MHz electrical signal in the photodetector.

The measurement beam is separated into the two component frequencies by a polarizing beam splitter. Light of the first frequency, $f_1$, is reflected by a fixed reflecting cube into a photodetector and polarizer, *B*. Light of the second frequency, $f_2$, is reflected by a movable reflecting cube and also enters *B*. The displacement to be measured is applied to the movable cube. With the movable cube in the null position, the light waves entering *B*

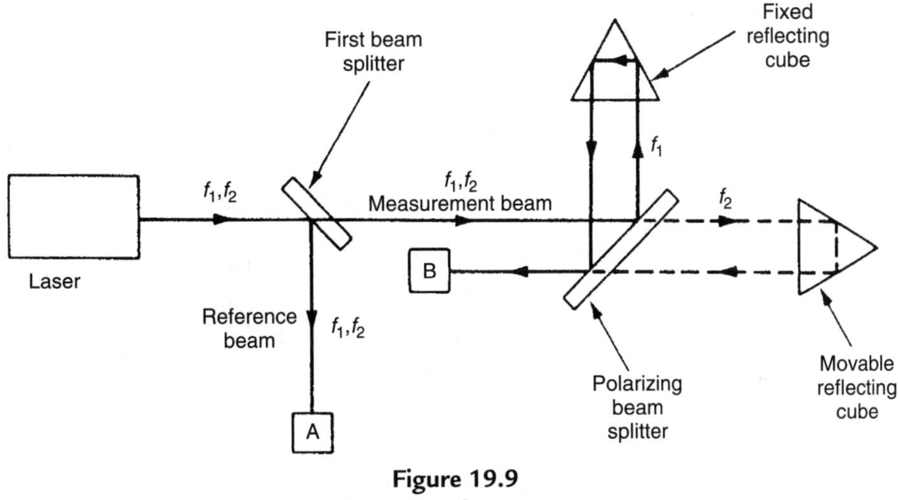


**Figure 19.9**
Laser interferometer.

produce an electrical signal output at a frequency of 2 MHz, which is the same frequency as the reference signal output from $A$. Any displacement of the movable cube causes a Doppler shift in the frequency, $f_2$, and changes the output from $B$. The frequency of the output signal from $B$ varies between 0.5 and 3.5 MHz according to the speed and direction of movement of the movable cube. Outputs from $A$ and $B$ are amplified and subtracted. The resultant signal is fed to a counter whose output indicates the magnitude of the displacement in the movable cube and whose rate of change indicates the velocity of motion.

This technique is used in applications requiring high-accuracy measurement, such as machine tool control. Such systems can measure displacements over ranges of up to 2 m with an inaccuracy of only a few parts per million. They are therefore an attractive alternative to the inductosyn, in having both high measurement resolution and a large measurement range within one instrument.

### *Fotonic sensor*

The fotonic sensor is one of many recently developed instruments that make use of fiber-optic techniques. It consists of a light source, a light detector, a fiber-optic light transmission system, and a plate that moves with the body whose displacement is being measured, as shown in Figure 19.10. Light from the outward fiber-optic cable travels across the air gap to the plate and some of it is reflected back into the return fiber-optic cable. The amount of light reflected back from the plate is a function of the air gap length, $x$, and hence of plate displacement. Measurement of the intensity of the light carried back along the return cable to the light detector allows displacement of the plate to be calculated. Common applications of fotonic sensors are measuring diaphragm displacements in pressure sensors and measuring the movement of bimetallic temperature sensors.

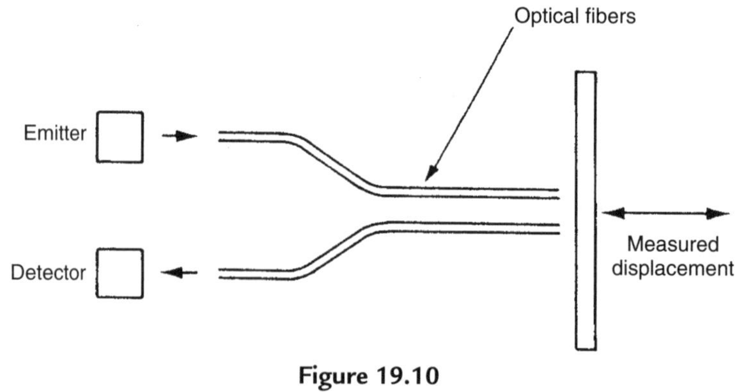


**Figure 19.10**
Fotonic sensor.

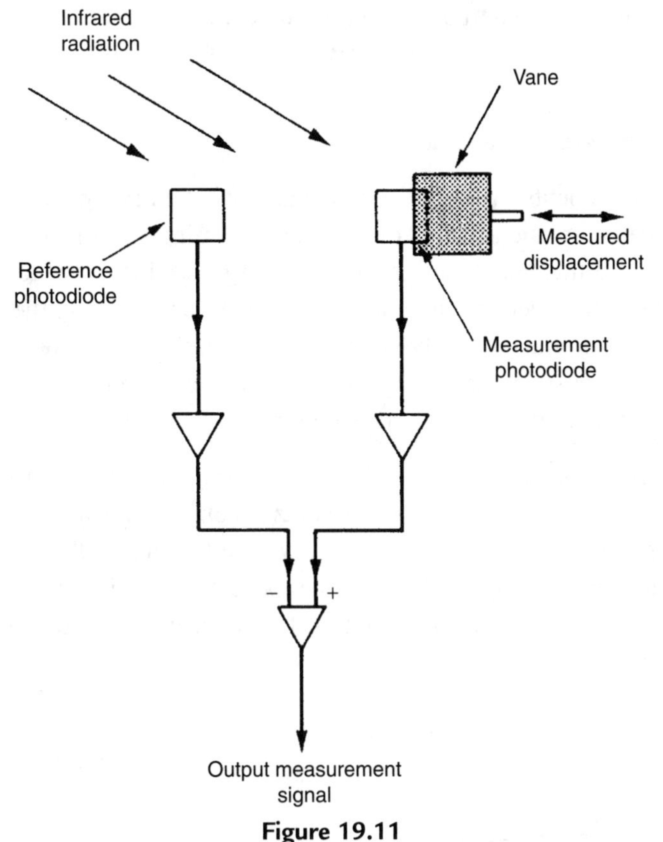


**Figure 19.11**
Noncontacting optical sensor.

*Noncontacting optical sensor*

Figure 19.11 shows an optical technique used to measure small displacements. The motion to be measured is applied to a vane, whose displacement progressively shades one of a pair of monolithic photodiodes that are exposed to infrared radiation. A displacement measurement is obtained by comparing the output of the reference (unshaded) photodiode with that of the shaded one. The typical range of measurement is $\pm 0.5$ mm with an inaccuracy of $\pm 0.1\%$ of full scale. Such sensors are used in some intelligent pressure-measuring instruments based on Bourdon tubes or diaphragms as described in Chapter 15.

### 19.2.9 Measurement of Large Displacements (Range Sensors)

One final class of instruments that has not been mentioned so far consists of those designed to measure relatively large translational displacements. These are usually known as range sensors and measure the motion of a body with respect to some fixed datum point. Most range sensors use an energy source and energy detector, but measurement using a rotary potentiometer and a spring-loaded drum provides an alternative method.

*Energy source/detector-based range sensors*

The fundamental components in energy source/detector-based range sensors are an energy source, an energy detector, and an electronic means of timing the time of flight of the energy between the source and the detector. The form of energy used is either ultrasonic or light. In some systems, both the energy source and the detector are fixed on the moving body and operation depends on the energy being reflected back from the fixed boundary as in Figure 19.12a. In other systems, the energy source is attached to the moving body and the energy detector is located within the fixed boundary, as shown in Figure 19.12b.

In ultrasonic systems, the energy is transmitted from the source in high-frequency bursts. A frequency of at least 20 kHz is usual, and 40 kHz is common for measuring distances up to 5 m. By measuring the time of flight of the energy, the distance of the body from the fixed boundary can be calculated, using the fact that the speed of sound in air is 340 m/s. Because of difficulties in measuring the time of flight with sufficient accuracy, ultrasonic systems are not

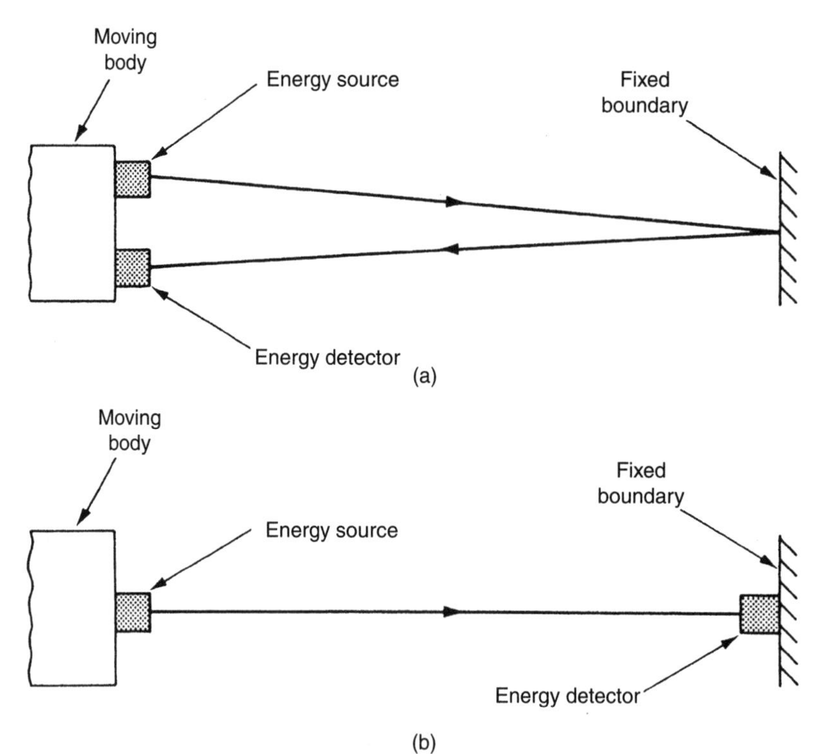


**Figure 19.12**
Range sensors.

suitable for measuring distances of less than about 300 mm. Measurement resolution is limited by the wavelength of the ultrasonic energy and can be improved by operating at higher frequencies. At higher frequencies, however, attenuation of the magnitude of the ultrasonic wave as it passes through air becomes significant. Therefore, only low frequencies are suitable if large distances are to be measured. The typical inaccuracy of ultrasonic range finding systems is $\pm 0.5\%$ of full scale.

Optical range-finding systems generally use a laser light source. The speed of light in air is about $3 \times 10^8 m/s$, so that light takes only a few nanoseconds to travel a meter. In consequence, such systems are only suitable for measuring very large displacements where the time of flight is long enough to be measured with reasonable accuracy.

#### *Rotary potentiometer and spring-loaded drum*

Another method for measuring large displacements that are beyond the measurement range of common displacement transducers is shown in Figure 19.13. This consists of a steel wire attached to the body whose displacement is being measured: the wire passes round a pulley and on to a spring-loaded drum whose rotation is measured by a rotary potentiometer. A multiturn potentiometer is usually required for this to give an adequate measurement resolution. With this measurement system, it is possible to reduce measurement uncertainty to as little as $\pm 0.01\%$ of full-scale reading.

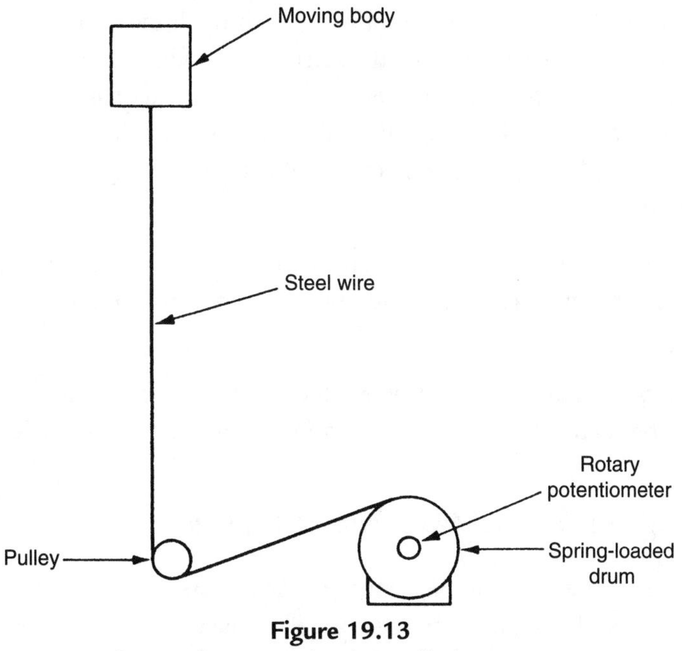


**Figure 19.13**
System for measuring large displacements.

### 19.2.10 Proximity Sensors

For the sake of completeness, it is proper to conclude this chapter on translational displacement transducers with consideration of proximity sensors. Proximity detectors provide information on the displacement of a body with respect to some boundary, but only insofar as to say whether the body is less than or greater than a certain distance away from the boundary. The output of a proximity sensor is thus binary in nature: the body is, or is not, close to the boundary.

Like range sensors, proximity detectors make use of an energy source and detector. The detector is a device whose output changes between two states when the magnitude of the incident reflected energy exceeds a certain threshold level. A common form of proximity sensor uses an infrared light-emitting diode (LED) source and a phototransistor. Light triggers the transistor into a conducting state when the LED is within a certain distance from a reflective boundary and the reflected light exceeds a threshold level. This system is physically small, occupying a volume of only a few cubic centimeters. If even this small volume is obtrusive, then fiber-optic cables can be used to transmit light from a remotely mounted LED and phototransistor. The threshold displacement detected by optical proximity sensors can be varied between 0 and 2 m.

Another form of proximity sensor uses the principle of varying inductance. Such devices are particularly suitable for operation in aggressive environmental conditions and can be made vibration and shock resistant by vacuum encapsulation techniques. The sensor contains a high-frequency oscillator whose output is demodulated and fed via a trigger circuit to an amplifier output stage. The oscillator output radiates through the surface of the sensor and, when the sensor surface becomes close to an electrically or magnetically conductive boundary, the output voltage is reduced because of interference with the flux paths. At a certain point, the output voltage is reduced sufficiently for the trigger circuit to change state and reduce the amplifier output to zero. Inductive sensors can be adjusted to change state at displacements in the range of 1 to 20 mm.

A third form of proximity sensor uses the capacitive principle. These can operate in similar conditions to inductive types. The threshold level of displacement detected can be varied between 5 and 40 mm.

*Fiber-optic proximity sensors* also exist where the amount of reflected light varies with the proximity of the fiber ends to a boundary, as shown earlier in Figure 13.2c.

### 19.2.11 Choosing Translational Measurement Transducers

Choice between the various translational motion transducers available for any particular application depends mainly on the magnitude of the displacement to be measured, although the operating environment is also relevant.

The requirement to measure displacements of less than 2 mm usually occurs as part of an instrument that is measuring some other physical quantity, such as pressure, and several types of devices have evolved to fulfill this task. The LVDT, strain gauges, fotonic sensor, variable capacitance transducers, and noncontacting optical transducer all find application in measuring diaphragm or Bourdon tube displacements within pressure transducers. Load cell displacements are also very small, which are commonly measured by nozzle flapper devices.

For measurements within the range of 2 mm to 2 m, the number of suitable instruments grows. Both the relatively inexpensive potentiometer and the LVDT, which is somewhat more expensive, are commonly used for such measurements. Variable inductance and variable capacitance transducers are also used in some applications. Additionally, strain gauges measuring the strain in two beams forced apart by a wedge (see Section 19.2.5) can measure displacements up to 50 mm. If very high measurement resolution is required, either the linear inductosyn or the laser interferometer is used.

Finally, range sensors are normally used if the displacement to be measured exceeds 2 meters.

As well as choosing sensors according to the magnitude of displacement to be measured, the measurement environment is also sometimes relevant. If the environmental operating conditions are severe (e.g., hot, radioactive or corrosive atmospheres), devices that can be protected easily from these conditions must be chosen, such as the LVDT, variable inductance, and variable capacitance instruments.

### *19.2.12 Calibration of Translational Displacement Measurement Transducers*

Most translational displacement transducers measuring displacements up to 50 mm can be calibrated at the workplace level using standard micrometers to measure a set of displacements and comparing the reading from the displacement transducer being calibrated when it is reading the same set of displacements. Such micrometers can provide a reference standard with an inaccuracy of $\pm 0.003\%$ of full-scale reading. If better accuracy is required, micrometer-based calibrators are available from several manufacturers that reduce the measurement inaccuracy down to $\pm 0.001\%$ of full-scale reading.

For sensors that measure displacements exceeding 50 mm (including those classified as range sensors), the usual calibration tool is a laser interferometer. This can provide measurement uncertainty down to $\pm 0.0002\%$ of full-scale reading. According to which laser interferometer model is chosen, a measurement range up to 50 meters is possible. Obviously, laser interferometers are expensive devices, which are also physically very large for a model measuring up to 50 meters, and therefore calibration services using these are usually devolved to specialist calibration companies or instrument manufacturers.

## 19.3 Velocity

Translational velocity cannot be measured directly and therefore must be calculated indirectly by other means as set out here.

### 19.3.1 Differentiation of Displacement Measurements

Differentiation of position measurements obtained from any of the translational displacement transducers described in Section 19.2 can be used to produce a translational velocity signal. Unfortunately, the process of differentiation always amplifies noise in a measurement system. Therefore, if this method has to be used, a low-noise instrument such as a d.c.-excited carbon film potentiometer or laser interferometer should be chosen. In the case of potentiometers, a.c. excitation must be avoided because of the problem that harmonics in the power supply would cause.

### 19.3.2 Integration of Output of an Accelerometer

Where an accelerometer is already included within a system, integration of its output can be performed to yield a velocity signal. The process of integration attenuates rather than amplifies measurement noise, and this is therefore an acceptable technique in terms of measurement accuracy.

### 19.3.3 Conversion to Rotational Velocity

Conversion from translational to rotational velocity is the final measurement technique open to the system designer and is the one used most commonly. This conversion enables any of the rotational velocity-measuring instruments described in Chapter 20 to be applied.

### 19.3.4 Calibration of Velocity Measurement Systems

Because translational velocity is never measured directly, the calibration procedure used depends on the system used for velocity measurement. If a velocity measurement is being calculated from a displacement or acceleration measurement, the traceability of system calibration requires that the associated displacement or acceleration transducer used is calibrated correctly. The only other measurement technique is conversion of the translational velocity into rotational velocity, in which case the system calibration depends on the calibration of the rotational velocity transducer used.

## 19.4 Acceleration

The only class of device available for measuring acceleration is the accelerometer. These are available in a wide variety of types and ranges designed to meet particular measurement requirements. They have a frequency response between zero and a high value, and have a form of output that can be integrated readily to give displacement and velocity measurements. The frequency response of accelerometers can be improved by altering the level of damping in the instrument. Such adjustment must be done carefully, however, because frequency response improvements are only achieved at the expense of degrading the measurement sensitivity. In addition to their use for general-purpose motion measurement, accelerometers are widely used to measure mechanical shocks and vibrations.

Most forms of accelerometer consist of a mass suspended by a spring and damper inside a housing, as shown in Figure 19.14. The accelerometer is fastened rigidly to the body undergoing acceleration. Any acceleration of the body causes a force, $F_a$, on the mass, $M$, given by $F_a = M\ddot{x}$.

This force is opposed by the restraining effect, $F_s$, of a spring with spring constant $K$, and the net result is that mass is displaced by a distance, $x$, from its starting position such that $F_s = Kx$.

In steady state, when the mass inside is accelerating at the same rate as the case of the accelerometer, $F_a = F_s$, and so

$$Kx = M\ddot{x} \quad \text{or} \quad \ddot{x} = (Kx)/M. \tag{19.4}$$

This is the equation of motion of a second-order system, and, in the absence of damping, the output of the accelerometer would consist of nondecaying oscillations. A damper is therefore included within the instrument, which produces a damping force, $F_d$, proportional to the

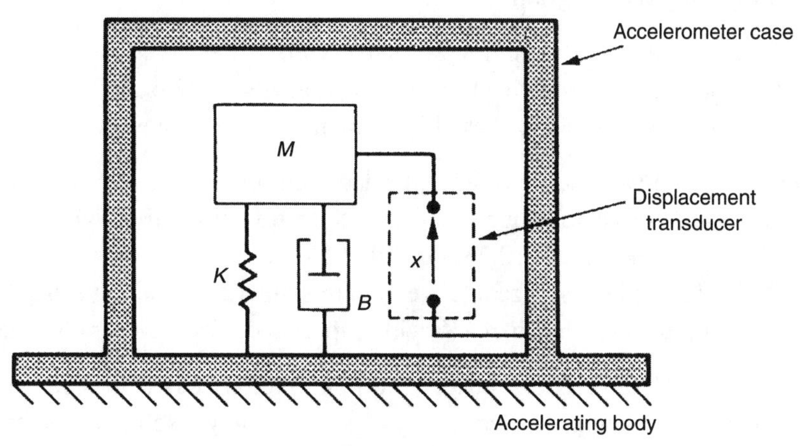


**Figure 19.14**
Structure of an accelerometer.

velocity of the mass, $M$, given by $F_d = B\dot{x}$. This modifies the previous equation of motion [Equation (19.4)] to the following:

$$Kx + B\dot{x} = M\ddot{x}. \tag{19.5}$$

One important characteristic of accelerometers is their sensitivity to accelerations at right angles to the sensing axis (the direction along which the instrument is designed to measure acceleration). This is defined as *cross-sensitivity* and is specified in terms of the output, expressed as a percentage of full-scale output, when an acceleration of some specified magnitude (e.g., $30g$) is applied at 90° to the sensing axis.

The acceleration reading is obtained from the instrument by measurement of the displacement of the mass within the accelerometer. Many different displacement-measuring techniques are used in the various types of accelerometers available commercially. Different types of accelerometers also vary in terms of the type of spring element and form of damping used.

Resistive potentiometers are one such displacement-measuring instrument used in accelerometers. These are used mainly for measuring slowly varying accelerations and low-frequency vibrations in the range of $0-50g$. The measurement resolution obtainable is about 1 in 400 and typical values of cross-sensitivity are $\pm 1\%$. Inaccuracy is about $\pm 1\%$ and life expectancy is quoted at two million reversals. A typical size and weight are 125 $cm^3$ and 500 gram.

Strain gauges and piezoresistive sensors are also used in accelerometers for measuring accelerations up to $200g$. These serve as the spring element as well as measuring mass displacement, thus simplifying the instrument's construction. Their typical characteristics are a resolution of 1 in 1000, inaccuracy of $\pm 1\%$, and cross-sensitivity of 2%. They have a major advantage over potentiometer-based accelerometers in terms of their much smaller size and weight (3 $cm^3$ and 25 gram).

Another displacement transducer found in accelerometers is the LVDT. This device can measure accelerations up to $700g$ with a typical inaccuracy of $\pm 1\%$ of full scale. They are of a similar physical size to potentiometer-based instruments but are lighter in weight (100 gram).

Accelerometers based on variable-inductance displacement-measuring devices have extremely good characteristics and are suitable for measuring accelerations up to $40g$. Typical specifications of such instruments are inaccuracy $\pm 0.25\%$ of full scale, resolution 1 in 10,000, and cross-sensitivity of 0.5%. Their physical size and weight are similar to potentiometer-based devices. Instruments with an output in the form of a varying capacitance also have similar characteristics.

The other common displacement transducer used in accelerometers is the piezoelectric type. The major advantage of using piezoelectric crystals is that they also act as the spring and damper within the instrument. In consequence, the device is quite small (15 $cm^3$) and low mass (50 gram), but because of the nature of a piezoelectric crystal operation, such instruments are not suitable for measuring constant or slowly time-varying accelerations. As the electrical impedance of a piezoelectric crystal is itself high, the output voltage must be measured with a very high-impedance instrument to avoid loading effects. Many recent piezoelectric crystal-based accelerometers incorporate a high impedance charge amplifier within the body of the instrument. This simplifies the signal conditioning requirements external to the accelerometer but can lead to problems in certain operational environments because these internal electronics are exposed to the same environmental hazards as the rest of the accelerometer. Typical measurement resolution of this class of accelerometer is 0.1% of full scale with an inaccuracy of $\pm 1\%$. Individual instruments are available to cover a wide range of measurements from $0.03g$ full scale up to $1000g$ full scale. *Intelligent accelerometers* are also now available that give even better performance through inclusion of processing power to compensate for environmentally induced errors.

Recently, very small microsensors have become available for measuring acceleration. These consist of a small mass subject to acceleration mounted on a thin silicon membrane. Displacements are measured either by piezoresistors deposited on the membrane or by etching a variable capacitor plate into the membrane.

Two forms of fiber-optic-based accelerometers also exist. One form measures the effect on light transmission intensity caused by a mass subject to acceleration resting on a multimode fiber. The other form measures the change in phase of light transmitted through a monomode fiber that has a mass subject to acceleration resting on it.

### 19.4.1 Selection of Accelerometers

In choosing between the different types of accelerometers for a particular application, the mass of the instrument is particularly important. This should be very much less than that of the body whose motion is being measured in order to avoid loading effects that affect the accuracy of the readings obtained. In this respect, instruments based on strain gauges are best.

### 19.4.2 Calibration of Accelerometers

The primary method of calibrating accelerometers is to mount them on a table rotating about a vertical axis such that the sensing axis of the accelerometer is pointing toward the axis of rotation of the table. Acceleration, $a$, is then given by

$$a = r(2\pi v)^2,$$

where $r$ is the radius of rotation measured from the center of the rotating table to the center of the accelerometer mass and $v$ is the velocity of rotation of the table (in revolutions per second).

This obviously requires that the rotational speed of the table is measured accurately by a calibrated sensor. Provided that this condition is met, various reference acceleration values can be generated by changing the rotational speed of the table.

## 19.5 Vibration

### 19.5.1 Nature of Vibration

Vibrations are encountered very commonly in the operation of machinery and industrial plants, and therefore measurement of the accelerations associated with such vibrations is extremely important in industrial environments. The peak accelerations involved in such vibrations can be $100g$ or greater in magnitude, while both the frequency of oscillation and the magnitude of displacements from the equilibrium position in vibrations have a tendency to vary randomly. Vibrations normally consist of linear harmonic motion that can be expressed mathematically as

$$X = X_o \sin(\omega t), \tag{19.6}$$

where $X$ is the displacement from the equilibrium position at any general point in time, $X_o$ is the peak displacement from the equilibrium position, and $\omega$ is the angular frequency of the oscillations. By differentiating Equation (19.6) with respect to time, an expression for the velocity $v$ of the vibrating body at any general point in time is obtained as

$$v = -\omega X_o \cos(\omega t). \tag{19.7}$$

Differentiating Equation (19.7) again with respect to time, we obtain an expression for the acceleration, $\alpha$, of the body at any general point in time as

$$\alpha = -\omega^2 X_o \sin(\omega t). \tag{19.8}$$

Inspection of Equation (19.8) shows that peak acceleration is given by

$$\alpha_{peak} = \omega^2 X_o. \tag{19.9}$$

This square law relationship between peak acceleration and oscillation frequency is the reason why high values of acceleration occur during relatively low-frequency oscillations. For example, an oscillation at 10 Hz produces peak accelerations of $2g$.

## Example 19.2

A pipe carrying a fluid vibrates at a frequency of 50 Hz with displacements of 8 mm from the equilibrium position. Calculate the peak acceleration.

■

## Solution

From Equation (19.9), $\alpha_{peak} = \omega^2 X_o = (2\pi 50)^2 \times (0.008) = 789.6 m/s^2$.

Using the fact that standard acceleration due to gravity, $g$, is $9.81 m/s^2$, this answer can be expressed alternatively as $\alpha_{peak} = 789.6/9.81 = 80.5g$.

■

### 19.5.2 Vibration Measurement

It is apparent that the intensity of vibration can be measured in terms of displacement, velocity, or acceleration. Acceleration is clearly the best parameter to measure at high frequencies. However, because displacements are large at low frequencies according to Equation (19.9), it would seem that measuring either displacement or velocity would be best at low frequencies. The amplitude of vibrations can be measured by various forms of displacement transducers. Fiber-optic-based devices are particularly attractive and can give measurement resolution as high as 1 μm. Unfortunately, there are considerable practical difficulties in mounting and calibrating displacement and velocity transducers and therefore they are rarely used. Because of this, vibration is usually measured by accelerometers at all frequencies. The most common type of transducer used is the piezoaccelerometer, which has typical inaccuracy levels of $\pm 2\%$.

The frequency response of accelerometers is particularly important in vibration measurement in view of the inherently high-frequency characteristics of the measurement situation. The bandwidth of both potentiometer-based accelerometers and accelerometers using variable inductance-type displacement transducers only goes up to 25 Hz. Accelerometers that include either the LVDT or strain gauges can measure frequencies up to 150 Hz, and the latest instruments using piezoresistive strain gauges have bandwidths up to 2 kHz. Finally, inclusion of piezoelectric crystal displacement transducers yields an instrument with a bandwidth that can be as high as 7 kHz.

When measuring vibration, consideration must be given to the fact that attaching an accelerometer to the vibrating body will significantly affect the vibration characteristics if the body has a small mass. The effect of such "loading" of the measured system can be quantified by the following equation:

$$a_1 = a_b \left( \frac{m_b}{m_b + m_a} \right),$$

where $a_1$ is the acceleration of the body with accelerometer attached, $a_b$ is the acceleration of the body without the accelerometer, $m_a$ is the mass of the accelerometer, and $m_b$ is the mass of the body. Such considerations emphasize the advantage of piezoaccelerometers for measuring vibration, as these have a lower mass than other forms of accelerometers and so contribute least to this system-loading effect.

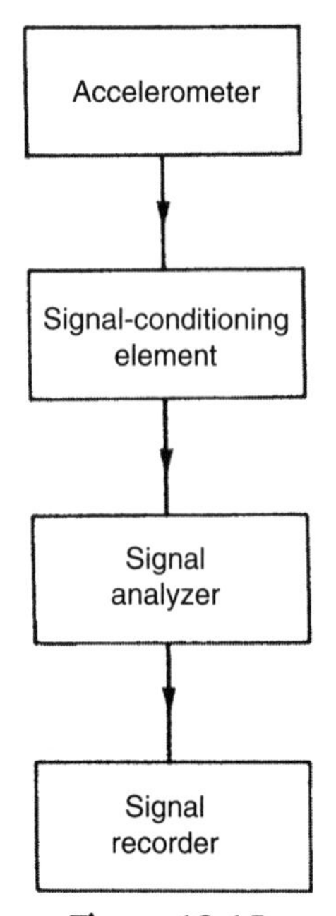


**Figure 19.15**
Vibration measurement system.

As well as an accelerometer, a vibration measurement system requires other elements to translate the accelerometer output into a recorded signal, as shown in Figure 19.15. The three other necessary elements are a signal conditioning element, a signal analyzer, and a signal recorder. The signal-conditioning element amplifies the relatively weak output signal from the accelerometer and also transforms the high output impedance of the accelerometer to a lower impedance value. The signal analyzer then converts the signal into the form required for output. The output parameter may be displacement, velocity, or acceleration, and this may be expressed as peak value, r.m.s. value, or average absolute value. The final element of the measurement system is the signal recorder. All elements of the measurement system, especially the signal recorder, must be chosen very carefully to avoid distortion of the vibration waveform. The bandwidth should be such that it is at least a factor of 10 better than the bandwidth of the vibration frequency components at both ends. Thus its lowest frequency limit should be less than or equal to 0.1 times the fundamental frequency of vibration and its upper frequency limit should be greater than or equal to 10 times the highest significant vibration frequency component.

If the frequency of vibration has to be known, the stroboscope is a suitable instrument to measure this. If the stroboscope is made to direct light pulses at the body at the same frequency as the vibration, the body will apparently stop vibrating.

### 19.5.3 Calibration of Vibration Sensors

Calibration of the accelerometer used within a vibration measurement system is normally carried out by mounting the accelerometer in a back-to-back configuration with a reference-calibrated accelerometer on an electromechanically excited vibrating table.

## 19.6 Shock

Shock describes a type of motion where a moving body is brought suddenly to rest, often because of a collision. This is very common in industrial situations, and usually involves a body being dropped and hitting the floor. Shocks characteristically involve large-magnitude decelerations (e.g., 500*g*) that last for a very short time (e.g., 5 ms). An instrument having a very high frequency response is required for shock measurement, and, for this reason, piezoelectric crystal-based accelerometers are commonly used. Again, other elements for analyzing and recording the signal are required as shown in Figure 19.16 and described in the last section. A storage oscilloscope is a suitable instrument for recording the output signal, as this allows time duration as well as acceleration levels in the shock to be measured. Alternatively, if a permanent record is required, the screen of a standard oscilloscope can be photographed.

### Example 19.3

A body is dropped from a height of 10 m and suffers a shock when it hits the ground. If the duration of the shock is 5 ms, calculate the magnitude of the shock in terms of *g*. ■

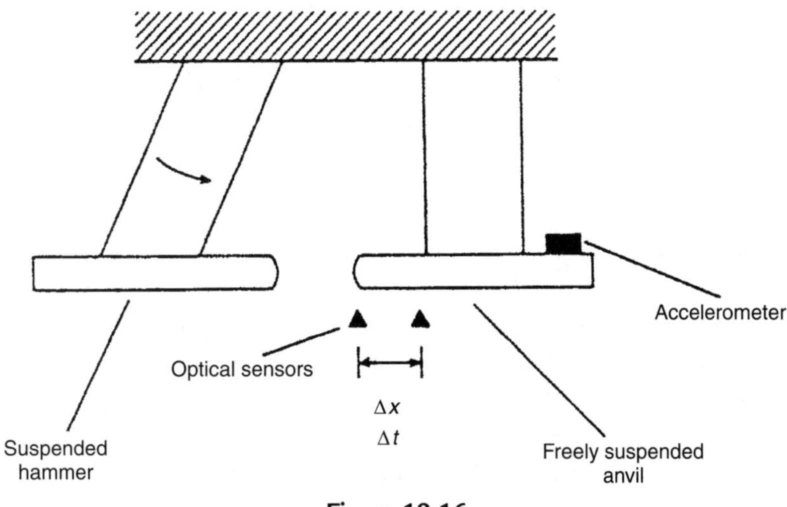


**Figure 19.16**
Shock measurement.

## Solution

The equation of motion for a body falling under gravity gives the following expression for terminal velocity, $v$:

$$v = \sqrt{2gx},$$

where $x$ is the height through which the body falls. Having calculated $v$, the average deceleration during the collision can be calculated as $\alpha = v/t$, where $t$ is the time duration of the shock. Substituting the appropriate numerical values into these expressions:

$$v = \sqrt{(2 \times 9.81 \times 10)} = 14.0m/s \quad ; \quad \alpha = 14.0/0.005 = 2801m/s = 286g.$$

■

### *19.6.1 Calibration of Shock Sensors*

Calibration of the accelerometer used within a shock sensor is carried out frequently using a pneumatic shock exciter. This device consists of a piston within a circular tube. High-pressure air is applied to one face of the piston, but it does not move initially because it is held at the end of the tube by a mechanical latching mechanism. When the latch is released, the piston accelerates at a high rate until it is brought to rest by a padded anvil at the other end of the tube. The accelerometer being calibrated and a calibrated reference accelerometer are both mounted on the anvil. By varying the characteristics of the padding, the deceleration level and hence magnitude of the shock produced on the anvil can be varied.

## 19.7 Summary

This chapter has been concerned with the measurement of translational (in a straight line) motion. This can take the form of displacement, velocity (rate of change of displacement), or acceleration (rate of change of velocity. We have looked at sensors for measuring each of these and, in the case of acceleration, we have also studied vibration and shock measurement, as both of these involve acceleration measurement.

Our study of displacement sensors started with the resistive potentiometer, where we learned that potentiometers come in three different forms: wire wound, carbon film, and plastic film. We then moved on to look at the linear variable differential transformer, variable capacitance, and variable inductance sensors. We noted that strain gauges were used to measure very small displacements (typically up to 50 μm in magnitude). We also noted that force-measuring piezoelectric sensors could also be regarded as displacement sensors, as their mode of operation is to generate an e.m.f. that is proportional to the distance by which it is compressed by the applied force. We also discussed the nozzle flapper, which measures displacements by converting them into a pressure change. We then moved on to summarize some other techniques for measuring small- and medium-sized displacements, including translating linear motion into rotational motion, integrating the output from velocity and acceleration sensors, and using specialist devices such as the linear inductosyn, laser interferometer, fotonic sensor, and noncontacting optical sensor. Moving the discussion on to the measurement of relatively large displacements, we noted that this could be achieved by several devices commonly called range sensors. We also included some mention of proximity sensors, as these belong properly within the classification of displacement sensors, although they are a special case in that their binary form of output merely indicates whether the sensor is, or is not, within some threshold distance of a boundary. Finally, before leaving the subject of displacement measurement, we looked at the techniques used to calibrate them.

Our discussion of translational velocity measurement introduced us to the fact that this cannot be measured directly. We then went on to look at the only three ways to measure it, these being differentiation of position measurements, integration of the output of an accelerometer, and conversion from translational to rotational velocity. Finally, we considered how measurements obtained via each of the techniques could be calibrated.

In the case of acceleration measurement, we observed that this could only be measured by some form of accelerometer. We noted that attributes such as frequency response and cross-sensitivity were important as well as measurement accuracy in accelerometers. We discovered that almost all accelerometers work on the principle that a mass inside them displaces when subject to acceleration. Accelerometers differ mainly in the technique used to measure this mass displacement, and we looked in turn at devices that use the resistive potentiometer, strain gauge, piezoresistive sensor, LVDT, variable inductance sensor, and variable capacitance sensor, respectively. We then looked at the one exception to the rule that accelerometers contain a moving mass. This is the piezoelectric accelerometer. Finally, we looked at the primary method of calibrating accelerometers using a rotating table.

We then concluded the chapter by looking at vibration and shock measurement. Both of these involve accelerations, and therefore both need an accelerometer to quantify their magnitude. Starting with vibration, we noted that this was a common phenomenon, especially in industrial situations. We learned that vibration consists of a linear harmonic in which the peak acceleration can exceed $100g$ and where the oscillation frequency and peak amplitude can vary randomly. We noted that the amplitude of vibration could be calculated from a measurement of the peak acceleration, and we went on to look at the suitability of various forms of accelerometers for such measurement.

Finally, we considered shock measurement. This revealed that very large magnitude decelerations are involved in the phenomenon of shock, which typically occurs when a falling body hits the floor or a collision occurs between two solid objects. A high-frequency response is particularly important in shock measurement, and the most suitable device to measure this is a piezoelectric crystal-based accelerometer.

## 19.8 Problems

19.1. Discuss the mode of operation and characteristics of a linear motion potentiometer.

19.2. What is an LVDT? How does it work?

19.3. Explain how the following two instruments work and discuss their main operating characteristics and uses: (a) variable capacitance transducer and (b) variable inductance transducer.

19.4. Sketch a linear imductosyn. How does it work? What are its main characteristics?

19.5. What is a laser interferometer and what are its principal characteristics? Explain how it works with the aid of a sketch.

19.6. What are range sensors? Describe two main types of range sensors.

19.7. Discuss the main types of proximity sensors available, mentioning particularly their suitability for operation in harsh environments.

19.8. What are the main considerations in choosing a translational motion transducer for a particular application? Give examples of some types of translational motion transducers and the applications that they are suitable for.

19.9. Discuss the usual calibration procedures for translational motion transducers.

19.10. What are the main ways of measuring translational velocities? How are such measurements calibrated?

19.11. What are the principles of operation of a linear motion accelerometer? What features would you expect to see in a high-quality accelerometer?

19.12. What types of displacement sensors are used within accelerometers? What are the relative merits of these alternative displacement sensors?

19.13. Write down a mathematical equation that describes the phenomenon of vibration. Explain briefly the three main ways of measuring vibration.

19.14. When an accelerometer is attached to a vibrating body, it has a loading effect that alters the characteristics of the vibration. Write down a mathematical equation that describes this loading effect. How can this loading effect be minimized?

CHAPTER 20

# Rotational Motion Transducers

## 20.1 Introduction

The different forms of motion have already been explained in the introduction to the last chapter. In that introduction, it was explained that motion occurred in two forms. These are translational motion, which describes the movement of a body along a single axis, and rotational motion, which describes the motion of a body about a single axis. Because the number of sensors involved in motion measurement is quite large, the review of them is divided into two chapters. The last chapter described translational motion sensors and now this chapter describes rotational motion sensors. Again, as for translational motion, rotational motion can occur in the form of displacement, velocity, or acceleration, which are considered separately in the following sections.

## 20.2 Rotational Displacement

Rotational displacement transducers measure the angular motion of a body about some rotation axis. They are important not only for measuring the rotation of bodies such as shafts, but also as a part of systems that measure translational displacement by converting the translational motion to a rotary form. The various devices available for measuring rotational displacements are presented here, and arguments for choosing a particular form in any given measurement situation are considered at the end of the chapter.

### 20.2.1 Circular and Helical Potentiometers

The *circular potentiometer* is the least expensive device available for measuring rotational displacements. It works on almost exactly the same principles as the translational motion potentiometer except that the track is bent round into a circular shape. The measurement range of individual devices varies from 0–10° to 0–360° depending on whether the track forms a full circle or only part of a circle. Where a greater measurement range than 0–360° is required, a *helical potentiometer* is used. The helical potentiometer accommodates multiple turns of the track by forming the track into a helix shape, and some devices are able to measure up to 60 full revolutions. Unfortunately, the greater mechanical complexity of a helical potentiometer makes the device significantly more expensive than a circular potentiometer. The two forms of devices are shown in Figure 20.1.

Both kinds of devices give a linear relationship between the measured quantity and the output reading because the output voltage measured at the sliding contact is proportional to the angular displacement of the slider from its starting position. However, as with linear track potentiometers, all rotational potentiometers can give performance problems if dirt on the track causes loss of contact. They also have a limited life because of wear between sliding surfaces. The typical inaccuracy of this class of devices varies from $\pm1\%$ of full scale for circular potentiometers down to $\pm0.002\%$ of full scale for the best helical potentiometers.

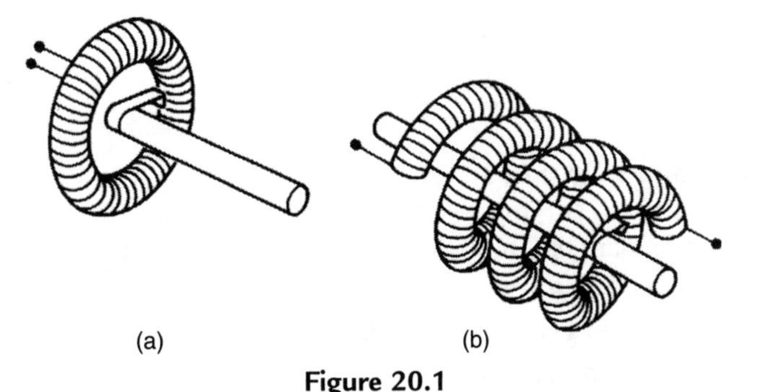


**Figure 20.1**
Rotary motion potentiometers: (a) circular and (b) helical.

### *20.2.2 Rotational Differential Transformer*

This is a special form of differential transformer that measures rotational rather than translational motion. The method of construction and connection of the windings is exactly the same as for the linear variable differential transformer, except that a specially shaped core is used that varies the mutual inductance between the windings as it rotates, as shown in Figure 20.2. Like its linear equivalent, the instrument suffers no wear in operation and therefore has a very long life with almost no maintenance requirements. It can also be modified for operation in harsh environments by enclosing the windings inside a protective enclosure. However, apart from the difficulty of avoiding some asymmetry between the secondary windings, great care has to be taken in these instruments to machine the core to exactly the right shape. In consequence, the inaccuracy cannot be reduced below $\pm 1\%$, and even this level of accuracy is only obtained for limited excursions of the core $\pm 40°$ away from the null position. For angular displacements of $\pm 60°$, the typical

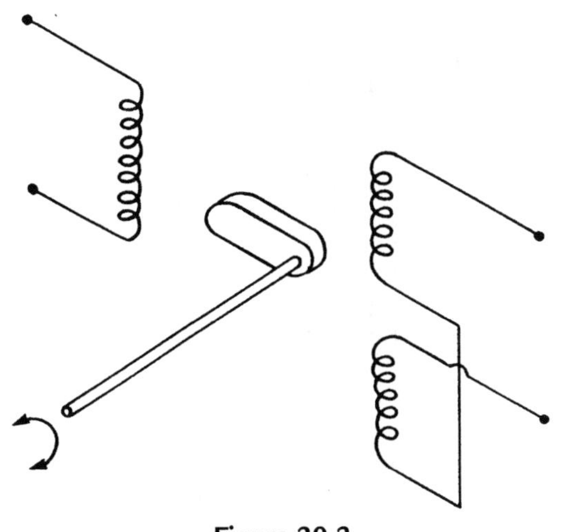

**Figure 20.2**
Rotary differential transformer.

inaccuracy rises to ±3%, and the instrument is unsuitable for measuring displacements greater than this.

### 20.2.3 Incremental Shaft Encoders

Incremental shaft encoders are one of a class of encoder devices that give an output in digital form. They measure the instantaneous angular position of a shaft relative to some arbitrary datum point, but are unable to give any indication about the absolute position of a shaft. The principle of operation is to generate pulses as the shaft whose displacement is being measured rotates. These pulses are counted and total angular rotation is inferred from the pulse count. The pulses are generated either by optical or by magnetic means and are detected by suitable sensors. Of the two, the optical system is considerably less expensive and therefore much more common. Such instruments are very convenient for computer control applications, as the measurement is already in the required digital form and therefore the analogue-to-digital signal conversion process required for an analogue sensor is avoided.

An example of an optical incremental shaft encoder is shown in Figure 20.3. It can be seen that the instrument consists of a pair of discs, one of which is fixed and one of which rotates with the body whose angular displacement is being measured. Each disc is basically opaque but has a pattern of windows cut into it. The fixed disc has only one window and the light source is aligned with this so that the light shines through all the time. The second disc has two tracks of windows cut into it that are spaced equidistantly around the disc, as

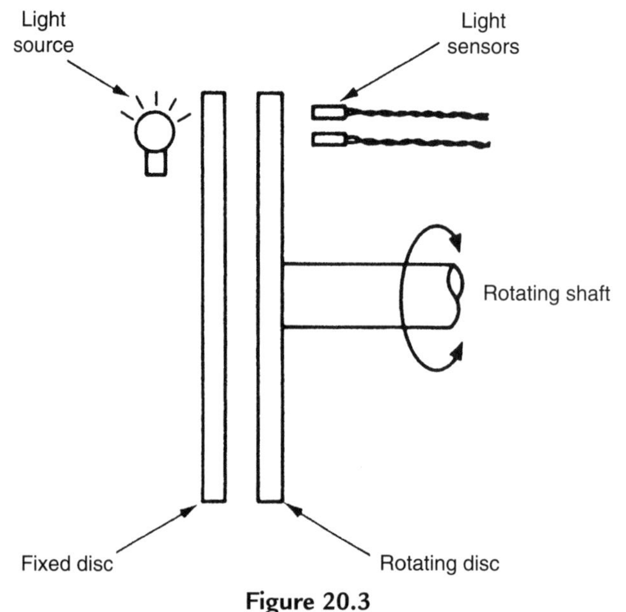


**Figure 20.3**
Optical incremental shaft encoder.

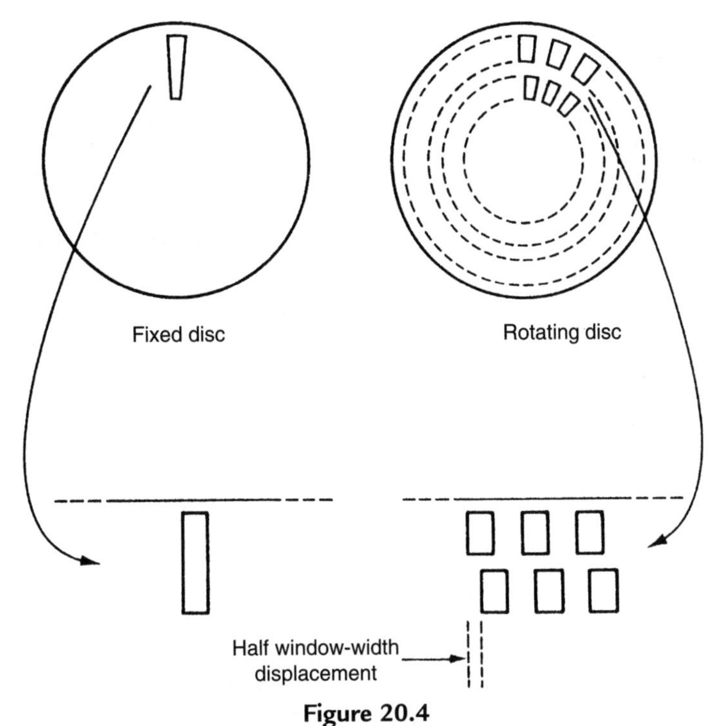


**Figure 20.4**
Window arrangement in incremental shaft encoder.

shown in Figure 20.4. Two light detectors are positioned beyond the second disc so that one is aligned with each track of windows. As the second disc rotates, light alternately enters and does not enter the detectors, as windows and then opaque regions of the disc pass in front of them. These pulses of light are fed to a counter, with the final count after motion has ceased corresponding to the angular position of the moving body relative to the starting position. The primary information about the magnitude of rotation is obtained by the detector aligned with the outer track of windows. However, the pulse count obtained from this gives no information about the direction of rotation. The necessary direction information is provided by the second, inner track of windows, which have an angular displacement with respect to the outer set of windows of half a window width. Pulses from the detector aligned with the inner track of windows therefore lag or lead the primary set of pulses according to the direction of rotation.

The maximum measurement resolution obtainable is limited by the number of windows that can be machined onto a disc. The maximum number of windows per track for a 150-mm-diameter disc is 5000, which gives a basic angular measurement resolution of 1 in 5000. By using more sophisticated circuits that increment the count on both the rising and the falling edges of the pulses through the outer track of windows, it is possible to double the resolution to a maximum of 1 in 10,000. At the expense of even greater complexity in the counting circuit, it is also possible to include pulses from the inner track of windows in the count, so giving a maximum measurement resolution of 1 in 20,000.

Optical incremental shaft encoders are a popular instrument for measuring relative angular displacements and are very reliable. Problems of noise in the system giving false counts can sometimes cause difficulties, although this can usually be eliminated by squaring the output from the light detectors. Such instruments are found in many applications where rotational motion has to be measured. Incremental shaft encoders are also used commonly in circumstances where a translational displacement has been transformed to a rotational one by suitable gearing. One example of this practice is in measuring the translational motions in numerically controlled drilling machines. Typical gearing used for this would give one revolution per millimeter of translational displacement. By using an incremental shaft encoder with 1000 windows per track in such an arrangement, a measurement resolution of 1 μm is obtained.

### 20.2.4 Coded Disc Shaft Encoders

Unlike the incremental shaft encoder that gives a digital output in the form of pulses that have to be counted, the digital shaft encoder has an output in the form of a binary number of several digits that provides an absolute measurement of shaft position. Digital encoders provide high accuracy and reliability. They are particularly useful for computer control applications, but have a significantly higher cost than incremental encoders. Three different forms exist, using optical, electrical, and magnetic energy systems, respectively.

*Optical digital shaft encoder*

The optical digital shaft encoder is the least expensive form of encoder available and is the one used most commonly. It is found in a variety of applications; one where it is particularly popular is in measuring the position of rotational joints in robot manipulators. The instrument is similar in physical appearance to the incremental shaft encoder. It has a pair of discs (one movable and one fixed) with a light source on one side and light detectors on the other side, as shown in Figure 20.5. The fixed disc has a single window, and the principal way in which the device differs from the incremental shaft encoder is in the design of the windows on the movable disc, as shown in Figure 20.6. These are cut in four or more tracks instead of two and are arranged in sectors as well as tracks. An energy detector is aligned with each track, and these give an output of "1" when energy is detected and an output of "0" otherwise. The measurement resolution obtainable depends on the number of tracks used. For a four-track version, the resolution is 1 in 16, with progressively higher measurement resolution being attained as the number of tracks is increased. These binary outputs from the detectors are combined together to give a binary number of several digits. The number of digits corresponds to the number of tracks on the disc, which, in the example shown in Figure 20.6, is four. The pattern of windows in each sector is cut

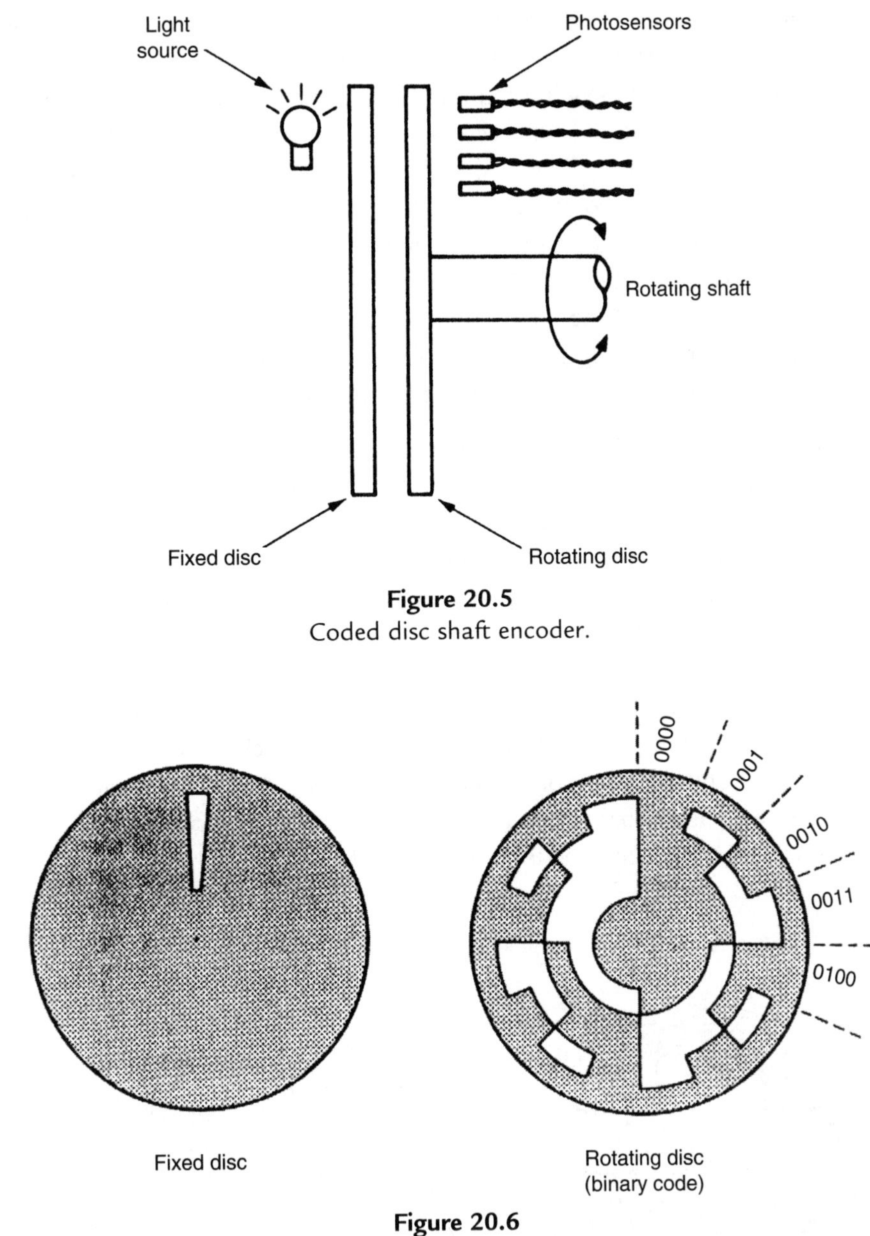


**Figure 20.5**
Coded disc shaft encoder.



**Figure 20.6**
Window arrangement for coded disc shaft encoder.

such that as that particular sector passes across the window in the fixed disc, the four energy detector outputs combine to give a unique binary number. In the binary-coded example shown in Figure 20.6, the binary number output increments by one as each sector in the rotating disc passes in turn across the window in the fixed disc. Thus the output from sector 1 is 0001, from sector 2 is 0010, from sector 3 is 0011, etc.

While this arrangement is perfectly adequate in theory, serious problems can arise in practice due to the manufacturing difficulty involved in machining the windows of the movable disc such that the edges of the windows in each track are aligned exactly with each other. Any misalignment means that, as the disc turns across the boundary between one sector and the next, the outputs from each track will switch at slightly different instants of time, and therefore the binary number output will be incorrect over small angular ranges corresponding to the sector boundaries. The worst error can occur at the boundary between sectors 7 and 8, where the output is switching from 0111 to 1000. If the energy sensor corresponding to the first digit switches before the others, then the output will be 1111 for a very small angular range of movement, indicating that sector 15 is aligned with the fixed disc rather than sector 7 or 8. This represents an error of 100% in the indicated angular position.

In practice, there are two ways that are used to overcome this difficulty. Both of these solutions involve an alteration to the manner in which windows are machined on the movable disc, as shown in Figure 20.7. The first method adds an extra outer track on the disc, known as an *antiambiguity track*, which consists of small windows that span a small angular range on either side of each sector boundary of the main track system. When energy sensors associated with this extra track sense energy, this is used to signify that the disc is aligned on a sector boundary and the output is unreliable. The second method is somewhat simpler and less expensive because it avoids the expense of machining the extra antiambiguity track. It does this by using a special code, known as the *Gray code*, to cut the tracks in each sector on the movable disk. The Gray code is a special binary representation where only one binary digit changes in moving from one decimal number representation to the next, that is, from one sector to the next in the digital shaft encoder. The code is illustrated in Table 20.1.

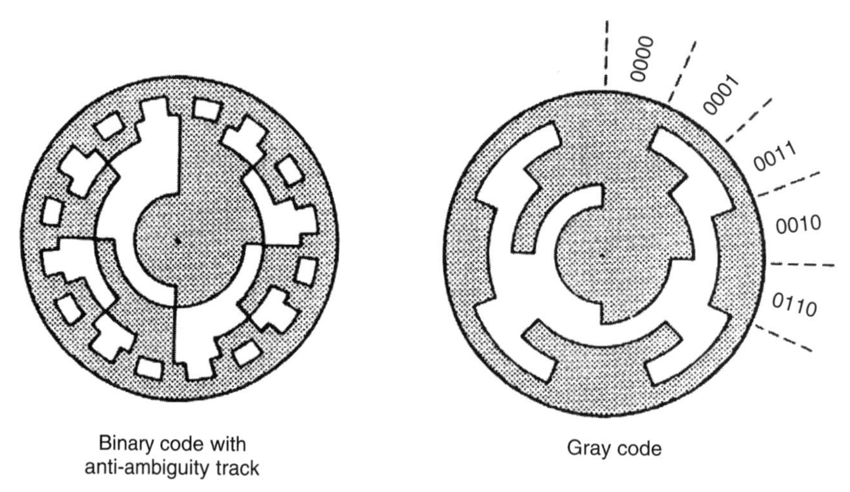


**Figure 20.7**
Modified window arrangements for a rotating disc.

**Table 20.1 The Gray Code**

| Decimal Number | Binary Code | Gray Code |
|---|---|---|
| 0 | 0000 | 0000 |
| 1 | 0001 | 0001 |
| 2 | 0010 | 0011 |
| 3 | 0011 | 0010 |
| 4 | 0100 | 0110 |
| 5 | 0101 | 0111 |
| 6 | 0110 | 0101 |
| 7 | 0111 | 0100 |
| 8 | 1000 | 1100 |
| 9 | 1001 | 1101 |
| 10 | 1010 | 1111 |
| 11 | 1011 | 1110 |
| 12 | 1100 | 1010 |
| 13 | 1101 | 1011 |
| 14 | 1110 | 1001 |
| 15 | 1111 | 1000 |

It is possible to manufacture optical digital shaft encoders with up to 21 tracks, which gives a measurement resolution of 1 part in $10^6$ (about 1 second of arc). Unfortunately, a high cost is involved in the special photolithography techniques used to cut the windows in order to achieve such a measurement resolution, and very high-quality mounts and bearings are needed. Hence, such devices are very expensive.

*Contacting (electrical) digital shaft encoder*

The contacting digital shaft encoder consists of only one disc, which rotates with the body whose displacement is being measured. The disc has conducting and nonconducting segments instead of the transparent and opaque areas found on the movable disc of the optical form of instrument, but these are arranged in an identical pattern of sectors and tracks. The disc is charged to a low potential by an electrical brush in contact with one side of the disc, and a set of brushes on the other side of the disc measures the potential in each track. The output of each detector brush is interpreted as a binary value of "1" or "0" according to whether the track in that particular segment is conducting or not and hence whether a voltage is sensed or not. As for the case of the optical form of instrument, these outputs are combined together to give a multibit binary number. Contacting digital shaft encoders have a similar cost to the equivalent optical instruments and have operational advantages in severe environmental conditions of high temperature or mechanical shock. They suffer from the usual problem of output ambiguity at the sector boundaries but this problem is overcome by the same methods used in optical instruments.

A serious problem in the application of contacting digital shaft encoders arises from their use of brushes. These introduce friction into the measurement system, and the combination of dirt and brush wear causes contact problems. Consequently, problems of intermittent output can occur, and such instruments generally have limited reliability and a high maintenance cost. Measurement resolution is also limited because of the lower limit on the minimum physical size of the contact brushes. The maximum number of tracks possible is 10, which limits the resolution to 1 part in 1000. Thus, contacting digital shaft encoders are normally only used where the environmental conditions are too severe for optical instruments.

#### *Magnetic digital shaft encoder*

Magnetic digital shaft encoders consist of a single rotatable disc, as in the contacting form of encoder discussed in the previous section. The pattern of sectors and tracks consists of magnetically conducting and nonconducting segments, and the sensors aligned with each track consist of small toroidal magnets. Each of these sensors has a coil wound on it that has a high or low voltage induced in it according to the magnetic field close to it. This field is dependent on the magnetic conductivity of that segment of the disc closest to the toroid.

These instruments have no moving parts in contact and therefore have a similar reliability to optical devices. Their major advantage over optical equivalents is an ability to operate in very harsh environmental conditions. Unfortunately, the process of manufacturing and accurately aligning the toroidal magnet sensors required makes such instruments very expensive. Their use is therefore limited to a few applications where both high measurement resolution and also operation in harsh environments are required.

### 20.2.5 The Resolver

The resolver, also known as a *synchro-resolver*, is an electromechanical device that gives an analogue output by transformer action. Physically, resolvers resemble a small a.c. motor and have a diameter ranging from 10 to 100 mm. They are frictionless and reliable in operation because they have no contacting moving surfaces; consequently, they have a long life. The best devices give measurement resolutions of 0.1%.

Resolvers have two stator windings, which are mounted at right angles to one another, and a rotor, which can have either one or two windings. As the angular position of the rotor changes, the output voltage changes. The simpler configuration of a resolver with only one winding on a rotor is illustrated in Figure 20.8. This exists in two separate forms that are distinguished according to whether the output voltage changes in amplitude or changes in phase as the rotor rotates relative to the stator winding.

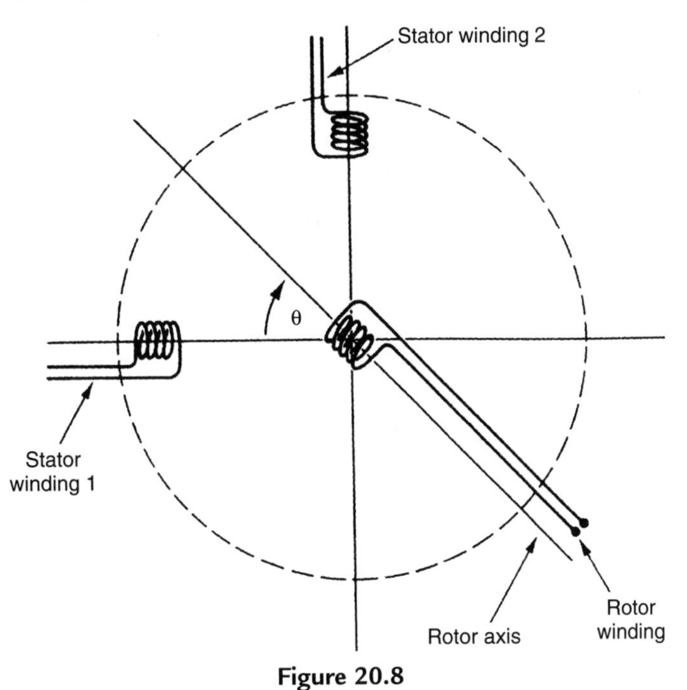


**Figure 20.8**
Schematic representation of resolver windings.

*Varying amplitude output resolver*

The stator of this type of resolver is excited with a single-phase sinusoidal voltage of frequency ω, where the amplitudes in the two windings are given by

$$V_1 = V\sin(\beta) \quad ; \quad V_2 = V\cos(\beta),$$

where $V = V_s \sin(\omega t)$.

The effect of this is to give a field at an angle of $(\beta + \pi/2)$ relative to stator winding 1.

Suppose that the angle of the rotor winding relative to that of the stator winding is given by θ. Then the magnetic coupling between the windings is a maximum for $\theta = (\beta + \pi/2)$ and a minimum for $(\theta = \beta)$. The rotor output voltage is of fixed frequency and varying amplitude given by

$$V_o = KV_s \sin(\beta - \theta)\sin(\omega t).$$

This relationship between shaft angle position and output voltage is nonlinear, but approximate linearity is obtained for small angular motions where $|\beta - \theta| < 15^{\circ}$.

Intelligent versions of this type of resolver are available that use a microprocessor to process the sine and cosine outputs. This can improve the measurement resolution to 2 minutes of arc.

*Varying phase output resolver*

This is a less common form of resolver that is only used in a few applications. The stator windings are excited with a two-phase sinusoidal voltage of frequency ω, and the instantaneous voltage amplitudes in the two windings are given by

$$V_1 = V_s \sin(\omega t) \quad ; \quad V_2 = V_s \sin(\omega t + \pi/2) = V_s \cos(\omega t).$$

The net output voltage in the rotor winding is the sum of the voltages induced due to each stator winding. This is given by

$$\begin{aligned} V_o &= KV_s \sin(\omega t)\cos(\theta) + KV_s \cos(\omega t)\cos(\pi/2 - \theta) \\ &= KV_s[\sin(\omega t)\cos(\theta) + \cos(\omega t)\sin(\theta)] \\ &= KV_s \sin(\omega t + \theta) \end{aligned}$$

This represents a linear relationship between shaft angle and the phase shift of the rotor output relative to the stator excitation voltage. The accuracy of shaft rotation measurement depends on the accuracy with which the phase shift can be measured. This can be improved by increasing the excitation frequency, ω, and it is possible to reduce inaccuracy down to $\pm 0.1\%$. However, increasing the excitation frequency also increases magnetizing losses. Consequently, a compromise excitation frequency of about 400 Hz is used.

### 20.2.6 The Synchro

Like the resolver, the synchro is a motor-like, electromechanical device with an analogue output. Apart from having three stator windings instead of two, the instrument is similar in appearance and operation to the resolver and has the same range of physical dimensions. The rotor usually has a dumbbell shape and, like the resolver, can have either one or two windings.

Synchros have been in use for many years for the measurement of angular positions, especially in military applications, and achieve similar levels of accuracy and measurement resolution to digital encoders. One common application is axis measurement in machine tools, where the translational motion of the tool is translated into a rotational displacement by suitable gearing. Synchros are tolerant to high temperatures, high humidity, shock, and vibration and are therefore suitable for operation in such harsh environmental conditions. Some maintenance problems are associated with the slip ring and brush system used to supply power to the rotor. However, the only major source of error in the instrument is asymmetry in the windings, and a reduction of measurement inaccuracy down to $\pm 0.5\%$ is easily achievable.

Figure 20.9 shows the simpler form of a synchro with a single rotor winding. If an a.c. excitation voltage is applied to the rotor via slip rings and brushes, this sets up a certain pattern

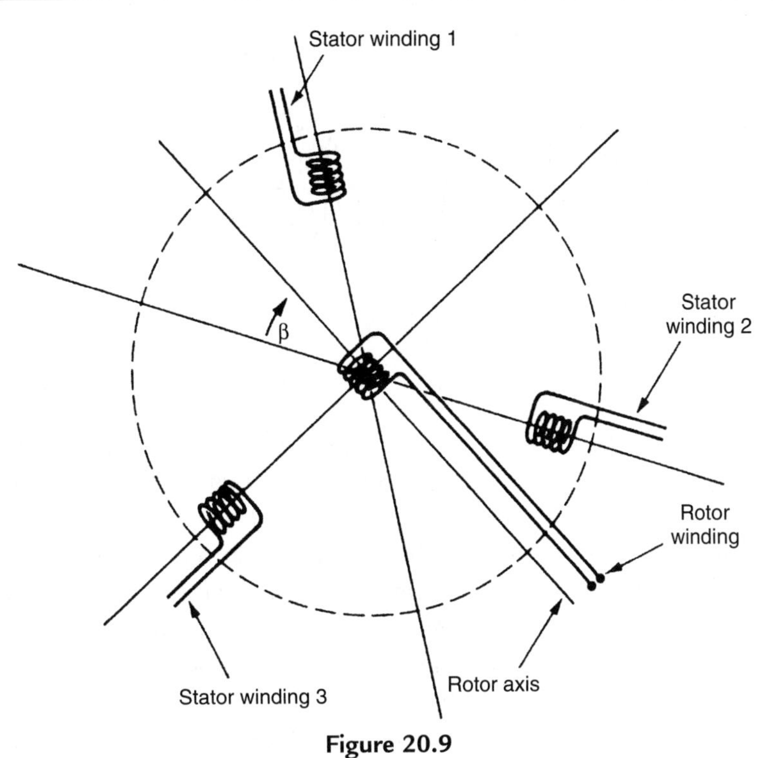


**Figure 20.9**
Schematic representation of synchro windings.

of fluxes and induced voltages in the stator windings by transformer action. For a rotor excitation voltage, $V_r$, given by

$$V_r = V\sin(\omega t)$$

the voltages induced in the three stator windings are

$$V_1 = V\sin(\omega t)\sin(\beta) \quad ; \quad V_2 = V\sin(\omega t)\sin(\beta + 2\pi/3) \quad ;$$
$$V_3 = V\sin(\omega t)\sin(\beta - 2\pi/3)$$

where β is the angle between the rotor and stator windings.

If the rotor is turned at constant velocity through one full revolution, the voltage waveform induced in each stator winding is as shown in Figure 20.10. This has the form of a carrier-modulated waveform, in which the carrier frequency corresponds to the excitation frequency, ω. It follows that if the rotor is stopped at any particular angle, β', the peak-to-peak amplitude of the stator voltage is a function of β'. If therefore the stator winding voltage is measured, generally as its root-mean-squared (r.m.s.) value, this indicates the magnitude of the rotor rotation away from the null position. The direction of rotation is determined by the phase difference between the stator voltages, which is indicated by their relative instantaneous magnitudes.

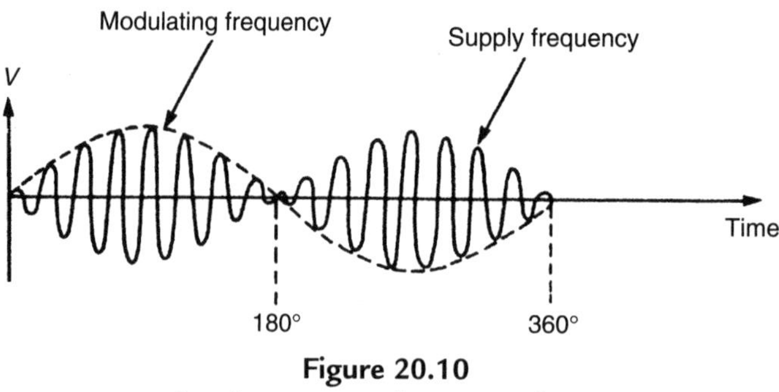


**Figure 20.10**
Synchro stator voltage waveform.

Although a single synchro is able to measure an angular displacement by itself, it is much more common to find a pair of them used for this purpose. When used in pairs, one member of the pair is known as the synchro transmitter and the other as the synchro transformer, and the two sets of stator windings are connected together, as shown in Figure 20.11. Each synchro is of the form shown in Figure 20.9, but the rotor of the transformer is fixed for displacement-measuring applications. A sinusoidal excitation voltage is applied to the rotor of the transmitter, setting up a pattern of fluxes and induced voltages in the transmitter stator windings. These voltages are transmitted to the transformer stator windings, where a similar flux pattern is established. This in turn causes a sinusoidal voltage to be induced in the fixed transformer rotor winding. For an excitation voltage, $V \sin(\omega t)$, applied to the transmitter rotor, the voltage measured in the transformer rotor is given by

$$V_o = V \sin(\omega t) \sin(\theta),$$

where $\theta$ is the relative angle between the two rotor windings.

Apart from their use as a displacement transducer, such synchro pairs are commonly used to transmit angular displacement information over some distance, for instance, to transmit gyro compass measurements in an aircraft to remote meters. They are also used for load positioning, allowing a load connected to the transformer rotor shaft to be controlled remotely by turning the

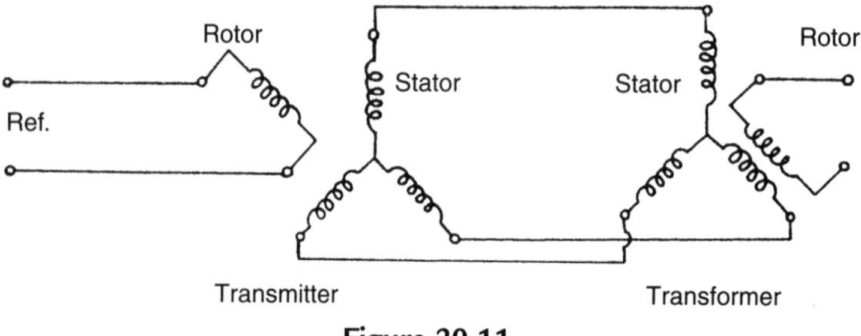


**Figure 20.11**
Synchro transmitter-transformer pair.

transmitter rotor. For these applications, the transformer rotor is free to rotate, and it is also damped to prevent oscillatory motions. In the simplest arrangement, a common sinusoidal excitation voltage is applied to both rotors. If the transmitter rotor is turned, this causes an imbalance in the magnetic flux patterns and results in a torque on the transformer rotor that tends to bring it into line with the transmitter rotor. This torque is typically small for small displacements, so this technique is only useful if the load torque on the transformer shaft is very small. In other circumstances, it is necessary to incorporate the synchro pair within a servomechanism, where the output voltage induced in the transformer rotor winding is amplified and applied to a servomotor that drives the transformer rotor shaft until it is aligned with the transmitter shaft.

### 20.2.7 The Induction Potentiometer

The induction potentiometer belongs to the same class of instruments as resolvers and synchros. It only has one rotor winding and one stator winding, but otherwise it is of similar size and appearance to other devices in this class of electromechanical, angular position measuring instruments. A single-phase sinusoidal excitation is applied to the rotor winding, which causes an output voltage in the stator winding through the mutual inductance linking the two windings. The magnitude of this induced stator voltage varies with rotation of the rotor. The variation of the output with rotation is naturally sinusoidal if the coils are wound such that their field is concentrated at one point, and only small excursions can be made away from the null position if the output relationship is to remain approximately linear. However, if the rotor and stator windings are distributed around the circumference in a special way, an approximately linear relationship for angular displacements of up to $\pm 90^\circ$ can be obtained.

### 20.2.8 The Rotary Inductosyn

This instrument is similar in operation to the linear inductosyn, except that it measures rotary displacements and has tracks that are arranged radially on two circular discs, as shown in Figure 20.12. Typical diameters of the instrument vary between 75 and 300 mm. The larger versions give a measurement resolution of up to 0.05 second of arc. However, like its linear equivalent, the rotary inductosyn has a very small measurement range. Therefore, a lower resolution, rotary displacement transducer with a larger measurement range must be used in conjunction with it.

### 20.2.9 Gyroscopes

Gyroscopes measure both absolute angular displacement and absolute angular velocity. Until recently, the mechanical, spinning-wheel gyroscope had a dominant position in the marketplace. However, this position is now being challenged by optical gyroscopes.

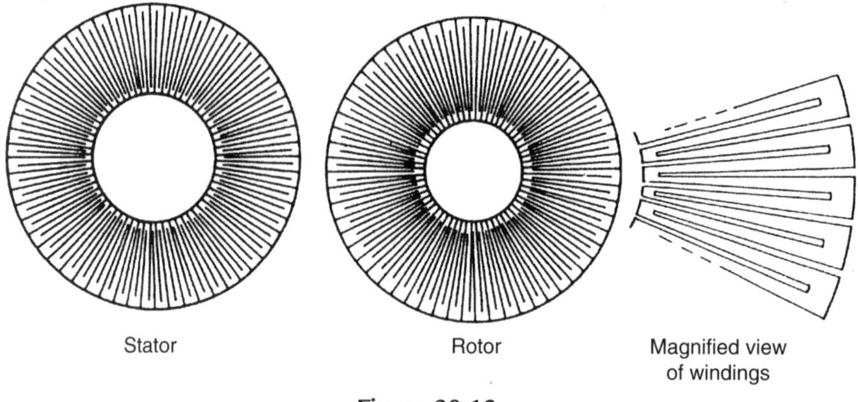


**Figure 20.12**
Rotary inductosyn.

*Mechanical gyroscopes*

Mechanical gyroscopes consist essentially of a large, motor-driven wheel whose angular momentum is such that the axis of rotation tends to remain fixed in space, thus acting as a reference point. The gyro frame is attached to the body whose motion is to be measured. The output is measured in terms of the angle between the frame and the axis of the spinning wheel. Two different forms of mechanical gyroscopes are used for measuring angular displacement—the free gyro and the rate-integrating gyro. A third type of mechanical gyroscope, the rate gyro, measures angular velocity and is described in Section 20.3.

The *free gyroscope* is illustrated in Figure 20.13. This measures the absolute angular rotation about two perpendicular axes of the body to which its frame is attached. Two alternative methods of driving the wheel are used in different versions of the instrument. One of these is to enclose the wheel in stator-like coils that are excited with a sinusoidal voltage. A voltage is applied to the wheel via slip rings at both ends of the spindle carrying the wheel. The wheel behaves as a rotor, and motion is produced by motor action. The other, less common, method is to fix vanes on the wheel, which is then driven by directing a jet of air onto the vanes.

The free gyroscope can measure angular displacements of up to 10° with a high accuracy. For greater angular displacements, interaction between the measurements on the two perpendicular axes starts to cause a serious loss of accuracy. The physical size of the coils in the motor-action-driven system also limits the measurement range to 10°. For these reasons, this type of gyroscope is only suitable for measuring rotational displacements of up to 10°. A further operational problem of free gyroscopes is the presence of angular drift (precession) due to bearing friction torque. This has a typical magnitude of 0.5° per minute and means that the instrument can only be used over short time intervals of say 5 minutes. This time duration can be extended if the angular momentum of the spinning wheel is increased.

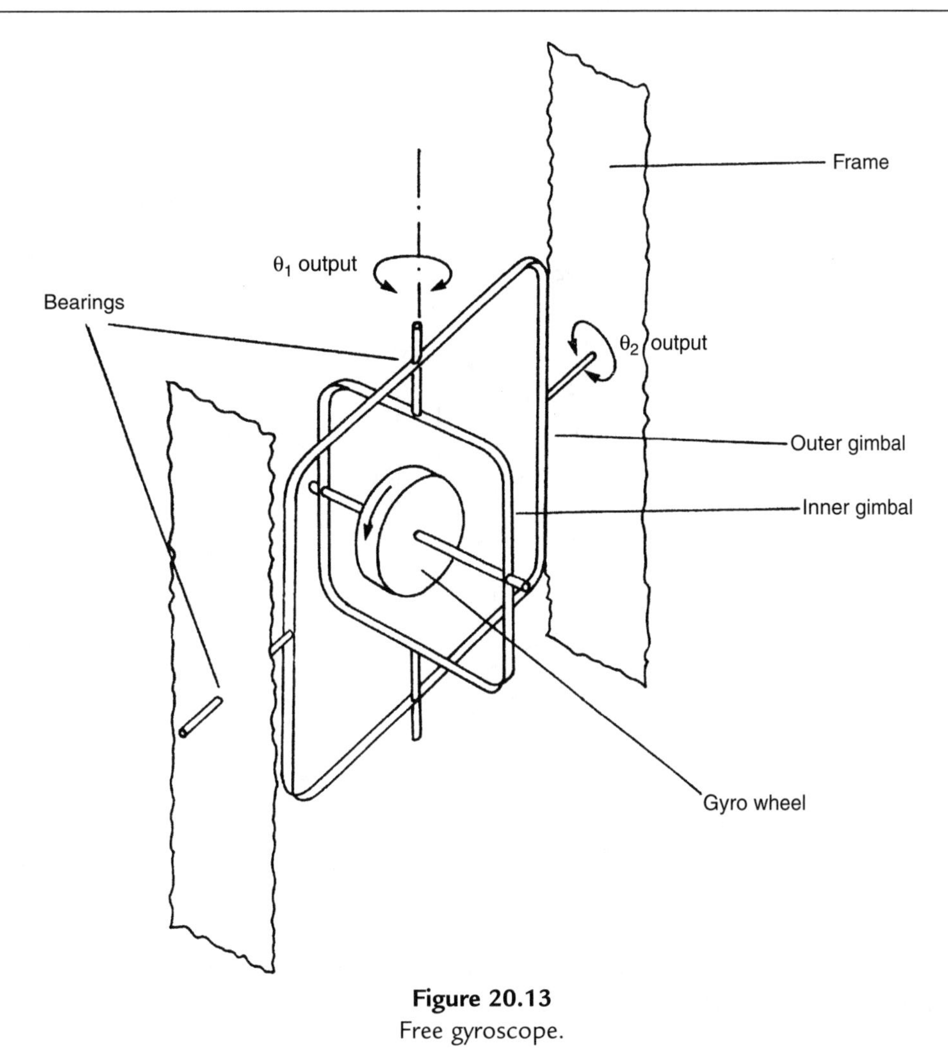


**Figure 20.13**
Free gyroscope.

A major application of the free gyroscope is in inertial navigation systems. Only two free gyros mounted along orthogonal axes are needed to monitor motions in three dimensions because each instrument measures displacement about two axes. The limited angular range of measurement is not usually a problem in such applications, as control action prevents the error in the direction of motion about any axis ever exceeding one or two degrees. However, precession is a much greater problem, and, for this reason, the rate-integrating gyro is used much more commonly.

The *rate-integrating gyroscope*, or *integrating gyro* as it is commonly known, is illustrated in Figure 20.14. It measures angular displacements about a single axis only, and therefore three instruments are required in a typical inertial navigation system. The major advantage

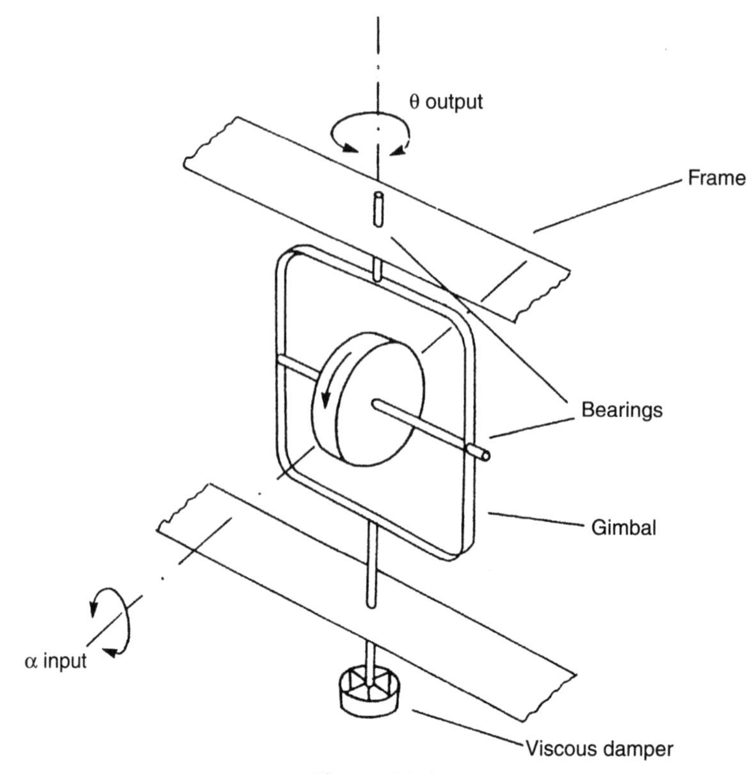


**Figure 20.14**
Rate-integrating gyroscope.

of the instrument over the free gyro is the almost total absence of precession, with typical specifications quoting drifts of only 0.01°/hour. The instrument has a first-order type of response given by

$$\frac{\theta_o}{\theta_i}(D) = \frac{K}{\tau D + 1}, \tag{20.1}$$

where $K = H/\beta$, $\tau = M/\beta$, $\theta_i$ is the input angle, $\theta_o$ is the output angle, $D$ is the D operator, $H$ is the angular momentum, $M$ is the moment of inertia of the system about the measurement axis, and β is the damping coefficient.

Inspection of Equation (20.1) shows that to obtain a high value of measurement sensitivity, $K$, a high value of $H$ and a low value of β are required. A large $H$ is normally obtained by driving the wheel with a hysteresis-type motor revolving at high speeds of up to 24,000 rpm. However, damping coefficient β can only be reduced so far because a small value of β results in a large value for the system time constant, τ, and an unacceptably low speed of system response. Therefore, the value of β has to be chosen as a compromise between these constraints.

In addition to their use as a fixed reference in inertial guidance systems, integrating gyros are also used commonly within aircraft autopilot systems and in military applications such as stabilizing weapon systems in tanks.

*Optical gyroscopes*

Optical gyroscopes are a relatively recent development and come in two forms—ring laser gyroscope and fiber-optic gyroscope.

The *ring laser gyroscope* consists of a glass ceramic chamber containing a helium–neon gas mixture in which two laser beams are generated by a single anode/twin cathode system, as shown in Figure 20.15. Three mirrors, supported by the ceramic block and mounted in a triangular arrangement, direct the pair of laser beams around the cavity in opposite directions. Any rotation of the ring affects the coherence of the two beams, raising the frequency of one and lowering the frequency of the other. The clockwise and anticlockwise beams are directed into a photodetector that measures the beat frequency according to the frequency difference, which is proportional to the angle of rotation. The advantages of the ring laser gyroscope over traditional, mechanical gyroscopes are considerable. The measurement accuracy obtained is substantially better than that afforded by mechanical gyros in a similar price range. The device is also considerably smaller physically, which is of considerable benefit in many applications.

The *fiber-optic gyroscope* measures angular velocity and is described in Section 20.3.

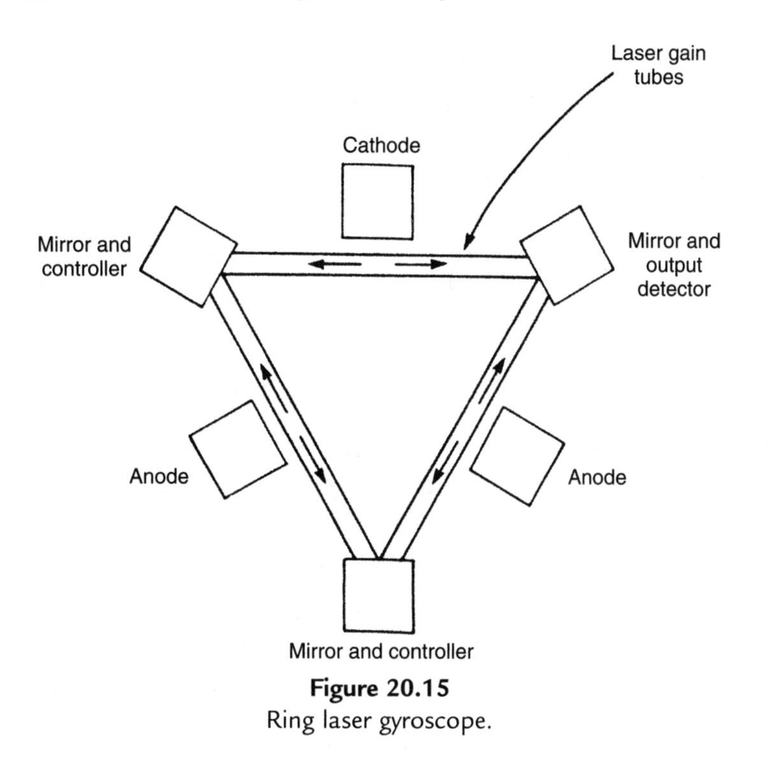


**Figure 20.15**
Ring laser gyroscope.

### 20.2.10 Choice between Rotational Displacement Transducers

Choice between the various rotational displacement transducers that might be used in any particular measurement situation depends first of all on whether absolute measurement of angular position is required or whether the measurement of rotation relative to some arbitrary starting point is acceptable. Other factors affecting the choice between instruments are the required measurement range, resolution of the transducer, and measurement accuracy afforded.

Where only measurement of relative angular position is required, the incremental encoder is a very suitable instrument. The best commercial instruments of this type can measure rotations to a resolution of 1 part in 20,000 of a full revolution, and the measurement range is an infinite number of revolutions. Instruments with such a high measurement resolution are very expensive, but much less expensive versions are available according to what lower level of measurement resolution is acceptable.

All the other instruments presented in this chapter provide an absolute measurement of angular position. The required measurement range is a dominant factor in the choice between these. If this exceeds one full revolution, then the only instrument available is the helical potentiometer. Such devices can measure rotations of up to 60 full turns, but are expensive because the procedure involved in manufacturing a helical resistance element to a reasonable standard of accuracy is difficult.

For measurements of less than one full revolution, the range of available instruments widens. The least expensive one available is the circular potentiometer, but much better measurement accuracy and resolution are obtained from coded-disc encoders. The least expensive of these is the optical form, but certain operating environments necessitate the use of the alternative contacting (electrical) and magnetic versions. All types of coded disc encoders are very reliable and are particularly attractive in computer control schemes, as the output is in digital form. A varying phase output resolver is yet another instrument that can measure angular displacements up to one full revolution in magnitude. Unfortunately, this instrument is expensive because of the complicated electronics incorporated to measure the phase variation and convert it to a varying amplitude output signal, and hence it is no longer in common use.

An even greater range of instruments becomes available as the required measurement range is reduced further. These include the synchro ($\pm 90^\circ$), the varying amplitude output resolver ($\pm 90^\circ$), the induction potentiometer ($\pm 90^\circ$), and the differential transformer ($\pm 40^\circ$). All these instruments have a high reliability and a long service life.

Finally, two further instruments are available for satisfying special measurement requirements—the rotary inductosyn and the gyroscope. The rotary inductosyn is used in applications where very high measurement resolution is required, although the measurement range afforded is extremely small and a coarser resolution instrument must be used in parallel with it to extend the measurement range. Gyroscopes, in both mechanical and optical forms, are used to measure small angular displacements up to $\pm 10°$ in magnitude in inertial navigation systems and similar applications.

### 20.2.11 Calibration of Rotational Displacement Transducers

The coded disc shaft encoder is normally used for the calibration of rotary potentiometers and differential transformers. A typical model provides a reference standard with a measurement uncertainty of $\pm 0.1\%$ of the full-scale reading. If greater accuracy is required, for example, in calibrating encoders of lesser accuracy, encoders with measurement uncertainty down to $\pm 0.0001\%$ of the full-scale reading can be obtained and used as a reference standard, although these have a very high associated cost.

## 20.3 Rotational Velocity

The main application of rotational velocity transducers is in speed control systems. They also provide the usual means of measuring translational velocities, which are transformed into rotational motions for measurement purposes by suitable gearing. Many different instruments and techniques are available for measuring rotational velocity as presented here.

### 20.3.1 Digital Tachometers

Digital tachometers or, to give them their proper title, digital *tachometric generators* are usually noncontact instruments that sense the passage of equally spaced marks on the surface of a rotating disk or shaft. Measurement resolution is governed by the number of marks around the circumference. Various types of sensors are used, such as optical, inductive, and magnetic ones. As each mark is sensed, a pulse is generated and input to an electronic pulse counter. Usually, velocity is calculated in terms of the pulse count in unit time, which of course only yields information about the mean velocity. If the velocity is changing, instantaneous velocity can be calculated at each instant of time that an output pulse occurs, using the scheme shown in Figure 20.16. In this circuit, each pulse from the transducer initiates the transfer of a train of clock pulses from a 1-MHz clock into a counter. Control logic resets the counter and updates the digital output value after receipt of each transducer pulse. The measurement resolution of this system is highest when the speed of rotation is low.

#### Optical sensing

Digital tachometers with optical sensors are often known as *optical tachometers*. Optical pulses can be generated by one of the two alternative photoelectric techniques illustrated in Figure 20.17. In the scheme shown in Figure 20.17a, pulses are produced as windows in a

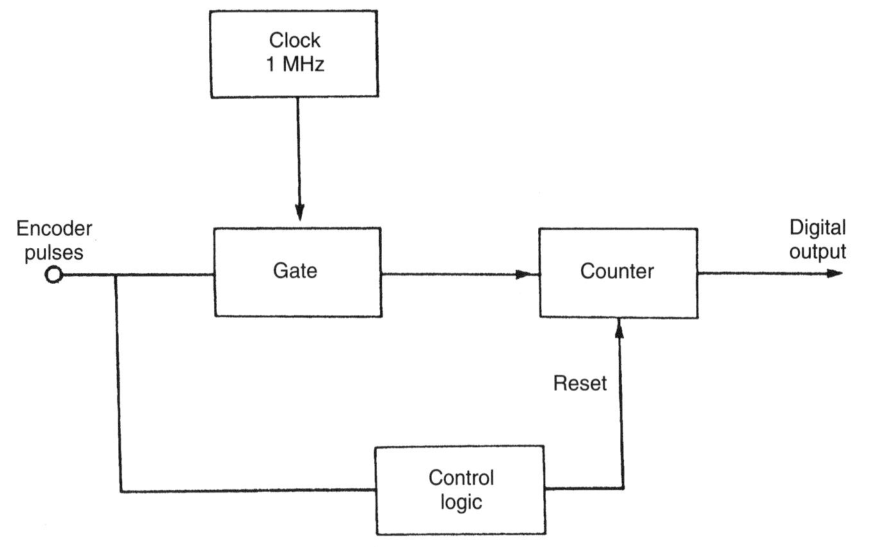

**Figure 20.16**

Scheme used to measure instantaneous angular velocities.

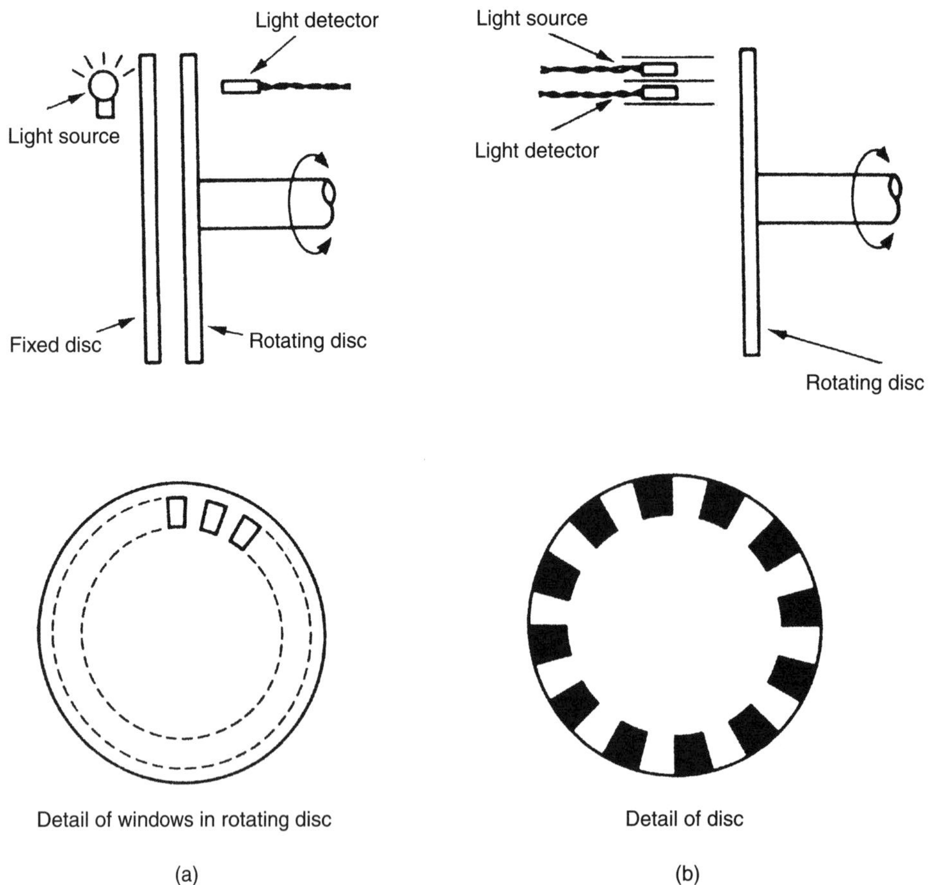


**Figure 20.17**

Photoelectric pulse generation techniques.

slotted disc pass in sequence between a light source and a detector. The alternative scheme, shown in Figure 20.17b, has both a light source and a detector mounted on the same side of a reflective disc that has black sectors painted onto it at regular angular intervals. Light sources are normally either lasers or light-emitting diodes, with photodiodes and phototransistors being used as detectors. Optical tachometers yield better accuracy than other forms of digital tachometers. However, they are less reliable than other forms because dust and dirt can block light paths.

#### *Inductive sensing*

*Variable reluctance velocity transducers*, also known as *induction tachometers*, are a form of digital tachometer that use inductive sensing. They are widely used in the automotive industry within antiskid devices, antilock braking systems, and traction control. One relatively simple and inexpensive form of this type of device was described earlier in Section 13.4 (Figure 13.2). A more sophisticated version, shown in Figure 20.18, has a rotating disc constructed from a bonded fiber material into which soft iron poles are inserted at regular intervals around its periphery. The sensor consists of a permanent magnet with a shaped pole piece, which carries a wound coil. The distance between the pickup and the outer perimeter of the disc is typically 0.5 mm. As the disc rotates, the soft iron inserts on the disc move in turn past the pickup unit. As each iron insert moves toward the pole piece, the reluctance of the magnetic circuit increases and hence the flux in the pole piece also increases. Similarly, the flux in the pole piece decreases as each iron insert moves away from the sensor. The changing magnetic flux inside the pickup coil causes a voltage to be induced in the coil whose magnitude is proportional to the rate of change of flux. This voltage is positive while the flux is increasing and negative while it is

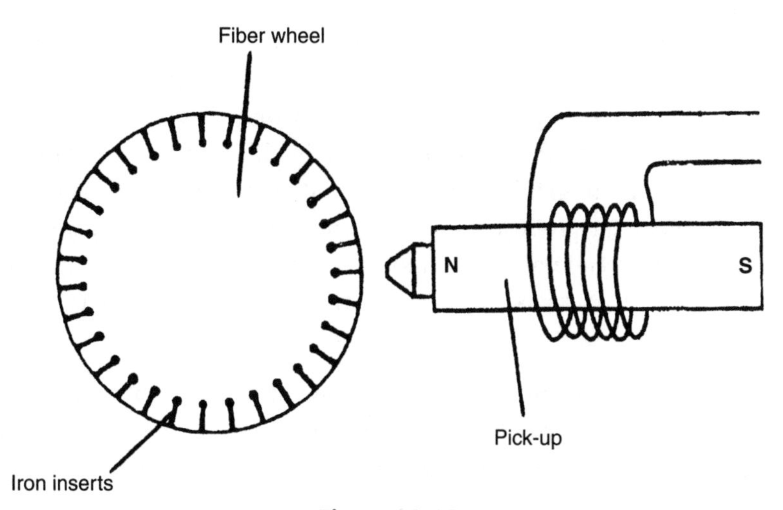


**Figure 20.18**
Variable reluctance transducer.

decreasing. Thus, the output is a sequence of positive and negative pulses whose frequency is proportional to the rotational velocity of the disc. The maximum angular velocity that the instrument can measure is limited to about 10,000 r.p.m. because of the finite width of the induced pulses. As the velocity increases, the distance between pulses is reduced; at a certain velocity, the pulses start to overlap. At this point, the pulse counter ceases to be able to distinguish separate pulses. The optical tachometer has significant advantages in this respect, as the pulse width is much narrower, allowing measurement of higher velocities.

A simpler and less expensive form of variable reluctance transducer also exists that uses a ferromagnetic gear wheel in place of a fiber disc. The motion of the tip of each gear tooth toward and away from the pickup unit causes a similar variation in the flux pattern to that produced by iron inserts in the fiber disc. However, pulses produced by these means are less sharp and, consequently, the maximum angular velocity measurable is lower.

*Magnetic (Hall-effect) sensing*

The rotating element in *Hall-effect* or *magnetostrictive tachometers* has a very simple design in the form of a toothed metal gear wheel. The sensor is a solid-state, Hall-effect device that is placed between the gear wheel and a permanent magnet. When an intertooth gap on the gear wheel is adjacent to the sensor, the full magnetic field from the magnet passes through it. Later, as a tooth approaches the sensor, the tooth diverts some of the magnetic field, and so the field through the sensor is reduced. This causes the sensor to produce an output voltage proportional to the rotational speed of the gear wheel.

### 20.3.2 Stroboscopic Methods

The stroboscopic technique of rotational velocity measurement operates on a similar physical principle to digital tachometers except that the pulses involved consist of flashes of light generated electronically and whose frequency is adjustable so that it can be matched with the frequency of occurrence of some feature on the rotating body being measured. This feature can either be some naturally occurring one, such as gear teeth or spokes of a wheel, or be an artificially created pattern of black and white stripes. In either case, the rotating body appears stationary when frequencies of the light pulses and body features are in synchronism. Flashing rates available in commercial stroboscopes vary from 110 up to 150,000 per minute according to the range of velocity measurement required, and typical measurement inaccuracy is $\pm 1\%$ of the reading. The instrument is usually in the form of a hand-held device that is pointed toward the rotating body.

It must be noted that measurement of the flashing rate at which the rotating body appears stationary does not automatically indicate the rotational velocity, because synchronism also occurs when the flashing rate is some integral submultiple of the rotational speed. The practical procedure followed is therefore to adjust the flashing rate until synchronism is obtained at the largest flashing rate possible, $R_1$. The flashing rate is then decreased carefully until synchronism is again achieved at the next lower flashing rate, $R_2$. The rotational velocity is then given by

$$V = \frac{R_1 R_2}{R_1 - R_2}.$$

### 20.3.3 Analogue Tachometers

Analogue tachometers are less accurate than digital tachometers but are nevertheless still used successfully in many applications. Various forms exist.

The *d.c. tachometer* has an output approximately proportional to its speed of rotation. Its basic structure is identical to that found in a standard d.c. generator used for producing power and is shown in Figure 20.19. Both permanent magnet types and separately excited field types are used. However, certain aspects of the design are optimized to improve its accuracy as a speed-measuring instrument. One significant design modification is to reduce the weight of the rotor by constructing the windings on a hollow fiberglass shell. The effect of this is to minimize any loading effect of the instrument on the system being measured. The d.c. output voltage from the instrument is of a relatively high magnitude, giving a high measurement sensitivity that is typically 5 volts per 1000 r.p.m. The direction of rotation is determined by the polarity of the output voltage. A common range of measurement is 0–6000 r.p.m. Maximum nonlinearity is usually about $\pm 1\%$ of the full-scale reading. One problem with these devices that can cause difficulties under some circumstances is the presence of an a.c. ripple in the output signal. The magnitude of this can be up to 2% of the output d.c. level.

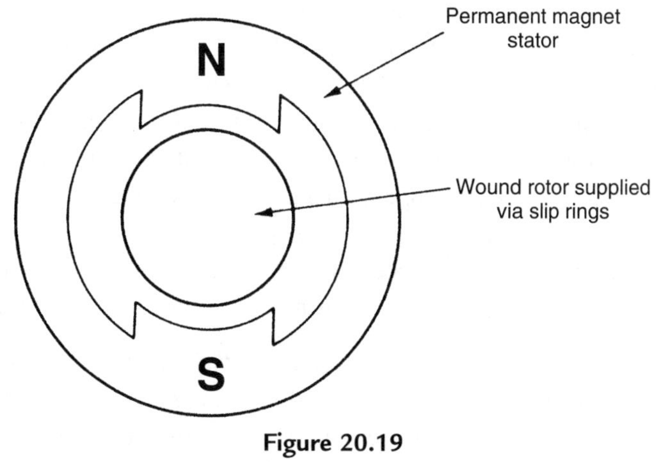


**Figure 20.19**
A d.c. tachometer.

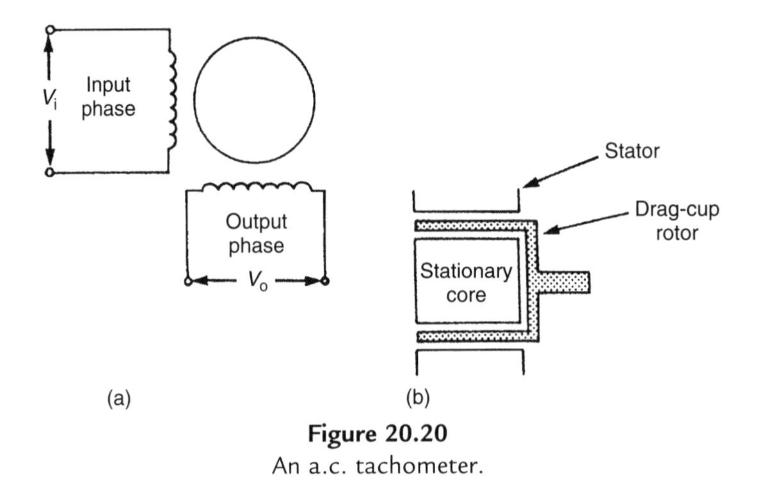


**Figure 20.20**
An a.c. tachometer.

The *a.c. tachometer* has an output approximately proportional to rotational speed like the d.c. tachogenerator. Its mechanical structure takes the form of a two-phase induction motor, with two stator windings and (usually) a drag-cup rotor, as shown in Figure 20.20. One of the stator windings is excited with an a.c. voltage, and the measurement signal is taken from the output voltage induced in the second winding. The magnitude of this output voltage is zero when the rotor is stationary, and otherwise is proportional to the angular velocity of the rotor. The direction of rotation is determined by the phase of the output voltage, which switches by 180° as the direction reverses. Therefore, both the phase and the magnitude of the output voltage have to be measured. A typical range of measurement is 0–4000 r.p.m., with an inaccuracy of ±0.05% of full-scale reading. Less expensive versions with a squirrel cage rotor also exist, but measurement inaccuracy in these is typically ±0.25%.

The *drag-cup tachometer*, also known as an *eddy-current tachometer*, has a central spindle carrying a permanent magnet that rotates inside a nonmagnetic drag cup consisting of a cylindrical sleeve of electrically conductive material, as shown in Figure 20.21. As the spindle and magnet rotate, voltage is induced that causes circulating eddy currents in the cup. These currents interact with the magnetic field from the permanent magnet and produce a torque. In response, the drag cup turns until the induced torque is balanced by the torque due to the restraining springs connected to the cup. When equilibrium is reached, angular displacement of the cup is proportional to the rotational velocity of the central spindle. The instrument has a typical measurement inaccuracy of ±0.5% and is used commonly in the speedometers of motor vehicles and also as a speed indicator for aero-engines. It is capable of measuring velocities up to 15,000 r.p.m.

Analogue output forms of the *variable reluctance velocity transducer* (see Section 20.3.1) also exist in which output voltage pulses are converted into an analogue, varying-amplitude, d.c. voltage by means of a frequency-to-voltage converter circuit. However, the measurement accuracy is inferior to digital output forms.

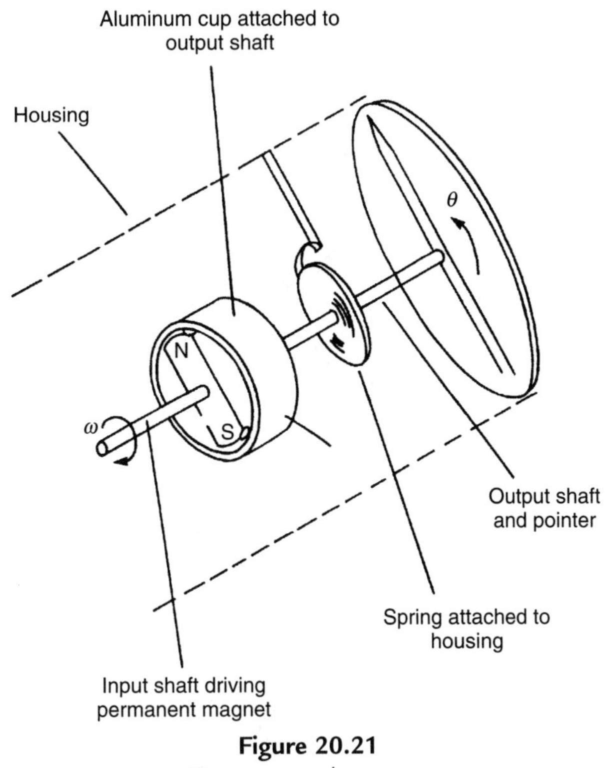


**Figure 20.21**
Drag-cup tachometer.

### 20.3.4 The Rate Gyroscope

The rate gyro, illustrated in Figure 20.22, has an almost identical construction to the rate-integrating gyro (Figure 20.14) and differs only by including a spring system that acts as an additional restraint on the rotational motion of the frame. The instrument measures the absolute angular velocity of a body and is widely used for generating stabilizing signals within vehicle navigation systems. The typical measurement resolution given by the instrument is 0.01°/s, and rotation rates up to 50°/s can be measured. The angular velocity, $\alpha$, of the body is related to the angular deflection of the gyroscope, $\theta$, by the equation:

$$\frac{\theta}{\alpha}(D) = \frac{H}{MD^2 + \beta D + K}, \tag{20.2}$$

where $H$ is the angular momentum of the spinning wheel, $M$ is the moment of inertia of the system, $\beta$ *is* the viscous damping coefficient, $K$ is the spring constant, and $D$ is the D operator.

This relationship [Equation (20.2)] is a second-order differential equation, and we must consequently expect the device to have a response typical of second-order instruments, as discussed in Chapter 2. Therefore, the instrument must be designed carefully so that the output

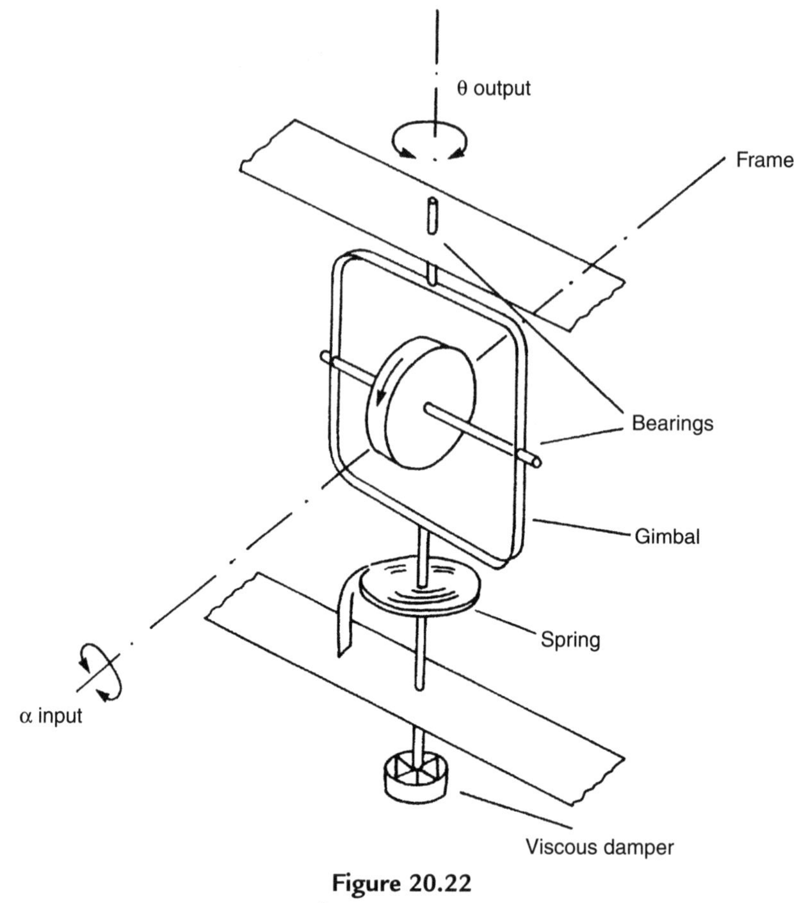


**Figure 20.22**
Rate gyroscope.

response is neither oscillatory nor too slow in reaching a final reading. To assist in the design process, it is useful to re-express Equation (20.2) in the following form:

$$\frac{\theta}{\alpha}(D) = \frac{K'}{D^2/\omega^2 + 2\xi D/\omega + 1}, \tag{20.3}$$

where $K' = H/K$, $\omega = \sqrt{K/M}$, and $\xi = \dfrac{\beta}{2\sqrt{KM}}$.

The static sensitivity of the instrument, $K'$, is made as large as possible by using a high-speed motor to spin the wheel and so make $H$ high. Reducing the spring constant $K$ further improves the sensitivity, but this cannot be reduced too far as it makes the resonant frequency ω of the instrument too small. The value of β is usually chosen such that the damping ratio ξ is as close to 0.7 as possible.

### 20.3.5 Fiber-Optic Gyroscope

This is a relatively new instrument that makes use of fiber-optic technology. Incident light from a source is separated by a beam splitter into a pair of beams, *a* and *b*, as shown in Figure 20.23. These travel in opposite directions around an fiber-optic coil (which may be several hundred meters long) and emerge from the coil as beams marked $a'$ and $b'$. The beams $a'$ and $b'$ are directed by the beam splitter into an interferometer. Any motion of the coil causes a phase shift between $a'$ and $b'$, which is detected by the interferometer.

### 20.3.6 Differentiation of Angular Displacement Measurements

Angular velocity measurements can be obtained by differentiating the output signal from angular displacement transducers. Unfortunately, the process of differentiation amplifies any noise in the measurement signal, and therefore this technique has been used only rarely in the past. However, the technique has become more feasible with the advent of intelligent instruments. For example, using an intelligent instrument to differentiate and process the output from a resolver can produce a velocity measurement with a maximum inaccuracy of ±1%.

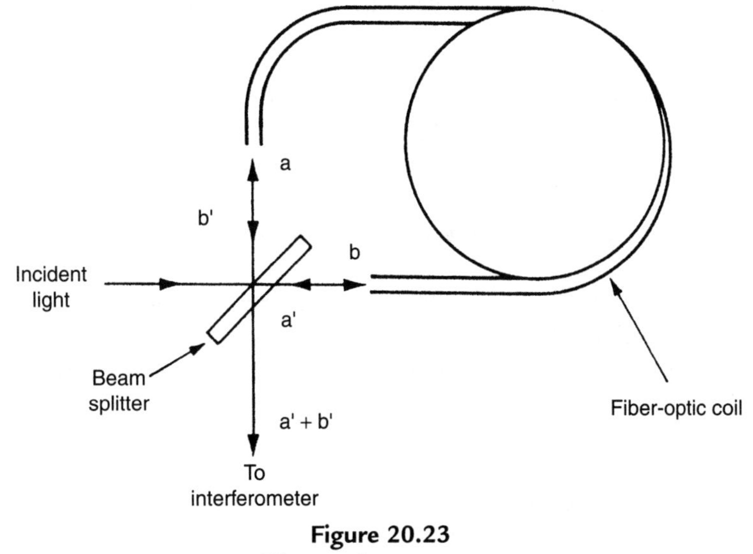


**Figure 20.23**
Fiber-optic gyroscope.

### 20.3.7 Integration of Output from an Accelerometer

In measurement systems that already contain an angular acceleration transducer, it is possible to obtain a velocity measurement by integrating the acceleration measurement signal. This produces a signal of acceptable quality, as the process of integration attenuates any measurement noise. However, the method is of limited value in many measurement situations because the measurement obtained is the average velocity over a period of time rather than a profile of the instantaneous velocities as motion takes place along a particular path.

### 20.3.8 Choice between Rotational Velocity Transducers

Choice between different rotational velocity transducers is influenced strongly by whether an analogue or digital form of output is required. Digital output instruments are now widely used and a choice has to be made among the variable reluctance transducer, devices using electronic light pulse-counting methods, and the stroboscope. The first two of these are used to measure angular speeds up to about 10,000 r.p.m. and the last one can measure speeds up to 25,000 r.p.m.

Probably the most common form of analogue output device used is the d.c. tachometer. This is a relatively simple device that measures speeds up to about 5000 r.p.m. with a maximum inaccuracy of $\pm 1\%$. Where better accuracy is required within a similar range of speed measurement, a.c. tachometers are used. The squirrel cage rotor type has an inaccuracy of only $\pm 0.25\%$, and drag-cup rotor types can have inaccuracies as low as $\pm 0.05\%$.

The drag-cup tachometer also has an analogue output, but has a typical inaccuracy of $\pm 5\%$. However, it is inexpensive and therefore suitable for use in vehicle speedometers where an inaccuracy of $\pm 5\%$ is normally acceptable.

### 20.3.9 Calibration of Rotational Velocity Transducers

The main device used as a calibration standard for rotational velocity transducers is a stroboscope. Provided the flash frequency of the reference stroboscope is calibrated properly, it is possible to provide velocity measurements where the inaccuracy is less than $\pm 0.1\%$.

## 20.4 Rotational Acceleration

Rotational accelerometers work on very similar principles to translational motion accelerometers. They consist of a rotatable mass mounted inside a housing attached to the accelerating, rotating body. Rotation of the mass is opposed by a torsional spring and damping. Any acceleration of the housing causes a torque, $J\ddot{\theta}$, on the mass. This torque is opposed by a backward torque due to the torsional spring and in equilibrium:

$$J\ddot{\theta} = K\theta \text{ and hence: } \ddot{\theta} = k\theta/J.$$

A damper is usually included in the system to avoid undying oscillations in the instrument. This adds an additional backward torque, $B\dot{\theta}$, to the system and the equation of motion becomes

$$J\ddot{\theta} = B\dot{\theta} + K\theta.$$

Different manufacturers produce accelerometers that measure the angular displacement of the mass within the accelerometer in different ways. However, it should be noted that the number of manufacturers producing rotational accelerometers is substantially less than the number manufacturing translational motion accelerometers because the requirement to measure rotational acceleration occurs much less frequently than requirements to measure translational acceleration.

### 20.4.1 Calibration of Rotational Accelerometers

This is normally carried out by comparison with a reference standard rotational accelerometer. The task is usually delegated to specialist calibration companies or accelerometer manufacturers because of the relatively small number of applications for rotational accelerometers and the corresponding shortage of personnel having the necessary calibration skills.

## 20.5 Summary

Having discussed sensors for measuring translational motion in the previous chapter, this chapter has been concerned with the measurement of the three aspects—displacement, velocity, and acceleration—of rotational motion. Starting with sensors for measuring rotation displacement, we first discussed circular and helical potentiometers. Next we considered the merits of the rotational differential transformer, incremental shaft encoder, coded disc shaft encoder, resolver, synchro, induction potentiometer, rotary inductosyn, and both free and rate-integrating gyroscopes.

Moving on to the measurement of rotational velocity, we first explored the various forms of digital tachometers available. Discussion then moved on to stroboscopic methods, followed by a review of analogue tachometers, which we noted were less accurate than digital tachometers but still in fairly widespread use. Next, we covered the two forms of gyroscopes that measure rotational velocity—rate gyro and fiber-optic gyro. Finally, we found that a velocity measurement could be obtained by differentiating an angular displacement measurement or by integrating an acceleration measurement. However, we noted that while the latter is acceptable because the process of integration attenuates any measurement noise, the differentiation technique is not used unless done within an intelligent instrument that can deal with the noise amplification that is inherent when measurements are obtained via differentiation.

Our final subject in the chapter was the measurement of rotational acceleration. We noted that rotational accelerometers worked on very similar principles to their translational motion counterparts, while observing that the requirement of measure rotational acceleration did not commonly arise.

## 20.6 Problems

20.1. Using simple sketches to support your explanation, explain the mode of operation and characteristics of the following devices: circular potentiometer, helical potentiometer, and rotary differential transformer.

20.2. Sketch an incremental shaft encoder. Explain what it measures and how it works. What special design features can be implemented to increase the measurement resolution of a disc of a given diameter?

20.3. What is a coded disc shaft encoder? How does its output differ from that of an incremental shaft encoder? What are the main types of coded disc shaft encoders?

20.4. Discuss the mode of operation of an optical coded disc shaft encoder, illustrating your discussion by means of a sketch.

20.5. What is the main consequence of any misalignment of the windows in an optical coded disc shaft encoder? Describe two ways in which the problem caused by window misalignment can be overcome.

20.6. Explain what a resolver is in the context of rotational position measurement. Discuss the two alternative forms of resolvers that exist.

20.7. How does a synchro work? Illustrate your explanation with a simple sketch.

20.8. What is a gyroscope? Discuss the characteristics and mode of operation of three kinds of gyroscopes that can measure angular position.

20.9. Explain the mode of construction and characteristics of each of the following: digital tachometer, optical tachometer, induction tachometer, and Hall-effect tachometer.

20.10. Discuss the characteristics of stroboscopic methods of measuring rotational velocity.

20.11. What are the main types of analogue tachometers available? Discuss the main characteristics of each.

20.12. How does a rate gyroscope work? What is its main application?

APPENDIX 1

# *Imperial–Metric–SI Conversion Tables*

## Length

SI units: mm, m, km
Imperial units: in, ft, mile

| | mm | m | km | in | ft | mile |
|---|---|---|---|---|---|---|
| mm | 1 | $10^{-3}$ | $10^{-6}$ | 0.039 3701 | $3.281 \times 10^{-3}$ | – |
| m | 1000 | 1 | $10^{-3}$ | 39.3701 | 3.280 84 | $6.214 \times 10^{-4}$ |
| km | $10^6$ | $10^3$ | 1 | 39 370.1 | 3280.84 | 0.621 371 |
| in | 25.4 | 0.0254 | – | 1 | 0.083 333 | – |
| ft | 304.8 | 0.3048 | $3.048 \times 10^{-4}$ | 12 | 1 | $1.894 \times 10^{-4}$ |
| mile | – | 1609.34 | 1.609 34 | 63 360 | 5280 | 1 |

## Area

SI units: $mm^2$, $m^2$, $km^2$
Imperial units: $in^2$, $ft^2$, $mile^2$

| | $mm^2$ | $m^2$ | $km^2$ | $in^2$ | $ft^2$ | $mile^2$ |
|---|---|---|---|---|---|---|
| $mm^2$ | 1 | $10^{-6}$ | – | $1.550 \times 10^{-3}$ | $1.076 \times 10^{-5}$ | – |
| $m^2$ | $10^6$ | 1 | $10^{-6}$ | 1550 | 10.764 | – |
| $km^2$ | – | $10^6$ | 1 | – | $1076 \times 10^7$ | 0.3861 |
| $in^2$ | 645.16 | $6.452 \times 10^{-4}$ | – | 1 | $6.944 \times 10^{-3}$ | – |
| $ft^2$ | 92 903 | 0.092 90 | – | 144 | 1 | – |
| $mile^2$ | – | $2.590 \times 10^6$ | 2.590 | – | $2.788 \times 10^7$ | 1 |

## Second Moment of Area

SI units: $mm^4$, $m^4$
Imperial units: $in^4$, $ft^4$

| | $mm^4$ | $m^4$ | $in^4$ | $ft^4$ |
|---|---|---|---|---|
| $mm^4$ | 1 | $10^{-12}$ | $2.4025 \times 10^{-6}$ | $1.159 \times 10^{-10}$ |
| $m^4$ | $10^{12}$ | 1 | $2.4025 \times 10^{6}$ | 115.86 |
| $in^4$ | 416 231 | $4.1623 \times 10^{-7}$ | 1 | $4.8225 \times 10^{-5}$ |
| $ft^4$ | $8.631 \times 10^{9}$ | $8.631 \times 10^{-3}$ | 20 736 | 1 |

## Volume

SI units: $mm^3$, $m^3$
Metric units: ml, l
Imperial units: $in^3$, $ft^3$, UK gallon

| | $mm^3$ | ml | l | $m^3$ | $in^3$ | $ft^3$ | UK Gallon |
|---|---|---|---|---|---|---|---|
| $mm^3$ | 1 | $10^{-3}$ | $10^{-6}$ | $10^{-9}$ | $6.10 \times 10^{-5}$ | – | – |
| ml | $10^{3}$ | 1 | $10^{-3}$ | $10^{-6}$ | 0.061 024 | $3.53 \times 10^{-5}$ | $2.2 \times 10^{-4}$ |
| l | $10^{6}$ | $10^{3}$ | 1 | $10^{-3}$ | 61.024 | 0.035 32 | 0.22 |
| $m^3$ | $10^{9}$ | $10^{6}$ | $10^{3}$ | 1 | 61 024 | 35.31 | 220 |
| $in^3$ | 16 387 | 16.39 | 0.0164 | $1.64 \times 10^{-5}$ | 1 | $5.79 \times 10^{-4}$ | $3.61 \times 10^{-3}$ |
| $ft^3$ | – | $2.83 \times 10^{4}$ | 28.32 | 0.028 32 | 1728 | 1 | 6.229 |
| UK gallon | – | 4546 | 4.546 | $4.55 \times 10^{-3}$ | 277.4 | 0.1605 | 1 |

Note: Additional unit: 1 US gallon = 0.8327 UK gallon.

## Density

SI unit: $kg/m^3$
Metric unit: $g/cm^3$
Imperial units: $lb/ft^3$, $lb/in^3$

| | $kg/m^3$ | $g/cm^3$ | $lb/ft^3$ | $lb/in^3$ |
|---|---|---|---|---|
| $kg/m^3$ | 1 | $10^{-3}$ | 0.062 428 | $3.605 \times 10^{-5}$ |
| $g/cm^3$ | 1000 | 1 | 62.428 | 0.036 127 |
| $lb/ft^3$ | 16.019 | 0.016 019 | 1 | $5.787 \times 10^{-4}$ |
| $lb/in^3$ | 27 680 | 27.680 | 1728 | 1 |

# Mass

SI units: g, kg, t
Imperial units: lb, cwt, ton

| | g | kg | t | lb | cwt | ton |
|---|---|---|---|---|---|---|
| g | 1 | $10^{-3}$ | $10^{-6}$ | $2.205 \times 10^{-3}$ | $1.968 \times 10^{-5}$ | $9.842 \times 10^{-7}$ |
| kg | $10^3$ | 1 | $10^{-3}$ | 2.204 62 | 0.019 684 | $9.842 \times 10^{-4}$ |
| t | $10^6$ | $10^3$ | 1 | 2204.62 | 19.6841 | 0.984 207 |
| lb | 453.592 | 0.453 59 | $4.536 \times 10^{-4}$ | 1 | $8.929 \times 10^{-3}$ | $4.464 \times 10^{-4}$ |
| cwt | 50 802.3 | 50.8023 | 0.050 802 | 112 | 1 | 0.05 |
| ton | $1.016 \times 10^6$ | 1016.05 | 1.01605 | 2240 | 20 | 1 |

# Force

SI units: N, kN
Metric unit: $kg_f$
Imperial units: pdl (poundal), $lb_f$, UK $ton_f$

| | N | $kg_f$ | kN | pdl | $lb_f$ | UK $ton_f$ |
|---|---|---|---|---|---|---|
| N | 1 | 0.1020 | $10^{-3}$ | 7.233 | 0.2248 | $1.004 \times 10^{-4}$ |
| $kg_f$ | 9.807 | 1 | $9.807 \times 10^{-3}$ | 70.93 | 2.2046 | $9.842 \times 10^{-4}$ |
| kN | 1000 | 102.0 | 1 | 7233 | 224.8 | 0.1004 |
| pdl | 0.1383 | 0.0141 | $1.383 \times 10^{-4}$ | 1 | 0.0311 | $1.388 \times 10^{-5}$ |
| $lb_f$ | 4.448 | 0.4536 | $4.448 \times 10^{-3}$ | 32.174 | 1 | $4.464 \times 10^{-4}$ |
| UK $ton_f$ | 9964 | 1016 | 9.964 | 72 070 | 2240 | 1 |

Note: Additional unit: 1 dyne = $10^{-5}$N = $7.233 \times 10^{-5}$ pdl.

# Torque (Moment of Force)

SI unit: N m
Metric unit: $kg_f$ m
Imperial units: pdl ft, $lb_f$ ft

| | N m | $kg_f$ m | pdl ft | $lb_f$ ft |
|---|---|---|---|---|
| N m | 1 | 0.1020 | 23.73 | 0.7376 |
| $kg_f$ m | 9.807 | 1 | 232.7 | 7.233 |
| pdl ft | 0.042 14 | $4.297 \times 10^{-3}$ | 1 | 0.031 08 |
| $lb_f$ ft | 1.356 | 0.1383 | 32.17 | 1 |

## Inertia

SI unit: N $m^2$
Imperial unit: $lb_f$ $ft^2$

$$1\ lb_f\ ft^2 = 0.4132\ N\ m^2$$
$$1\ N\ m^2 = 2.420\ lb_f\ ft^2$$

## Pressure

SI units: mbar, bar, N/$m^2$ (pascal)
Imperial units: lb/$in^2$, in Hg, atm

| | mbar | bar | N/$m^2$ | lb/$in^2$ | in Hg | atm |
|---|---|---|---|---|---|---|
| mbar | 1 | $10^{-3}$ | 100 | 0.014 50 | 0.029 53 | $9.869 \times 10^{-4}$ |
| bar | 1000 | 1 | $10^5$ | 14.50 | 29.53 | 0.9869 |
| N/$m^2$ | 0.01 | $10^{-5}$ | 1 | $1.450 \times 10^{-4}$ | $2.953 \times 10^{-4}$ | $9.869 \times 10^{-6}$ |
| lb/$in^2$ | 68.95 | 0.068 95 | 6895 | 1 | 2.036 | 0.068 05 |
| in Hg | 33.86 | 0.033 86 | 3386 | 0.4912 | 1 | 0.033 42 |
| atm | 1013 | 1.013 | $1.013 \times 10^5$ | 14.70 | 29.92 | 1 |

## Additional Conversion Factors

1 inch water = 0.073 56 in Hg = 2.491 mbar
1 torr = 1.333 mbar
1 pascal = 1 N/$m^2$

## Energy, Work, Heat

SI unit: J
Metric units: $kg_f$ m, kW h
Imperial units: ft $lb_f$, cal, Btu

| | J | $kg_f$ m | kW h | ft $lb_f$ | cal | Btu |
|---|---|---|---|---|---|---|
| J | 1 | 0.1020 | $2.778 \times 10^{-7}$ | 0.7376 | 0.2388 | $9.478 \times 10^{-4}$ |
| $kg_f$ m | 9.8066 | 1 | $2.724 \times 10^{-6}$ | 7.233 | 2.342 | $9.294 \times 10^{-3}$ |
| kW h | $3.600 \times 10^6$ | 367 098 | 1 | $2.655 \times 10^6$ | 859 845 | 3412.1 |
| ft $lb_f$ | 1.3558 | 0.1383 | $3.766 \times 10^{-7}$ | 1 | 0.3238 | $1.285 \times 10^{-3}$ |
| cal | 4.1868 | 0.4270 | $1.163 \times 10^{-6}$ | 3.0880 | 1 | $3.968 \times 10^{-3}$ |
| Btu | 1055.1 | 107.59 | $2.931 \times 10^{-4}$ | 778.17 | 252.00 | 1 |

## Additional Conversion Factors

1 therm = $10^5$ Btu = $1.0551 \times 10^8$ J
1 thermie = $4.186 \times 10^6$ J
1 hp h = 0.7457 kW h = $2.6845 \times 10^6$ J
1 ft pdl = 0.042 14 J
1 erg = $10^{-7}$ J

## Power

SI units: W, kW
Imperial units: HP, ft $lb_f$/s

| | W | kW | HP | ft $lb_f$/s |
|---|---|---|---|---|
| W | 1 | $10^{-3}$ | $1.341 \times 10^{-3}$ | 0.735 64 |
| kW | $10^3$ | 1 | 1.341 02 | 735.64 |
| HP | 745.7 | 0.7457 | 1 | 548.57 |
| ft $lb_f$/s | 1.359 35 | $1.359 \times 10^{-3}$ | $1.823 \times 10^{-3}$ | 1 |

## Velocity

SI units: mm/s, m/s
Metric unit: km/h
Imperial units: ft/s, mile/h

| | mm/s | m/s | km/h | ft/s | mile/h |
|---|---|---|---|---|---|
| mm/s | 1 | $10^{-3}$ | $3.6 \times 10^{-3}$ | $3.281 \times 10^{-3}$ | $2.237 \times 10^{-3}$ |
| m/s | 1000 | 1 | 3.6 | 3.280 84 | 2.236 94 |
| km/h | 277.778 | 0.277 778 | 1 | 0.911 344 | 0.621 371 |
| ft/s | 304.8 | 0.3048 | 1.097 28 | 1 | 0.681 818 |
| mile/h | 447.04 | 0.447 04 | 1.609 344 | 1.466 67 | 1 |

## Acceleration

SI unit: $m/s^2$
Other metric unit: $cm/s^2$
Imperial unit: $ft/s^2$
Other unit: $g$

| | $m/s^2$ | $cm/s^2$ | $ft/s^2$ | $g$ |
|---|---|---|---|---|
| $m/s^2$ | 1 | 100 | 3.281 | 0.102 |
| $cm/s^2$ | 0.01 | 1 | 0.0328 | 0.001 02 |
| $ft/s^2$ | 0.3048 | 30.48 | 1 | 1 |
| $g$ | 9.81 | 981 | 32.2 | 1 |

## Mass Flow Rate

SI unit: g/s
Metric units: kg/h, tonne/d
Imperial units: lb/s, lb/h, ton/d

| | g/s | kg/h | tonne/d | lb/s | lb/h | ton/d |
|---|---|---|---|---|---|---|
| g/s | 1 | 3.6 | 0.086 40 | $2.205 \times 10^{-3}$ | 7.937 | 0.085 03 |
| kg/h | 0.2778 | 1 | 0.024 00 | $6.124 \times 10^{-4}$ | 2.205 | 0.023 62 |
| tonne/d | 11.57 | 41.67 | 1 | 0.025 51 | 91.86 | 0.9842 |
| lb/s | 453.6 | 1633 | 39.19 | 1 | 3600 | 38.57 |
| lb/h | 0.1260 | 0.4536 | 0.010 89 | $2.788 \times 10^{-4}$ | 1 | 0.010 71 |
| ton/d | 11.76 | 42.34 | 1.016 | 0.025 93 | 93.33 | 1 |

## Volume Flow Rate

SI unit: $m^3/s$
Metric units: l/h, ml/s
Imperial units: gal/h, $ft^3/s$, $ft^3/h$

| | l/h | ml/s | $m^3/s$ | gal/h | $ft^3/s$ | $ft^3/h$ |
|---|---|---|---|---|---|---|
| l/h | 1 | 0.2778 | $2.778 \times 10^{-7}$ | 0.2200 | $9.810 \times 10^{-6}$ | 0.035 316 |
| ml/s | 3.6 | 1 | $10^{-6}$ | 0.7919 | $3.532 \times 10^{-5}$ | 0.127 14 |
| $m^3/s$ | $3.6 \times 10^6$ | $10^6$ | 1 | $7.919 \times 10^5$ | 35.31 | $1.271 \times 10^5$ |
| gal/h | 4.546 | 1.263 | $1.263 \times 10^{-6}$ | 1 | $4.460 \times 10^{-5}$ | 0.160 56 |
| $ft^3/s$ | $1.019 \times 10^5$ | $2.832 \times 10^4$ | 0.028 32 | $2.242 \times 10^4$ | 1 | 3600 |
| $ft^3/h$ | 28.316 | 7.8653 | $7.865 \times 10^{-6}$ | 6.2282 | $2.778 \times 10^{-4}$ | 1 |

## Specific Energy (Heat per Unit Volume)

SI units: J/m$^3$, kJ/m$^3$, MJ/m$^3$
Imperial units: kcal/m$^3$, Btu/ft$^3$, therm/UK gal

| | J/m$^3$ | kJ/m$^3$ | MJ/m$^3$ | kcal/m$^3$ | Btu/ft$^3$ | therm/UK gal |
|---|---|---|---|---|---|---|
| J/m$^3$ | 1 | $10^{-3}$ | $10^{-6}$ | $1.388 \times 10^{-4}$ | $2.684 \times 10^{-5}$ | – |
| kJ/m$^3$ | 1000 | 1 | $10^{-3}$ | 0.2388 | 0.02684 | – |
| MJ/m$^3$ | $10^6$ | 1000 | 1 | 238.8 | 26.84 | $4.309 \times 10^{-5}$ |
| kcal/m$^3$ | 4187 | 4.187 | $4.187 \times 10^{-3}$ | 1 | 0.1124 | $1.804 \times 10^{-7}$ |
| Btu/ft$^3$ | $3.726 \times 10^4$ | 37.26 | 0.03726 | 8.899 | 1 | $1.605 \times 10^{-6}$ |
| therm/UK gal | – | – | $2.321 \times 10^4$ | $5.543 \times 10^6$ | $6.229 \times 10^5$ | 1 |

## Dynamic Viscosity

SI unit: N s/m$^2$
Metric unit: cP (centipoise), P (poise) [1 P = 100 g/m s]
Imperial unit: lb$_m$/ft h

| | lb$_m$/ft h | P | cP | N s/m$^2$ |
|---|---|---|---|---|
| lb$_m$/ft h | 1 | $4.133 \times 10^{-3}$ | 0.4134 | $4.134 \times 10^{-4}$ |
| P | 241.9 | 1 | 100 | 0.1 |
| cP | 2.419 | 0.01 | 1 | $10^{-3}$ |
| N s/m$^2$ | 2419 | 10 | 1000 | 1 |

Note: Additional unit: 1 pascal second = 1 N s/m$^2$.

## Kinematic Viscosity

SI unit: m$^2$/s
Metric unit: cSt (centistokes), St (stokes)
Imperial unit: ft$^2$/s

| | ft$^2$/s | m$^2$/s | cSt | St |
|---|---|---|---|---|
| ft$^2$/s | 1 | 0.0929 | $9.29 \times 10^4$ | 929 |
| m$^2$/s | 10.764 | 1 | $10^6$ | $10^4$ |
| cSt | $1.0764 \times 10^{-5}$ | $10^{-6}$ | 1 | 0.01 |
| St | $1.0764 \times 10^{-3}$ | $10^{-4}$ | 100 | 1 |

APPENDIX 2

# Thévenin's Theorem

Thévenin's theorem is extremely useful in the analysis of complex electrical circuits. It states that any network that has two accessible terminals, A and B, can be replaced, as far as its external behavior is concerned, by a single e.m.f. acting in series with a single resistance between A and B. The single equivalent e.m.f. is that e.m.f. measured across A and B when the circuit external to the network is disconnected. The single equivalent resistance is the resistance of the network when all current and voltage sources within it are reduced to zero. To calculate this internal resistance of the network, all current sources within it are treated as open circuits and all voltage sources as short circuits. The proof of Thévenin's theorem can be found in Skilling (1967).

Figure A2.1 shows part of a network consisting of a voltage source and four resistances. As far as its behavior external to terminals A and B is concerned, this can be regarded as a single voltage source, $V_t$, and a single resistance, $R_t$. Applying Thévenin's theorem, $R_t$ is found first of all by treating $V_1$ as a short circuit, as shown in Figure A2.2. This is simply two resistances, $R_1$ and $(R_2 + R_4 + R_5)$, in parallel. The equivalent resistance, $R_t$, is thus given by

$$R_t = \frac{R_1(R_2 + R_4 + R_5)}{R_1 + R_2 + R_4 + R_5},$$

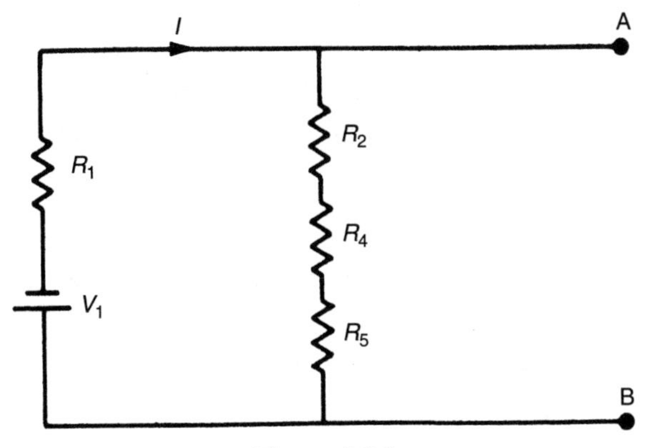


Figure A2.1

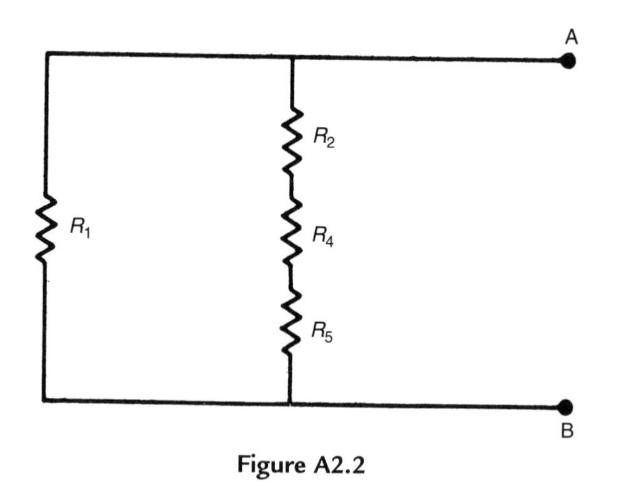


**Figure A2.2**

where $V_t$ is the voltage drop across AB. To calculate this, it is necessary to carry out an intermediate step of working out the current flowing, $I$. Referring to Figure A2.1, this is given by

$$I = \frac{V_1}{R_1 + R_2 + R_4 + R_5}.$$

Now, $V_t$ can be calculated from

$$V_t = I(R_2 + R_4 + R_5)$$
$$= \frac{V_1(R_2 + R_4 + R_5)}{R_1 + R_2 + R_4 + R_5}.$$

The network of Figure A2.1 has thus been reduced to the simpler network shown in Figure A2.3.

Let us now proceed to the typical network problem of calculating the current flowing in resistor $R_3$ of Figure A2.4. $R_3$ can be regarded as an external circuit or load on the rest of the

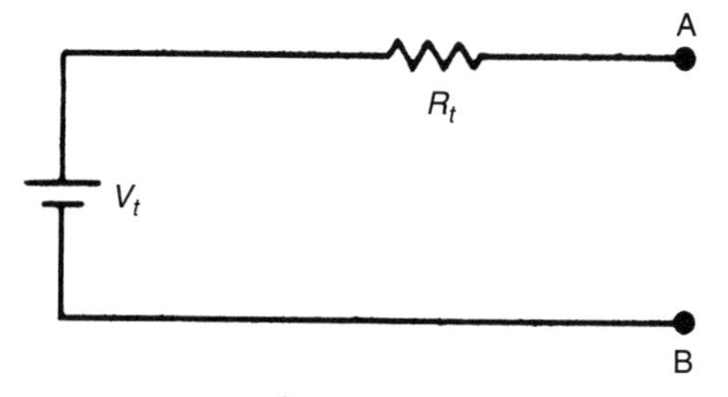


**Figure A2.3**

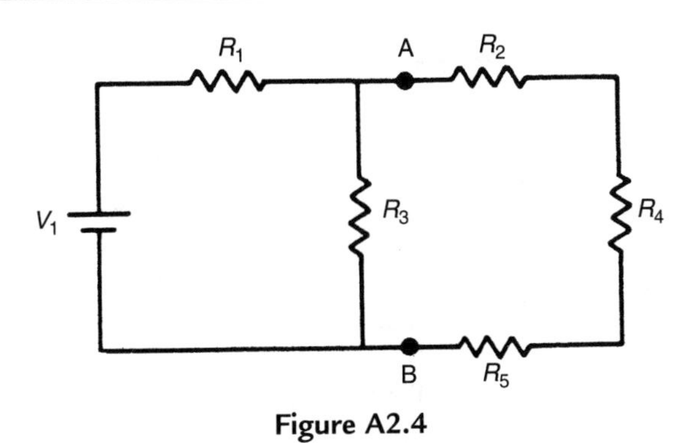


Figure A2.4

network consisting of $V_1, R_1, R_2, R_4$, and $R_5$, as shown in Figure A2.5. This network of $V_1, R_1$, $R_2$, $R_4$, and $R_5$ is that shown in Figure A2.6. This can be rearranged to the network shown in Figure A2.1, which is equivalent to the single voltage source and resistance, $V_t$ and $R_t$, calculated earlier. The whole circuit is then equivalent to that shown in Figure A2.7, and the current flowing through $R_3$ can be written as

$$I_{AB} = \frac{V_t}{R_t + R_3}.$$

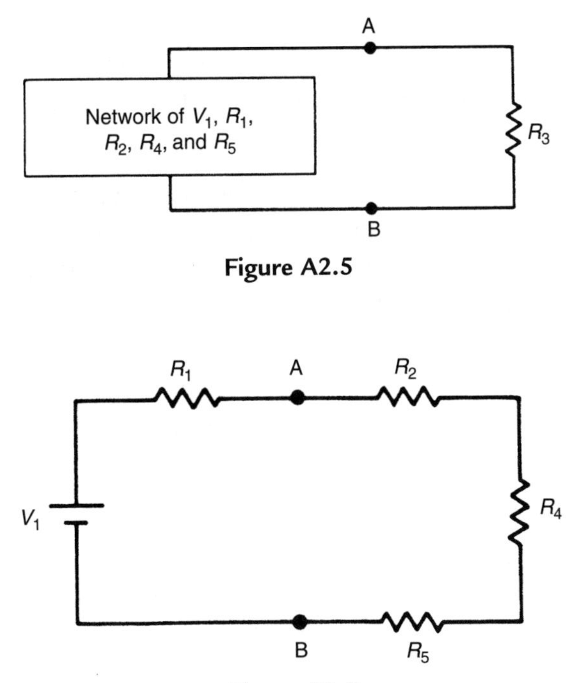


Figure A2.5

Figure A2.6

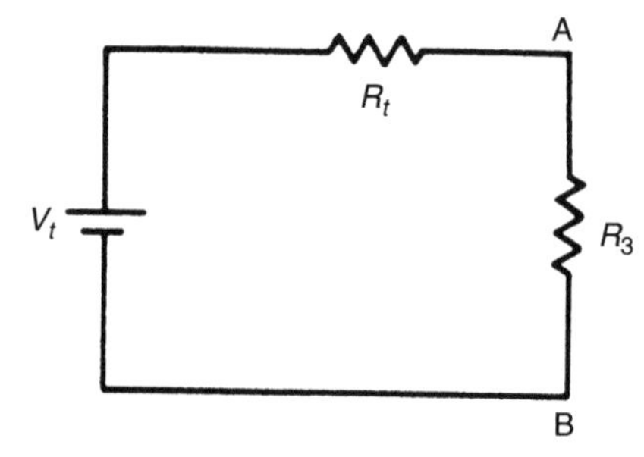


Figure A2.7

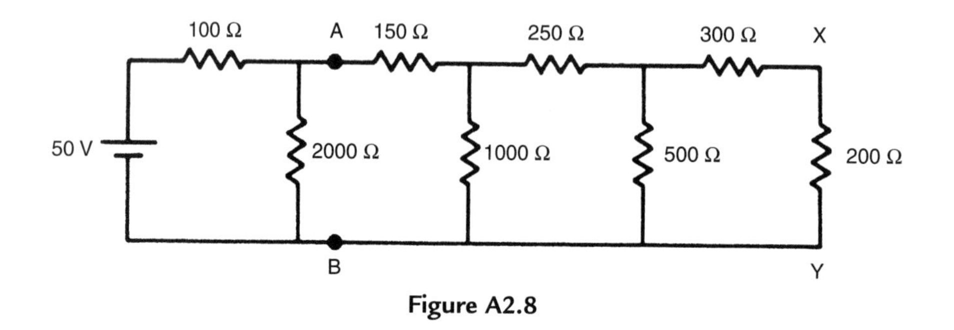


Figure A2.8

Thévenin's theorem can be applied successively to solve ladder networks of the form shown in Figure A2.8. Suppose in this network that it is required to calculate the current flowing in branch XY.

The first step is to imagine two terminals, A and B, in the circuit and regard the network to the right of AB as a load on the circuit to the left of AB. The circuit to the left of AB can be reduced to a single equivalent voltage source, $E_{AB}$, and resistance, $R_{AB}$, by Thévenin's theorem. If the 50-V source is replaced by its zero internal resistance (i.e., by a short circuit), then $R_{AB}$ is given by

$$\frac{1}{R_{AB}} = \frac{1}{100} + \frac{1}{2000} = \frac{2000 + 100}{200,000}.$$

Hence,

$$R_{AB} = 95.24\ \Omega.$$

When AB is open circuit, the current flowing round the loop to the left of AB is given by

$$I = \frac{50}{100 + 2000}.$$

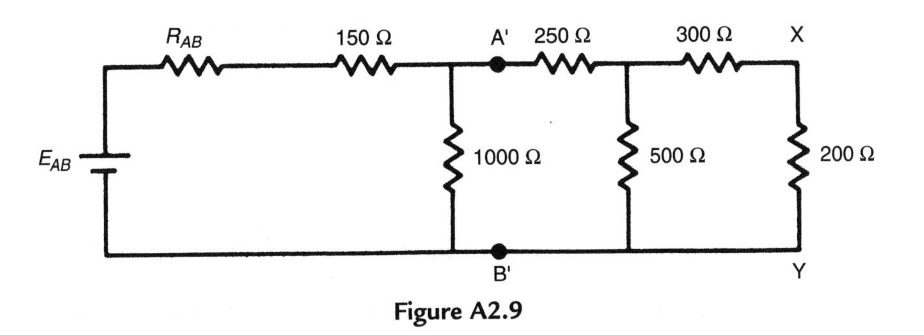


**Figure A2.9**

Hence, $E_{AB}$, the open circuit voltage across AB, is given by

$$E_{AB} = I \times 2000 = 47.62 \text{ volts.}$$

We can now replace the circuit shown in Figure A2.8 by the simpler equivalent circuit shown in Figure A2.9.

The next stage is to apply an identical procedure to find an equivalent circuit consisting of voltage source $E_{A'B'}$ and resistance $R_{A'B'}$ for the network to the left of points A′ and B′ in Figure A2.9:

$$\frac{1}{R_{A'B'}} = \frac{1}{R_{AB} + 150} + \frac{1}{1000} = \frac{1}{245.24} + \frac{1}{1000} = \frac{1245.24}{245,240}.$$

Hence,

$$R_{A'B'} = 196.94\ \Omega$$

$$E_{A'B'} = \frac{1000}{R_{AB} + 150 + 1000} \times E_{AB} = 38.24\ volts.$$

The circuit can now be represented in the yet simpler form shown in Figure A2.10. Proceeding as before to find an equivalent voltage source and resistance, $E_{A''B''}$ and $R_{A''B''}$, for the circuit to the left of A″ and B″ in Figure A2.10:

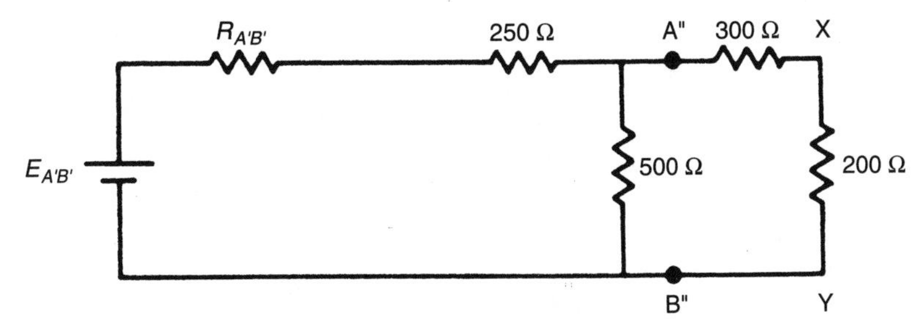


**Figure A2.10**

$$\frac{1}{R_{A''B''}} = \frac{1}{R_{A'B'} + 250} + \frac{1}{500} = \frac{500 + 446.94}{223,470}.$$

Hence,

$$R_{A''B''} = 235.99\ \Omega$$

$$E_{A''B''} = \frac{500}{R_{A'B'} + 250 + 500} E_{A'B'} = 20.19\ volts.$$

The circuit has now been reduced to the form shown in Figure A2.11, where the current through branch XY can be calculated simply as

$$I_{XY} = \frac{E_{A''B''}}{R_{A''B''} + 300 + 200} = \frac{20.19}{735.99} = 27.43\ \text{mA}.$$

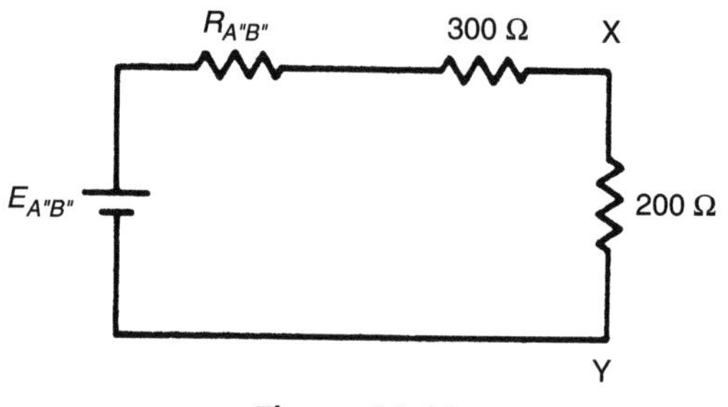


**Figure A2.11**

## Reference

Skilling, H.H. (1967). *Electrical engineering circuits.* Wiley: New York.

APPENDIX 3

# Thermocouple Tables

Type E: chromel–constantan
Type J: iron–constantan
Type K: chromel–alumel
Type N: nicrosil–nisil
Type S: platinum/10% rhodium–platinum
Type T: copper–constantan

| Temp. (°C) | Type E | Type J | Type K | Type N | Type S | Type T |
|---|---|---|---|---|---|---|
| −270 | −9.834 | | −6.458 | −4.345 | | |
| −260 | −9.795 | | −6.441 | −4.336 | | |
| −250 | −9.719 | | −6.404 | −4.313 | | |
| −240 | −9.604 | | −6.344 | −4.277 | | −6.105 |
| −230 | −9.456 | | −6.262 | −4.227 | | −6.003 |
| −220 | −9.274 | | −6.158 | −4.162 | | −5.891 |
| −210 | −9.063 | −8.096 | −6.035 | −4.083 | | −5.753 |
| −200 | −8.824 | −7.890 | −5.891 | −3.990 | | −5.603 |
| −190 | −8.561 | −7.659 | −5.730 | −3.884 | | −5.438 |
| −180 | −8.273 | −7.402 | −5.550 | −3.766 | | −5.261 |
| −170 | −7.963 | −7.122 | −5.354 | −3.634 | | −5.070 |
| −160 | −7.631 | −6.821 | −5.141 | −3.491 | | −4.865 |
| −150 | −7.279 | −6.499 | −4.912 | −3.336 | | −4.648 |
| −140 | −6.907 | −6.159 | −4.669 | −3.170 | | −4.419 |
| −130 | −6.516 | −5.801 | −4.410 | −2.994 | | −4.177 |
| −120 | −6.107 | −5.426 | −4.138 | −2.807 | | −3.923 |
| −110 | −5.680 | −5.036 | −3.852 | −2.612 | | −3.656 |
| −100 | −5.237 | −4.632 | −3.553 | −2.407 | | −3.378 |
| −90 | −4.777 | −4.215 | −3.242 | −2.193 | | −3.089 |
| −80 | −4.301 | −3.785 | −2.920 | −1.972 | | −2.788 |
| −70 | −3.811 | −3.344 | −2.586 | −1.744 | | −2.475 |
| −60 | −3.306 | −2.892 | −2.243 | −1.509 | | −2.152 |
| −50 | −2.787 | −2.431 | −1.889 | −1.268 | −0.236 | −1.819 |
| −40 | −2.254 | −1.960 | −1.527 | −1.023 | −0.194 | −1.475 |
| −30 | −1.709 | −1.481 | −1.156 | −0.772 | −0.150 | −1.121 |
| −20 | −1.151 | −0.995 | −0.777 | −0.518 | −0.103 | −0.757 |
| −10 | −0.581 | −0.501 | −0.392 | −0.260 | −0.053 | −0.383 |

| Temp. (°C) | Type E | Type J | Type K | Type N | Type S | Type T |
|---|---|---|---|---|---|---|
| 0 | 0.000 | 0.000 | 0.000 | 0.000 | 0.000 | 0.000 |
| 10 | 0.591 | 0.507 | 0.397 | 0.261 | 0.055 | 0.391 |
| 20 | 1.192 | 1.019 | 0.798 | 0.525 | 0.113 | 0.789 |
| 30 | 1.801 | 1.536 | 1.203 | 0.793 | 0.173 | 1.196 |
| 40 | 2.419 | 2.058 | 1.611 | 1.064 | 0.235 | 1.611 |
| 50 | 3.047 | 2.585 | 2.022 | 1.339 | 0.299 | 2.035 |
| 60 | 3.683 | 3.115 | 2.436 | 1.619 | 0.365 | 2.467 |
| 70 | 4.329 | 3.649 | 2.850 | 1.902 | 0.432 | 2.908 |
| 80 | 4.983 | 4.186 | 3.266 | 2.188 | 0.502 | 3.357 |
| 90 | 5.646 | 4.725 | 3.681 | 2.479 | 0.573 | 3.813 |
| 100 | 6.317 | 5.268 | 4.095 | 2.774 | 0.645 | 4.277 |
| 110 | 6.996 | 5.812 | 4.508 | 3.072 | 0.719 | 4.749 |
| 120 | 7.683 | 6.359 | 4.919 | 3.374 | 0.795 | 5.227 |
| 130 | 8.377 | 6.907 | 5.327 | 3.679 | 0.872 | 5.712 |
| 140 | 9.078 | 7.457 | 5.733 | 3.988 | 0.950 | 6.204 |
| 150 | 9.787 | 8.008 | 6.137 | 4.301 | 1.029 | 6.702 |
| 160 | 10.501 | 8.560 | 6.539 | 4.617 | 1.109 | 7.207 |
| 170 | 11.222 | 9.113 | 6.939 | 4.936 | 1.190 | 7.718 |
| 180 | 11.949 | 9.667 | 7.338 | 5.258 | 1.273 | 8.235 |
| 190 | 12.681 | 10.222 | 7.737 | 5.584 | 1.356 | 8.757 |
| 200 | 13.419 | 10.777 | 8.137 | 5.912 | 1.440 | 9.286 |
| 210 | 14.161 | 11.332 | 8.537 | 6.243 | 1.525 | 9.820 |
| 220 | 14.909 | 11.887 | 8.938 | 6.577 | 1.611 | 10.360 |
| 230 | 15.661 | 12.442 | 9.341 | 6.914 | 1.698 | 10.905 |
| 240 | 16.417 | 12.998 | 9.745 | 7.254 | 1.785 | 11.456 |
| 250 | 17.178 | 13.553 | 10.151 | 7.596 | 1.873 | 12.011 |
| 260 | 17.942 | 14.108 | 10.560 | 7.940 | 1.962 | 12.572 |
| 270 | 18.710 | 14.663 | 10.969 | 8.287 | 2.051 | 13.137 |
| 280 | 19.481 | 15.217 | 11.381 | 8.636 | 2.141 | 13.707 |
| 290 | 20.256 | 15.771 | 11.793 | 8.987 | 2.232 | 14.281 |
| 300 | 21.033 | 16.325 | 12.207 | 9.340 | 2.323 | 14.860 |
| 310 | 21.814 | 16.879 | 12.623 | 9.695 | 2.414 | 15.443 |
| 320 | 22.597 | 17.432 | 13.039 | 10.053 | 2.506 | 16.030 |
| 330 | 23.383 | 17.984 | 13.456 | 10.412 | 2.599 | 16.621 |
| 340 | 24.171 | 18.537 | 13.874 | 10.772 | 2.692 | 17.217 |
| 350 | 24.961 | 19.089 | 14.292 | 11.135 | 2.786 | 17.816 |
| 360 | 25.754 | 19.640 | 14.712 | 11.499 | 2.880 | 18.420 |
| 370 | 26.549 | 20.192 | 15.132 | 11.865 | 2.974 | 19.027 |
| 380 | 27.345 | 20.743 | 15.552 | 12.233 | 3.069 | 19.638 |
| 390 | 28.143 | 21.295 | 15.974 | 12.602 | 3.164 | 20.252 |
| 400 | 28.943 | 21.846 | 16.395 | 12.972 | 3.260 | 20.869 |
| 410 | 29.744 | 22.397 | 16.818 | 13.344 | 3.356 | |
| 420 | 30.546 | 22.949 | 17.241 | 13.717 | 3.452 | |
| 430 | 31.350 | 23.501 | 17.664 | 14.091 | 3.549 | |
| 440 | 32.155 | 24.054 | 18.088 | 14.467 | 3.645 | |
| 450 | 32.960 | 24.607 | 18.513 | 14.844 | 3.743 | |
| 460 | 33.767 | 25.161 | 18.938 | 15.222 | 3.840 | |
| 470 | 34.574 | 25.716 | 19.363 | 15.601 | 3.938 | |

| *Temp. (°C)* | *Type E* | *Type J* | *Type K* | *Type N* | *Type S* | *Type T* |
|---|---|---|---|---|---|---|
| 480 | 35.382 | 26.272 | 19.788 | 15.981 | 4.036 | |
| 490 | 36.190 | 26.829 | 20.214 | 16.362 | 4.135 | |
| 500 | 36.999 | 27.388 | 20.640 | 16.744 | 4.234 | |
| 510 | 37.808 | 27.949 | 21.066 | 17.127 | 4.333 | |
| 520 | 38.617 | 28.511 | 21.493 | 17.511 | 4.432 | |
| 530 | 39.426 | 29.075 | 21.919 | 17.896 | 4.532 | |
| 540 | 40.236 | 29.642 | 22.346 | 18.282 | 4.632 | |
| 550 | 41.045 | 30.210 | 22.772 | 18.668 | 4.732 | |
| 560 | 41.853 | 30.782 | 23.198 | 19.055 | 4.832 | |
| 570 | 42.662 | 31.356 | 23.624 | 19.443 | 4.933 | |
| 580 | 43.470 | 31.933 | 24.050 | 19.831 | 5.034 | |
| 590 | 44.278 | 32.513 | 24.476 | 20.220 | 5.136 | |
| 600 | 45.085 | 33.096 | 24.902 | 20.609 | 5.237 | |
| 610 | 45.891 | 33.683 | 25.327 | 20.999 | 5.339 | |
| 620 | 46.697 | 34.273 | 25.751 | 21.390 | 5.442 | |
| 630 | 47.502 | 34.867 | 26.176 | 21.781 | 5.544 | |
| 640 | 48.306 | 35.464 | 26.599 | 22.172 | 5.648 | |
| 650 | 49.109 | 36.066 | 27.022 | 22.564 | 5.751 | |
| 660 | 49.911 | 36.671 | 27.445 | 22.956 | 5.855 | |
| 670 | 50.713 | 37.280 | 27.867 | 23.348 | 5.960 | |
| 680 | 51.513 | 37.893 | 28.288 | 23.740 | 6.064 | |
| 690 | 52.312 | 38.510 | 28.709 | 24.133 | 6.169 | |
| 700 | 53.110 | 39.130 | 29.128 | 24.526 | 6.274 | |
| 710 | 53.907 | 39.754 | 29.547 | 24.919 | 6.380 | |
| 720 | 54.703 | 40.382 | 29.965 | 25.312 | 6.486 | |
| 730 | 55.498 | 41.013 | 30.383 | 25.705 | 6.592 | |
| 740 | 56.291 | 41.647 | 30.799 | 26.098 | 6.699 | |
| 750 | 57.083 | 42.283 | 31.214 | 26.491 | 6.805 | |
| 760 | 57.873 | 42.922 | 31.629 | 26.885 | 6.913 | |
| 770 | 58.663 | 43.563 | 32.042 | 27.278 | 7.020 | |
| 780 | 59.451 | 44.207 | 32.455 | 27.671 | 7.128 | |
| 790 | 60.237 | 44.852 | 32.866 | 28.063 | 7.236 | |
| 800 | 61.022 | 45.498 | 33.277 | 28.456 | 7.345 | |
| 810 | 61.806 | 46.144 | 33.686 | 28.849 | 7.454 | |
| 820 | 62.588 | 46.790 | 34.095 | 29.241 | 7.563 | |
| 830 | 63.368 | 47.434 | 34.502 | 29.633 | 7.672 | |
| 840 | 64.147 | 48.076 | 34.908 | 30.025 | 7.782 | |
| 850 | 64.924 | 48.717 | 35.314 | 30.417 | 7.892 | |
| 860 | 65.700 | 49.354 | 35.718 | 30.808 | 8.003 | |
| 870 | 66.473 | 49.989 | 36.121 | 31.199 | 8.114 | |
| 880 | 67.245 | 50.621 | 36.524 | 31.590 | 8.225 | |
| 890 | 68.015 | 51.249 | 36.925 | 31.980 | 8.336 | |
| 900 | 68.783 | 51.875 | 37.325 | 32.370 | 8.448 | |
| 910 | 69.549 | 52.496 | 37.724 | 32.760 | 8.560 | |
| 920 | 70.313 | 53.115 | 38.122 | 33.149 | 8.673 | |
| 930 | 71.075 | 53.729 | 38.519 | 33.538 | 8.786 | |
| 940 | 71.835 | 54.341 | 38.915 | 33.926 | 8.899 | |
| 950 | 72.593 | 54.949 | 39.310 | 34.315 | 9.012 | |

| Temp. (°C) | Type E | Type J | Type K | Type N | Type S | Type T |
|---|---|---|---|---|---|---|
| 960 | 73.350 | 55.553 | 39.703 | 34.702 | 9.126 | |
| 970 | 74.104 | 56.154 | 40.096 | 35.089 | 9.240 | |
| 980 | 74.857 | 56.753 | 40.488 | 35.476 | 9.355 | |
| 990 | 75.608 | 57.349 | 40.879 | 35.862 | 9.470 | |
| 1000 | 76.357 | 57.942 | 41.269 | 36.248 | 9.585 | |
| 1010 | | 58.533 | 41.657 | 36.633 | 9.700 | |
| 1020 | | 59.121 | 42.045 | 37.018 | 9.816 | |
| 1030 | | 59.708 | 42.432 | 37.402 | 9.932 | |
| 1040 | | 60.293 | 42.817 | 37.786 | 10.048 | |
| 1050 | | 60.877 | 43.202 | 38.169 | 10.165 | |
| 1060 | | 61.458 | 43.585 | 38.552 | 10.282 | |
| 1070 | | 62.040 | 43.968 | 38.934 | 10.400 | |
| 1080 | | 62.619 | 44.349 | 39.315 | 10.517 | |
| 1090 | | 63.199 | 44.729 | 39.696 | 10.635 | |
| 1100 | | 63.777 | 45.108 | 40.076 | 10.754 | |
| 1110 | | 64.355 | 45.486 | 40.456 | 10.872 | |
| 1120 | | 64.933 | 45.863 | 40.835 | 10.991 | |
| 1130 | | 65.510 | 46.238 | 41.213 | 11.110 | |
| 1140 | | 66.087 | 46.612 | 41.590 | 11.229 | |
| 1150 | | 66.664 | 46.985 | 41.966 | 11.348 | |
| 1160 | | 67.240 | 47.356 | 42.342 | 11.467 | |
| 1170 | | 67.815 | 47.726 | 42.717 | 11.587 | |
| 1180 | | 68.389 | 48.095 | 43.091 | 11.707 | |
| 1190 | | 68.963 | 48.462 | 43.464 | 11.827 | |
| 1200 | | 69.536 | 48.828 | 43.836 | 11.947 | |
| 1210 | | | 49.192 | 44.207 | 12.067 | |
| 1220 | | | 49.555 | 44.577 | 12.188 | |
| 1230 | | | 49.916 | 44.947 | 12.308 | |
| 1240 | | | 50.276 | 45.315 | 12.429 | |
| 1250 | | | 50.633 | 45.682 | 12.550 | |
| 1260 | | | 50.990 | 46.048 | 12.671 | |
| 1270 | | | 51.344 | 46.413 | 12.792 | |
| 1280 | | | 51.697 | 46.777 | 12.913 | |
| 1290 | | | 52.049 | 47.140 | 13.034 | |
| 1300 | | | 52.398 | 47.502 | 13.155 | |
| 1310 | | | 52.747 | | 13.276 | |
| 1320 | | | 53.093 | | 13.397 | |
| 1330 | | | 53.438 | | 13.519 | |
| 1340 | | | 53.782 | | 13.640 | |
| 1350 | | | 54.125 | | 13.761 | |
| 1360 | | | 54.467 | | 13.883 | |
| 1370 | | | 54.807 | | 14.004 | |
| 1380 | | | | | 14.125 | |
| 1390 | | | | | 14.247 | |
| 1400 | | | | | 14.368 | |
| 1410 | | | | | 14.489 | |
| 1420 | | | | | 14.610 | |
| 1430 | | | | | 14.731 | |

| *Temp. (°C)* | *Type E* | *Type J* | *Type K* | *Type N* | *Type S* | *Type T* |
|---|---|---|---|---|---|---|
| 1440 | | | | | 14.852 | |
| 1450 | | | | | 14.973 | |
| 1460 | | | | | 15.094 | |
| 1470 | | | | | 15.215 | |
| 1480 | | | | | 15.336 | |
| 1490 | | | | | 15.456 | |
| 1500 | | | | | 15.576 | |
| 1510 | | | | | 15.697 | |
| 1520 | | | | | 15.817 | |
| 1530 | | | | | 15.937 | |
| 1540 | | | | | 16.057 | |
| 1550 | | | | | 16.176 | |
| 1560 | | | | | 16.296 | |
| 1570 | | | | | 16.415 | |
| 1580 | | | | | 16.534 | |
| 1590 | | | | | 16.653 | |
| 1600 | | | | | 16.771 | |
| 1610 | | | | | 16.890 | |
| 1620 | | | | | 17.008 | |
| 1630 | | | | | 17.125 | |
| 1640 | | | | | 17.243 | |
| 1650 | | | | | 17.360 | |
| 1660 | | | | | 17.477 | |
| 1670 | | | | | 17.594 | |
| 1680 | | | | | 17.711 | |
| 1690 | | | | | 17.826 | |
| 1700 | | | | | 17.942 | |
| 1710 | | | | | 18.056 | |
| 1720 | | | | | 18.170 | |
| 1730 | | | | | 18.282 | |
| 1740 | | | | | 18.394 | |
| 1750 | | | | | 18.504 | |
| 1760 | | | | | 18.612 | |

# Index

Note: Page numbers followed by *f* indicate figures and *t* indicate tables.

## D

## F

## J



## K



## L

## M

## Q



## R

# S

## W